The Red Cell Membrane

Contemporary Biomedicine

The Red Cell Membrane
Edited by ***B. U. Raess*** and ***Godfrey Tunnicliff,*** 1989

Leukolysins and Cancer
Edited by ***Janet H. Ransom*** and ***John R. Ortaldo***, 1988

Methods of Hybridoma Formation
Edited by ***Arie H. Bartal*** and ***Yashar Hirshaut,*** 1987

Monoclonal Antibodies in Cancer
Edited by ***Stewart Sell*** and ***Ralph A. Reisfeld,*** 1985

Calcium and Contractility
Edited by ***A. K. Grover*** and ***E. E. Daniel,*** 1984

Carcinogenesis and Mutagenesis Testing
Edited by ***J. F. Douglas,*** 1984

The Human Teratomas
Edited by ***Ivan Damjanov, Barbara B. Knowles,*** and ***Davor Solter,*** 1983

Human Cancer Markers
Edited by ***Stewart Sell*** and ***Britta Wahren,*** 1982

Cancer Markers
Edited by ***Stewart Sell,*** 1980

The Red Cell Membrane

A Model for Solute Transport

Edited by

B. U. Raess

and

Godfrey Tunnicliff

Indiana University School of Medicine
Evansville, Indiana

Humana Press • Clifton, New Jersey

Library of Congress Cataloging in Publication Data
Main entry under title:

The Red Cell Membrane: a model for solute transport / edited by B.U. Raess and Godfrey Tunnicliff.

p. cm. — (Contemporary biomedicine)
Includes bibliographies and index.
1. Biological transport. 2. Erythrocytes. 3. Cell membrances. 4. biological models. I. Raess, B. U. II. Tunnicliff, Godfrey. III. Series.
[DNLM: 1. Biological Transport. 2. Erythrocyte Membrane. 3. Models, Molecular. WH 150 R3057]
QH507.R43 1989
574.87'5—dc20
DNLM/DLC
for Library of Congress
ISBN 0-89603-158-6 **89-15438**

Crescent Manor
PO Box 2148
Clifton, NJ 07015

Printed in the United States of America

CONTENTS

Part 2. Biochemistry and Biophysics

Catalytic Mechanisms of the Ca-Pump ATPase
Alcides F. Rega and Patricio J. Garrahan

Regulation of the Plasma Membrane Ca^{2+}-Pump
Frank F. Vincenzi

The Anion Transport Protein: *Structure and Function*
Michael L. Jennings

Kinetics of Anion Transport

Philip A. Knauf

Chloride-Dependent Cation Transport in Human Erythrocytes

Gordon W. Stewart and J. Clive Ellory

Part 3. Pharmacology

Pharmacological Modification of the Red Cell Ca^{2+}-Pump

B. U. Raess

Irreversible Modification of the Anion Transporter

P. J. Bjerrum

Drug Actions on Potassium Fluxes in Red Cells
Balázs Sarkadi and George Gárdos

Part 4. Pathophysiology

Membrane Lipid Changes in the Abnormal Red Cell
Paolo Luly

Calcium Fluxes in Pathologically Altered Red Cells
Ole Scharff and Birthe Foder

Preface

"After being frequently urged to write upon this subject, and as often declining to do it, from apprehension of my own inability, I am at length compelled to take up the pen, however unqualified I may still feel myself for the task."

William Withering, M.D.[1]

I have yet to find a description or a quote that better summarizes my initial ambivalence towards embarking on such an endeavor as participating in putting together this monograph. The impetus for *The Red-Cell Membrane: A Model for Solute Transport* has been a simple, genuine desire to bring together an authoritative account of the "state of the art and knowledge" in the red-cell-membrane transport field. In particular, it seems important to emphasize the pivotal role the red cell has played for several decades in the discovery and the elucidation of mechanisms of plasma-membrane transport processes. It is only with such knowledge that we can hope to push ahead and make progress in this exciting, multifaceted area. Eventually, one hopes to not only further our knowledge of red cells, but apply the newly gained insights to any other cell with the common denominator of the plasma membrane.

In this compendium of reviews, the reader will find that the term model will take on a variety of gists and meanings. In some chapters, the red cell has been chosen as a model membrane solely on the basis of its preeminent design and simplicity. In other chapters, the model is chosen for its predictive powers by using a more sophisticated approach, i.e., the description and mathematical treatment of a set of experimentally generated data. To each his own! Our hope is that the combination of the two applications will help us visualize and explain, in a simplified way, the intricacies and complexities of life in and around the plasma membrane.

Clearly, one must realize that, within the constraints of a manageable review of the field's development in the past decade or so, many important topics and questions had to be omitted. Making the choices and decisions of what to include or, by contrast, omit or, for that matter, whose expertise to call upon was a thoroughly humbling and occasionally frustrating experience. Unfortunately, many excellent investigators could not be included, but their views are recognized in articles treating their particular disciplines.

Perhaps more importantly, this project has given me the opportunity to step off the teacher's dais (or daze, as the case may be on any given day) and become a listener to old mentors and new experts once again. It is my sincere hope that readers and students in "transportology" will enjoy and learn as much as I have since joining Goff Tunnicliff in organizing the present work.

B. U. Raess

[1]In *An Account of the Foxglove, and Some of Its Medical Uses; with Practical Remarks on Dropsy, and Other Diseases*. Birmingham MDCCLXXXV.

Contributors

JAVIER ALVAREZ • *Departamento de Bioquimica y Biologia Molecular y Fisiologia, Facultad de Medicina, Universidad de Valladolid, Valladolid, Spain*

LEE ANN BAXTER-LOWE • *Clinical Laboratories, The Blood Center of Southeastern Wisconsin, Milwaukee, Wisconsin*

P. J. BJERRUM • *Department of Clinical Chemistry, University Hospital, Copenhagen, Denmark*

RHODA BLOSTEIN • *Departments of Medicine and Biochemistry, Montreal General Hospital Research Institute, McGill University, Montreal, Quebec, Canada*

ROBERT M. BOOKCHIN • *Department of Medicine, Albert Einstein College of Medicine, Bronx, New York*

ANTHONY CARRUTHERS • *Department of Biochemistry, University of Massachusetts Medical Center, Worcester, Massachusetts*

J. CLIVE ELLORY • *University Laboratory of Physiology, University of Oxford, Oxford, UK*

BIRTHE FODER • *Department of Clinical Physiology and Nuclear Medicine, University Hospital, Copenhagen, Denmark*

CAROL J. FREEMAN • *Physiological Laboratory, Cambridge University, Cambridge, UK*

JAVIER GARCÍA-SANCHO • *Departamento de Bioquimica y Biologia Molecular y Fisiologia, Facultad de Medicina, Universidad de Valladolid, Valladolid, Spain*

GEORGE GÁRDOS • *Department of Cell Metabolism, National Institute of Haematology and Blood Transfusion, Budapest, Hungary*

PATRICIO J. GARRAHAN • *Instituto de Quimica y Fisicoquimica Biologicas, Facultad de Farmacia y Bioquimica, Buenos Aires, Argentina*

LOWELL E. HOKIN • *Department of Pharmacology, University of Wisconsin Medical School, Madison, Wisconsin*

MICHAEL L. JENNINGS • *Department of Physiology and Biophysics, University of Texas Medical Branch at Galvaston, Galvaston, Texas*

PHILIP A. KNAUF • *Department of Biophysics, University of Rochester School of Medicine and Dentistry, Rochester, New York*

VIRGILIO L. LEW • *Physiological Laboratory, Cambridge University, Cambridge, UK*

PAOLO LULY • *Dipartimento di Biologia, II Universita Degli Studi di Roma, Roma, Italy*

OLGA E. ORTIZ • *Department of Medicine, Albert Einstein College of Medicine, Bronx, New York*

B. U. RAESS • *Department of Pharmacology and Toxicology, Indiana University School of Medicine, Evansville, Indiana*

ALCIDES F. REGA • *Instituto de Quimica y Fisicoquimica Biologicas, Facultad de Farmacia y Bioquimica, Buenos Aires, Argentina*

BASIL D. ROUFOGALIS • *Department of Biochemistry, University of Sydney, Sydney, Australia*

BALÁZS SARKADI • *Department of Cell Metabolism, National Institute of Haematology and Blood Transfusion, Budapest, Hungary*

OLE SCHARFF • *Department of Clinical Physiology and Nuclear Medicine, University Hospital, Copenhagen, Denmark*

H. J. SCHATZMANN • *Veterinär Pharmakologisches Institut, Universität Bern, Bern, Switzerland*

GORDON W. STEWART • *Department of Medicine, University College London, London, UK*

GODFREY TUNNICLIFF • *Department of Biochemistry, Indiana University School of Medicine, Evansville, Indiana*

ANTONIO VILLALOBO • *Instituto de Investigaciones Biomedicas, Facultad de Medicina U.A.M., Madrid, Spain*

FRANK F. VINCENZI • *Department of Pharmacology, University of Washington, Seattle, Washington*

1
INTRODUCTION

Why Red Cells?

H. J. Schatzmann

1. Introduction

This is a book about the passage of matter across the plasma membrane of the mature and, for the most part, nonnucleated red blood cell (of mammals). It seems appropriate at the outset to try to answer the question of why one wishes to know much about such processes (although the reader certainly has his or her own answer). The textbook truth is that the only obvious function of mature red cells in the body's economy is to transport oxygen (why it is advantageous to have haemoglobin within a membrane-bounded space rather than in free solution must not concern us here). Yet, necessarily, there is a red-cell membrane. In point of fact, there is too much of it for the volume of the cell. Human red cells can swell by about 50 per cent (increasing their water content by 70 per cent) before reaching the volume of a sphere having the surface of the original disc. Since water-permeable bags filled with a 5 m*M* protein solution will swell and eventually burst by colloid-osmotic lysis, keeping water out of the cell is a vital function of the plasma membrane. On aging in the circulation, red cells do not swell but become even further dehydrated (and therefore heavier). Lew and Bookchin (1986) offer an astute explanation for this all but self-evident fact.

At the dawn of what we call cellular physiology, the red cell was cherished for having a semipermeable membrane, allowing the verification of van't Hoff's theory of solutions on one hand, and to demonstrate the validity of physicochemical laws in living things on the other (Hamburger, 1886, 1890, 1897a, b; Höber, 1914). When in later years people looked more carefully into the matter, it turned out that red cells do not quite behave like perfect osmometers, as was originally postulated, and the explanation for the deviation from ideality is still controversial (Freedman and Hoffman, 1979; Sawitz et al., 1964; Solomon et al., 1986). Semipermeability means simply that water passes much faster across the membrane than any solute. One-tenth of the osmotic water flow occurs in between the lipids of the membrane, whereas nine-tenths is by way of proteinacious water channels that are narrow enough to exclude ions and even small organic solutes (Macey, 1982; Ojcius and Solomon, 1988). Today, the high water permeability ($P_{H_2O} \sim 10^{-2}$ cm/s; Whittembury, personal communication) of cell membranes (and, incidentally, of artificial lipid bilayers) is an accepted fact but still a riddle, and all the interest revolves around the originally negligible side of the problem, namely the "slow" permeation across the membrane of all sorts of solutes.

In 1948, Eric Ponder wrote in the introduction to his book *Hemolysis and Related Phenomena*: "I have been told that I tend to speak of the red cell as if it were a microcosm, and as if an understanding of its nature and properties would include an understanding of nearly everything else in the cellular world. To some extent this is true. . . ." The properties of the red cell membrane that came to light since certainly do not depreciate this statement. The red cell membrane contains many transport systems that are both basic for the life of a cell and universal in the sense that they are found in all animal cells, and some even in plant and prokaryotic cells.

In addition, the red-cell membrane is a typical lipid bilayer that is very well characterized with respect to its phospholipid and cholesterol components, and with respect to the role of fatty acids in phospholipids and the dynamics of phospholipids. The physicochemical influence of lipid-soluble compounds on the structure of, and their passage through, the bilayer are also well studied. It is true that simple artificial lipid bilayers may give clearer answers to specific questions concerning physicochemistry or actions of xenobiotic substances, but the red-cell membrane is invaluable for the verification of a simple model notion under realistic conditions prevailing in biological, complex membranes.

What mainly attracts the interest, however, are the specific transport proteins in the membrane. Nonetheless, before compiling an inventory of the specific transport properties of red-cell membranes, it ought to be said what red cells do *not* have. Mature red cells lack rapid, voltage or transmitter gated channels for Na, K, or Ca^{2+}, such as those found in muscle and nerve. Further, they are not blatantly responsive to transmitters or hormones in any of their transport functions (except avian nucleated red cells). At least, there has not been presented any utterly convincing evidence for receptor-controlled mechanisms yet. The case of β-adrenergic substances is interesting. Binding studies in mature human red cells show sites with the requisite high affinity that is characteristic of receptors (Bree et al., 1984; Sager, 1982), but there is no production of cAMP upon exposure to β-adrenergic agents (Farfel and Cohen, 1984), although a coupling *N*-protein (guanosine-phosphate-binding protein) is present (Nielsen et al., 1980). The accepted explanation is that there are α- and β-receptors (left over from the erythroblast stage), but that adenylate cyclase is degraded upon maturation (Kaiser et al., 1978).

Another case in point is the insulin receptor. A binding site at the cell surface with a dissociation constant in the nanomolar range is found that might well be an insulin receptor (Gambhir et al., 1978; Herzberg et al., 1980; Robinson et al., 1979; Dows et al., 1981; Im et al., 1983), since it even induces tyrosine-specific protein kinase activity (Grigorescu et al., 1983). However, there is only one report showing that insulin (at unphysiologically high concentration) increases glucose flux (Zipper and Mawe, 1972), and another intimating that insulin stimulates the Na-pump and increases ouabain binding (Baldini et al., 1985, 1986). Attempts have been made in defining something that might be a muscarinic cholinergic receptor (Mantione and Hanin, 1980), but again, evidence for any cholinergic effect is scarce and barely convincing (Tang et al., 1984).

At present, the most likely view is that precursor cells (erythroblasts) embody a whole array of signal receiving systems, but that, on maturation of the red cell, nature does away with the machinery for the generation of second messengers, thus letting the receptors facing the outside medium fall into obsolescence. The ways in which the cells dismantle the internal signaling devices might in themselves be an interesting topic to study.

2. Transport Systems in Mature Red Cells

2.1. Transport of Inorganic Ions

2.1.1. The Anion Transport System

The red cell is notorious for an extraordinarily high density of the anion exchange system in the membrane, although other cells undoubtedly also display it. It is a protein called capnophorin (band 3 in SDS-PAGE*), which has sites that move across the membrane only if charged by an anion. The physiological significance of this exchange system resides in its allowing rapid exchange of chloride for bicarbonate (Wieth and Brahm, 1985). Were it not for this exchange, the red cell could not fully exploit the large pH buffer capacity of haemoglobin for rapid buffering of carbonic acid in the tissue and in the lungs. Haemoglobin is present at approximately 5 mmol/l cells and has a buffer capacity of 50–

*Sodium dodecylsulfate polyacrylamide gel electrophoresis.

60 mmol/l cells (Macey et al., 1978) (the buffer capacity is defined as the steepness of the titration curve, i.e., the concentration of neutralizing equivalents/pH unit, near the isoelectric point). Upon exposure to CO_2, the cellular $KHCO_3$ concentration rises because CO_2 enters rapidly across the cell membrane and is transformed into H_2CO_3 by the carbonic anhydrase. Haemoglobin buffers the carbonic acid, such that one proton disappears and is replaced by potassium. HCO_3^- now is no longer at equilibrium inside and outside the cell and, since there is a nonexchange anion transport system, will diffuse out of the cell setting up a positive potential inside, and thus eventually driving K out of the cell. This is prevented by the HCO_3^-–Cl^- exchange (Hamburger-shift), which balances the charge without net salt movement. Conversely, if the membrane were tight to HCO_3^-, the plasma volume could not participate in diluting the HCO_3^-. Moreover, the passage of carbonic acid in the form of CO_2 in one direction and of HCO_3^- in the opposite direction allows very rapid pH equilibration across the membrane (Jacobs-Stewart cycle) such that $H_o:H_i = OH_i:OH_o$ always equals $Cl_i:Cl_o$ whatever the membrane potential may be. The effect of the exchange system is that seemingly hydrochloric acid moves in the direction of the CO_2 movement. Even at atmospheric CO_2 tension (0.3 mm Hg), a pH difference across the membrane is dissipated in minutes (Macey et al., 1978). The system also mediates Cl^-–Cl^- self-exchange and accepts other anions as partners in the exchange; nitrate passes rapidly, whereas sulphate moves much more slowly than Cl^- or HCO_3^- and is probably accompanied by one proton.

The rate constant for chloride exchange is 10^6 times larger than the rate constant for K permeation. This must not be confused with the Cl^- permeability (nonexchange-restricted chloride movement), which is "only" about 100 times higher than the K permeability ($P'_{Cl} = 0.036$ min^{-1}; Hunter, 1977) [P′ is related to the permeability constant (P) in $cm \cdot min^{-1}$ by $P' = P \times$ cell membrane area in cm^2/mL of cell H_2O]. It is a matter of debate whether nonexchange-restricted (net) chloride permeability is mediated by a separate system or is the reflexion of occasional slips in the exchange system, passing a Cl^- anion once in a while, without any compensating exchange movement of another anion taking place.

2.1.2. Na-K-Transport

In most species, the Na-K-pump is the contrivance that protects red cells from colloid-osmotic swelling, but its study in red cells had much wider implications. The basis for an understanding of its mechanism was laid by the work of Glynn, Post, Whittam and Hoffman in red cells (the important papers are too numerous to be quoted here). It is the paradigm of a mechanism in red cells from which the widest generalizations to other cells are possible. In the red cell, it was possible to demonstrate that Na-K-stimulated ATPase (Skou, 1957, 1962) required Na at the inside and K at the outside, which, together with its ouabain sensitivity, made it likely to be the pump protein.

The red cell membrane is of course permeable to Na and K, and the steady-state internal concentrations of Na (7 mmol/l cells) and K (90 mmol/l cells) in human red cells result from the leak-pump principle: In the steady state (passive), influx of sodium equals (active) efflux, and the opposite is true for potassium. The leak is much less than, for instance, in muscle, but one ought to be careful in not overlooking the complexity of the passive fluxes. There is a true leak (electrodiffusional leak) that is similar in magnitude to that found in lipid bilayers. In addition, there is a saturable component of K influx and efflux that comprises a Na_o-dependent K efflux and a Na-dependent K influx (Beaugé & Lew, 1977). This holds for the case when the pump mechanism is blocked by ouabain. Otherwise, abnormal (passive) fluxes through the pump can complicate the picture. Since there is also an uphill Na efflux that is dependent on a K gradient in the outward direction, the proposal of a Na-K-cotransport requiring chloride is attractive (Wiley and Cooper, 1974; Chipperfield, 1986). This must be distinguished from the K-Cl-cotransport emerging in swollen cells of many species (possibly including those of humans; Kaji, 1986). A discrimination is possible, since the former is sensitive to loop diuretics (bumetanide, K_{diss} $3 \cdot 10^{-6}M$) and the latter is not (K_{diss} $10^{-3}M$) (Ellory et al., 1985). Na (more easily than K) can pass through the anion exchange system as the ion pair $NaCO_3^-$, a process that is obviously more rapid in alkalosis (Funder and Wieth, 1967).

In human (but not in ruminant) red cells, a comparatively rapid K pathway opens upon entry of traces of Ca^{2+} to the point that P_K approaches or exceeds P_{Cl} (Gardos, 1958; Simons, 1976; Hoffman and Blum, 1977). Recently, Vestergaard-Bogind et al. (1985) demonstrated that the "Gardos channel" obeys single-file kinetics. This is a fine example of a mechanism first discovered in red cells that later on was found to be of importance in other cells. It seems quite likely that, in excitable cells with a slow inward Ca^{2+} current during the action potential, the Ca_i^{2+}-sensitive K-channel accelerates the repolarization. In the red cell itself, opening of the Gardos channel induces shrinking, and it has been claimed (Lew and Bookchin, 1986) that this might have the following biological meaning: under shear stress in the blood capillaries, leak fluxes may increase. By way of incoming Ca^{2+}, the K efflux will be made larger than the Na influx, thus protecting the cell from a volume gain at a moment when swelling is particularly risky.

2.1.3. Ca-Transport

The red cell of humans displays an ATP-driven Ca^{2+} extrusion pump, which is another example of a system first discovered in red cells (Schatzmann, 1966, 1986) and later demonstrated in many other cells (including plant and prokaryotic cells). Here again, the very low internal free Ca^{2+} concentration reflects the steady state between passive influx and active efflux. The passive fluxes seem to be mostly saturable processes (Ferreira and Lew, 1977).

2.1.4. Mg-Transport

The latest addition to the list is a Mg^{2+} efflux dependent on an inwardly directed Na gradient and on adenosine triphosphate in human (Féray and Garay, 1986; Lüdi and Schatzmann, 1987) and bird (Günther et al., 1984) red cells. This may turn out to offer an explanation as to why in, for example, muscle cells the intracellular free Mg^{2+} concentration falls short of the expected equilibrium value.

The ion transport systems enumerated above set the steady-state values of volume, internal pH, cation distribution, and so on, that are shown in Table 1 for a given composition of the medium (the source of the membrane potential is a moot point). Lew and Bookchin (1986) have recently demonstrated that the existing knowledge is sufficient for predicting, by mathematical modeling, the transients leading from one steady state to the next when some perturbation in the form of a change of medium or increase of permeability for a particular ion is imposed (*see* Chapter 13).

2.2. Transport of Organic Molecules

2.2.1. Hexose Transport

The red-cell membrane is equipped with a hexose transfer system whose nature and specificity for hexoses and pentoses in the pyranose form was well studied already in the 1950s (LeFevre, 1948, 1954; Wilbrandt, 1961; Miller, 1968a,b). Many of the fundamental properties of a carrier-mediated transport have been exemplified in the red-cell glucose transport system (Miller, 1968a,b; Wilbrandt and Rosenberg, 1961). It saw the rise and fall of the "mobile carrier hypothesis." It is an equilibrating system mediating facilitated diffusion and, yet, displays an asymmetry in that the affinity for glucose is unequal on the two sides of the membrane. This means that the transfer rate in the two directions must

Table 1
Steady-State Composition of Human Red Cells in a Plasma-Like Medium*

Volume	82–107 μm^3	(1)
Surface area	134–145 μm^2	(1)
Water	0.7 L/lc	
Haemoglobin:		
Concentration	5mmol/lc	
Buffer capacity	50–60 meq/lc	(2,3)
Isoelectric point	6.85	
pH_i	7.26	
$A_o^-:A_i^- = H_i^+:H_o^+$	1.4	
E_m	−9 mV	
Impermeant anions (including Hb)	60 meq/lcw	(3)
Cl_i^-	95 mmol/lcw	
K_i^+	140 mmol/lcw	
Na_i^+	10 mmol/lcw	
Mg_i; (Mg_i^{2+})	2.4 mmol/lc; (0.4 mmol/lcw)	(4)
Ca_i; (Ca_i^{2+})	0.01 mmol/lc; ($<10^{-7}$mol/lcw)	(5)
Na pump flux = Na − leak	2.5 mmol/lc·h	
Na flux:K flux	1.5	(6,7,8)
Ca pump flux = Ca − leak	0.05 mmol/lc·h	(5)

*(mM) Na_o^+ 140, K_o^+ 5, Ca_o^{2+} 1, Mg_o^{2+} 0.5, Cl_o^- 131, impermeant anionic buffer 10, impermeant anions 10, glucose 10; pH 7.4. lc = liter cells; lcw = liter cell water; A^- = permeant anions, E_m = membrane potential.

(1) Weinstein, 1974
(2) Macey et al., 1978
(3) Lew and Bookchin, 1986
(4) Flatman and Lew, 1980
(5) Lew et al., 1982
(6) Whittam and Wheeler, 1970
(7) Glynn and Karlish, 1975
(8) Garay and Garrahan, 1973

also be unequal, if the system does not consume energy and fails to create a non-equilibrium distribution of glucose (Baker and Widdas, 1973; Widdas, 1980). The glucose transport is thought not to be coupled to Na movement and, thus, to mediate only downhill flow of glucose. However, it was claimed (Mahatma and Thomas, 1979) that a moderate glucose gradient is maintained in red cells. Zipper and Mawe (1972) demonstrated that, within 20 s after exposure of packed cells to mmol insulin concentrations, net glucose flux was increased by 47%. The physiological bearing of this is doubtful, but the finding is interesting in view of more recent binding studies demonstrating the existence of insulin-receptors on mature red cells.

Interestingly, pig red cells lack both the glycolytic enzymes and the glucose transport system. They might sustain their activity on inosine coming from liver cells (Kim et al, 1980, Jarvis et al., 1980).

2.2.2. Transport of Amino Acids

Except for glutathione red cells do not synthesize peptides and, therefore, do not have a need for many amino acids. Nevertheless, the fact that amino-acid permeability is much higher in red cells than in liposomes, and that there are large differences in the rates of penetration between different amino acids and between D- and L-amino acids, clearly shows that there are specific transport systems present. As with glucose, the accepted view is that the systems do not create gradients and are, therefore, in general not coupled Na-amino-acid transporting systems (for review, *see* Young and Ellory, 1977), but out of the ten known specific systems, a major and three minor ones were found to depend on Na (Ellory et al., 1981; Ellory 1987).

An interesting finding (Srivastava, 1977) is that red cells are able to export oxidized glutathione (GSSG) against a gradient, obviously by an ATP-fueled active mechanism. This might be a protective mechanism in case glutathione reduction fails.

2.2.3. Small Hydrophilic Organic Molecules

Urea, methyl-urea, thio-urea, glycerol, erythritol, and formamide were taken to penetrate mamalian red cell membranes by the water pathway. However, in bird red cells, carrier kinetics were found for glycerol (Hunter, 1970), and in mammalian, amphibian, and reptilian, but not bird red cells, even urea shows characteristics of facilitated diffusion (Kaplan et al., 1974).

2.2.4. Organic Acids

Organic acids penetrating in ionized form apparently use mechanisms that implicate a binding site at the membrane. Therefore, these are probably facilitated diffusion systems and not aqueous pores (with positive charges) for passive free diffusion (for review, *see* Motais, 1977).

2.3. Comparative Aspects

The red cell also offers some vistas in comparative physiology. The pig has been mentioned. Dog and cat (and possibly other carnivore) red cells display no measurable Na-K-pump activity and are high-Na cells, yet avoid swelling. They maintain their volume by a combination of the Ca-pump with a Ca^{2+}/Na exchange system, which leads to Na extrusion on account of the energy expenditure in the Ca^{2+} pumping (Parker, 1978,

1979; Parker et al., 1975). Ruminant red cells show genetic dimorphism: In each breed, many individuals have high-Na (low-K) cells and a few have low-Na cells, although all are born with low-Na cells. High-Na cells have fewer pump sites and their Na-pump has a high affinity for K at the internal surface. The "ruminant type" is expressed in the adult cell and a "normal type" in foetal cells (Ellory and Tucker, 1983; Christinaz and Schatzmann, 1972; Stucki and Schatzmann, 1983).

2.4. Conclusion

Some of these transport systems are vital for volume stability and for the proper function of the red-cell gas transport. This is certainly true for the Na-K-pump, the Ca-pump, the K-Cl-cotransport, the glucose transport, and the anion exchange. Others, like the amino-acid transport, may be remnants from the earlier life of the cell that have not been dismantled upon maturation.

3. Advantages of Red Cells

The red-cell membrane, as a transporting hydrophobic boundary around the cell content, is prodigiously complex and the rates of transport often a mere pittance compared to those of more lively cells. The convenience that makes the red cell a favorite object for studying transport phenomena stems from the morphological simplicity of red cells and the ease with which they can be handled.

1. They are single cells that, when incubated, come immediately into contact with the medium from which they can easily be separated by centrifugation and washing procedures.
2. There is only one type of membrane, and the whole cell is one single compartment, which eliminates the formidable difficulties of multicompartment analysis necessary in more complete cells.
3. They behave mechanically as (nonideal) osmometers, such that permeation of solutes can be followed by osmotic methods (Ponder, 1948; Wilbrandt, 1955) in cases where chemical measurements are not available and radio-tracer methods unsuitable.
4. It is not difficult to prepare the membranes (with some spectrin) in pure form, either as vesiculated preparation (inside-out) for permeation studies in the absence of cell contents, or for ATPase assays, phosphorylation studies, and chemical analysis in disrupted form.

5. Finally, the intracellular composition can be manipulated comparatively easily by cold storage, starvation, reversal of haemolysis, ionophores, PCMBS-treatment, or salicylate treatment. This is not to say that everything is plain sailing, but methodology is relatively simple.

6. A great asset is that red cells can be obtained in large quantities without very much disturbing the donor. Therefore, they are the most easily available living human cells, and their study allows direct inferences to human physiology and pathology, which is not a negligible advantage, since the most noble pursuit of physiological research still is to explain human nature in health and disease.

4. Conclusion

For the student of biological transport phenomena, the red cell is a comparatively easy to handle object whose membrane displays very many of the vitally important translocation systems seen in other surface membranes and can be studied in unadulterated form. Its morphological simplicity often makes possible rigorously quantitative measurements of such phenomena.

In contrast, one who is interested in movement, regulation, growth, and development of cells will not find much satisfaction in red-cell studies, for in these respects the red cell is indeed "very nearly dead." We should endeavor to appreciate the subtle humor in Ponder's phrase quoted above, stating that it is *partly* true that we *wholly* understand everything else after having taken the red cell's lesson.

References

Baker, G. F. and Widdas, W. F. (1973) The asymmetry of the facilitated transfer system for hexoses in human red cells and the simple kinetics of a two-compound model. *J. Physiol.* **231,** 143–165.

Baldini, P., Incerpi, S., Luly, P., and Verna, R. (1985) Insulin responsiveness of human erythrocyte plasma membrane. *J. Physiol.* **369,** 112P.

Baldini, P., Incerpi, S., Pascale F., Rinaldi, C., Verna, R., Luly, P. (1986) Insulin effects on human red blood cells. *Molec. Cell. Endocrinology* **46,** 93–102.

Beaugé, L. and Lew, V. L. (1977) Passive fluxes of sodium and potassium across the red cell membrane, in *Membrane Transport in Red Cells* (Ellory, J. C. and Lew, V. L., eds.) Academic, London, New York, pp. 39–51.

Bree, F., Goult, J., d'Athis, P., and Tillement, J. P. (1984) Beta adrenoceptors of human red blood cells; determination of their subtypes. *Biochem. Pharmacol.* **33,** 4045–4050.

Chipperfield, A. R. (1986) The ($Na^+ - K^+ - ce^-$) co-transport system. *Clin. Sci.* **71,** 465–476.

Christinaz, P. and Schatzmann, H. J. (1972). High potassium and low potassium erythrocytes in cattle. *J. Physiol.* **224,** 391–406

Dows, R. F., Corash, L. M., and Gorden, P. (1981) The insulin receptor is an age-dependent integral component of the human erythrocyte membrane. *J. Biol. Chem.* **256,** 2982–2987.

Ellory, J. C. (1987) Amino acid transport systems in mammalian red cells, in *Amino Acid Transport in Animal Cells.* (Yudilevich, D. L. and Boyd, C. A. R., eds.), Manchester University Press, pp. 106–119.

Ellory, J. C. and Tucker, E. M. (1983) Cation transport in red blood cells, in *Red Blood Cells of Domestic Mammals* (Agar, N. S. and Board, P. G., eds.) Elsevier, Amsterdam, pp. 291–312.

Ellory, J. C., Hall, A. C., and Stewart, G. W. (1985) Volume-sensitive cation fluxes in mammalian red cells. *Molecular Physiol.* **8,** 235–246.

Ellory, J. C., Jones, S. E. M., and Young, J. D. (1981) Glycine transport in human erythrocytes. *J. Physiol.* **320,** 403–422.

Farfel, Z. and Cohen, Z. (1984) Adenylate cyclase in the maturing human reticulocyte. Selective loss of the catalytic unit but not of the receptor-cyclase coupling protein. *Eur. J. Clin. Invest.* **14,** 79–82.

Féray, J. C. and Garay, R. (1986) A Na^+-stimulated Mg^{2+} transport system in human red blood-cells. *Biochim. Biophys. Acta* **856,** 76–84.

Ferreira, H. G. and Lew, V. L. (1977) Passive Ca transport and cytoplasmic Ca buffering in intact red cells, in *Membrane Transport in Red Cells* (Ellory, J. C. and Lew, V. L., eds.) Academic, London, San Francisco, pp 53–91.

Flatman, P. W. and Lew V. L. (1980). Magnesium buffering in intact human red blood cells measured using the ionophore A 23187. *J. Physiol.* **305,** 13–20.

Freedman, J. C. and Hoffman, J. F. (1979) Ionic osmotic equilibria of human red blood cells treated with nystatin. *J. Gen. Physiol.* **74,** 157–185.

Funder, J. and Wieth, J. O. (1967) Effects of some monovalent anions on fluxes of Na and K, and on glucose metabolism of ouabain treated human red cells. *Acta Physiol. Scand.* **71,** 168–185.

Gambhir, K. K., Archer, J. A., and Bradley, C. J. (1978) Characteristics of human erythrocyte insulin receptors. *Diabetes* **27**, 701–708

Garay, R. P. and Garrahan, P. J. (1973) The interaction of sodium and potassium with the sodium pump in red cells. *J. Physiol.* **231,** 297–325.

Gardos, G. (1958) The function of calcium in the potassium permeability of human red cells. *Biochim. Biophys. Acta* **30,** 653–654.

Glynn I. M. and Karlish, S. J. D. (1975) The sodium pump. *Ann Rev. Physiol.* **37,** 13–55.

Grigorescu, F., White, M. F., and Kalm, C. R. (1983) Insulin binding and insulin-dependent phosphorylation of the insulin receptor from human erythrocytes. *J. Biol. Chem.* **258,** 13708–13716.

Günther, T., Vormann, J., and Förster, R. (1984) Regulation of intracellular magnesium by Mg^{2+}-efflux. *Biochem. Biophys. Res. Comm.* **119,** 124–131.

Hamburger, H. J. (1886) Ueber den Einfluss chemischer Verbindungen auf Blutkörperchen in Zusammenhang mit ihren Molekulargewichten. *Arch. Physiol.* 476–487.

Hamburger, H. J. (1890) Die isotonischen Koeffizienten und die rothen Blutkörperchen. *Z. Physik. Chemie* **6,** 319–333.

Hamburger, H. J. (1897a) Die Gefrierpunktserniedrigung des lackfarbenen Blutes und das Volumen der Blutkörperchenschatten. *Arch. Physiol.* 486–496.

Hamburger, H. J. (1897b) Die Blutkörperchenmethode für die Bestimmung des osmotischen Druckes von Lösungen und für die Bestimmung der Resistenzfähigkeit der rothen Blutkörperchen. *Arch. Physiol.* 144–145.

Herzberg, V., Boughter, J. M., Carlisle, S., and Hill, D. E. (1980) Evidence for two insulin receptor populations on human erythrocytes. *Nature* **286,** 279–281.

Höber, R. (1914) Physikalische Chemie der Zellen und der Gewebe, 4. Aufl. W. Engelmann, Leipzig-Berlin.

Hoffman, J. F. and Blum, R. M. (1977). On the nature of the pathway used for Ca-dependent K movement in human red blood cells, in *Membrane Toxicity* (Miller, M. W. and Shamoo, A. E., eds.), Plenum, New York, pp. 381–405.

Hunter, F. R. (1970) Facilitated diffusion in pigeon erythrocytes. *Am. J. Physiol.* **218,** 1765–1772.

Hunter, M. J. (1977). Human erythrocyte anion permeabilities measured under conditions of net charge transfer. *J. Physiol.* **268,** 35–49.

Im, J. H., Meezan, E., Rackley, C. E., and Kim, H. D. (1983) Isolation and characterization of human erythrocyte insulin receptors. *J. Biol. Chem.* **258,** 5021–5026.

Jarvis, S. M., Young, J. D., Ansay, M., Archibald, A. L., Harkness, R. A., and Simmonds, R. J. (1980) Is inosine the physiological energy source of pig erythrocytes? *Biochim. Biophys. Acta* **597,** 183–188.

Kaiser, G., Wiener, G., Kremer, G., Dietz, J. Helwich, M., and Palm, D. (1978) Correlation between isoprenaline-stimulated synthesis of cyclic AMP and occurence of β-adrenoceptors in immature erythrocytes from rats. *Eur. J. Pharmocol.* **48,** 253–262.

Kaji, D. (1986) Volume sensitive K transport in human erythrocytes. *J. Gen. Physiol.* **88,** 719–738.

Kaplan, M. A., Hays, L., and Hays, R. (1974) Evolution of facilitated diffusion pathway for amides in erythrocytes. *Am. J. Physiol.* **226,** 1327–1337.

Kim, H. D., Watts, R. P., Luthra, M. G., Schwalbe, C. R., Comer, R., T., and Brendel, K. (1980) A symbiotic relationship of energy metabolism between a "non-glycolytic" mammalian red cell and the liver. *Biochim. Biophys. Acta* **589,** 256–263.

LeFevre, P. G. (1948) Evidence of active transfer of certain non-electrolytes across the human red cell membrane. *J. Gen. Physiol.* **31,** 505–527.

LeFevre, P. G. (1954) The evidence for active transport of monosaccharides across the red cell membrane. *Symp. Soc. exptl. Biol.* **8,** 118–135.

Lew, V. L. and Bookchin, R. M. (1986) Volume, pH and ion content regulation in human red cells: analysis of transient behaviour with integrated models. *J. Membr. Biol.* **92,** 57–74.

Lew, V. L., Tsien, R. Y., Miner, C., and Bookchin, R. M. (1982) Physiological $(Ca^{2+})_i$ level and pump-leak in intact red cells measured using an incorporated Ca chelator. *Nature* (London) **298,** 478–481.

Lüdi, H. and Schatzmann, H. J. (1987) Some properties of a system for Na-dependent outward movement of Mg^{2+} from metabolizing human red blood cells. *J. Physiol.* **390,** 367–382.

Macey, R. I. (1982) Water transport in red blood cells, in *Membranes and Transport* (Martonosi, A. N., ed.), Plenum, New York, pp. 461–466.

Macey, R. I., Adorante, J. S., and Orme, F. W. (1978) Erythrocyte membrane potentials determined by hydrogen ion distribution. *Biochim. Biophys. Acta* **512,** 284–295.

Mahatma, M. and Thomas, H. W. (1979). Can glucose transport across the human erythrocyte membrane be sustained against a concentration gradient? *J. Physiol.* **296,** 104P.

Mantione, C. R. and Hanin, I. (1980). Further characterization of man red blood cell cholinergic receptor. *Molec. Pharmacol.* **18,** 28–32.

Miller, D. M. (1968a) The kinetics of selective biological transport. III Erythrocyte-monosaccharide transport data. *Biophys. J.* **8,** 1329–1339.

Miller, D. M. (1968b) The kinetics of selective biological transport. IV Assessment of three carrier systems using the erythrocyte-monosaccharide data. *Biophys. J.* **8,** 1339–1352.

Motais, R. (1977) Organic anion transport in red blood cells, in *Membrane Transport in Red Cells* (Ellory, J. C. and Lew, V. L., eds.) Academic, pp. 197–220.

Nielsen, T. B., Lad, P. M., Preston, S., and Rodbell, M. (1980) Characteristics of the guanine nucleotide regulation component of adenylate cyclase in human erythrocyte membranes. *Biochim. Biophys. Acta* **629,** 143–155.

Ojcius, D. M. and Solomon, A. K. (1988) Sites of p-chloromercuribenzene sulfonate inhibition of red cell urea and water transport. *Biochim. Biophys. Acta* **942,** 73–82.

Parker, J. C. (1978) Sodium and calcium movements in dog red blood cells. *J. Gen. Physiol.* **71,** 1–17.

Parker, J. C. (1979) Active and passive Ca movements in dog red blood cells and resealed ghosts. *Am. J. Physiol.* **237,** C10–C16.

Parker, J. C., Gitelman, H. J., Glosson, P. S., and Leonard, D. L. (1975) Role of calcium in volume regulation by dog red blood cells. *J. Gen. Physiol.* **65,** 84–96.

Ponder, E. (1948) *Hemolysis and Related Phenomena.* Grune & Stratton, New York.

Robinson, T. J., Archer, J. A., Gambhir, K. K., Hollis, V. W. Jr., Carter, L., and Bradley, C. (1979) Erythrocytes: A new cell type for the evaluation of insulin receptor defects in diabetic humans. *Science* **205,** 200–202.

Sager, G. (1982) Receptor binding sites for β-adrenergic ligands on human erythrocytes. *Biochem. Pharmacol.* **31,** 99–104.

Sawitz, D., Sidel, V. S., and Solomon, A. K. (1964) Osmotic properties of human red cells. *J. Gen. Physiol.* **48,** 79–91.

Schatzmann, H. J. (1966) ATP-dependent Ca^{++}-extrusion from human red cells. *Experientia* **22,** 364–368.

Schatzmann, H. J. (1986) The human red blood cell calcium pump, in: *Membrane Control of Cellular Activity* (Lüttgau, H. C., ed.) Fischer-Verlag, Stuttgart. Fortschritte der Zoologie **33,** 435–442.

Simons, T. J. B. (1976) Calcium dependent potassium exchange in human red cell ghosts. *J. Physiol.* **256,** 227–244.

Skou, J. C. (1957) The influence of some cations on an adenosine triphosphatase from peripheral nerves. *Biochim. Biophys. Acta* **23,** 394–401.

Skou, J. C. (1962) Preparation from mammalian brain and kidney of the enzyme system involved in active transport of Na^+ and K^+. *Biochim. Biophys. Acta* **58,** 314–325.

Solomon, A. K., Toon, M. R., and Dix, J. A. (1986) Osmotic properties of human red cells. *J. Membr. Biol.* **91,** 259–273.

Srivastava, S. K. (1977) Glutathione movements, in: *Membrane Transport in Red Cells* (Ellory, J. C. and Lew V. L., eds.) Academic, London, New York, San Francisco, pp. 327–335.

Stucki, P. and Schatzmann, H. J. (1983) The response to potassium of the Na-K-pump ATPase in low-K red blood cells from cattle at birth and in later life. *Experientia* **39,** 535–536.

Tang, L. C., Shoomaker, E., and Wiesman, W. P. (1984). Cholinergic agonists stimulate calcium uptake and cAMP formation in human erythrocytes. *Biochim. Biophys. Acta* **772,** 235–238.

Vestergaard-Bogind, B., Stampe, P., and Christophersen, P. (1985) Single file diffusion through the Ca-activated K^+-channel of human red cells. *J. Membrane Biol.* **88,** 67–75.

Weinstein, R. S. (1974) The morphology of adult cells, in *The Red Blood Cell* (Surgenor, D. M., ed.) Academic, New York, London, pp. 213–268.

Whittam, R. and Wheeler, K. P. (1970) Transport across cell membranes. *Ann. Rev. Physiol.* **32,** 21–60.

Widdas, W. F. (1980) The asymmetry of the hexose transfer system in the human red cell membrane. *Curr. Topics. Membr. Transport* **14,** 165–223.

Wieth, J. O. and Brahm, J. (1985). Cellular anion transport, in *The Kidney.* (Seldin, D. W. and Giebisch, G., eds.) Raven, New York, pp. 49–89.

Wilbrandt, W. (1955) *Osmotische Erscheinungen und osmotische Methoden an Erythrozyten. Handbuch physiol.-pathol.-chem. Analyse* Hoppe-Seyler/Thierfelder, 10. Aufl., **Vol. II,** pp. 49–71.

Wilbrandt, W. (1961) Zuckertransporte. 12th Colloq. Ges. physiol. Chemie, Mosbach-Baden, pp. 113–136.

Wilbrandt, W. and Rosenberg, T. C. (1961) The concept of carrier transport and its corollaries in pharmacology. *Pharmacol. Rev.* **13,** 109–183.

Wiley, J. S. and Cooper, R. A. (1974) A furosemide-sensitive cotransport of sodium and potassium in the human red cell. *J. Clin. Invest.* **53,** 745–755.

Young, J. D. and Ellory, J. C. (1977). Red cell amino acid transport, in *Membrane Transport in Red Cells* (Ellory, J. C. and Lew, V. L., eds.) Academic, London, New York, San Francisco, pp. 301–325.

Zipper, H. and Mawe, R. C. (1972) The exchange and maximal net flux of glucose across the human erythrocyte. I. The effect of insulin, insulin derivatives and small proteins. *Biochim. Biophys. Acta* **282,** 311–325.

Need and Applications of Integrated Red Cell Models

Virgilio L. Lew, Carol J. Freeman, Olga E. Ortiz, and Robert M. Bookchin

1. Introduction

The behavior of a cell within a tissue, organ, or organism is the result of direct or indirect interactions among diverse, functional molecular units. The methodological advances of the last few decades have provided much information on the identity and function of a large variety of cell components, and on the chemical structure and operation of isolated, purified, or reconstituted molecular entities. If the understanding of a cell, organ, and organismal physiology is ever to be derived from the integration of elemental functions into progressively higher order mathematical representations, a general approach to integrated modeling must first be explored.

In this chapter, we will analyze the steps followed in the development of an integrated red-cell model aimed at explaining the volume, pH, and ion content regulatory functions of human red cells (Lew and Bookchin, 1986). Full details of the model may be found in the original paper. Here we concentrate on those aspects of its development that may have more general relevance. The predictive power of the model-building process and of the finished product will be illustrated with two representative examples at the end of the chapter.

2. General Considerations

2.1. Purpose and Requirements of an Integrated Red-Cell Model

The model of Lew and Bookchin (1986) was primarily developed to answer specific questions about the effects of ion transport on cell volume and pH, about the behavior of normal and abnormal hemoglobins, and about many puzzling or unexplained experimental observations in the literature. Without such a model, it was not possible to answer with any precision apparently simple questions such as by how much, and at what rates, should red cells be expected to swell, shrink, or change pH or ionic composition when suspended in different media, permeabilized to ions, exposed to transport inhibitors, or metabolically depleted. The fundamental information required to answer these questions had been available for years. This information concerned the nonideal osmotic behavior of hemoglobin, the composition of and charge distribution on the impermeant cell ions, the proton buffering behavior of the red-cell cytoplasm, the equality of permeant-anion and proton distribution ratios across the red-cell membrane in steady state, the operation of the Jacobs-Stewart mechanism, and the identity and kinetic properties of the main ion transporters of the red-cell membrane. What was lacking was the kind of mathematical framework with which this information could be put to use to answer those questions.

Earlier red-cell models were either steady-state models with simplified treatments of ion transport, or nonsteady-state models with explicit formulations of ion transport kinetics, but with oversimplified treatments of impermeant ion structure and no consideration of proton buffering. They answered the specific purpose that prompted their formulation, but lacked the general applicability that was now sought. The fundamental characterization and critique of model components, as well as the pioneer modeling efforts, are described in papers by Warburg (1922), Van Slyke et al. (1923), Adair (1929), Maizels and Paterson (1937), Jacobs and Stewart (1947), Harris and Maizels (1952), Dick (1959), Tosteson and Hoffman (1960), Tosteson (1964), McConaghey and Maizels (1961), Fitzsimmons and Sendroy (1961), Funder and Wieth (1966), Deuticke et al. (1971), Glynn and Warner (1972), Cass and Dalmark (1973), Duhm (1976), Hunter (1977), Hladky and Rink (1977), Freedman and Hoffman (1979), Lew and Beauge (1979), Lew et al. (1979), Lauf and Theg (1980), Haas et al. (1982), Civan and Bookman (1982), Bookchin et al. (1984), Brugnara et al. (1984), Lauf (1985a),

Larsen and Rasmussen (1985), Werner and Heinrich (1985), and Solomon et al. (1986).

The model that was now required had to:

1. Integrate the diverse functions described above
2. Be modular in its program structure, to allow easy incorporation and removal of components
3. Be flexible in its mathematical handling of the phenomenological equations representing the function of the various components, so that critical comparisons between predicted and measured behavior could be used to test hypotheses and refine kinetic options and
4. Be able to predict nonsteady-state behavior, for only so could rate processes be modeled, realistic simulations performed, feasible experiments designed, and transition mechanisms between steady states be studied and explained.

Incorporation of proton transport and buffering also required that we treat the red cell and its environment as a closed, two-compartment system. Such a treatment also optimizes comparison between simulated and observed behavior in most experimental conditions. For when the cell fraction is relatively high, transfer of water and ions between cells and medium would alter the composition of both compartments, and the extracellular environment may no longer be considered an infinite reservoir.

A final condition was that the model should be unique and minimal. This means that the choice of structures and number of parameters required to explain and predict a defined domain of functions with equal precision is reduced to one or a small number of similar alternatives. For if modeling of a system that has room to accommodate fundamentally different starting points or arbitrary parameters is attempted, the modeling effort is likely to be premature and predict only what is already known.

2.2. Components of an Integrated Red-Cell Model

The requisites above determine the minimal set of functions and components required in an integrated model expected to explain and predict red-cell behavior concerned with volume, pH, and ion content regulation. These functions are:

1. Proton buffering in cells and medium
2. The nonideal osmotic behavior of hemoglobin and
3. Ion transport.

Intracellular proton buffering within the pH range 6–8.5 was found to be determined mainly by proton binding residues on hemoglobin, and was well described over that range by a simple two-parameter equation (Maizels and Paterson, 1937; Dalmark, 1975; Freedman and Hoffman, 1979). The nonideal osmotic behavior of hemoglobin had been characterized by many investigators and found to be remarkably similar across species. A quadratic equation in hemoglobin concentration, with two virial coefficients, was found to offer a good empirical fit of experimental findings.

Ion transport will be determined by the choice of transporters and by the equations used to define their function. This exercise may seem open to arbitrary choice and infinite variety. The rule of thumb here is to list the known transporters, see if they account for most of the ion fluxes across the cell membrane, and then see if they provide the overall balance required by electroneutrality and zero net flux for each ion in the steady state. If they do, the representation of the transport function in the model cannot stray far from real life. If they do not, the exercise of formulating the specific requirements for electroneutrality may still lead to specific insights and predictions, as we hope to illustrate with an example below, on the design of a reticulocyte model. The reason why the mere attempt to apply this rule may be useful is that it places isolated and fragmentary experimental observations in a new context, different from that in which they are usually considered. This generates unexpected associations and derived questions of potential value.

The main ion transporters that operate in the mature human red cell under physiological conditions, listed in order of decreasing traffic, are the Jacobs-Stewart mechanism with the anion exchange carrier and CO_2 shunt (Jacobs and Stewart, 1942, 1947; Hladky and Rink, 1977; Fortes, 1977); the diffusible anion channel; the Na pump; a mixed bag of ouabain-insensitive carriers that mediate largely electroneutral and variously linked co- and countertransports of Na, K, Cl, and protons; and minute electrodiffusive channels for Na, K, and perhaps protons. In red cells from other species, some of the ouabain-insensitive fluxes may play much more prominent roles. Nonphysiological ion transport pathways may become important in abnormal conditions in vivo or in various experimental conditions. Activation of normally inactive transporters such as Ca^{2+}-sensitive K channels (Gardos, 1958) or K:Cl cotransporters (Lauf et al., 1984; Hall and Ellory, 1986; Canessa et al., 1986, 1987; Brugnara and Tosteson, 1987) occur in certain pathological conditions. Ionophores incorporate easily into the red-cell membrane and are also often used experimentally to induce, directly or indirectly, selective permeability changes.

Once the transporters are identified, the next challenge is in the choice of kinetic equations used to represent them in the model. Simple phenomenological expressions of complex processes have a distinguished tradition in physiology (Hill, 1910; Hodgkin and Huxley, 1952). The equations describing the nonideal osmotic behavior of hemoglobin and its proton buffering capacity are good examples. Each of these expressions was developed for a specific purpose. There is no general rule of how to choose the simplest kinetic representation of a transporter, but analysis of the Jacobs-Stewart mechanism provides a good example of a practical approach. The Jacobs-Stewart mechanism comprises the kinetics of CO_2 hydration inside and outside the cell, CO_2 diffusion across the membrane, and the kinetics and pH dependence of the anion exchange carrier. A detailed and explicit formulation of all these steps is a formidable task, and would generate a cumbersome computational burden in any model. Analysis of the effects of the Jacobs-Stewart mechanism, however, shows that in most conditions it cannot be distinguished from the operation of a simple electroneutral H:A cotransporter (*see* equation 12 in Lew and Bookchin, 1986). A hypothetical H:A cotransporter would therefore suffice to represent the operation of the Jacobs-Stewart mechanism over a wide range of conditions. Such phenomenological representations are, in any case, initial proposals, open to correction and refinement through experimental feedback. It is the computation strategy that must now provide the flexibility to change kinetics as we would want to do when testing alternative hypotheses.

2.3. The Computation Strategy

Experiments with fresh red cells usually start by taking blood, washing the cells, resuspending them, and doing something to them while measuring this or that. If we want a model capable of simulating such experiments, we must first define a reference steady state that represents the condition of the cell at the start of the experiment. We then ought to be able to introduce a perturbation representative of the experimental treatment and let the computations report the evolution in time of the system variables. A similar sequence will apply for simulations in vivo. In any case, the computations proceed in two stages: (1) the computation of a reference state, and (2.) the computation of transients.

To compute any reference state, we ought to choose an initial set of parameters, variables, and experimental or physiological conditions from which all others may be derived. The smaller the number of arbitrary decisions at this stage, the more realistic the model is likely to be. For the reference state of the human red cell, the initial values can be taken from well-established measurements (*see* Table 1 in Lew and Bookchin, 1986).

A perturbation may be a single, sequential or periodic change in some experimental condition (e.g., external pH or Cl concentration) or in any of the cell parameters (e.g., increased K permeability, inhibition of the anion carrier, or Na pump). After the perturbation, the evolution of the system variables in time is computed for successive small time increments by executing the following steps within each time interval. We compute:

1. The new membrane potential by applying the electroneutrality condition that the sum of all ionic currents through the membrane should be zero (capacitative transients may often be neglected)
2. The individual ion fluxes using the new membrane potential value
3. The change in the amount of each ion by multiplying the new flux by the incremental time interval; add this to the amounts at the beginning of the interval to calculate the new amounts
4. The new cell volume from the new amounts and from the isotonicity condition (this assumes rapid water equilibration across the red-cell membrane)
5. The new cell concentrations
6. The new extracellular volume and concentrations from the redistribution and proton buffering equations.

We then repeat the cycle beginning at 1. using the new variable values, and any computed or externally introduced change in parameter values.

Computation of the membrane potential requires solution of a transcendental equation; cell volume may also be solved with the same routine. There are a number of numerical methods for this, based on the secant iteration or the Newton-Raphson algorithms. To preserve the freedom to add, remove, or change the kinetic definitions of fluxes after comparing predicted and observed results, it is essential to avoid any procedure that demands precise analytical expressions for the derivatives of the flux equations used to compute the membrane potential. For if the first derivative needs to be recalculated each time a flux equation is modified, flexibility will be lost. The numerical solution currently used in the integrated red-cell model applies the Newton-Raphson method with cord approximations instead of true tangents. This requires some careful adjustments between the size of the cord interval, the precision imposed on the solution of the implicit equations, and the size of the incremental interval, but it preserves the speed advantages of the Newton-Raphson method and provides the desired flexibility.

A final consideration in computing strategy concerns the size of the incremental time interval. In theory, this should be infinitely small. In practice, economy of computing time recommends the use of the largest size interval with which the final results will differ from those obtained at smaller intervals by less than a specified tolerance limit. In red cells, the turnover of the various transporters may differ by six to seven orders of magnitude. It is therefore impossible to fix a single interval appropriate for all possible simulations. An elegant rule that simplifies the computations and the output of partial results is to relate the size of the incremental interval in each cycle inversely to the sum of the absolute value of all fluxes in that interval. This tailors the size of the interval to the size of traffic and change in the system.*

3. Examples

In this section, we will illustrate the predictive power of integrated modeling with two examples. In the first example, we will try to model a reticulocyte and show how the conditions required to define a reference steady state lead to specific predictions of a link between Na and Cl fluxes in these cells. The second example will report the experimental test of new and unexpected behavior predicted by the integrated red-cell model. In addition, in the chapter in this volume by Bookchin et al., detailed simulations with the integrated red-cell model describes the scope and limitations of the Na-pump hypothesis on the mechanism of dehydration of sickle cell anemia red cells.

3.1. Computing a Reference Steady State for a Reticulocyte

Reticulocytes from the sheep and the human were found to have a large capacity pH- and volume-sensitive K:Cl cotransporter (Lauf and Theg, 1980; Dunham et al., 1980; Dunham and Ellory, 1981; Lauf, 1983, 1985a,b; Brugnara et al., 1984, 1985; Brugnara and Tosteson, 1987; Canessa et al., 1986, 1987; Lauf and Bauer, 1987). This transporter persists in a functionally inactive state in mature HK-type sheep red cells. It can be partially reactivated by treatment of the red cells with *N*-ethylmaleimide or with high pressure in vitro (Lauf et al., 1984; Hall and Ellory, 1986). At physiological external pH, the K:Cl cotransport in

*The program (written in BASIC) is available from the authors on request.

human reticulocytes, and possibly in other young or abnormal, hemoglobinopathic red cells mediates a K(Rb) efflux of at least 10 mmol/(L cells·h) probably 20 mmol/(L cells·h) or more if the proportion of cells with highly active K:C1 cotransport in the top density cell fraction, where the relevant measurements were performed (Brugnara et al., 1985), was 50% or less. If unbalanced, such a flux could shrink the cells by about 20% or more in an hour or two. Since normal reticulocytes evolve to mature red cells over a period of about 24 h in the peripheral circulation (Rapoport, 1986), and in the process their volume is reduced by only about 17% (from about 106 to 88 fl.) (Killman, 1964; Clarkson and Moore, 1976), it may be assumed that circulating reticulocytes depart little from a quasi steady state of volume and composition as they mature.

The attempt to describe the quasi steady state of a young reticulocyte immediately poses the question of how the fluxes of K and Cl through the K:C1 cotransporter are balanced. The only known path for uphill K uptake is through the Na pump. For the pump to bring in at least 10 mmol/(L cells·h) of K, it must pump out more than 15 mmol/(L cells·h) of Na. In turn, this means that Na influx must also be highly increased in reticulocytes relative to mature red cells, where it has a value of only about 2.5–3 mmol/(L cells·h). This increased Na traffic in reticulocytes has been documented by Wiley and Shaller (1977), but the pathways involved remain unknown. Overall, this simple analysis already points out an important difference between reticulocytes and mature red cells concerning their volume stability. The reticulocyte, with a volume only slightly higher than that of the erythrocyte, has a cation traffic at least ten times higher. This alone may create volume instability, in the sense that a small perturbation can cause large ionic shifts, with consequent large volume and pH changes. In addition, the introduction of pH- and volume-controlling factors in the flux equation describing the K:C1 cotransporter may not yield unique steady states for equivalent sets of parameters. Drift and hysteresis must therefore be kept in mind as potentially true reticulocyte responses to transient perturbations.

What had never been considered before, and what the attempt to describe a reticulocyte steady state brings sharply into focus, is the need to balance the net Cl efflux through the K:C1 cotransporter. There are three known Cl transport pathways in the mature red cell: the anion exchange carrier, the diffusional anion channel (often considered to be a minor leak through the same anion carrier), and the K:C1Na:2Cl cotransporter. Net Cl entry through any of these pathways requires a suitable gradient. This represents a departure from the equilibrium conditions known to exist in the mature red cell, where the net flux through each of these transporters in a normal steady state is negligible.

Net Cl entry in reticulocytes may be through any of these pathways or through different, as yet undiscovered, transporters. The only available ion gradient able to energize Cl entry would seem to be that of Na. Three main possibilities exist:

1. Na entry is largely through a diffusional channel; this would depolarize the membrane potential away from the Cl equilibrium potential and drive Cl in
2. Na entry is mainly through an electroneutral Na:Cl cotransporter
3. Na entry is via a H:Na exchanger; this would generate a secondary proton gradient that, in turn, would drive Cl in through the Jacobs-Stewart mechanism, functionally equivalent to an H:Cl cotransport (Lew and Bookchin, 1986).

In all instances, Cl entry is directly or indirectly Na-dependent. Each of these possibilities generates many novel, previously unexplored predictions that are easily tested. A comprehensive analysis of all the options is outside the scope of the present chapter. We have chosen the example in Fig. 1 to illustrate the predictions derived from a potential driven Cl entry model.

If the inward Cl flux that balances the net Cl efflux through the K:Cl cotransport is mainly through a diffusional pathway, the driving force must be the membrane potential. This has to be depolarized relative to the Cl equilibrium potential in order to drive Cl in. The only ion whose diffusional flux can provide such depolarization is Na. A reticulocyte steady-state model designed to analyze potential-driven Cl entry predicts the relations between membrane potential and Na/Cl permeabilities given in Fig. 1. In this model, the diffusional Na permeability has to be increased relative to its value in mature red cells. The minimal increase is obtained if one assumes that all passive Na influx is diffusional, as in the figure. The first prediction then is that, even when minimally increased, the diffusional Na permeability has to be at least two orders of magnitude higher in reticulocytes than in mature red cells to provide the potential required to drive Cl in. In this model, the inverse proton and Cl concentration ratios remain equal. However, unlike mature red cells, the membrane potential is depolarized relative to the proton or Cl equilibrium potential. The figure tells us that the potential may be experimentally indistinguishable from the Cl equilibrium potential if the diffusional Cl permeability is about ten or more times higher than in the mature cells. Such an increase should be easy to detect. Alternatively, the reticulocytes' Cl permeability may be similar to that in the mature cells (within experimental error). In that case, membrane depolarization should be large and easily measurable. This simple example illustrates

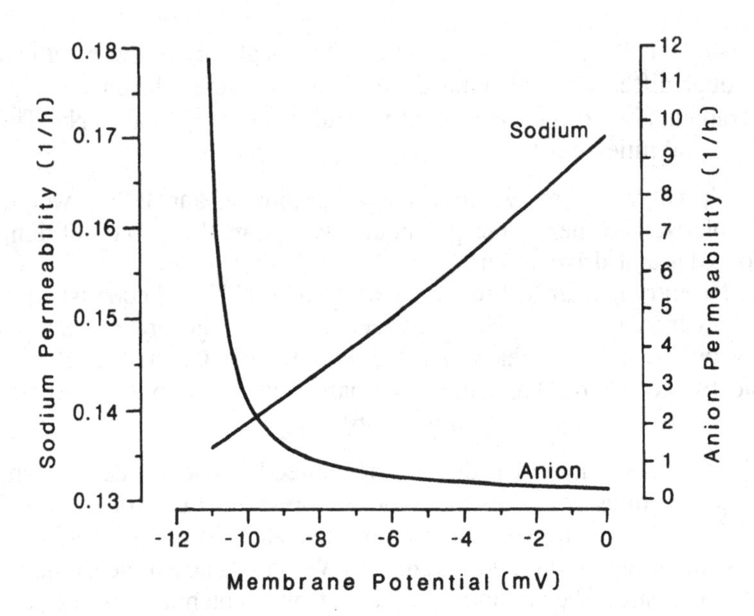

Fig. 1. Predicted relation between membrane potential, sodium, and anion permeabilities in a reticulocyte model in which the net influx of Cl required to balance net Cl efflux through the K:Cl cotransporter is driven by Na-dependent membrane depolarization.

how the mere attempt to organize available information within the conservation constraints of a model generates specific questions and predictions that can guide research in the field.

3.2. Experimental Test of Predictions of the Behavior of K-Permeabilized Red Cells

A surprising prediction of the integrated red-cell model was that the fluid and electrolyte loss from K-permeabilized red cells, suspended in low-K media, would be alkaline and hypertonic to the medium, with an excess of K over accompanying anions, that cell dehydration would precede medium alkalinization, and that partial inhibition of anion transport would delay alkalinization further relative to cell dehydration and K loss (Lew and Bookchin, 1986). It was important to investigate this prediction, because it represented a critical test of the integrated model, and also because, if it was true, it could account for some unexplained properties of dehydrated sickle cell anemia and other abnormal red cells, i.e., a decrease in total monovalent cation concentrations in cell water, and acidification of dense cells (Brugnara et al., 1984; Clark et al., 1978;

Glader et al., 1974, 1978; Kaperonis et al., 1979). It could also provide alternative interpretations of the evidence for a K:C1H exchanger in amphibian red cells (Cala, 1980, 1983a,b).

The mechanism of these effects, as explained by the model, is as follows. The increase in K permeability hyperpolarizes the cells and generates a driving force for net anion efflux through diffusional pathways. This initially results in the loss of an isotonic fluid containing only K and diffusible anions. As the cells dehydrate, the rising cell concentration of impermeant solutes leads to intracellular dilution of K and diffusible anion. The consequent increase in inward anion concentration ratio induces an inward proton flux through the Jacobs-Stewart mechanism. This, in turn, acidifies the cells and alkalinizes the medium. The extent of the proton shift in K permeabilized cells is determined by the titration properties of the cell buffers. Part of the K lost from the cells would then appear as K:C1H exchange or net KOH loss. The inward anion concentration ratio, which constitutes the driving force for proton entry, increases progressively with time as the internal anion concentration falls. This explains the prediction of a delay in pH change relative to cell dehydration. The curves in Fig. 2 illustrate the time-dependent changes in the fluxes of K, diffusible anion, and protons predicted by the model for human red cells at nonlimiting anion permeability. It can be seen that the predicted proton flux increases to a maximum by the time the K flux is less than 30% of its initial maximum. Thereafter, all net fluxes decline, as the driving gradients become vanishingly small. As the cells dehydrate, the increase in osmotic coefficient of hemoglobin "creates" osmotic particles that equilibriate between cells and medium. The cell effluent thus becomes progressively more hypertonic.

These predictions were recently investigated in human red cells by Freeman et al. (1987). In order to explore the effects of K permeabilization with a minimal number of critical measurements, preliminary simulations were used to optimize the design of the experiments. These were performed with red cells suspended at about 30% hematocrit in an initially K-free Na-saline, and permeabilized to K by the addition of valinomycin. Cl anions were partially replaced by the more permeable thiocyanate anions, so as to prevent rate-limiting effects of the anion on the rate of K loss and cell dehydration (Lew and Garcia-Sancho, 1985). The expected fluxes were therefore those shown in Fig 2. When present, the anion transport inhibitor SITS (4-acetamido-4′-isothiocyano-2,2′-stilbene disulphonic acid) was used at a concentration of 300 μM in the suspension. At the high hematocrit of these experiments, that concentration was expected to inhibit anion transport by about 80%.

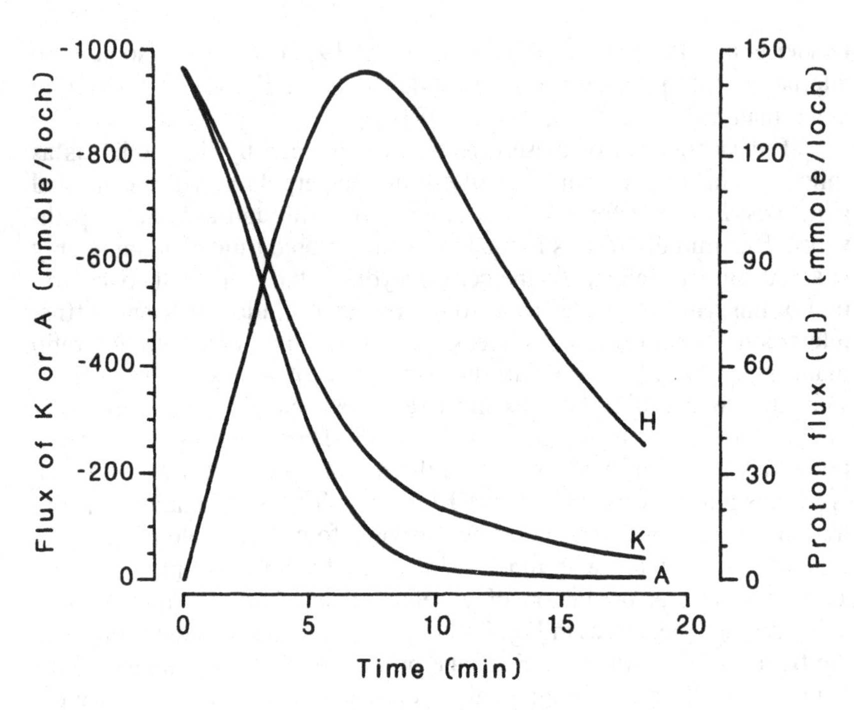

Fig. 2. Relation between time-dependent changes in the fluxes of potassium (K), anion (A), and hydrogen (H) ions in K-permeabilized human red cells, as predicted by the integrated red-cell model for nonlimiting anion permeability.

The results showed that, by the time a quasi steady state had been reached, the cells had lost the equivalent of a hypertonic fluid containing over 350 imOsmol/L of solutes (initial medium osmolarity was 295 imOsmol/L), about 180 m*M* KCl(SCN), and 10 m*M* KOH. Cell dehydration did precede alkalinization of the medium, more so in the presence of SITS. The three parameters of the model required to fit the time course of the observed changes in cell volume, external K concentration, and pH were the valinomycin-induced K permeability, the turnover of the Jacobs-Stewart mechanism, and the diffusional anion permeability. Only one set of values would prove an adequate fit. The effects of SITS could be accounted for by a similar reduction in anion diffusion and exchange turnover of between 70 and 80%, without change in the K permeability. These results therefore document good agreement between theory and experiment.

Acknowledgments

We wish to thank the Wellcome Trust, the Medical Research Council (GB), and the National Institute of Health (USA) (HL 28018 and HL 21016) for funds.

References

Adair, G. S. (1929) Thermodynamic analysis of the observed osmotic pressures of protein salts in solutions of finite concentration. *Proc. R. Soc. London A* **126,** 16–24.

Bookchin, R. M., Lew, D. J., Balazs, T., Ueda, Y., and Lew, V. L. (1984) Dehydration and delayed proton equilibria of red cells suspended in isosmotic buffers. Implications for studies of sickle cells. *J. Lab. Clin. Med.* **104,** 855–866.1

Brugnara, C. and Tosteson, D. C. (1987) Cell volume, K transport, and cell density in human erythrocytes. *Am. J. Physiol.* **252,** C269–C276.

Brugnara, C., Kopin, A. S., Bunn, H. F., and Tosteson, D. C. (1984) Electrolyte composition and equilibrium in hemoglobin CC red blood cells. *Trans. Assoc. Am. Phys.* **97,** 104–112.

Brugnara, C., Kopin, A. S., Bunn, H. F., and Tosteson, D. C. (1985) Regulation of cation content and cell volume in erythrocytes from patients with homozygous hemoglobin C disease, *J. Clin. Invest.* **75**, 1608–1617.

Cala, P. M. (1980) Volume regulation by Amphiuma red blood cells: The membrane potential and its implications regarding the nature of the ion-flux pathways. *J. Gen. Physiol.* **76,** 683–708.

Cala, P. M. (1983a) Volume regulation by red blood cells: Mechanisms of ion transport. *Mol. Physiol.* **4,** 33–52.

Cala, P. M. (1983b) Cell volume regulation by Amphiuma red blood cells. *J. Gen. Physiol.* **82,** 761–784.

Canessa, M., Spalvins, A., and Nagel R. L. (1986) Volume-dependent and NEM-stimulated K^+,Cl^- transport is elevated in oxygenated SS, SC and CC human red cells. *FEBS Lett.* **200,** 197–202.

Canessa, M., Fabry, M. E., Blumenfeld, N., and Nagel, R. L. (1987) Volume-stimulated, Cl^--dependent K^+ efflux is highly expressed in young human red cells containing normal hemoglobin or HbS. *J. Membr. Biol.* **97,** 97–105.

Cass, A. and Dalmark, M. (1973) Equilibrium dialysis of ions in nystatin-treated cells. *Nature New Biol* **244,** 47–49.

Civan, M. M. and Bookman, R. J. (1982) Transepithelial Na^+ transport and the intracellular fluids: A computer study. *J. Membr. Biol.* **65,** 63–80.

Clark, M. R., Ungar, R. C., and Shohet, S. B. (1978) Monovalent cation composition and ATP and lipid content of irreversibly sickled cells. *Blood* **51,** 1169–1178.

Clarkson, D. R. and Moore, E. M. (1976) Reticulocyte size in nutritional anemias. *Blood* **48,** 669–677.

Dalmark, M. (1975) Chloride and water distribution in human red cells. *J. Physiol.* **250,** 65–84.

Deuticke, B., Duhm, J., and Dierkesmann, R. (1971) Maximal elevation of 2,3-diphosphoglycerate concentrations in human erythrocytes: Influence on glycolytic metabolism and intracellular pH. *Pflugers Arch.* **326,** 15–34.

Dick, D. A. T. (1959) Osmotic properties of living cells. *Int. Rev. Cytol.* **8,** 387–448.

Duhm, J. (1976) Influence of 2,3-diphosphoglycerate on the buffering properties of human blood. Role of the red cell membrane. *Pflugers Arch.* **363,** 61–67.

Dunham, P. B. and Ellory, J. C. (1981) Passive potassium transport in low potassium sheep red cells: Dependence upon cell volume and chloride. *J. Physiol.* **318,** 511–530.

Dunham, P. B., Stewart, G. W., and Ellory, J. C. (1980) Chloride-activated passive potassium transport in human erythrocytes. *Proc. Nat. Acad. Sci. USA* **77,** 1711–1715.

Fitzsimons, E. J. and Sendroy, J., Jr. (1961) Distribution of electrolytes in human blood. *J. Biol. Chem.* **236,** 1595–1601.

Fortes, P. A. G. (1977) Anion movements in red cells. *Membrane Transport in Red Cells* (Ellory, J. C. and Lew, V. L., eds.), Academic, New York, pp. 175–195.

Freedman, J. C. and Hoffman, J. F. (1979) Ionic and osmotic equilibria of human red blood cells treated with nystatin. *J. Gen. Physiol.* **74,** 157–185.

Freeman, C. J., Bookchin, R. M., Ortiz, O. E., and Lew, V. L. (1987) K-permeabilized human red cells lose an alkaline, hypertonic fluid containing an excess K over diffusible anions. *J. Membr. Biol.* **96,** 235–242.

Funder, J. and Wieth, J. O. (1966) Chloride and hydrogen ion distribution between human red cells and plasma. *Acta Physiol. Scand.* **68,** 234–235.

Gardos, G. (1958) The function of calcium in the potassium permeability of human erythrocytes. *Biochim. Biophys. Acta* **30,** 653–654.

Glader, B. E., Fortier, N., Albala, M. M., and Nathan, D. G. (1974) Congenital hemolytic anemia associated with dehydrated erythrocytes and increased potassium loss. *N. Eng. J. Med.* **291,** 491–496.

Glader, B. E., Lux, S. E., Muller-Soyano, A., Platt, O. S., Propper, R. D., and Nathan, D. G. (1978) Energy reserve and cation composition of irreversibly sickled cells in vivo. *Br. J. Haematol.* **40,** 527–532.

Glynn, I. M., and Warner, A. E. (1972) Nature of the calcium dependent potassium leak induced by (+)-propranolol, and its possible relevance to the drug's antiarrhythmic effect. *Br. J. Pharmacol.* **44,** 271–278.

Haas, M., Schmidt, W. F., III, and McManus, T. J. (1982) Catecholamine-stimulated ion transport in duck red cells. *J. Gen. Physiol.* **80,** 125–147.

Hall, A. C. and Ellory, J. C. (1986) Evidence for the presence of volume-sensitive KCl transport in "young" human red cells. *Biochim. Biophys. Acta* **858,** 317–320.

Harris, E. J. and Maizels, M. (1952) Distribution of ions in suspensions of human erythrocytes. *J. Physiol.* **118,** 40–53.

Hill, A. V. (1910) The possible effects of the aggregation of the molecules of haemoglobin on its dissociation curve. *J. Physiol.* **40,** iv.

Hladky, S. B. and Rink, T. J. (1977) pH equilibrium across the red cell membrane. *Membrane Transport in Red Cells* (Ellory, J. C. and Lew, V. L., eds.), Academic, London, pp. 115–135.

Hodgkin, A. L. and Huxley, A. F. (1952) Currents carried by sodium and potassium ions through the membrane of the giant axon of Loglio. *J. Physiol.* **116,** 449–472.

Hunter, M. J. (1977) Human erythrocyte anion permeabilities measured under conditions of net charge transfer. *J. Physiol.* **268,** 35–49.

Jacobs, M. H. and Stewart, D. R. (1942) The role of carbonic anhydrase in certain ionic exchanges involving the erythrocyte. *J. Gen. Physiol.* **25,** 539–552.

Jacobs, M. H. and Stewart, D. R. (1947) Osmotic properties of the erythrocyte. XII. Eonic and osmotic equilibria with a complex external solution. *J. Cell. Comp. Physiol.* **30,** 79–103.

Kaperonis, A. A., Bertles, J. F., and Chien, S. (1979) Variability of intracellular pH within individual populations of SS and AA erythrocytes. *Br. J. Haemat.* **43,** 391–400.

Killman, S -A. (1964) On the size of normal human reticulocytes. *Acta Med. Scand.* **176,** 529–533.

Larsen, E. H. and Rasmussen, B. E. (1985) A mathematical model of amphibian skin epithelium with two types of transporting cellular units. *Pflugers Arch.* **405(Suppl 1),** S50–S58.

Lauf, P. K. (1983) Thiol-dependent passive K/Cl transport in sheep red cells: II. Loss of Cl^- and *N*-ethylmaleimide sensitivity in maturing high K^+ cells. *J. Membr. Biol.* **73,** 247–256.

Lauf, P. K. (1985a) K^+:Cl^- cotransport: Sulfhydryls, divalent cations, and the mechanism of volume activation in a red cell. *J. Membr. Biol.* **88,** 1–13.

Lauf, P. K. (1985b) Passive K^+-Cl^- fluxes in low-K^+ sheep erythrocytes: modulation by A23187 and bivalent cations. *Am. J. Physiol.* **249,** C271–C278.

Lauf, P. K. and Bauer, J. (1987) Direct evidence for chloride-dependent volume reduction in macrocytic sheep reticulocytes. *Biochem. Biophys. Res. Comm.* **144,** 849–855.

Lauf, P. K. and Theg, B. E. (1980) A chloride dependent K^+ flux induced by *N*-ethylmaleimide in genetically low K^+ sheep and goat erythrocytes. *Biochem. Biophys. Res. Comm.* **92,** 1422–1428.

Lauf, P. K., Adragna, N. C., and Garay, R. P. (1984) Activation by *N*-ethylmaleimide of a latent K^+-Cl^- flux in human red blood cells. *Am. J. Physiol.* **246,** C385–C390.

Lew, V. L., and Beauge, L. A. (1979) Passive cation fluxes in the red cell membranes. *Transport across Biological Membranes*, **vol. II.**, (Giebisch, G., Tosteson, D. C., and Ussing, H. H., eds.), Springer-Verlag, Berlin, pp. 85–115.

Lew, V. L. and Bookchin, R. M. (1986) Volume, pH and ion content regulation in human red cells: analysis of transient behavior with an intregated model. *J. Membr. Biol.* **92,** 57–74.

Lew, V. L. and Garcia-Sancho, J. (1985) Use of the ionophore A23187 to measure and control cytoplasmic Ca^{2+} levels in intact red cells. *Cell Calcium* **6,** 15–23.

Lew, V. L., Ferreira, H. G., and Moura, T. (1979) The behavior of transporting epithelial cells. I. Computer analysis of a basic model. *Proc. R. Soc. London B* **206,** 53–83.

Maizels, M. and Paterson, J. L. H. (1937) CCVII. Base binding in erythrocytes. *Biochem. J.* **31,** 1642–1656.

McConaghey, P. D. and Maizels, M. (1961) The osmotic coefficients of haemoglobin in red cells under varying conditions. *J. Physiol.* **155,** 28–45.

Rapoport, S. M. (1986) *The Reticulocyte*, CRC Press, Boca Raton, p. 35.

Solomon, A. K., Toon, M. R., and Dix, J. A. (1986) Osmotic properties of human red cells. *J. Membr. Biol.* **91,** 259–273.

Tosteson, D. C. (1964) Regulation of cell volume by sodium and potassium transport. *The Cellular Functions of Membrane Transport* (Hoffman, J. F., ed.), Prentice Hall, Englewood Cliffs, pp. 3–22.

Tosteson, D. C. and Hoffman, J. F. (1960) Regulation of cell volume by active cation transport in high and low potassium sheep red cells. *J. Gen. Physiol.* **44,** 169–194.

Van Slyke, D. D., Wu, H., and McLean, F. C. (1923) Studies of gas and electrolyte equilibria in the blood. V. Factors controlling the electrolyte and water distribution in the blood. *J. Biol. Chem.* **56,** 765–849.

Warburg, E. J. (1922) XXII. Studies on carbonic acid compounds and hydrogen ion activities in blood and salt solutions. A contribution to the theory of the equation of Lawrence J. Henderson and K. A. Hasselbalch. *Biochem. J.* **16,** 153–340.

Werner, A. and Heinrich, R. (1985) A kinetic model for the interaction of energy metabolism and osmotic states of human erythrocytes. Analysis of the stationary "in vivo" state and time dependent variations under blood preservation conditions. *Biomed. Biochim. Acta* **44,** 185–212.

Wiley, J. S. and Shaller, C. C. (1977) Selective loss of calcium permeability on maturation of reticulocytes. *J. Clin. Invest.* **59,** 1113–1119.

2
BIOCHEMISTRY AND BIOPHYSICS

The Molecular Structure of the Na,K-ATPase

Lee Ann Baxter-Lowe and Lowell E. Hokin

1. Introduction

The sodium- and potassium-activated ATPase, or the Na,K-pump (Na,K-ATPase), maintains a low intracellular concentration of sodium and a high intracellular concentration of potassium in all animal cells. These concentration gradients are necessary for a variety of physiological functions, including regulation of cell volume, maintenance of osmotic pressure, Na^+-coupled transport of certain organic and inorganic molecules, and electrical excitability of nerve and muscle (for review, *see* Stekhoven and Bonting, 1981). The Na,K-ATPase effects the coupled transports of three sodium ions out of the cell and two potassium ions into the cell. The energy required for this active transport is derived from the hydrolysis of ATP (for review, *see* Wallick et al., 1979; Kaplan, 1985).

The red blood cell is an excellent model system to investigate the transport activity of the Na,K-ATPase. Red blood cells provide a fairly homogeneous cell population without intracellular compartments. These cells can be easily lysed and reconstituted into semipermeable ghosts that are a very powerful tool for the study of the kinetics of the Na,K-pump. Red blood cells have also been utilized in clinical research to investigate potential involvement of the activity of the Na,K-ATPase in certain pathological conditions.

There has also been considerable effort directed toward the determination of the structure of the Na,K-ATPase. Many of these studies have required isolation of pure Na,K-ATPase. It has been easiest to isolate the Na,K-ATPase from specialized tissues that have high levels of the protein, such as rectal salt glands of dogfish (Hokin et al., 1973; Perrone et al., 1975), electric organs of the electric eel (Perrone et al., 1975), brain (Uesugi et al., 1971), and kidney (Kyte, 1972; Jorgensen,

1974). The Na,K-ATPases in membranes of brain and kidney are enriched 170 and 400 ×, respectively, over the Na,K-ATPase in membranes of the red blood cell (Bader et al., 1968).

The Na,K-ATPases that have been isolated from this wide range of sources are strikingly similar. In all cases, the Na,K-ATPase consists of two subunits, designated α and β. There is some evidence for the possible existence of a third subunit, termed γ. The structures of these subunits have been probed by a variety of methods, including the derivation of the amino-acid sequences from the nucleotide sequences of cDNA clones. These studies, regardless of tissue source, will be included in this review in order to present the most up-to-date and detailed information concerning the structure of the Na,K-ATPase.

2. The Subunits and Their Function

2.1. α-Subunit

The general properties presented below have been described in detail in previous reviews (Wallick et al., 1979; Jorgensen, 1982; Kaplan, 1985). The α-subunit of the Na,K-ATPase is the largest of the two subunits, with an approximate mol wt of 110,000. It is often described as the catalytic subunit, because it contains the binding sites for K^+, Na^+, and ATP. This subunit undergoes conformational changes throughout the transport cycle, which are hypothesized to be involved in the transport of ions across the membrane. In addition, the α-subunit contains binding sites for cardiac glycosides and vanadate, compounds that are known to inhibit the activity of the Na,K-ATPase.

2.2. β-Subunit

The β-subunit, which has a mol wt of about 50,000, is a sialoglycoprotein whose function is yet to be established. Its proximity to the α-subunit has been demonstrated by copurification, chemical cross-linking studies, and labeling with photoactive cardiac glycosides.

It is not yet clear whether the β-subunit plays a catalytic role. Studies have suggested that the β-subunit is functionally important. Involvement of the β-subunit is indicated by inhibition of enzymatic activity by a polyclonal antibody raised against purified β-subunit (Rhee and Hokin, 1975). The β-subunit may also play a role in the binding of cardiac glycosides, since the β-subunit can be photolabeled by certain cardiac glycoside analogs (Hall and Ruoho, 1980; Lee and Fortes, 1985).

Perhaps the strongest evidence for the involvement of the β-subunit in the function of the Na,K-ATPase is the failure so far to demonstrate full Na,K-ATPase activity in the absence of the β-subunit. The α-subunit can carry out certain partial reactions, but full enzymatic and transport activities have only been demonstrated in enzymes containing both α- and β-subunits (Perrone et al., 1975; Hilden and Hokin, 1975; Goldin, 1977). Although one study did suggest that the β-subunit may not be required for enzymatic activity (Freytag, 1983), subsequent studies have suggested that these observations were the result of experimental artifact (Kamimura et al., 1985).

2.3. γ-Subunit

The existence of a third subunit, γ, was first proposed by Rivas et al. (1972), who suggested that a proteolipid component of the Na,K-ATPase was involved in forming the cardiac glycoside binding site. The relationship between this proteolipid and the Na,K-ATPase has been controversial. This proteolipid has not been consistently seen in highly purified preparations of Na,K-ATPase (Perrone et al., 1975). There is data suggesting that it may be a proteolytic fragment of the α-subunit (Ball et al., 1983).

Support for the existence of the γ-subunit was provided by use of a photoactive ouabain derivative that labeled the α-subunit and a small protein (Forbush et al., 1978). In addition, a small protein has been isolated from a Na,K-ATPase preparation, and a partial amino-acid sequence has been determined. This sequence was different from that of the α- and β-subunits of the Na,K-ATPase (Collins and Leszyk, 1987).

3. The Structure of the α-Subunit

3.1. Molecular Weight

During the past two decades, there have been many attempts to determine the mol wt of the α-subunit. A variety of physical methods have been utilized to arrive at estimates of the mol wt, which range between 84,000 and 120,000 (for review, *see* Jorgensen, 1982). Since the primary sequences of α-subunits from several species have now been determined, the mol wt of the protein component of the α-subunit is known to be about 112,000 (Shull et al., 1985, 1986a; Kawakami, et al., 1985, 1986b; Ovchinnikov et al., 1986; Baxter-Lowe et al., submitted). Several studies have demonstrated that the α-subunit may also contain carbohydrate (Table 1). However, the amounts of carbohydrate reported vary (i.e., 0–3.6%).

3.2. Isoforms

In 1978, high resolution SDS-PAGE was utilized to demonstrate that there are two forms of the α-subunit (designated α_1 and α_2) in the brine shrimp (Peterson et al., 1978, 1982). Subsequently, two forms of α-subunit [designated α and α(+)] were detected in several rat tissues including brain (Sweadner, 1979), retinal cells (McGrail and Sweadner, 1986), muscle (Matsuda et al., 1984), and adipose cells (Lytton et al., 1985).

There is considerable evidence to suggest that the existence of the isoforms of the α-subunit is physiologically important. The two isoforms show differences in a variety of functional properties, including phosphorylation (Churchill, 1984), sensitivity to cardiac glycosides (Sweadner, 1979; Churchill et al., 1984), sensitivity to trypsin digestion (Sweadner, 1979), reactivity of sulfhydryl groups (Sweadner, 1979), and affinity for intracellular Na^+ (Lytton, 1985).

The expression of these isoforms is under physiological regulation. Tissue-specific expression has been demonstrated, and there are examples of cells that express only one of the two forms or both forms (for review, *see* Stahl, 1986). The expression of the isoforms also varies as a function of development (Peterson et al., 1982; Specht, 1984; Wolitzky and Fambrough, 1986), and endocrine status (Lo et al., 1976; Lytton, 1985).

Recently, examination of the sequences of cDNAs from rat brain has revealed that there are at least three unique mRNAs encoding α-subunits, and these have been designated α, α(+), and αIII (Shull et al., 1986a). The nomenclature of the cDNAs encoding the two rat isoforms, α and α(+), follows the designations that were made when the rat isoforms were first isolated (Shull et al., 1986a; Lytton, 1985). The initial designation of α and α(+) was based upon the relative migration of the two isoforms on SDS-PAGE (Sweadner, 1979). In this system, the α(+) form migrated more slowly than the α form, suggesting that it was the larger of the two subunits (Sweadner, 1979). Surprisingly, the derivation of the amino-acid sequences from the nucleotide sequences has revealed that the α(+)-subunit is actually smaller than the α-subunit.

It is possible that the number of isoforms of the α-subunit exceeds the two or three that have been detected. At least five or six genes encoding the α-subunits have been detected in the human genome (Shull

Table 1
Carbohydrate Composition of the Subunits of the Na,K-ATPase
(Expressed as Mol Carbohydrate/100 Mol Amino Acid)

	α	β	Reference
Brine shrimp	3.6	6.8	Peterson and Hokin, 1980
Eel electroplax	1.7	19.9	Churchill et al., 1979
Rabbit kidney	4.4	20.1	Peters et al., 1981
Lamb kidney	1.6	7.9	Munakata et al., 1982

and Lingrel, 1987). However, it is not known whether all of these genes are expressed.

3.3. α-Subunit in the Red Blood Cell

The α-subunit in red blood cells has not been examined by high resolution SDS-PAGE, and cDNA encoding the α-subunit(s) of red blood cells has not been sequenced. Therefore, the exact structure of the α-subunit of the red blood cell remains unknown. However, there is some indirect evidence available concerning the structure of the Na,K-ATPase in the red blood cell.

Antibodies raised against pig kidney Na,K-ATPase (α form) inhibited the enzymatic activity in human red blood cell membranes, indicating some common antigenic determinants between the α form of kidney and an isoform present in human red blood cells (Jorgensen et al., 1973). Additional evidence was provided by the observation that antibodies raised against dog kidney Na,K-ATPase (α form) were more effective inhibitors of enzymatic activity of human red-cell Na,K-ATPase than those of dog heart or dog brain, which contain α and α(+) isoforms (McCans et al., 1975). These results could imply that the α-isoform that is expressed in the kidney may also be expressed in red blood cells. However, one must bear in mind that there are multiple antigenic determinants that may be recognized by these polyclonal antisera (on both α- and β-subunits), making it impossible to make accurate comparisons of the α-subunits. A monoclonal antibody that recognizes the α-subunit of chicken kidney demonstrated little or no binding to red blood cells (Fambrough and Bayne, 1983), but it is not clear whether this failure to detect binding was caused by low levels or complete absence of the antigenic determinant.

A cDNA encoding the human α-subunit (derived from placenta) failed to hybridize to mRNA isolated from reticulocytes, but did hybridize to mRNA isolated from immature human erythroleukemia cells (Chehab et al., 1987). It is possible the $mRNA_{\alpha}$ is difficult to detect in reticulocytes because its synthesis may be terminated at an early stage of maturation. Alternatively, the lack of hybridization to $cDNA_{\alpha}$ may indicate the presence of a different isoform in the reticulocytes. Additional investigation will be required to resolve these questions.

3.4. Primary Structure

Pairwise comparisons of the primary sequences of the α, α(+), and αIII isoforms from the rat show that a high level of homology exists between these isoforms (85–86% identity). Pairwise comparisons of the amino-acid sequences of the α-subunits from rat, pig, lamb, and HeLa cells show very high levels of identity (96–99%). Pairwise comparisons of the amino-acid sequences of the α-subunit of torpedo electroplax with the mammalian sequences show somewhat less homology (83–87%). This degree of similarity is quite amazing, considering the diversity of the organisms. The homology between the brine shrimp and mammalian Na,K-ATPases is approximately 70%. The brine shrimp dates back to +500,000,000 years, where it branched off from echinoderms shortly (in the evolutionary scale) after echinoderms and platyhelminthes diverged (Field et al., 1988). Since the brine shrimp is a "fast-clock" organism (Field et al., 1988), 70% homology with the mammalian species attests to the high conservation of the Na,K-ATPase.

There are a few regions of the α-subunits that show high variability, whether one compares species or isoforms. These include the amino terminus, a few regions believed to be in the cytoplasmic domain (residues 460–480, 495–505, 520–530, and 560–590), and the carboxyl terminus of the protein. It is not known whether these differences arose because of a lack of selective forces (i.e., these regions are not essential for enzymatic activity) or the introduction of physiologically advantageous alterations. The latter argument would be particularly gratifying if it explained the existence and physiological significance of the isoforms of the α-subunit.

The greatest similarities occur around the phosphorylation sites (about 100 amino acids), the hydrophobic regions that are believed to form transmembrane segments, and about 200 residues of the cytoplasmic loop leading to the putative fifth transmembrane domain (Fig. 1).

3.5. Secondary and Tertiary Structure

The primary structure of the α-subunits can be utilized to make some predictions about the secondary and tertiary structures. Hydrophobicity

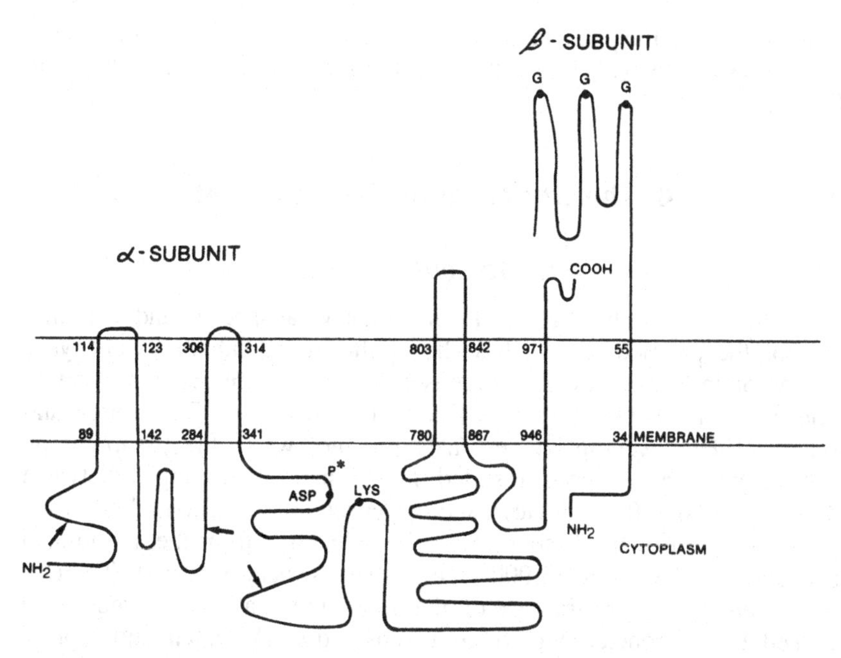

Fig. 1. Model for the structure of the Na,K-ATPase. The amino-acid residues forming the transmembrane segments of the α- and β-subunits are indicated by numbers corresponding to the points of entry and departure from the membrane. The position that is phophorylated during the enzymatic cycle is indicated by "ASP,P*." The lysine that reacts with FITC is indicated by "LYS." Potential glycosyation sites of the β-subunit are shown by "G." Proteolytic cleavage sites described in "Structure Function Relationships" are shown by arrows.

plots have been utilized to determine the location of potential transmembrane fragments. Hydrophobicity plots of each of the eight available primary sequences show a similar pattern, but there has been considerable variation in the interpretation of the patterns. Estimates of the number of transmembrane fragments range between six and ten (Shull et al., 1985; Ovchinnikov et al., 1986). Analysis of the peptides produced by selective tryptic digestion of membrane-bound Na,K-ATPase (Arzamazova et al., 1988) suggested that there are seven transmembrane segments in the α-subunit, as shown in Fig. 1.

In all models that have been presented, there are four transmembrane helices located at the amino terminal end of the α-subunit (Fig. 1). The central segment of the α-subunit forms a large cytoplasmic domain that is believed to be involved in the hydrolysis of ATP. This portion of the protein contains the phosphorylation site and binding sites for fluorescein 5′-isothiocyanate (FITC) and 5′-(p-fluorosulfonyl)-benzoyladenosine (FSBA).

4. The Structure of the β-Subunit

4.1. Molecular Weight

The β-subunit has been isolated from several species, and examination of the purified protein has shown that the β-subunit is always a glycoprotein (for review, *see* Jorgensen, 1982). The degree of glycosylation does vary between species (Table 1) and tissues (Fambrough and Bayne, 1983). Attempts to estimate the mol wt of the β-subunit by several physical methods resulted in estimates that ranged between 37,000 and 56,000 (for review, *see* Jorgensen, 1982). The mol wt of the protein portion of the β-subunit can be calculated from the amino-acid sequences to be about 35,000 (*see* primary structure below). There is heterogeneity in preparations of the β-subunit that is believed to be caused by microheterogeneity in glycosylation (Marshall and Hokin, 1979).

4.2. Primary Structure

The amino-acid sequences of the β-subunits have been derived from the nucleotide sequences of cDNA clones from five different species (Kawakami et al., 1986a; Mercer et al., 1986; Noguchi et al., 1986; Ovchinnikov et al., 1986; Shull et al., 1986b; Brown et al., 1987). In all cases, the β-subunit cDNAs encode a peptide with a mol wt of about 35,000, consisting of 303–305 amino-acid residues.

Pairwise comparisons of the amino-acid sequences of the β-subunits from dog, human, pig, rat, and sheep show that the proteins are at least 91% homologous. However, comparison of the sequence of the torpedo β-subunit with these sequences shows only about 60% homology. The amino- and carboxyl-terminal segments of the β-subunit are most highly conserved. Residues 1–94 are identical in all mammalian β-subunits with the exception of three conservative substitutions in the sheep and one semi-conservative substitution in the pig. Residues 235–302 are also conserved, with the exception of a single amino-acid substitution in the pig sequence (Brown et al., 1987). This conservation of sequence is suggestive that these domains may be important for the function of the β-subunit.

4.3. Secondary and Tertiary Structure

Proteolytic fragmentation (Chin, 1985; Farley et al., 1986) and chemical labeling (Jorgensen and Brunner, 1983) have suggested that the β-subunit is a transmembrane protein. A hydropathy plot based upon the amino-acid sequences of the β-subunits shows that the β-subunit probably has a single transmembrane domain. This domain is located near the amino-terminus and is in a highly conserved region of the β-subunit. The conservation of sequence in this region has led to the suggestion that this region is important for interaction between the α- and β-subunits and may play additional functional roles.

If the β-subunit is anchored in the membrane by a single transmembrane segment encompassing positions 34–55, the amino-terminus will consist of a short cytoplasmic domain and, with the exception of the transmembrane segment, the remaining portion of the subunit will be located extracellularly. This is consistent with the presence of potential glycosylation sites in the region that is predicted to be extracellular (Fig. 1).

5. Stoichiometry

5.1. Mass Ratios

There is fairly good agreement that the α:β molar ratio is one. Estimates of the α:β mass ratio of 2.3:2.9 have been made by scanning the gels and quantitating the protein by several different methods (Jorgensen, 1974; Peterson and Hokin, 1981). An α:β mass ratio of three was derived from amino-acid analysis (Peters et al., 1981). These values correspond to a molar ratio of one if the mol wt of the α-subunit is about 112,000 and the mol wt of the β-subunit is about 35,000.

Other studies that have examined Na,K-ATPase preparations with specific activities below 25 μmol Pi/min/mg protein have produced lower values for the α:β mass ratio (1.6 and 2.3); these values would correspond to one or two β-subunits per α-subunit (Lane et al., 1973; Kyte, 1972). It is possible that these lower estimates were caused by the presence of impurities or proteolytic fragments in the preparations of the Na,K-ATPase.

5.2. Chemical Cross-Linking Studies

Another technique that has been utilized to study the subunit composition of the Na,K-ATPase in the membrane is chemical cross-linking.

These studies have also suggested that the $\alpha:\beta$ ratio is one (Craig and Kyte, 1980; Peterson et al., 1982; Periyasamy et al., 1983).

5.3. Functional Unit

Although most evidence supports the presence of an equimolar ratio of the α- and β-subunits, there is still some question regarding minimal composition of a unit capable of ion transport activity and ATPase activity. Several approaches have been utilized to address this problem, including examination of soluble Na,K-ATPase, radiation inactivation, ligand binding, cross-linking, and electron microscopy.

Examination of soluble Na,K-ATPase has been accomplished by dissolving purified Na,K-ATPase in a solution containing a nonionic detergent, and measuring both enzymatic activity and mol wt of the soluble protein. One limitation of this approach is that it is not possible to examine the process of transport of ions across membranes. However, it can be argued that experimental results obtained with soluble preparations can be extrapolated to active transport in membranes provided certain characteristics of the enzyme are identical in membranes and in solution (Craig, 1982). Enzymatic activity has been demonstrated for soluble preparations of αβ monomers (Craig, 1982; Brotherus et al., 1983; Nakao et al., 1983) that would indicate that this unit satisfies the minimum requirement for activity.

Radiation inactivation studies have produced variable results that appear to be dependent upon the method of sample preparation. Radiation inactivation of lyophilized enzyme preparations resulted in an estimated target size for Na,K-ATPase activity that corresponds with an αβ dimer (Ottolenghi and Ellroy, 1983). Radiation inactivation of frozen aqueous suspensions of the enzyme suggested that the minimal functional unit is an αβ monomer (Karlish and Kempner, 1984). The *in situ* state of the Na,K-ATPase has been analyzed by radiation inactivation of the enzyme in human erythrocytes (Hah et al., 1985). When intact cells were used, the target sizes for transport and ATPase activities were 620,000 daltons, which would be consistent with an αβ tetramer or possibly an αβ dimer that is in close association with other protein(s) and/or lipids (Hah et al., 1985).

6. Structure–Function Relationships

6.1. Amino-Terminal Domain

A comparison of the sequences of the α-subunit of the Na,K-ATPase, the H,K-ATPase, and the Ca-ATPase suggests that these transport ATPases have evolved from a common ancestor. Comparison of these primary sequences shows that the highest degree of diversity occurs in the amino and carboxyl termini of the proteins. One possible interpretation of this observation is that these domains may be involved in the ion selectivity that differentiates these transport ATPases.

Shull et al. (1986a) have noted that comparison of the N-terminal sequences of all α-subunits of the Na,K-ATPase reveals a high variability between isoforms and species, but there is a distinctive feature of a lysine-rich region. This lysine-rich region is adjacent to a highly conserved region.

The amino terminus has also been implicated in both binding of ions and conformational changes that occur during the transport cycle by studies involving limited proteolysis of the Na,K-ATPase. In a medium containing NaCl, tryptic cleavage of the Na,K-ATPase of pig kidney results in a fast phase of inactivation that is associated with cleavage of the α-subunit between Lys-30 and Glu-31 (Jorgensen and Collins, 1986). The release of the 30 amino-terminal residue results in increases in the apparent affinity for Na^+ at low pH and shifts the equilibrium between the two conformational forms of the enzyme. The role that the amino-terminal residue might play in the function of the enzyme is curious because certain α-subunits of the Na,K-ATPase [rat α(+) and brine shrimp] do not contain this amino-terminal residue, which is present in all the other α-subunits sequenced to date.

6.2. Conformational Transitions and Ion Binding

A limited digestion of the membrane-bound enzyme with trypsin or chymotrypsin results in specific cleavage of the α-subunit that varies as a function of the ionic environment of the enzyme (Jorgensen and Collins, 1986). Cleavage between Leu-266 and Ala-267 does not alter binding sites for cations or nucleotides. However, cleavage at this site does abolish cation exchange reactions and conformational transitions. Cleavage between Arg-438 and Ala-439 does not interfere with conformational transitions or cation exchange reactions. These observations suggest that the first 438 amino-acid residues in the α-subunit may be sufficient for ion binding and exchange reactions when it remains associated with the remainder of the Na,K-ATPase in the membrane.

6.3. ATP Binding and Hydrolysis

At least two amino-acid residues located in the α-subunit are believed to be essential for binding and hydrolysis of ATP. The first is Asp-376, which is phosphorylated by ATP. The second is Lys-507, which binds to FITC; this binding results in inactivation of the Na,K-

ATPase, which can be prevented by the addition of ATP (Farley et al., 1984). Another ATP analog, FSBA, also shows ATP-protectable inhibition of the Na,K-ATPase. FSBA labels two peptides that are located in the large cytoplasmic loop of the α-subunit (Ohta et al., 1986). These results strongly implicate the large cytoplasmic group in the binding and hydrolysis of ATP. This is consistent with the observation that these regions are highly conserved in other transport ATPases, including the Ca-ATPase and the H,K-ATPase. Conformational changes that are associated with the hydrolysis of ATP can be detected by attaching a fluorescent label, *N*-[*p*-(2-benzimidazolyl)phenyl]maleimide (BIPM), to Cys-964 of the sheep kidney α-subunit (Nagai et al., 1986).

6.4. Binding of Cardiac Glycosides

It is generally believed that cardiac glycosides (e.g., ouabain, digitalis) exert their positive inotropic effects by inhibiting the activity of the Na,K-ATPase (for review, *see* Hansen, 1984). The interaction between the Na,K-ATPase and cardiac glycosides is greatly influenced by the tertiary structure of the enzyme. For example, the conformational state of the enzyme has dramatic effects on the binding of cardiac glycosides (Schwartz et al., 1968). In addition, the sensitivity of the Na,K-ATPase to cardiac glycosides varies between isoforms and the species from which the enzyme was derived. This has led to attempts to identify the binding site by comparison of sequences from sensitive and insensitive forms of the α-subunits. Such comparisons have pointed to the extracellular portion of the α-subunit, which is located between the first and second transmembrane segments (Shull et al., 1986a).

7. Summary

The Na,K-ATPase is comprised of two or possibly three subunits (α, β, and possibly γ). The largest subunit, α, has been implicated in several important functions of the Na,K-ATPase, including binding of ions, hydrolysis of ATP, binding of cardiac glycosides, and comformational changes. There are several isoforms of the α-subunit, and the expression of these isoforms is tissue-specific and regulated as a function of developmental and endocrine status of the organism. The expression of the various isoforms may provide an important mechanism for regulation of the activity of the Na,K-ATPase.

References

Arzamazova, N. M., Arystarkhova, E. A., and Gevondyan, N. M. (1988) Sequence analysis of exposed domains of membrane-bound Na,K-ATPase. A model of transmembrane arrangement, in *Progress in Clinical and Biological Research.* Vol. 268 B. The Na^+,K^+-Pump Part A: Molecular Aspects (Skou, J. C., Norby, J. G., Maunsbach, A. B., and Esman, M. eds.), Alan R. Liss, New York, pp. 57–64.

Bader, H., Post, R. L., and Bond, G. H. (1968) Comparison of sources of a phosphorylated intermediate in transport of ATPase. *Biochim. Biophys. Acta* **150,** 41–46.

Ball, W. J., Jr., Collins, J. H., Land, L., and Schwartz, A. (1983) Antigenic properties of the α-, β-, and γ-subunits of Na,K-ATPase. *Curr. Topics Membr. Transp.* **19,** 781–785.

Brotherus, J. R., Jacobsen, L., and Jorgensen, P. L. (1983) Soluble and enzymatically stable (Na^+ + K^+)-ATPase from mammalian kidney consisting of predominantly αβ-subunits. *Biochim. Biophys. Acta* **731,** 290–303.

Brown, T., Horowitz, B., Miller, R., McDonough, A., and Farley, R. (1987) Molecular cloning and sequence analysis of the (Na^+ + K^+)-ATPase β-subunit from dog kidney. *Biochim. Biophys. Acta* **912,** 244–253.

Chehab, F., Kan, Y., Law, M., Hartz, J., Kao, F., and Blostein, R. (1987) Human placental Na^+,K^+-ATPase α-subunit: cDNA cloning, tissue expression, DNA polymorphism, and chromosomal localization. *Proc. Natl. Acad. Sci. USA* **84,** 7901–7905.

Chin, G. J. (1985) Papain fragmentation of the (Na^+,K^+)-ATPase β-subunit reveals multiple membrane-bound domains. *Biochemistry* **24,** 5943–5947.

Churchill, L. (1984) Differences in phosphorylation of the two large subunits of brine shrimp Na,K-ATPase. *J. Exp. Zool.* **231,** 335–341.

Churchill, L., Peterson, G. L., and Hokin, L. E. (1979) The large subunit of (sodium + potassium)-activated adenosine triphosphatase from the electroplax of *Electrophorus electricus* is a glycoprotein. *Biochem. Biophys. Res. Commun.* **90,** 488–490.

Churchill, L., Hall, C., Peterson, G. L., Ruoho, A., and Hokin, L. E. (1984) Photoaffinity labeling of the ouabain binding site in Na,K-ATPase in developing brine shrimp. *J. Exp. Zool.* **231,** 343–350.

Collins, J. H. and Leszyk, J. (1987) The "γ-subunit" of Na,K-ATPase: A small, amphiphilic protein with a unique amino acid sequence. *Biochemistry* **26,** 8665–8668.

Craig, W. S. (1982) Monomer and sodium and potassium ion activated adenosinetriphosphatase displays complete enzymatic function. *Biochemistry* **21,** 5707–5717.

Craig, W. S. and Kyte, J. (1980) Stoichiometry and molecular weight of the minimum asymmetric unit of canin renal sodium and potassium-activated adenosine triphosphatase. *J. Biol. Chem.* **255,** 6262–6269.

Fambrough, D. M. and Bayne, E. K. (1983) Multiple forms of $(Na^+ + K^+)$-ATPase in the chicken. *J. Biol. Chem.* **258,** 3926–3935.

Farley R. A., Miller, R. D., and Kudrow, A. (1986) Orientation of the β-subunit polypeptide of $(Na^+ + K^+)$-ATPase in cell membrane. *Biochim. Biophys. Acta* **873,** 136–142.

Farley, R. A., Tran, C. M., Carilli, C. T., Hawke, D., and Shively, J. E. (1984) The amino acid sequence of a fluorescein-labeled peptide from the active site of (Na,K)-ATPase. *J. Biol. Chem.* **259,** 9532–9535.

Field, K. G., Olsen, G. J., Lane, D. J., Giovannoni, S. J., Ghiselin, M. T., Raff, E. C., Pace, N. R., and Raff, R. A. (1988) Molecular phylogeny of the animal kingdom. *Science* **239,** 748–753.

Forbush, B. III, Kaplan, J. H., and Hoffman, J. F. (1978) Characterization of a new photoaffinity derivative of ouabain: Labeling of the large polypeptide and of a proteolipid component of the Na,K-ATPase. *Biochemistry* **17,** 3667–3676.

Freytag, J. W. (1983) The (Na^+,K^+)-ATPase exhibits enzymatic activity in the absence of the glycoprotein subunit. *FEBS Lett.* **159,** 280–284.

Goldin, S. (1977) Active transport of sodium and potassium ions by the sodium and potassium ion-activated adenosine triphosphatase from renal medulla. Reconstitution of the purified enzyme into a well defined *in vitro* transport system. *J. Biol. Chem.* **252,** 5630–5642.

Hah, J., Goldinger, J. M., and Jung, C. Y. (1985) *In situ* assembly states of (Na^+,K^+)-pump ATPase in human erythrocytes. *J. Biol. Chem.* **260,** 14016–14019.

Hall, C. and Ruoho, A. (1980) Ouabain-binding-site photoaffinity probes that label both subunits of Na^+,K^+-ATPase. *Proc. Natl. Acad. Sci. USA* **77,** 4529–4533.

Hansen, O. (1984) Interaction of cardiac glycosides with $(Na^+ + K^+)$-activated ATPase. A biochemical link to digitalis-induced inotropy. *Pharmacol. Rev.* **36,** 143–163.

Hilden, S. and Hokin, L. E. (1975) Active potassium transport coupled to active sodium transport in vesicles reconstituted from purified sodium and potassium ion-activated adenosine triphosphatase from the rectal gland of *Squalus acanthias. J. Biol. Chem.* **250,** 6296–6303.

Hokin, L. E., Dahl. J. L., Deupree, J. D., Dixon, J. F., Hackney, J. F., and Perdue, J. F. (1973) Studies on the characterization of the sodium-potassium transport adenosine triphosphatase. Purification of the rectal gland of *Squalus acanthias. J. Biol. Chem.* **248,** 2503-2605.

Jorgensen, P. L. (1974) Purification and characterization of the $(Na^+ + K^+)$-ATPase. III. Purification from the outer medulla from mammalian kidney after selective removal of membrane components by sodium dodecylsulphate. *Biochim. Biophys. Acta* **356,** 36–52.

Jorgensen, P. L. (1982) Mechanism of the Na^+,K^+-pump protein structure and conformations of the pure $(Na^+ + K^+)$-ATPase. *Biochim. Biophys. Acta* **694,** 27–68.

Jorgensen, P. L. and Brunner, J. (1983) Labeling of intramembrane segments of the α-subunit and β-subunit of pure membrane-bound (Na^+,K^+)-ATPase with 3-trifluoromethyl-3-(*m*-[^{125}I]iodophenyl)diazimine. *Biochim. Biphys. Acta* **735,** 291–296.

Jorgensen, P. L. and Collins, J. (1986) Tryptic and chymotryptic cleavage sites in sequence of alpha-subunit of (Na^+ + K^+)-ATPase from outer medulla of mammalian kidney. *Biochim. Biophys. Acta* **860,** 570–576.

Jorgensen, P. L., Hansen, O., Glynn, I. M., and Cavieres, J. D. (1973) Antibodies to pig kidney (Na^+ + K^+)-ATPase inhibit the Na^+ pump in human red cells provided thay have access to the inner surface of the cell membrane. *Biochim. Biophys. Acta* **291,** 795–800.

Kamimura, K., Morohoshi, M., and Kawamura, M. (1985) Papain digestion of the subunits of (Na,K)-ATPase. *FEBS Lett.* **187,** 135–140.

Kaplan, J. (1985) Ion movements through the sodium pump. *Annu. Rev. Physiol.* **47,** 535–544.

Karlish, S. J. D. and Kempner E. S. (1984) Minimal functional unit for transport and enzyme activities of (Na^+ + K^+)-ATPase as determined by radiation inactivation. *Biochim. Biophys. Acta* **776,** 288–298.

Kawakami, K., Noguchi, S., Noda, M., Takahashi, H., Ohta, T., Kawamura, M., Nojima, H., Nagano, K., Hirose, T., Inayama, S., Hayashida, H., Miyata, T., and Numa, S. (1985) Primary structure of the α-subunit of *Torpedo californica* (Na^+ + K^+)-ATPase deduced from cDNA sequence. *Nature* **316,** 733–736.

Kawakami, K., Nojima, H., Ohta, T., and Nagano, K. (1986a) Molecular cloning and sequence analysis of human Na,K-ATPase β-subunit. *Nucl. Acids Res.* **14,** 2833–2844.

Kawakami, K., Ohta, T., Nojima, H., and Nagano, K. (1986b) Primary structure of the α-subunit of human Na,K-ATPase deduced from cDNA sequence. *J. Biochem.* **100,** 389–397.

Kyte, J. (1972) Properties of the two polypeptides of sodium- and potassium-dependent adenosine triphosphatase. *J. Biol. Chem.* **247,** 7642–7649.

Lane, L. K., Copenhaver, J. H., Lindenmayer, G. E., and Schwartz, A. (1973) Purification and characterization of and [^{3}H]ouabain binding to the transport adenosine triphosphatase from outer medulla of canine kidney. *J. Biol. Chem.* **248,**7197–7200.

Lee, J. A. and Fortes, P. A. G. (1985) Labeling of the glycoprotein subunit of (Na,K)ATPase with fluorescent probes. *Biochemistry* **24,** 322–330.

Lo, C. S., August, T. R., Liberman, U. A., and Edelman, I. S. (1976) Dependence of renal (Na^+ + K^+)-adenosine triphosphatase activity on thyroid status. *J. Biol. Chem.* **251,** 7826–7833.

Lytton, J. (1985) The catalytic subunits of the (Na^+ + K^+)-ATPase α and α(+) isozymes are the products of different genes. *Biochem. Biophys. Res. Comm.* **132,** 764–769.

Lytton, J., Lin, J., and Guidotti, G. (1985) Identification of two molecular forms of (Na^+ + K^+)-ATPase in rat adipocytes. Relation to insulin stimulation of the enzyme. *J. Biol. Chem.* **260,** 1177–1184.

Marshall, P. J. and Hokin, L. E. (1979) Microheterogeneity of the glycoprotein subunit of the (sodium + potassium)-activated adenosine triphosphatase from the electroplax of *Electrophorus electricus*. *Biochem. Biophys. Res. Comm.* **87,** 476–482.

Matsuda, T., Iwata, H., and Cooper, J. R. (1984) Specific inactivation of α(+) molecular form of (Na^+ + K^+)-ATPase by pyrithiamin. *J. Biol. Chem.* **259,** 3858–3863.

McCans, J. L., Lindenmayer, G. E., Pitts, B. J. R., Ray, M. V., Raynor, B. D., Butler, V. P., Jr., and Schwartz, A. (1975) Antigenic differences in (Na^+ + K^+)-ATPase preparations isolated from various organs and species. *J. Biol. Chem.* **250,** 7257–7265.

McGrail, K. M. and Sweadner, K. J. (1986) Immunofluorescent localization of two different Na,K-ATPase in the rat retina and in identified dissociated retinal cells. *J. Neurosci.* **6,** 1272–1283.

Mercer, R. W., Schneider, J. W., Savitz, A., Emanuel, J., Benz, E. J., Jr., and Levensen, R. (1986) Rat-brain Na,K-ATPase β-chain gene: primary structure, tissue specific expression, and amplification in ouabain resistant HeLa C^+ cells. *Mol. Cell. Biol.* **6,** 3884–3890.

Munakata, H., Schmid, K., Collins, J. H., Zot, A. S., Lane L. K., and Schwartz, A. K. (1982) The α and β subunits of lamb kidney Na,K-ATPase are both glycoproteins. *Biochem. Biophys. Res. Comm.* **107,** 229–231.

Nagai, M., Taniguchi, K., Kangawa, K., Matsuo, H., Nakamura, S., and Iida, S. (1986) Identification of *N*-[*p*-(2-benzimidazolyl)phenyl]maleimide-modified residue participating in dynamic fluorescence changes accompanying Na^+,K^+-dependent ATP hydrolysis. *J. Biol. Chem.* **261,** 13197–13202.

Nakao, T., Ohno, T., Nakao, M., Maeki, G., Tsukita, S., and Ishikawa, H. (1983) Monomeric and trimeric structures of active Na,K-ATPase in $C_{12}E_8$ solution. *Biochem. Biophys. Res. Comm.* **113,** 361–367.

Noguchi, S., Noda, M., Takahashi, H., Kawakami, K., Ohta, T., Nagano, K., Hirose, T., Inayama, S., Kawamura, M., and Numa, S. (1986) Primary structure of the β-subunit of *Torpedo californica* (Na^+ + K^+)-ATPase deduced from the cDNA sequence. *FEBS Lett.* **196,** 315–320.

Ohta, T., Nagano, K., and Yoshida, M. (1986) The active site structure of Na^+/K^+-transporting ATPase: Location of the 5′-(*p*-fluorosulfonyl)-benzoyladenosine binding site and soluble peptides released by trypsin. *Proc. Natl. Acad. Sci. USA* **83,** 2071–2075.

Ottolenghi, P. and Ellroy, J. C. (1983) Radiation inactivation of (Na,K)-ATPase, an enzyme showing multiple radiation-sensitive domains. *J. Biol. Chem.* **258,** 14895–14907.

Ovchinnikov, Y. A., Modyanov, N. N., Broude, N. E., Petrukhin, K. E., Grishin, A. V., Arzamazova, N. M., Aldanova, N. A., Monastyrskaya, G. S., and Sverdlov, E. D. (1986) Pig kidney (Na^+ + K^+)-ATPase primary structure and spatial organization. *FEBS Lett.* **201,** 237–245.

Periyasamy, S. M., Huang, W.-H., and Askari, A. (1983) Subunit associations of (Na^+ + K^+)-dependent adenosine triphosphatase. *J. Biol. Chem.* **258,** 9878–9885.

Perrone, J., Hackney, J., Dixon, J., and Hokin, L. E. (1975) Molecular properties of purified (sodium + potassium)-activated adenosine triphosphatases and their subunits from the rectal gland of *Squalus acanthias* and the electric organ of *Electrophorus electricus*. *J. Biol. Chem.* **250,** 4178–4184.

Peters, W. H. M., DePont, J. J. H. H. M., Koppers, A., and Bonting, S. L. (1981) Studies on (Na^+ + K^+)-activated. ATPase XLVII. Chemical composition, molecular weight and molar ratio of the subunits of the enzyme from rabbit kidney outer medulla. *Biochim. Biophys. Acta* **641,** 55–70.

Peterson, G. L. and Hokin, L. E. (1980) Improved purification of brine-shrimp (*Artemia salina*) (Na^+ + K^+)-activated adenosine triphosphatase and amino acid and carbohydrate analyses of the isolated subunits. *Biochem. J.* **192,** 107–118.

Peterson, G. L. and Hokin, L. E. (1981) Molecular weight and stoichiometry of the sodium- and potassium-activated adenosine triphosphatase subunits. *J. Biol. Chem.* **256,** 3751–3761.

Peterson, G. L., Churchill, L., Fisher, J. A., and Hokin, L. E. (1982) Structural and biosynthetic studies on the two molecular forms of the (Na^+ + K^+)-activated adenosine triphosphatase large subunit in *Artemia salina* nauplii. *J. Exp. Zool.* **221,** 295–308.

Peterson, G. L., Ewing, R. D., Hootman, S. R., and Conte, F. P. (1978) Large-scale partial purification and molecular and kinetic properties of the Na + K-activated adenosine triphosphatase in *Artemia salina* nauplii. *J. Biol. Chem.* **253,** 4762–4770.

Rhee, H. M. and Hokin, L. E. (1975) Inhibition of the purified sodium-potassium activated adenosinetriphosphatase from the rectal gland of *Squalus acanthias* by antibody against the glycoprotein subunit. *Biochem. Biophys. Res. Comm.* **63,** 1139–1145.

Rivas, E., Lew, V., and DeRobertis, E. (1972) [^{3}H]Ouabain binding to a hydrophobic protein from electroplax membranes. *Biochim. Biophys. Acta* **290,** 419–423.

Schwartz, A., Matsui, H., and Laughter, A. H. (1968) Tritiated digoxin binding to (Na^+ + K^+)-activated adenosine triphosphatase: possible allosteric site. *Science* **160,** 323–325.

Shull, G., Greeb, J., and Lingrel, J. (1986a) Molecular cloning of three distinct forms of the Na^+,K^+-ATPase α-subunit from rat brain. *Biochemistry* **25,** 8125–8132.

Shull, G. E., Lane, L. K., and Lingrel, J. B. (1986b) Amino acid sequence of the β-subunit of the (Na^+ + K^+)-ATPase deduced from a cDNA. *Nature* **321,** 429–431.

Shull, M. and Lingrel, J. (1987) Multiple genes encode the human Na^+, K^+-ATPase catalytic subunit. *Proc. Natl. Acad. Sci. USA* **84,** 4039–4043.

Shull, G. E., Schwartz, A., and Lingrel, J. B. (1985) Amino-acid sequence of the catalytic subunit of the (Na^+ + K^+)ATPase deduced from a complementary DNA. *Nature* **316,** 691–695.

Specht, S. C. (1984) Development and regional distribution of two molecular forms of the catalytic subunit of the Na,K-ATPase in rat brain. *Biochem. Biophys. Res. Comm.* **121,** 208–212.

Stahl, W. (1986) The Na,K-ATPase of nervous tissue. *Neurochem. Int.* **8,** 449–476.

Stekhoven, F. and Bonting, S. (1981) Transport adenosine triphosphatases: properties and functions. *Physiol. Rev.* **61,** 1–76.

Sweadner, K. J. (1979) Two molecular forms of (Na^+ + K^+)-stimulated ATPase in brain. Separation and difference in affinity for strophanthidin. *J. Biol. Chem.* **254,** 6060–6067.

Uesugi, S., Dulak, N., Dixon, J., Hexum, T., Dahl, J., Perdue, J., and Hokin, L. (1971) Studies on the characterization of the sodium-potassium transport adenosine triphosphatase. VI. Large scale partial purification and properties of a lubrol-solubilized bovine brain enzyme. *J. Biol. Chem.* **246,** 531–543.

Wallick, E., Lane, L., and Schwartz, A. (1979) Biochemical mechanism of the sodium pump. *Annu. Rev. Physiol.* **41,** 397–411.

Wolitzky, B. A. and Fambrough, D. M. (1986) Regulation of the (Na^+ + K^+)-ATPase in cultured chick skeletal muscle. *J. Biol. Chem.* **261,** 9990–9999.

Enzymatic and Functional Aspects of Na^+/K^+ Transport*

Rhoda Blostein

1. Introduction

It is almost 50 years since Harris (1941) and Danowski (1941) showed that glycolysis energizes net extrusion of Na^+ and uptake of K^+ in red blood cells, thus providing convincing evidence for the existence of an energy-driven active alkali cation transport system. The most direct demonstration that ATP fuels active cation transport was achieved in subsequent experiments with resealed red-cell ghosts, whereby it was shown that Na^+ extrusion was dependent on the presence of intracellular ATP (Gárdos, 1954; Hoffman, 1962).

Schatzmann's discovery of the highly specific inhibition of energy-dependent Na^+ and K^+ transport by cardiotonic steroids such as ouabain or its aglycone, strophanthidin, provided a means to quantify and characterize "pump"-mediated Na^+ and K^+ fluxes, as well as the associated ATP utilization (Schatzmann, 1953). Thus, the ouabain-inhibitable fraction of Na^+ efflux, K^+ influx, and ATP hydrolysis are attributed specifically to the normal reactions of the sodium pump. From studies of intact human red cells, it was shown that the efflux of three Na^+ ions is coupled to the influx of two K^+ ions (Post and Jolly, 1957) and that this process is energized by the hydrolysis of one molecule of ATP (Sen and Post, 1964; Garrahan and Glynn, 1967). The Na,K-ATPase, first discovered in peripheral crab nerve microsomes by Skou (1957), was later described and characterized in red-cell ghosts (Dunham and Glynn, 1961; Post et al., 1960). However, the unambiguous identification of the ouabain-sensitive

*Throughout this article, reviews rather than original reports are frequently cited. I apologize to those authors not credited directly for their contributions.

Na,K-ATPase as the molecular structure responsible for active transport was achieved by reconstitution experiments in which the Na,K-ATPase, purified from much more active tissues and reconstituted with phospholipids into proteoliposomes, was shown to catalyze ATP-dependent Na/K exchange (Goldin and Tong, 1974; Hilden et al., 1974).

Several reviews of Na,K-ATPase structure and mechanism have appeared over the past two decades, including recent accounts by Kaplan (1985), Glynn (1985), Stein (1986), and Hoffman (1986). The latter provides a concise overview of active transport of Na^+ and K^+ derived mainly from studies of red blood cells. This chapter describes selected features of the red-cell Na,K-ATPase with the hope of providing further insights into the diversity of Na,K pump behavior and the regulation of its activity as it exists *in situ*.

2. Modes of Pump Behavior and Their Enzymic Correlates

The 3:2 stoichiometry of Na^+ efflux:K^+ influx during pump activity generates net transfer of charge in various tissues (Thomas, 1972), as well as red cells (Hoffman et al., 1979). The fit of the flux data to the Hill equation whereby $n = 3$ for Na^+ and $n = 2$ for K^+ (*see* Cavieres, 1977) is consistent with both the directly measured stoichiometry and the idea of multiple binding sites for these ions at the two membrane surfaces.

The molecular mechanism of the Na,K-ATPase has been delineated from studies in several laboratories, notably those of Post and Albers and their coworkers in the early 1960s (Post et al., 1972; Taniguchi and Post, 1975; Fahn et al., 1966a, b; Siegel and Albers, 1967). From studies of the overall hydrolytic reaction, the partial reactions, as well as side-specificity of cation requirements, it is apparent that the mechanism is characterized by a complex series of partial reaction steps. The minimal reaction sequence comprises Na_i^+-dependent† phosphorylation and K_o^+-activated† dephosphorylation of the enzyme linked to conformational transitions of dephospho- and phosphoforms of the enzyme. The simplified reaction mechanism depicted in Fig. 1 is a modified version of the Post-Albers model. It encompasses ion translocation modes associated with the reaction steps or sequence of steps of the Na,K-ATPase and is based, to a considerable extent, on results of experiments carried out with human red cells. The fundamental question as to whether the

†Subscripts i and o refer to the normal intracellular and extracellular sides of the membrane.

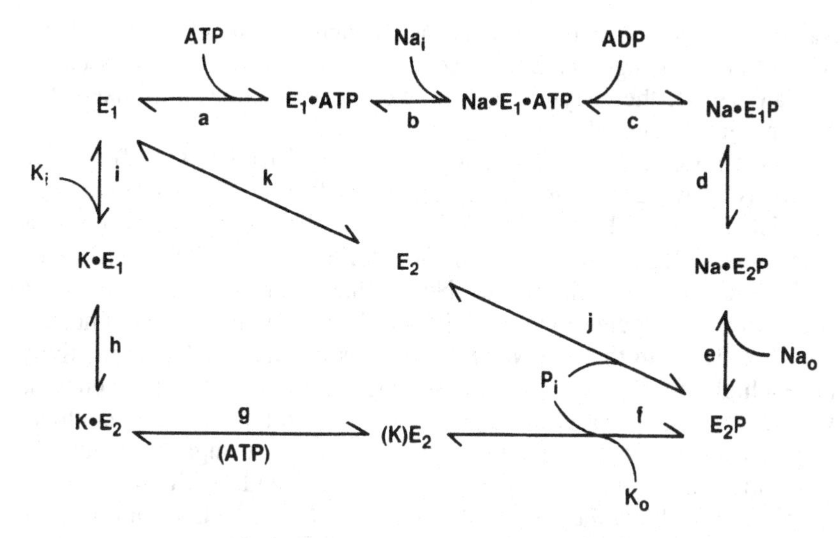

Fig. 1. A simplified Na,K-ATPase reaction mechanism. The scheme is a modified version of the Post-Albers model and is described in the text.

reaction pathway is ping-pong (Cleland, 1963), whereby the enzyme alternately binds and releases Na^+ and K^+, or is one in which both ions are bound, at least at some phase of the cycle, has been evaluated in detail by Sachs (1986a). His results give strong support to the predominance of a ping-pong Na/K exchange mechanism.

2.1. Na/Na Exchange, K/K Exchange, Uncoupled Na^+ Efflux and Uncoupled K^+ Efflux

In addition to normal Na/K exchange, the pump can mediate other modes of transport depending on the intra- and extracellular cation composition and on the intracellular composition of substrates, namely ATP, ADP, and inorganic phosphate. The characteristics of Na/Na exchange, K/K exchange, and uncoupled Na^+ efflux of red cells have been reviewed elsewhere (Kaplan, 1985; Glynn, 1985; Hoffman, 1986). Briefly, exchange of Na_i^+ for Na_o^+ is observed with Na^+ in each of the two compartments labeled specifically ($^{22}Na^+$ and $^{24}Na^+$). Accordingly, an ouabain-inhibitable influx of Na^+ appears in the absence of extracellular $K^+(K_o)$ and diminishes as a function of increasing K_o^+ and also if Na_i^+ is removed. This flux requires intracellular ADP as well as ATP, and can be accounted for by the sequence of steps denoted by the forward and reverse reactions of the sequence a, b, c, d and e in Fig. 1. Although no net hydrolysis of ATP occurs, the reaction is associated with

Na^+-dependent phosphorylation of the enzyme as well as Na^+-dependent ADP–ATP exchange, namely reactions a, b, and c. As discussed by Kaplan (1985), the enzymic and ion exchange reactions are linked, but only partially coupled.

In the absence of alkali cations, Glynn and Karlish (1967) noted a relatively slow ouabain-sensitive downhill efflux of Na^+ from red cells associated with ATP hydrolysis. Originally, the absence of an obvious countercation led these investigators to refer to this behavior as uncoupled Na^+ efflux; for convenience, this Na^+ efflux will be referred to here as a Na/O flux. Its association with ATP hydrolysis is similar to that of normal Na/K exchange in that the Na/O flux and associated Na-ATPase activity have a high affinity for ATP, representing a slow turnover of the enzyme via the sequence of reactions a, b, c, d, e, j, and k (Fig. 1). In contrast, during normal Na/K exchange, pump turnover is high provided the concentration of ATP is sufficiently high to bind, with much lower affinity, to facilitate the release at step i of the K^+, which becomes bound during K^+-stimulated dephosphorylation of E_2P (Hegyvary and Post, 1971). Thus, Na/K exchange is represented by the sequence of reactions a, b, c, d, e, f, g, h, and i.

Several years ago, Dissing and Hoffman (1983) showed that the so-called uncoupled Na^+ efflux is an electroneutral Na^+-anion coupled efflux from red cells, at least at physiological pH. To observe this flux, they studied Na^+ efflux from sulphate-loaded red cells pretreated with the anion exchange inhibitor DIDS.‡ Under this condition, they observed an ouabain-sensitive $SO_4^=$ efflux stoichiometrically equivalent to the efflux of two Na^+ ions. This anion-coupled Na^+ efflux as well as Na/Na exchange are inhibited, albeit incompletely, by low amounts of extracellular Na^+ (2–5 m*M*). It now appears that the behaviour characterized originally as "uncoupled" Na efflux is further complicated as indicated by the following observations: (1) in the presence of 2–5 m*M* Na_o^+, the residual ouabain-sensitive, Na_o^+-insensitive efflux is characterized by another type of anion-coupled transport, namely the cotransport of Na^+ with P_i derived *directly* from the ATP hydrolyzed by Na,K-ATPase (Marin and Hoffman, 1988); (2) a new ouabain-sensitive ATP-inhibited Na^+ efflux has also been described; it is coupled to P_i efflux, supported by ADP plus P_i and is completely inhibited by 10 m*M* Na_o or by 20 m*M* K_o provided intracellular K^+ is absent (Marin and Hoffman, 1986).

The model shown in Fig. 1 is essentially similar to the one depicted elsewhere (Fig. 10 of the paper of Sachs, 1986b). Not only does it account for the Na/O flux coupled to hydrolysis of E_2P, but it predicts the

‡DIDS: 4,4′-diisothiocyano-2,2′-stilbene disulfonate.

observed uncoupled K^+ efflux from K^+-filled human red cells suspended in alkali cation-free medium (K/O efflux) via the sequence i, h, g, f (Sachs, 1986b).

Another mode of pump operation evident only in the red cells of certain species (human and goat) is ouabain-sensitive K/K exchange (Glynn et al., 1970). Analogous to Na/Na exchange, this mode is not associated with hydrolysis of ATP, but is accounted for by the sequence of reversible steps denoted by f, g, h, and i in Fig. 1, requiring both ATP or one of its nonhydrolyzable analogues (Simons, 1974 and 1975), or ADP (Kaplan and Kenney, 1982) and either inorganic phosphate, or arsenate (Kenney and Kaplan, 1985). Moreover, as in the case of ouabain-sensitive ADP-ATP exchange and Na^+/Na^+ exchange, the biochemical reaction (P_i—water oxygen exchange) is more rapid than the K/K exchange, indicating coupling without tight linkage (*see* Kaplan, 1985).

Although the various pump modes were first described in intact human red blood cells, most are evident in hemoglobin-depleted resealed ghosts and/or inside–out vesicles, the latter providing the advantage of allowing direct access to the cytoplasmic surface. Thus, with resealed ghosts, the sidedness of Na^+ and K^+ effects on overall hydrolysis was established (*see* Hoffman, 1986). With inside-out vesicles, it was shown that cytoplasmic (extravesicular) Na^+ promotes phosphorylation and extracellular (intravesicular) K^+ promotes dephosphorylation of the enzyme (Blostein and Chu, 1977; Blostein et al., 1979). Furthermore, K^+ interactions at both surfaces of inside-out vesicles were evidenced in studies of K^+-activation of strophanthidin-sensitive *p*-nitrophenylphosphatase (Drapeau and Blostein, 1980a): in the absence of Na^+, K_i is required, whereas in the presence of both Na_i and ATP, K_o^+ promotes activity, consistent with the reversible sequence of steps f, g, h, and i (c.f. K/K exchange).

In the past, fluxes in mammalian tissues other than red cells have been hard to assess. More recently, proteoliposomes comprising detergent-purified Na,K-ATPase from active sources such as the kidney have provided further insights into certain aspects of pump behavior not readily apparent in the weakly active red-cell system. These include, for example, the voltage dependence of the E_1P to E_2P conformational transition reaction (Raphaeli et al., 1986) and characterization of cation translocation kinetics during a single turnover of the pump (Forbush, 1984). However, it remains difficult to compare the enzyme of the red-cell and kidney proteoliposomes quantitatively or to know whether apparent differences in pump behavior reflect fundamental differences between tissue Na,K-ATPases. Anion-coupled Na efflux, for instance, has been described only in the red cell. Another problem is the fact that, relative to the red cell or even inside-out red-cell membrane vesicle, the other (purified) preparations comprise enzyme subjected to detergent treatment during purifications; following reconstitution of the purified enzyme into lipids, the resulting proteoliposomes have small surface-to-volume ratios, which has usually limited the assessment as well as quantification of the various pump modes. In certain experiments, namely those concerned with normal Na/K exchange, the latter problem has been elegantly circumvented by using an optical method for detecting K^+ fluxes, thus enabling the accurate measurement of fluxes on a rapid time scale (Marcus et al., 1986).

In spite of the low pump density of human red blood cells, small ouabain-sensitive facilitated diffusion of cations such as the K^+ congener Rb^+, first described in experiments with purified kidney Na,K-ATPase proteoliposomes (Karlish and Stein, 1982), have also been observed in red-cell ghosts (Kenney and Kaplan, 1985). These small fluxes of Rb^+ occur in the absence of reactants that normally interact at substrate (ATP, ADP, P_i) sites. The implications regarding mechanism are interesting. Thus, the finding that extracellular Rb^+ can bind with high affinity provides evidence of interactions of Rb^+ with unphosphorylated E_2 as well as E_2P.

2.2. Affinities for Cations and the Role of Protons

Originally, various pump modes were characterized by the alternative behavior observed when the substrates ATP, ADP, and P_i were changed, and also when intracellular Na^+ was substituted for K^+, or extracellular K^+, for Na^+. It is also clear that the affinities of the pump for Na^+ at the cytoplasmic surface and for K^+ at the extracellular surface are not absolute. Other cations, such as Li^+, Cs^+, Rb^+, as well as K^+, were shown to compete with Na^+ at the intracellular loading site(s) and others, such as Rb^+, Cs^+, and Li^+, for K^+ at the extracellular discharge site(s). (For a more detailed discussion, *see* Hoffman, 1986). Therefore, it may not be surprising to find that pump modes that appear different from normal Na/K exchange may reflect the lack of absolute selectivity of the pump for intracellular Na^+ and extracellular K^+. An example is Na/Na exchange, not dependent upon ADP, and associated with net hydrolysis of ATP in a manner consistent with extracellular Na^+ acting as a substitute for extracellular K^+ (Lee and Blostein, 1981; Blostein, 1983) although its maximal activity is approximately an order of magnitude lower than normal Na/K exchange.

Results from experiments with inside-out vesicles support the notion that protons can replace *both* Na^+ at the cytoplasmic and K^+ at the extracellular surfaces (Blostein, 1985; Polvani and Blostein, 1988a,b). In these studies, $^{86}Rb^+$ efflux (equivalent to K^+ or Rb^+ influx into cells) was assayed with inside-out membrane vesicles containing $^{86}Rb^+$ and suspended in alkali cation-free medium; alternatively, $^{22}Na^+$ influx (equivalent to efflux from cells) was assayed with alkali cation-free vesicles suspended in $^{22}Na^+$-containing medium. Under these two conditions, respectively, an O/K flux (actually Rb^+ efflux), almost absent at physiological pH, was clearly evident when the pH was lowered to about 6.0; the Na/O flux normally observed at physiological pH was increased by lowering the pH to 6.0. In contrast, normal Na/K exchange was decreased more than 50% under these conditions.

These fluxes of Rb^+ (presumably K^+ also) and Na^+ are associated with H^+ translocations in the opposite directions consistent with the notion of pump-mediated Na/H and H/K exchanges. To detect proton movements across the red-cell membrane, inside-out vesicles derived from cells pretreated with DIDS to inhibit anion permeation were filled, during vesiculation, with FITC-dextran.§ Under the conditions described above in which the O/K and Na/O fluxes were enhanced, intravesicular acidification and alkalinization, respectively, were observed (Polvani and Blostein, 1988b). An Na^+-like role of protons has been reported in similar experiments carried out with K^+-filled kidney Na,K-ATPase proteoliposomes (Hara and Nakao, 1986), although counterfluxes of K^+ or Rb^+ have not been described. Whether the putative Na/H exchange of red cells replaces Na/O flux occurring at physiological pH (*see above*) or is an additional mode is not known at present.

There is also indirect evidence for H^+ influx and efflux, even in the presence of both Na_i^+ and K_o^+ when the pH is reduced. Thus, with inside-out vesicles, it was observed that the number of Na^+ as well as K^+ ions transported/mol ATP hydrolyzed is reduced about 50% when the pH is reduced from 7.6 to 6.2 (Polvani and Blostein, 1988a). Whether this reduced "efficiency" reflects H/H exchange and/or Na/H and H/K exchanges in addition to normal Na/K exchange is a moot question. Relevant to the possibility of H/H exchange are reports that cardiac glycoside-sensitive ATP hydrolysis is markedly enhanced in the absence of both Na^+ and K^+ when the pH is reduced. This has been observed with red-cell membranes (Polvani and Blostein, 1988a) as well as purified kidney Na,K-ATPase (Plesner and Plesner, 1985).

§FITC: Fluorescein isothiocyanate.

Finally, it would be interesting to know whether the sodium pump's efficiency in transporting Na^+ and K^+ is reduced when acid-base balance is disturbed. Only a small change in coupling could have critical consequences for the organism's cation homeostasis.

3. Are There Characteristics of the Na,K-ATPase Unique to Red Cells?

Experiments with red cells and membrane preparations derived from them have supported the conclusion that the Na,K-ATPase reaction cycle is fundamentally similar to that of the enzyme from other tissues, such as the Na,K-ATPase-rich kidney. Nevertheless, several differences have been noted, including differences in temperature sensitivities and ligand-induced ouabain binding. In addition, certain features of pump behavior, most notably the anion-coupled Na^+ efflux have been described only in red cells. As described subsequently, experiments with red-cell ghosts have provided information about specific pump interactions with, as well as modulation by, component(s) of the membrane environment. Another aspect of distinctive red cell Na,K-ATPase behavior concerns the different kinetic behavior of Na,K-ATPase among the genetically distinct red cells of certain species.

3.1. Effects of Temperature

The steady-state composition of phosphoenzyme intermediates depicted in Fig. 1 comprises two major classes: ADP-sensitive, K^+-insensitive phosphoenzyme (predominantly E_1P) and ADP-insensitive K^+-sensitive (predominantly E_2P) phosphoenzyme. When the Na,K-ATPase reaction is studied at low temperature (0°C), it is observed that the kinetic behavior of the red-cell and kidney enzymes are markedly different. Thus, the steady-state distribution of phosphoenzyme forms of the red-cell enzyme comprise predominantly K^+-insensitive E_1P, whereas kidney phosphoenzyme is predominantly K^+-sensitive E_2P (White and Blostein, 1982: Kaplan and Kenney, 1984). It may be relevant that alterations in lipid composition of kidney Na,K-ATPase proteoliposomes, most notably the cholesterol content, have been shown to change its steady-state phosphoenzyme composition and kinetic behavior (Yoda and Yoda, 1987). It remains to be determined whether differences in lipid composition are the basis for the distinctive red-cell kinetics observed at low temperature.

3.2. Interactions with Ouabain

Cardiac glycosides such as ouabain bind to Na,K-ATPase at the extracellular surface (Hoffman, 1980; Perrone and Blostein, 1973), and the rate of binding is modulated by ligands present at both the cytoplasmic and extracellular surfaces. (For a more detailed description of the effects and their plausible interpretations, *see* Hoffman, 1978). A major, yet unexplained difference between ouabain binding to mature red-cell membranes compared to other systems (cf. Schwartz et al., 1968) is an absolute requirement of the latter for Na^+ (Hoffman, 1978). This feature, taken together with the marked differences, at least at 0°C, in conformational states of phosphoenzyme of the red-cell vs kidney Na,K-ATPase discussed in 3.1. points to the possibility that the relative rates of one (or more) steps of the reaction cycle may be different for the two systems, not only at 0°C, but also at higher temperatures, thus altering the steady-state distribution of form(s) to which ouabain preferentially binds.

3.3. Intracellular Ca^{2+}

It is well known that intracellular Ca^{2+}, even at relatively low concentrations, inhibits Na,K-ATPase. This sensitivity to Ca^{2+} inhibition was shown to be increased more than an order of magnitude (30-fold) by material present in human red-cell hemolysate (Yingst, 1983). The material named calnaktin is a potentially important regulator of Na,K-ATPase activity. The material has been identified as a ~30,000-dalton peptide, distinct from calmodulin (Yingst, 1988).

3.4. Membrane-Bound Glycolytic Enzymes and Their Interactions with Na,K-ATPase via Compartmentalized ATP

During the course of experiments with red-cell ghosts, it became apparent that the membrane environment of the Na,K-ATPase comprises a membrane-bound pool of ATP. This pool is preferentially used by the sodium pump and can be filled by membrane-bound glyceraldehyde-3 phosphate dehydrogenase and phosphoglycerate kinase (Parker and Hoffman, 1967; Proverbio and Hoffman, 1977; Mercer and Dunham, 1981). Although the physiological relevance of compartmentalization of ATP turnover has been enigmatic, its possible relevance to pumped regulation and efficiency can be readily appreciated. Moreover, this aspect of *in situ* pump regulation may not be confined to the red cell: evidence for pump energization by compartmentalized energy production, albeit less direct, has been reported in brain slices (Lipton and Robacker, 1983), vascular smooth muscle (Paul et al., 1979), and in cardiac muscle cells (Saks et al., 1977). The phenomenon of compartmentalized ATP has even broader implications for regulation of cation homeostasis. Thus, in the heart, glycolysis appears to be a preferential source of the ATP required to prevent ATP-sensitive K^+-channels from opening (Weiss and Lamp, 1987).

4. Functional Variants of the Na,K-ATPase: Sodium Pump Behavior in Genetically Dimorphic Sheep Red Blood Cells

Whereas the mature red cells of certain species, such as humans, have high intracellular K^+, others, such as those of dogs and cats, have low K^+. In addition, there is a genetic dimorphism in the cation content of certain species, most notably ruminants, such as sheep, goats, and cattle. In sheep, for example, the cells of one type (high-K^+ or HK) have a higher intracellular K^+ content than the other type (low-K^+ or LK) (Abderhalden, 1898; Kerr, 1937; Evans, 1957). These differences are associated with higher active Na/K pump activity and lower passive fluxes in HK compared to LK sheep cells (Tosteson and Hoffman, 1960). In addition, the *kinetic* behavior of the pumps of these two types of cells are distinct with respect to affinities for specific ligands, particularly Na^+ and K^+. This is examplified by the ouabain-sensitive pump activity of intact HK and LK cells examined as a function of varying intracellular K^+ concentration (Hoffman and Tosteson, 1971). Kinetic differences between HK and LK are also revealed when the Na^+-activated ATP hydrolysis of permeable ghosts is assayed, particularly at low ATP concentrations, to minimize nonspecific ATP hydrolysis (Whittington and Blostein, 1971).

As illustrated in Fig. 2, in both types of functional assays, the activity of LK, but not HK is markedly inhibited by K^+. With preparations having sidedness, namely sealed inside-out membrane vesicles, it is evident that the apparent affinity for K_i as an inhibitor and for Na_o as an activator of Na_i-activated ATP hydrolysis is greater in LK than HK (Drapeau and Blostein, 1980b), whereas the affinity for Na_i as an activator is less in LK. Another striking difference is that the ATP concentration-dependent K_o-activation of Na-ATPase is marked in HK, but clearly weak in LK (ATP added in the range of 0.2–2.0 μM). The latter kinetic difference may remind us that relative reactivities may not reflect true affinities, and it is entirely possible that the LK and HK

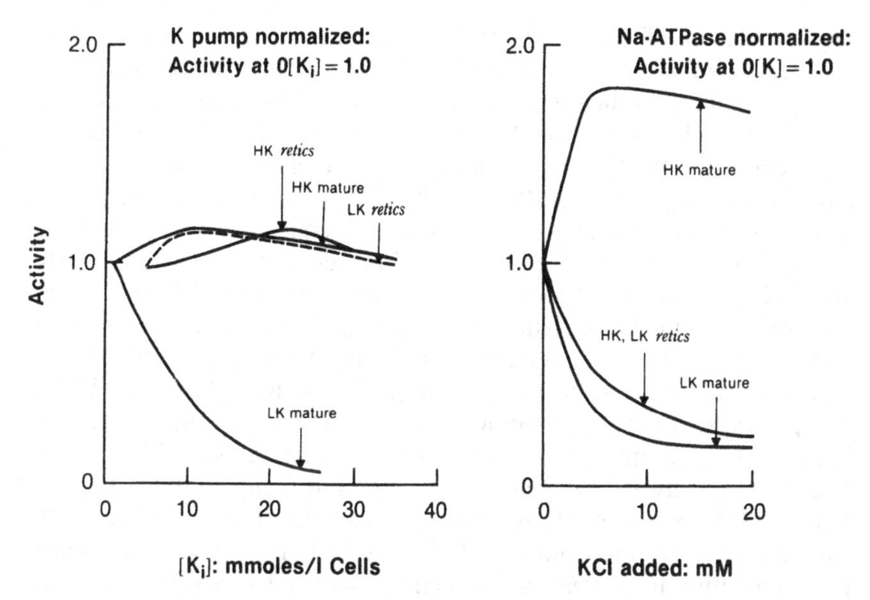

Fig. 2. Kinetic differences in Na,K-pump and Na-ATPase of mature and immature red cells (reticulocytes) of HK and LK sheep. Data shown schematically are taken from studies of Hoffman and Tosteson (1971), Blostein et al. (1974, 1983), Whittington and Blostein (1971), and Dunham and Blostein (1976).

enzymes differ in their ability to catalyze reaction step(s) other than one directly involved in ion binding and release, for example, the interaction with and modulation by ATP. Moreover, from a consideration of the sequence of reactions involved in K^+ binding and occlusion (Fig. 1), it is likely that the relative rate of transition of E_1K to $E_2(K)$ is slowed, whether K binds initially to E_1 at the cytoplasmic face or to E_2P or even E_2 at the extracellular side. Thus, one explanation for HK-LK differences would be a lack of modulation of the $E_2K \rightleftharpoons E_1K$ transition by ATP. Such a basic difference would likely be evidenced in other functional differences such as the lower pump turnover (K^+ pump activity/ouabain binding site) of LK compared to HK cells (Joiner and Lauf, 1978).

In contrast to mature cells, immature cells, mostly reticulocytes, of anemic LK as well as HK sheep have high intracellular K^+ and much higher pump activity than mature cells (Lee et al., 1966, Blostein et al., 1974). The kinetic behavior of their pumps is distinct from that of either mature LK or HK cells. Thus, as illustrated in Fig. 2, when K^+-pump activity is measured as a function of intracellular K^+ concentration, the kinetic behavior of reticulocyte-type pumps, whether from an HK or LK animal, is similar to that of mature HK pumps (Dunham and Blostein, 1976). In contrast, when Na-ATPase is examined as a function of added K^+, the reticulocyte enzyme has characteristics of the LK-type enzyme (Blostein et al., 1974). It can be concluded, therefore, that maturation of reticulocytes of genetically HK and LK sheep is attended by quantitative as well as kinetic changes, as discussed subsequently.

In several respects, the kinetic differences associated with the HK and LK phenotypes as well as the maturation-associated functional changes may reflect modulation of pump behavior resulting from interactions with specific surface antigen(s). Thus, the polymorphism of the pump is associated with the polymorphism of two classes of blood groups, M and L: HK cells are M type whereas LK cells are L type, the LK allele being dominant. (For a comprehensive discussion of the genetic relationships between antigens and the sodium pump of red cells, *see* Lauf, 1982). Interestingly, interaction of specific anti-L antibodies with LK cells increases the number of functioning pumps, partly by reducing noncompetitive inhibition by intracellular K^+ and partly by somehow increasing V_{max} (Dunham and Anderson, 1987). Furthermore, the pumps of mature LK cells treated with anti-L resemble those of reticulocytes in that the pump kinetics are HK-like (Dunham and Blostein, 1976), whereas their Na,K-ATPase kinetics are LK-like (Blostein et al., 1974; Blostein and Whittington, 1973). At the present time, the molecular basis for the apparent multifaceted behavioral differences of the Na,K-ATPase of reticulocytes and of mature HK and LK cells remains unknown. One possibility is that reticulocyte-type pumps not only diminish during cell maturation, but also change to mature LK or HK pumps. A simple accounting for the LK maturation changes would be the interaction of the enzyme with the L antigen. Another possibility is that more than one isoform of the enzyme complex exists. Accordingly, cell maturation-associated losses in one type, perhaps in addition to posttranslational modifications, ensue. Particularly relevant to the latter possibility are earlier experiments that showed developmentally regulated pump changes. Thus, the pumps of mature red cells of LK neonates resemble those of mature HK cells. During ontogeny, these cells are gradually replaced by cells with the characteristics of mature HK cells (Blostein et al., 1974; Valet et al., 1978).

Taken together, these observations indicate that the genetic dimorphism evidenced in the distinct HK and LK cells of sheep is the result of selective changes that occur during ontogeny and during reticulocyte maturation. These changes proceed differently in the cells of the two genotypes. In both, functional decreases are associated with losses in the catalytic subunit detected on immunoblots of the electrophoresed red-cell membranes, not only in HK and LK sheep (Blostein and Grafova, 1988), but also in a strain of HK and LK dogs (Inaba and Maede, 1986). We have observed, however, that the decreases in total specific ouabain binding sites do not necessarily coincide with the decrease in catalytic subunit. In HK animals in particular, the decrease in ouabain binding sites is considerably greater than the decrease in catalytic subunit. This may be relevant to the report of pump heterogeneity in mature HK cells, in which about 70% of the pump sites exhibit almost all of the pump activity (Joiner and Lauf, 1978). It is also intriguing to note that the distinctive molecular architecture of the pump as it exists *in situ* may dictate not only its behavior, but also its turnover (Blostein and Grafova, 1987, 1988).

5. Conclusion

This brief account illustrates the unique usefulness of the red cell for studies of the ubiquitous Na,K-ATPase as it exists in its virtually native environment. Devoid of a nucleus, this simple parcel containing mainly hemoglobin has, nevertheless, a dynamic membrane physiology. The mature cell and its immediate precursor, the reticulocyte, may enable us to gain important insights into the role of the membrane in regulating Na,K-ATPase behavior.

Acknowledgment

I thank Pierre Drapeau for the helpful criticisms of the manuscript. This work was supported by a grant from the Medical Research Council of Canada.

References

Abderhalden, E. (1898) Zur Quantitativen vergleichenden Analyse des Blutes. *Hoppe Seylers Z. Physiol. Chem.* **25,** 65–115.

Bolstein, R. (1983) Sodium pump catalyzed sodium-sodium exchange associated with ATP hydrolysis. *J. Biol. Chem.* **248,** 7948–7953.

Blostein, R. (1985) Proton-activated rubidium transport catalyzed by the sodium pump. *J. Biol. Chem.* **260,** 839–843.

Blostein, R. and Chu, L. (1977) Sidedness of (sodium, potassium)—adenosine triphosphatase of inside-out red cell membrane vesicles. *J. Biol. Chem.* **205,** 3035–3043.

Blostein, R. and Grafova, E. (1987) Characterization of membrane transport losses during reticulocyte maturation. *Biochem. Cell Biol.* **65,** 869–875.

Blostein, R. and Grafova, E. (1988) Loss of Na,K-ATPase during sheep reticulocyte maturation, in *The Na^+,K^+-Pump, Part B: Cellular Aspects* (Alan R. Liss, Inc., New York), pp. 357–364.

Blostein, R., Drapeau, P., Benderoff, S., and Weigensberg, A. (1983) Changes in Na^+-ATPase during maturation of sheep reticulocytes. *Can. J. Biochem. Cell Biol.* **61,** 23–28.

Blostein, R. and Whittington, E. S. (1973) Studies of high and low potassium sheep erythrocyte membrane sodium-adenosine triphosphatase. Interactions with oligomycin, ATP, sodium and potassium. *J. Biol. Chem.* **248,** 1772–1777.

Blostein, R., Whittington, E. S., and Kuebler, E. S. (1974) Na^+-ATPase of mammalian erythrocyte membranes: kinetic changes associated with postnatal development and following active erythropoiesis. *Ann. N.Y. Acad. Sci.* **242,** 305–316.

Blostein, R., Pershadsingh, H. A., Drapeau, P., and Chu, L. (1979) Side-specificity of alkali cation interactions with Na,K-ATPase, in *Na,K-ATPase.* (Skou, J. C. and Norby, J. G. eds.), Academic Press, New York. pp. 233–245.

Cavieres, J. D. (1977) The sodium pump in human red cells, in *Membrane Transport in Red Cells.* (Ellory, J. C., ed.), Academic Press, New York, pp. 1–38.

Cleland, W. W. (1963) The kinetics of enzyme-catalyzed reactions with two or more substrates or products. I. Nomenclature and rate equations. *Biochim. Biophys. Acta* **67,** 104–137.

Danowski, T. S. (1941) The transfer of potassium across the human blood cell membrane. *J. Biol. Chem.* **139,** 693–705.

Dissing, S. and Hoffman, J. F. (1983) Anion-coupled Na efflux mediated by the Na/K pump in human red blood cells. *Curr. Top. Membr. Trans.* **19,** 693–695.

Drapeau, P. and Blostein, R. (1980a) Interactions of K^+ with (Na,K)-ATPase: Orientation of K^+ phosphatase sites studied with inside-out red cell membrane vesicles. *J. Biol. Chem.* **255,** 7827–7834.

Drapeau, P. and Blostein, R. (1980b) Sodium and potassium interactions with Na^+-ATPase of inside-out membrane vesicles from high-K^+ and low-K^+ sheep red cells. *Biochim. Biophys. Acta* **596,** 372–380.

Dunham, E. T. and Glynn, I. M. (1961) Adenosine triphosphatase activity and the active movements of alkali metal ions. *J. Physiol. (London)* **156,** 274–293.

Dunham, P. B. and Anderson, G. (1987) On the mechanism of stimulation of the Na/K pump of LK sheep erythrocytes by anti-L antibody. *J. Gen. Physiol.* **90,** 3–25.

Dunham, P. B. and Blostein, R. (1976) Active potassium transport in reticulocytes of high K^+ and low K^+ sheep. *Biochim. Biophys. Acta* **455,** 749–758.

Evans, J. V. (1957) The stability of the potassium concentration in the erythrocytes of the individual sheep compared with the variability between different sheep. *J. Physiol. (London)* **136,** 41–59.

Fahn, S., Koval, G. J., and Albers, R. W. (1966a) Sodium-potassium-activated adenosine triphosphatase of Electrophorus electric organ. I. An associated sodium activated transphosphorylation. *J. Biol. Chem.* **241,** 1882–1889.

Fahn, S., Hurley, M. R., Koval, G. J. and Albers, R. W. (1966b) Sodium-potassium-activated adenosine triphosphatase of Electrophorus electric organ. II. Effects of *N*-ethylmaleimide and other sulfhydryl reagents. *J. Biol. Chem.* **241,** 1890–1895.

Forbush, B. (1984) Na^+ movement in a single turnover of the Na pump. *Proc. Natl. Acad. Sci. (USA)* **81,** 5310–5314.

Gardos, G. (1954) Akumulation der Kaliumionen durch menschliche Blutkörperchen. *Acta Physiol. Acad. Sci. Hung.* **6,** 191–199.

Garrahan, P. J. and Glynn, I. M. (1967) The stoichiometry of the sodium pump. *J. Physiol. (London)* **192,** 217–235.

Glynn, I. M. (1985) The Na,K-transporting adenosine triphosphatase, in *The Enzymes of Biological Membranes*, Vol. 3, 2nd Ed. (Martonosi, A., ed.), Plenum Press, New York, pp. 35–114.

Glynn, I. M. and Karlish, S. J. D. (1967) ATP hydrolysis associated with an uncoupled Na efflux through the sodium pump: evidence for allosteric effects of intracellular ATP and extracellular Na. *J. Physiol. (London)* **256,** 465–496.

Glynn, I. M., Lew, V. L., and Lüthi, U. (1970) Reversal of the potassium entry mechanism in red cells, with and without reversal of the entire pump cycle. *J. Physiol. (London)* **207,** 371–391.

Goldin, S. M. and Tong, S. W. (1974) Reconstitution of active transport by the purified Na and K ion-stimulated adenosine triphosphatase from canine renal medulla. *J. Biol. Chem.* **249,** 5907–5915.

Hara, Y. and Nakao, M. (1986) ATP-dependent proton uptake by proteoliposomes reconstituted with purified Na^+,K^+-ATPase. *J. Biol. Chem.* **261,** 12655–12660.

Harris, J. E. (1941) The influence of the metabolism of human erythroocytes on their potassium content. *J. Biol. Chem.* **141,** 579–595.

Hegyvary, C. and Post, R. L. (1971) Binding of adenosine triphosphate to sodium and potassium-ion stimulated adenosine triphosphatase. *J. Biol. Chem.* **246,** 5234–5240.

Hilden, S., Rhee, H. M., and Hokin, L. E. (1974) Sodium transport by phospholipid vesicles containing purified sodium and potassium-ion activated adenosine triphosphatase. *J. Biol. Chem.* **249,** 7432–7440.

Hoffman, J. F. (1962) The active transport of sodium by ghosts of human red blood cells. *J. Gen. Physiol.* **45,** 837–859.

Hoffman, J. F. (1978) Asymmetry and the mechanism of the red cell Na-K pump, determined by ouabain binding, in *Molecular Specialization and Symmetry in Membrane Function*. Solomon, A. K. and Karnovsky, M., eds.), Harvard Univ. Press, Mass. pp. 191–211.

Hoffman, J. F. (1980) The link between metabolism and active transport of sodium in human red cell ghosts. *J. Membr. Biol.* **57,** 143–161.

Hoffman, J. F. (1986) Active transport of Na^+ and K^+ by red blood cells, in *Physiology of Membrane Disorders*, 2nd Ed. (Andreoli, T. E., Hoffman, J. F., Fanestil, D. F., and Schultz, S. G., eds.), Plenum Press, New York, pp. 221–231.

Hoffman, J. F., Kaplan, J. H., and Callahan, T. J. (1979) The Na:K pump in red cells is electrogenic. *Fed. Proc.* **38,** 2440–2441.

Hoffman, P. G. and Tosteson, D. C. (1971) Active sodium and potassium transport in high potassium and low potassium sheep red cells. *J. Gen. Physiol.* **58,** 438–466.

Inaba, M. and Maeda, Y. (1986) Na,K-ATPase in dog red cells. Immunological identification and maturation-associated degradation by the proteolytic system. *J. Biol. Chem.* **261,** 16099–16105.

Joiner, C. H. and Lauf, P. K. (1978) The correlation between ouabain binding and potassium pump inhibition in human and sheep erythrocytes. *J. Physiol. (London)* **283,** 155–175.

Kaplan, J. H. (1985) Ion movements through the sodium pump. *Annu. Rev. Physiol.* **47,** 535–544.

Kaplan, J. H. and Kenney, L. J. (1982) ADP supports ouabain-sensitive K-K exchange in human red blood cells. *Ann. N.Y. Acad. Sci.* **402,** 292–295.

Kaplan, J. H. and Kenney, L. J. (1984) Temperature effects on sodium pump phosphoenzyme distribution in human red blood cells. *J. Gen. Physiol.* **85,** 123–136.

Karlish, S. J. D. and Stein, W. D. (1982) Passive rubidium fluxes mediated by the Na,K-ATPase reconstituted into phospholipid vesicles when ATP- and phosphate-free. *J. Physiol. (London)* **328,** 295–316.

Kenney, L. J. and Kaplan, J. H. (1985) Arsenate replaces phosphate in ADP-dependent and ADP-independent Rb^+-Rb^+ exchange mediated by the red cell sodium pump, in *The Sodium Pump*. (Glynn, I. and Ellory, C., eds.) The Company of Biologists, Cambridge, pp. 535–539.

Kerr, S. E. (1937) Studies on the inorganic composition of blood. IV. The relationship of potassium to the acid-soluble phosphorus fractions. *J. Biol. Chem.* **117,** 227–235.

Lauf, P. K. (1982) Active and passive cation transport and its association with Membrane antigens in sheep erythrocytes, in *Membranes and Transport*, Vol. 1 Martonosi, A., ed.) Plenum, New York, pp. 553–558.

Lee, K. H. and Blostein, R. (1981) Red cell sodium fluxes catalyzed by the sodium pump in the absence of K^+ and ADP. *Nature* **285,** 338–339.

Lee, P., Woo, A., and Tosteson, D. C. (1966) Cytodifferentiation and membrane transport properties in LK sheep red cells. *J. Gen. Physiol.* **50,** 379–390.

Lipton, P. and Robacker, K. (1983) Glycolysis and brain function: $(K^+)_o$ stimulation of protein synthesis and K^+ uptake require glycolysis. *Fed. Proc.* **42,** 2875–2880.

Marcus, M. M., Apell, H.-J., Roudna, M., Schwendener, R. A., Weder, H. G., and Lauger, P. (1986) $(Na^+ + K^+)$-ATPase in artificial lipid vesicles: influence of lipid structure on pumping rate. *Biochim. Biophys. Acta* **854,** 270–278.

Marin, R. and Hoffman, J. F. (1986) Ouabain-sensitive Na/K pump mediated efflux of Na and inorganic phosphate (P_i) stimulated by ADP and inhibited by ATP in human red cell ghosts. *Biophys. J.* **49,** 582a.

Marin, R. and Hoffman J. F. (1988) Two different types of ATP-dependent anion coupled Na transport are mediated by the human red blood cell Na/K pump, in *The $Na^+_,K^+$-Pump Part B: Cellular Aspects* (Alan R. Liss, Inc., New York) pp. 539–544.

Mercer, R. W. and Dunham, P. B. (1981) Membrane-bound ATP fuels the Na/K pump: Studies on membrane bound glycolytic enzymes on inside-out vesicles from human red cell membranes. *J. Gen. Physiol.* **78,** 547–568.

Parker, J. C. and Hoffman, J. F. (1967) Thr role of membrane phosphoglycerate kinase in the control of glycolytic rate by cation transport in human red blood cells. *J. Gen. Physiol.* **50,** 893–916.

Paul, R. J., Bauer, M., and Peake, W. (1979) Vascular smooth muscle: aerobic glycolysis linked to sodium and potassium transport processes. *Science* **206,** 1414–1416.

Perrone, J. F. and Blostein, R. (1973) Asymmetric interaction of inside-out and right-side out erythrocyte membrane vesicles with ouabain. *Biochim. Biophys. Acta* **291,** 680–689.

Plesner, L. and Plesner, I. W. (1985) Ouabain-inhibited ATPase activity in the absence of Na^+ and K^+, in *The Sodium Pump.* (Glynn, I. and Ellory, C., eds.), The Company of Biologists, Cambridge, pp. 469–474.

Polvani, C. and Blostein, R. (1988a) Proton effects on the sodium pump, in *The $Na^+_,K^+$-Pump Part B: Molecular Aspects* (Alan R. Liss, Inc., New York) pp. 553–559.

Polvani, C. and Blostein, R. (1988b) Protons as substitutes for sodium and potassium in the sodium pump reaction. *J. Biol. Chem.* **293,** 16757–16763.

Post, R. L. and Jolly, P. C. (1957) The linkage of sodium, potassium and ammonium active transport across the human erythrocyte membrane. *Biochim. Biophys. Acta* **25,** 118–128.

Post, R. L., Hegyvary, C., and Kume, S. (1972). Activation of adenosine triphosphate in the phosphorylation kinetics of sodium and potassium ion transport adenosine triphosphatase. *J. Biol. Chem.* **247,** 6530–6540.

Post, R. L., Merritt, L. C., Kinsolving, C. R., and Albright, C. D. (1960) Membrane adenosine triphosphatase as a participant in the active transport of sodium and potassium in the human erythrocyte. *J. Biol. Chem.* **235,** 2796–2802.

Proverbio, F. and Hoffman, J. F. (1977) Membrane compartmentalized ATP and its preferential use by the Na,K-ATPase of human red cell ghosts. *J. Gen. Physiol.* **69,** 605–632.

Raphaeli, A., Richards, D. E., and Karlish, S. J. D. (1986) Electrical potential accelerates the $E_1P(Na)$—E_2P conformational transition of (Na,K)-ATPase in reconstituted vesicles. *J. Biol. Chem.* **261,** 12437–12440.

Sachs, J. R. (1986a) The order of addition of sodium and release of potassium at the inside of the sodium pump of the human red cell. *J. Physiol. (London)* **381,** 149–168.

Sachs, J. R. (1986b) Potassium-potassium exchange as part of the over-all reaction mechanism of the sodium pump of the human red blood cell. *J. Physiol. (London)* **374,** 221–244.

Saks, V. A., Lipina, N. V., Sharov, V. G., Smirnov, V. N., Chazov, E., and Grosse, R. (1977) The localization of the MM isozyme of creatine phosphokinase on the surface membrane of myocardial cells and its functional coupling to ouabain-inhibited (Na^+, K^+)-ATPase. *Biochim. Acta.* **465,** 292–313.

Schatzmann, H. J. (1953) Herzglykoside als Hemmstoff für den aktiven Kalium und Natrium-transport durch die Erythrocyten-membrane. *Helv. Physiol. Pharmacol. Acta* **11,** 346–354.

Schwartz, A., Matsui, H., and Laughter, A. H. (1968) Tritiated digoxin binding to (Na^+K^+)-activated adenosine triphosphatase: Possible allosteric site. *Science* **160,** 323–325.

Sen, A. K. and Post, R. L. (1964) Stoichiometry and localization of adenosine triphosphate-dependent sodium and potassium transport in the erythrocyte. *J. Biol. Chem.* **239,** 345–352.

Siegel, G. J. and Albers, R. W. (1967) Sodium-potassium-activated adenosine triphosphatase of Electrophorus electric organ. IV. Modification of responses to sodium and potassium by arsenite plus 2,3-dimercaptopropanol. *J. Biol. Chem.* **242,** 4972–4979.

Simons, T. J. B. (1974) Potassium-potassium exchange catalyzed by the sodium pump in human red cells. *J. Physiol. (London)* **237,** 123–155.

Simons T. J. B. (1975) The interaction of ATP-analogues possessing a blocked phosphate group with the sodium pump in human red cells. *J. Physiol. (London)* **244,** 731–739.

Skou, J. C. (1957) The influence of some cations on an adenosine triphosphatase form peripheral nerve. *Biochim. Biophys. Acta* **23,** 394–401.

Stein, W. D. (1986) *Transport and Diffusion Across Cell Membranes.* Academic Press, New York, Chapter 6.

Taniguchi, K. and Post, R. L. (1975) Synthesis of adenosine triphosphate and exchange between inorganic phosphate and adenosine triphosphate in sodium and potassium ion transport adenosine triphosphatase. *J. Biol. Chem.* **250,** 3010–3018.

Thomas, R. C. (1972) Electrogenic sodium pump in nerve and muscle cells. *Physiol. Rev.* **52,** 563–594.

Tosteson, D. C. and Hoffman, J. F. (1960) Regulation of cell volume by active cation transport in high and low potassium sheep red cells. *J. Gen. Physiol.* **44,** 169–194.

Valet, G., Franz, G., and Lauf, P. K. (1978) Different red cell population in new born, genetically low potassium sheep: relationship to hematopoietic, immunologic and physiologic differentiation. *J. Cell Physiol.* **94,** 215–228.

Weiss, J. N. and Lamp, S. T. (1987) Glycolysis preferentially inhibits ATP-sensitive K^+ channels in isolated guinea pig cardiac myocytes. *Science* **238,** 67–69.

White, B. and Blostein, R. (1982) Comparison of kidney and red cell Na,K-ATPase at 0°C. *Biochim. Biophys. Acta* **688,** 685–690.

Whittington, E. S. and Blostein, R. (1971) Comparative properties of high potassium and low potassium sheep erythrocyte membrane sodium-activated adenosine triphosphatase. *J. Biol. Chem.* **246,** 3518–3523.

Yingst, D. R. (1983) Hemolysate increases calcium inhibition of the Na^+, K^+-pump of resealed human red cell ghosts. *Biochim. Biophys. Acta* **732,** 312–315.

Yingst, D. R. (1988) Modulation of the Na,K-ATPase by calcium and intracellular proteins. *Ann. Rev. Physiol.* **50,** 291–303.

Yoda, A. and Yoda, S. (1987) Two different phosphorylation-dephosphorylation cycles of Na,K-ATPase proteoliposomes accompanying Na^+ transport in the absence of K^+. *J. Biol. Chem.* **262,** 110–115.

The (Ca^{2+} + Mg^{2+})-ATPase

Purification and Reconstitution

Basil D. Roufogalis and Antonio Villalobo

1. Introduction

Solubilization and partial purification of membrane ATPases was achieved in the late 1960s. In the early 1970s, hopes were expressed that the Ca^{2+}-translocating ATPase, whose transport function was first described by Schatzmann (1966), could be purified and its active transport function reconstituted asymmetrically in a lipid bilayer. Serious difficulties were experienced in achieving these aims because of the multiplicity of organelles containing Ca^{2+}-transport activities in most cells, and in the case of red blood cells, because of the low content of the Ca^{2+}-transport ATPase in the plasma membrane. Developments in the better understanding of the amphiphilic nature of membrane proteins, in the utilization of mild detergents for protein solubilization with minimal denaturation, and the discovery of calmodulin as a potent activator of the (Ca^{2+} + Mg^{2+})-ATPase (Gopinath and Vincenzi, 1977; Jarret and Penniston, 1977) led to the successful purification of this transport enzyme in an active and calmodulin-sensitive form. This occurred about seven years ago in the laboratories of Carafoli, Penniston, and Wolf, and was followed by the asymmetric reconstitution of the pure enzyme into liposomes a few years later. In this review, we consider the principles behind the successful purification of the Ca^{2+}-translocating ATPase from a plasma membrane source, and we discuss the kinetic properties of the purified enzyme and its transport properties in the reconstituted state.

2. Purification of (Ca^{2+} + Mg^{2+})-ATPase

2.1. Early Attempts

Early efforts to purify the (Ca^{2+} + Mg^{2+})-ATPase from the human erythrocyte membrane were plagued by difficulties in maintaining the enzyme in an active form once it was removed from its natural membrane environment and by the unavailability of a method specific for the isolation of this protein from the remaining 99.7% or more of other protein components in the membrane. Nevertheless, remarkable progress was made in this direction by various groups. The laboratory of Wolf reported the early solubilization of (Ca^{2+} + Mg^{2+})-ATPase with Triton X-100 (Wolf and Gietzen, 1974; Wolf et al., 1976; Wolf et al., 1977). The Triton X-100-solubilized enzyme was inactivated in all subsequent attempts to purify it by gel filtration or ion exchange chromatography (Wolf et al., 1977). This problem of instability was solved by the addition of suspensions of phosphatidylcholine, but a nonaggregated form of the enzyme was obtained only when the enzyme was dispersed in mixed micelles of Triton X-100, or Tween 20 and sonicated phospholipids (Wolf et al., 1977). When combined with the addition of protease inhibitors, diisopropylfluorophosphate (DFP), and N^{α}-*p*-tosyl-L-lysine chloromethyl ketone hydrochloride (TLCK), during cell hemolysis and membrane isolation, and with the addition of cysteine, Ca^{2+}, Mg^{2+}, and K^+ as enzyme effectors, the (Ca^{2+} + Mg^{2+})-ATPase was eluted from a Sepharose CL-6B gel filtration column as a complex that largely excluded spectrin. A yield of 45–50% of the starting membrane bound activity and an increase in specific activity from 0.02 U/mg protein to about 2.2 U/mg protein was obtained. The peak fraction with the eluted (Ca^{2+} + Mg^{2+})-ATPase activity consisted of three major bands (denoted α, β, and γ), of which the 145,000 MW band (α) was shown to be phosphorylated by [γ-^{32}P]ATP in a Ca^{2+}-dependent manner (Wolf et al., 1977). The enzyme activity so isolated was sensitive to Ca^{2+} in the range of 0.1–30 μM free Ca^{2+}, with a Hill coefficient for Ca^{2+} activation of 1.25 $\pm$ 0.11.

Following on from their earlier observations (Ronner et al., 1977) and from the work of Roelofsen and Schatzmann (1977) using lipid depletion by phospholipase digestion, the laboratory of Carafoli further characterized the requirement of the (Ca^{2+} + Mg^{2+})-ATPase for the presence of phospholipids for its activity. This group found that maximal activity was achieved when the Triton X-100-solubilized enzyme was combined with the acidic phospholipid, phosphatidylserine, or the unsaturated fatty acid oleic acid (Peterson et al., 1978). Whereas the various purification protocols used by these workers resulted in a significant increase in specific activity of the enzyme, the enzyme preparation was reported to be contaminated by many protein bands, particularly bands 4.5, 6, and 7, as well as band 3. Using the solubilization and mixed micelle chromatography procedures of Wolf et al. (1977) for the solubilization and purification of the (Ca^{2+} + Mg^{2+})-ATPase from pig erythrocytes, Haaker and Racker (1979) reconstituted the Triton X-100-solubilized ATPase into vesicles following detergent depletion by phenyl-Sepharose 4B hydrophobic chromatography. These authors also reported activation of the Ca^{2+}-ATPase activity in washed ghosts by both Triton X-100 and Tween 20, although above a detergent/protein ratio of 1.2 (w/w), the enzyme was inhibited by Triton X-100. Activation by the modulator protein, calmodulin, was observed both in the presence and absence of activating concentrations of Tween 20. The purity of the solubilized Ca^{2+}-ATPase used in the reconstitution experiments was not reported in detail in this study. The enzyme-mixed micelle complex was eluted from gel chromatography columns with an apparent size of about 700,000 (Haaker and Racker, 1979), which is similar to the size reported by Wolf et al. (1977). However, the micelle of the Triton-X-100-solubilized enzyme was found to elute with an apparent mol wt of 155,000 ± 20,000 by Niggli et al. (1979a).

2.2. Calmodulin-Affinity Chromatography

2.2.1. Principles

The possibility of purifying the (Ca^{2+} + Mg^{2+})-ATPase by calmodulin-affinity chromatography arose from the discovery that calmodulin activates the enzyme in a Ca^{2+}-dependent manner. Furthermore, the solubilized enzyme in the detergent–lipid complex was found to remain responsive to calmodulin binding in the solubilized state (Lynch and Cheung, 1979; Niggli et al., 1979a), as long as calmodulin was depleted from the preparations during the enzyme isolation.

The first report of the successful purification of (Ca^{2+} + Mg^{2+})-ATPase using immobilized calmodulin was by Verena Niggli in collaboration with John Penniston and Ernesto Carafoli (Niggli et al., 1979b). Soon after, Gietzen et al. (1980b) reported a similar result, with the exception that the enzyme resulting from the method they used retained its calmodulin sensitivity. Subsequently, Niggli et al. (1981b) showed that the enzyme solubilized and bound to the calmodulin affinity column could be eluted in a calmodulin-sensitive form by exchanging the phosphatidylserine used initially with phosphatidylcholine. This work established that the Ca^{2+}-transport ATPase existed in a low Ca^{2+} affinity state under certain conditions, as had been reported earlier by workers studying the enzyme in the erythrocyte ghost membrane (Schatzmann and Rossi, 1971; Quist and Roufogalis, 1975). It was shown further that the addition of acidic phospholipids increased the Ca^{2+} sensitivity of the enzyme in a manner apparently similar to that of calmodulin (Niggli et al., 1981a). However, the reduction in the Triton X-100 concentration from 0.4 to 0.05% may have also contributed to the recovery of calmodulin sensitivity of the purified preparation of Niggli et al. (1979b), since Triton X-100 was shown to maximally activate the (Ca^{2+} + Mg^{2+})-ATPase activity at a detergent:protein ratio of 1.24 (w/w), at which point the enzyme activity was no longer activated by calmodulin (Gietzen et al., 1980b). Triton X-100 has been shown to inhibit calmodulin activation of other calmodulin-dependent enzymes, probably by binding to hydrophobic sites on calmodulin (Sharma and Wang, 1981).

2.2.2. Procedures

The procedures for the purification of (Ca^{2+} + Mg^{2+})-ATPase by calmodulin-affinity chromatography have been published in detail in the original papers referred to above. The following is a summary of the protocol used in our laboratory (Villalobo et al., 1986). Calmodulin-depleted erythrocyte ghosts (150–200 mg protein) are prepared either by lysis and repeated washing by centrifugation or by continuous washing by filtration in a Millipore Pellicon Cassette system, in the presence of 0.5 m*M* phenylmethylsulphonylfluoride (PMSF). The pellet of a 125,000 × g_{max} centrifugation of these membranes is solubilized with 0.5% (w/v) Triton X-100 in a medium of 300 m*M* KCl, 10 m*M* potassium-HEPES, 1 m*M* $MgCl_2$, 100 μ*M* $CaCl_2$, and 2 m*M* dithiothreitol (pH 7.4) at 0°C. To the supernatant from the centrifugation of the solubilized material at 125,000 × g_{max} is added 0.1% (w/v) sonicated asolectin (or other phospholipids, as required), 2 m*M* dithiothreitol, and additional 0.05% (w/v) Triton X-100. This preparation is passed through a column of calmodulin-Agarose (11.3 cm × 1.5 cm), obtained commercially from the Sigma Chemical Co. and containing 18 mg covalently bound calmodulin, previously equilibrated with the calcium medium containing 0.1% (w/v) asolectin and 0.05% (w/v) Triton X-100. The fraction not bound is determined by collecting the material that passes through the column during this loading step. The column containing the bound activity is washed extensively (usually overnight) with approximately 700 mL of a similar medium to remove contaminating protein (Graf et al., 1982). The Ca^{2+}-translocating ATPase is then eluted from the column with a solution of 2 m*M* EDTA (pH 7.4) in 0.05% (w/v) Triton X-100, 300 m*M* KCl, 10 m*M* potassium-HEPES, 2 m*M* dithiothreitol, and 0.1% (w/v) sonicated asolectin, in order to chelate the calcium. Fractions

(≈2.0 mL) are collected, and the eluted protein is determined by UV-absorption at 280 nm. The fractions with maximum activity are pooled, and small aliquots are frozen at − 196°C and stored at − 70°C until used.

A tendency to form aggregates of 200,000 MW (as seen on SDS-gel electrophoresis) has been reported (Graf. et al., 1982), particularly after freeze-thawing, exposure to temperatures greater than 30°C, and after prolonged storage at − 76°C. In some preparations, we also obtain a diffuse Coomassie blue R-250-staining band of 145,000 MW, devoid of Ca^{2+}-dependent phosphorylating activity, just above the active (Ca^{2+} + Mg^{2+})-ATPase band of 132,000 MW (Villalobo et al., 1986). Other lower mol wt bands constitute trace contaminants, many of which are introduced with the asolectin used in the purification.

2.2.3. Recovery of Enzyme Activity

A typical recovery of (Ca^{2+} + Mg^{2+})-ATPase activity is presented in Fig. 1. Recoveries of 22–43% of the membrane-associated activity, or 19–49% of the total solubilized activity are obtained (Gietzen et al., 1980b; Gietzen and Kolandt, 1982; Niggli et al., 1979b; Niggli et al., 1981b). The specific activity of the enzyme usually reported varies from 1–6 μmol Pi/mg protein/min, but activity as high as 20 μmol/mg protein/min has been reported (Graf et al., 1982).

3. Kinetic Properties of the Purified Enzyme

Purification of the (Ca^{2+} + Mg^{2+})-ATPase from human erythrocytes allowed the study of its kinetic properties and particularly a comparison of the behavior of the pure enzyme with that of the membrane-associated enzyme. Differences in the properties of the enzyme in the two environments might yield important clues about the modulation of the enzyme function.

An immediate contribution was the confirmation that the Ca^{2+} transport ATPase exists in a functionally significant low Ca^{2+}-affinity form, which had been persistently reported in a number of studies on the membrane-bound enzyme (Quist and Roufogalis 1975, 1977; Scharff 1972, 1976). In addition to modulation of the Ca^{2+} affinity of the enzyme by calmodulin, which emerged from the work of Gopinath and Vincenzi (1977) and Jarrett and Penniston (1977), the early work on the purified enzyme established conclusively that, in the presence of acidic phospholipids, particularly phosphatidylserine, the Ca^{2+}-translocating ATPase was maintained in a fully active high Ca^{2+}-affinity form, largely insensitive to calmodulin. By contrast, in the presence of neutral phospholi-

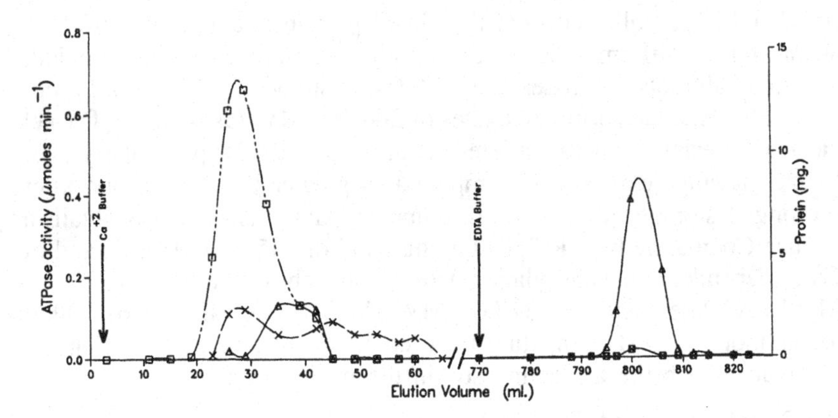

Fig. 1. Distribution of ATPase activities and protein during purification by calmodulin-affintiy chromatography. ATPase activity and protein were determined as in Villalobo et al. (1986). (□-□) protein; (x-x) Mg^{2+}-ATPase, (△-△) Ca^{2+}-ATPase assayed in the presence of calmodulin.

pids, particularly phosphatidylcholine, the enzyme exists in a low Ca^{2+}-affinity form that is sensitive to calmodulin (Niggli et al., 1981b; Gietzen et al., 1980b).

Since magnesium concentration has been shown to influence the Ca^{2+} sensitivity of the membrane associated Ca^{2+}-translocating ATPase profoundly under certain conditions (Katz et al., 1979), it is of interest to consider the effect of Mg^{2+} on the purified enzyme. The purified Ca^{2+} transport ATPase is devoid of a small but consistent Mg^{2+}-ATPase activity (10–20%) found in the erythrocyte membrane, indicating that the Mg^{2+}-ATPase is a separate enzyme and the effects of Mg^{2+} on the Ca^{2+} transport ATPase are modulatory (Niggli et al., 1981b). The interrelationship between Mg^{2+} and other enzyme effectors on the Ca^{2+} sensitivity of the purified Ca^{2+}-ATPase is illustrated in Fig. 2. For the enzyme purified in the presence of acidic phospholipids, as found in asolectin, calmodulin activation is more apparent at 5 m*M* $MgCl_2$ than at 1 m*M* $MgCl_2$, where the Ca^{2+} affinity is already high (Fig. 2A). However, in the presence of phosphatidylcholine, calmodulin increases the Ca^{2+} sensitivity both at high and low $MgCl_2$ concentrations (Fig. 2B). Kinetic analysis indicated that Mg^{2+} competes with Ca^{2+} activation of the pure enzyme (Villalobo et al., 1986), as found also for the membrane-associated enzyme (Penniston, 1982). In the millimolar concentration range, free Mg^{2+} inhibits the turnover of the enzyme (Villalobo et al., 1986).

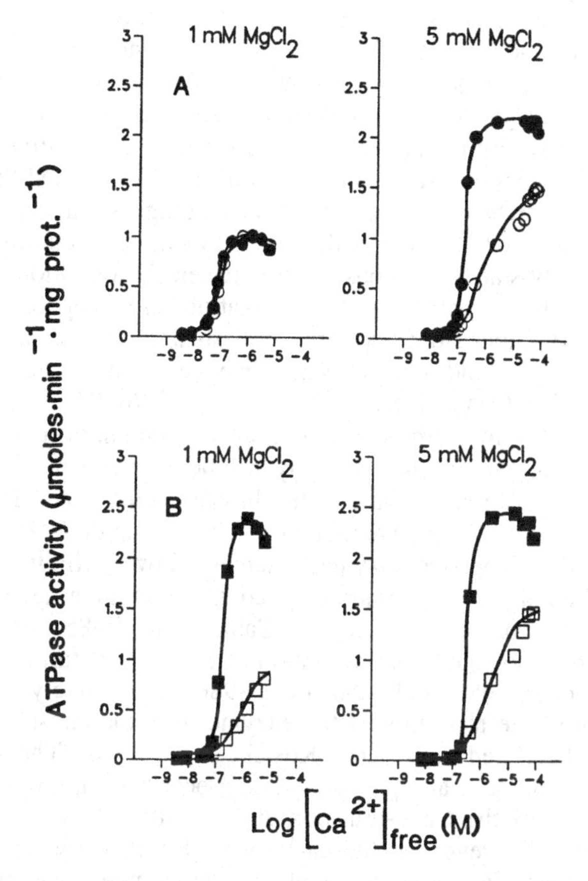

Fig. 2. Calcium sensitivity of (Ca^{2+} + Mg^{2+})-ATPase purified and assayed in asolectin (A) or phosphatidylcholine (soybean, Sigma Type II-S) (B). ATPase activity and purification were performed as in Villalobo et al. (1986). Closed symbols are data obtained in the presence of calmodulin, whereas open symbols are data without added calmodulin.

A feature of the Ca^{2+} activation of the purified enzyme is its cooperativity, which changes from 2–4 in the absence and presence of calmodulin, respectively (Villalobo et al., 1986). A high degree of cooperativity of Ca^{2+} activation is maintained in the presence of phosphatidylcholine (*see* Fig. 2), whereas the cooperativity in the presence of calmodulin is reduced from 4–2 as the pH is decreased from pH 8–pH 6 (Villalobo et al., 1986). Cooperativity of Ca^{2+} activation was also noted in the membrane-associated enzyme, the extent of which varied with the nature of the preparation (Scharff, 1981; Downes and Michell, 1981).

The cooperativity observed is consistent with the presence of oligomeric association of the enzyme monomers (Minocherhomjee et al., 1983; Kosk-Kosicka and Inesi, 1985), and was also observed by direct equilibrium binding analysis on the nonphosphorylated enzyme (Kosk-Kosicka and Inesi, 1985). From an analysis of the kinetics of the partial reactions of the purified enzyme. Kosk-Kosicka et al. (1986) reported that, in the fully activated state of the enzyme, the binding of Ca^{2+} to the high affinity site is rapid. Slow and therefore possibly rate-controling steps under nonsteady-state conditions occur between EP formation and cleavage (as shown by an overshoot of EP formation before dephosphorylation was activated). Other slow steps are the dissociation of calcium from the E_2 phosphoenzyme and reversal of the enzyme transition from the E_1 to E_2 phosphorylated state (Kosk-Kosicka et al., 1986). The rate and equilibrium constants for some of these steps are summarized in Table 1.

Another feature of the ATP hydrolytic mechanism found to be retained in the soluble enzyme is the biphasic effect of ATP. A high affinity binding constant presumably reflects binding of ATP at the ATP hydrolytic site of the free enzyme, whereas a lower affinity component represents binding of ATP either at a second site or at the same site in the phosphorylated form of the enzyme (Cable et al., 1985). This binding appears to be associated with activation of the enzyme turnover (Stieger and Luterbacher, 1981; Villalobo et al., 1986), which likely results from acceleration of the transition of the enzyme from the phosphorylated to the dephosphorylated form (Kosk-Kosicka et al., 1986). The nonsteady-state binding of Ca^{2+} and phosphorylation occurred rapidly in the fully activated state of the Ca^{2+}-transport ATPase studied by Kosk-Kosicka et al. (1986). However, in the calmodulin-depleted erythrocyte membrane, both the Ca^{2+} sensitivity of enzyme turnover and the rate of Ca^{2+}-dependent phosphorylation of the enzyme were enhanced by calmodulin (Allen et al., 1987). It is therefore possible that Ca^{2+} binding and Ca^{2+}-dependent phosphorylation may be rate-controling steps under some conditions.

3.1. Requirement for Phospholipids

Although the earlier studies on purification of the enzyme clearly established the importance of the addition of phospholipids for the successful purification of the Ca^{2+}-transport ATPase (*see above*), recent studies indicate that the enzyme may be stabilized by the presence of appropriate detergents alone. Thus, Bridges and Katz (personal communication) purified the Triton X-100-solubilized enzyme by calmodulin-affinity chromatography in the absence of added phospholipids, using

0.05% Tween 20 during washing and elution of the enzyme from the affinity matrix. Although the enzyme activity was negligible when assayed in these conditions, it could be restored by the addition of phospholipids to the assay mixture. A similar preparation was reported by Kosk-Kosicka and Inesi (1985), who used the subunit-dissociating detergent $C_{12}E_8$ for purification in the absence of added phospholipids and found the enzyme to be equiactive to that obtained by the original procedures employing phospholipid, as long as phospholipids were reintroduced in the assay medium. It thus appears that phospholipids are essential for the preservation of the catalytic sequence, but not for stabilizing the protein in the solubilized state.

3.2. Calmodulin-Insensitive Forms

Recently Krug and Wolf (1982) reported the presence of a lower mol wt form of the (Ca^{2+} + Mg^{2+})-ATPase in human erythrocyte membranes that was independent of calmodulin, but it remains to be shown that the form isolated was not derived from a single Ca^{2+}-translocating ATPase by proteolytic cleavage. Improved recoveries of enzyme will be required before the significance of other enzyme forms, if any, can be assessed.

4. Reconstitution of (Ca^{2+} + Mg^{2+})-ATPase

The transport function and mode of operation of the Ca^{2+}-translocating ATPase from human erythrocytes was first studied in inside-out plasma membrane vesicles. However, the presence of other transport systems in the plasma membrane made it difficult to ascertain the mode of operation of this enzyme. In particular, at least three important parameters must be known for the understanding of the transport mechanism that could easily be misinterpreted when inside-out plasma membrane vesicles are used for the transport measurements. These parameters are:

1. the mechanistic stoichiometric ratio of moles of Ca^{2+} translocated/mole of ATP hydrolyzed
2. the electrogenicity or electroneutrality of the transport process and
3. the possible direct cotransport or countertransport of other ions coupled to Ca^{2+} translocation during the catalytic cycle of the enzyme.

Table 1
Kinetic Constants of the Purified Ca^{2+}-Translocating ATPase of Human Erythrocytes

Property	Constant		Reference
	$-$CaM	+CaM	
$K_{d(Ca^{2+})}$ (direct binding to nonphosphorylated enzyme E_1)	1×10^{-7}M[a]	–	Kosk-Kosicka and Inesi (1985)
$K_{0.5(Ca^{2+})}$ (phosphatidylcholine)	$2 \times 10^{-5}M$ $> (3\text{–}3.5) \times 10^{-6}M$[b]	$5 \times 10^{-7}M$ $(1.8\text{–}3.5) \times 10^{-7}M$[b]	Niggli et al. (1981b); Villalobo and Roufogalis (unpublished)
$K_{0.5(Ca^{2+})}$ (acidic phospholipids)	$5 \times 10^{-7}M$ $(0.3\text{–}8) \times 10^{-7}M$[b]	$5 \times 10^{-7}M$ $(0.2\text{–}2) \times 10^{-7}M$[b]	Niggli et al. (1981b) Villalobo et al. (1986)
$K_{0.5(Ca^{2+})}(E_2{\cdot}P)$	$2 \times 10^{-3}M$	–	Kosk-Kosicka et al. (1986)
k_1 (forward rate constant for Ca^{2+} binding)	$10^6\ s^{-1}M^{-1}$	–	Kosk-Kosicka and Inesi (1985)

k_{-1} (rate constant for Ca^{2+} dissociation)	$0.1\ s^{-1}$	–	Kosk-Kosicka and Inesi (1985)
k_p (forward rate constant for phosphorylation)	$220\ s^{-1}$	–	Kosk-Kosicka et al. (1986)
$k_{(cat)}$ (overall enzyme turnover)			
at 4 μM ATP	$1\ s^{-1}$	–	Kosk-Kosicka et al. (1986)
at 720 μM ATP	$4\ s^{-1}$	–	
$K_{0.5(CaM)}$	$6.3 \times 10^{-9}M$[c]	–	Villalobo et al. (1986)
$K_{m(MgATP)}$			
High affinity site	$7 \times 10^{-6}M$	[d]	Villalobo et al. (1986)
	$3.5 \times 10^{-6}M$		Stieger and Luterbacher (1981)
Low affinity site	$1.4 \times 10^{-4}M$	[d]	Villalobo et al. (1986)
	$1.2 \times 10^{-4}M$		Stieger and Luterbacher (1981)

[a] At the protein concentration used, the enzyme activity is fully activated and calmodulin insensitive.
[b] The apparent affinity constant varies with the free Mg^{2+} concentration used.
[c] This value was independent of Mg^{2+} concentrations from 0.5–5mM.
[d] The constants were independent of the presence of calmodulin.

The possible errors obtained when inside-out vesicles are used in the determination of those parameters arise from the operation of other primary and/or secondary ion transport systems of the plasma membrane during the Ca^{2+}-translocation. This could result in recycling of Ca^{2+} across the membrane, collapse, and/or generation of electrical gradients because of the movement of other charge species, and/or apparent coupling of other ionic species to the Ca^{2+} transport that in reality are secondarily translocated by independent carrier systems. One practical solution to these possible problems is to reconstitute the purified (Ca^{2+} + Mg^{2+})-ATPase in artificial phospholipid vesicles (proteoliposomes), where other plasma membrane transport systems are absent.

4.1. Early Attempts

In 1978 a partially purified (Ca^{2+} + Mg^{2+})-ATPase from human erythrocyte plasma membrane was reconstituted in artificial phospholipid vesicles of different composition using a sonication procedure (Peterson et al., 1978). The reconstituted enzyme gave the highest ATP hydrolytic activity in vesicles of phosphatidylserine. In contrast, reconstitution of the enzyme in phosphatidylcholine or phosphatidylethanolamine vesicles resulted in no or negligible ATP hydrolytic activity. Reconstitution in vesicles made of pure oleic acid also showed significant ATP hydrolytic activity, although this was 2.3–2.5-fold lower than that with phosphatidylserine. However, none of the reconstituted vesicles described above were able to sustain Ca^{2+} transport activity (Peterson et al., 1978).

The first successful reconstitution of a plasma membrane Ca^{2+}-translocating ATPase was attained one year later with the enzyme from pig erythrocyte, using asolectin and a freeze-thaw/sonication procedure to prepare the liposomes (Haaker and Racker, 1979). In these experiments, two criteria to ascertain the reconstitution of a functional Ca^{2+}-translocating ATPase were met: (1) direct demonstration of ATP-dependent $^{45}Ca^{2+}$ uptake by proteoliposomes having the ATP catalytic site on the outside of the vesicles, and (2) stimulation of the ATP hydrolytic activity (nearly threefold) by the addition of the $Ca^{2+}/2H^{+}$ exchanger A23187 in the assay system, so preventing the generation of a Ca^{2+} concentration gradient and resulting in the full expression of the ATP hydrolytic activity (Haaker and Racker, 1979). Reconstitution of Ca^{2+}-translocating activity using a partially purified (Ca^{2+} + Mg^{2+})-ATPase from human erythrocytes and a detergent dialysis method was reported by Gietzen et al. (1980a).

4.2. Principle of the Methods

In the first attempts to reconstitute the Ca^{2+}-transport activity of the enzyme, a sonication procedure was employed (Peterson et al., 1978), since sonication of a suspension of phospholipids results in the formation of liposomes. The semipurified enzyme preparations used in those studies contained from 0.3–0.5% (w/v) Triton X-100. Consequently, after mixing the phospholipid suspension with the enzyme preparation a significant amount of Triton X-100 was present and was later removed by the use of a column of Bio-Beads (Bio-Rad). The exact concentration of Triton X-100 present in the mixture was not reported in his work; however, it was probably not optimal for reconstitution of Ca^{2+} transport following subsequent removal of the detergent. The crude preparation used for reconstitution could also contain other ion carriers resulting in Ca^{2+} recycling across the membrane.

The use of a mixture of soybean phospholipids (asolectin) and a freeze-thaw sonication procedure was the key to obtaining Ca^{2+} transport in proteoliposomes that were loaded with 0.2*M* potassium oxalate (Haaker and Racker, 1979). The procedure was followed by hydrophobic chromatography in phenyl-Sepharose 4B to remove detergent present in the preparation. The sonication/Triton X-100 removal method (using a Bio-Beads SM-2 column) was further refined by the use of a highly purified ATPase and the inclusion of 2% (w/v) Triton X-100 as the initial concentration (higher than before), so preventing the problems of earlier studies (Niggli et al., 1981b). Reconstitution of (Ca^{2+} + Mg^{2+})-ATPase able to support net Ca^{2+} transport was attained using this last method in various phospholipid vesicles, including asolectin, phosphatidylserine, phosphatidylcholine, or phosphatidylinositol. Competent Ca^{2+} transport in those systems was indirectly demonstrated by stimulation of the ATP hydrolytic activity of the reconstituted enzyme by A23187 (Niggli et al, 1981a,b) and by direct demonstration of Ca^{2+} transport with the metallochromic dye Arsenazo III (Niggli et al., 1981b).

In further work, the reconstitution was performed by a cholate/dialysis method, using enzyme purified not only from erythrocyte plasma membrane (Niggli et al., 1981a; Carafoli and Zurini, 1982; Niggli et al., 1982; Carafoli et al., 1982; Clark and Carafoli, 1983; Verma and Penniston, 1985; Villalobo and Roufogalis, 1986), but from calf heart sarcolemma (Carafoli and Zurini, 1982; Caroni et al., 1983; Caroni et al., 1982) and pig antrum smooth muscle plasma membrane (Verbist et al., 1984). This method essentially consists of the sonication of a suspension of phospholipids in the presence of a salt of cholic acid, followed by extensive dialysis to remove the detergent. The cholate/dialysis method has the advantage of producing proteoliposomes more tightly coupled than those obtained with the other previously described methods.

4.3. Characterization of the Proteoliposomes

Regardless of the method used for reconstitution, and the ionic composition of the different buffers employed, it was observed that to attain an active reconstituted enzyme the presence of a reducing agent, usually dithiothreitol, was required in the reconstitution media, particularly when long reconstitution procedures were used (Villalobo and Roufogalis, unpublished results).

When the vesicles were prepared by the Triton X-100 removal procedure with Bio-Beads, it was estimated that the final concentration of the detergent was reduced to 0.01% (*see* Holloway, 1973). The reconstitution of the enzyme by a sonication/Triton X-100 removal procedure, using Bio-Beads, results in the formation of monolamellar vesicles varying in size from 50–100 nm in diameter, when the procedure was performed at 4°C (Peterson et al., 1978). However, larger vesicles (200–400 nm diameter) were obtained when the reconstitution was performed at 23°C. In this work, the determination of vesicle size was done by thin section electron micrography. Preparation of liposomes by the cholate/dialysis method also results in the formation of small unilamellar vesicles, with a diameter ranging from 30–100 nm. The size of the vesicles was determined by gel filtration chromatography using Sephacryl S-100, and by negative staining electron micrography (Verbist et al., 1984). However, larger unilamellar liposomes could be prepared by inducing fusion of the preformed vesicles by a freeze-thaw procedure (Holloway, 1973). When a mixture of different phospholipids is used in the preparation of liposomes, it could result in an asymmetric incorporation of the different phospholipids in the membrane. For example, when a mixture of phosphatidylcholine and phosphatidylserine was used, the latter lipid tended to be incorporated on the inside of the bilayer, probably because of the less bulky polar head (*see* Niggli et al., 1981a for discussion).

We have determined the size of the liposomes prepared by the cholate/dialysis method used in our laboratory by Dynamic Light Scattering of a laser beam, as described by Perevucnik et al. (1985). The results (presented in Fig. 3A) show that there are two populations of vesicles. The smaller ones are of diameters ranging from 45–60 nm, with an average value of 53 nm, and the larger ones are of diameters ranging from 130–450 nm, with an average value of 225 nm. In addition, a small

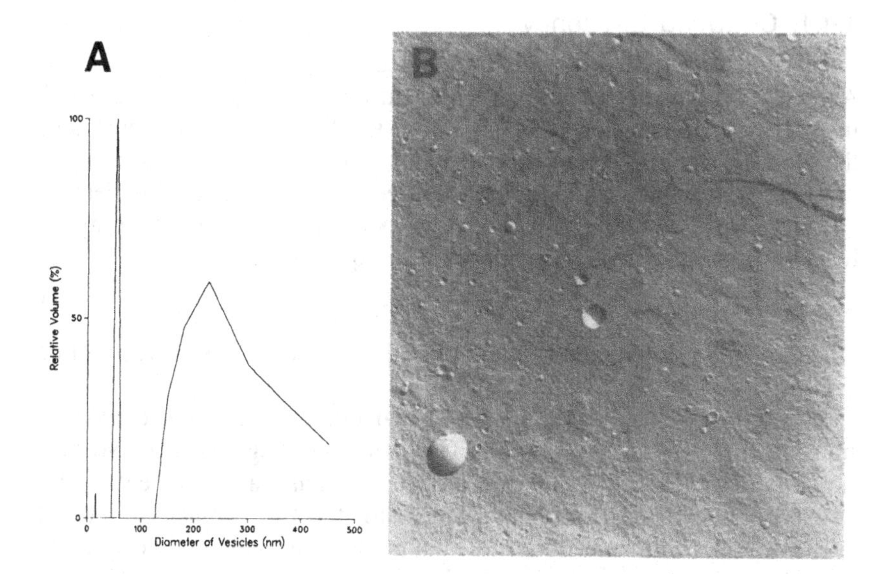

Fig. 3. Size distribution of proteoliposomes containing the Ca^{2+}-translocating ATPase prepared by a cholate-dialysis method (*see* Villalobo and Roufogalis, 1986) measured by Dynamic Light Scattering. B. Freeze-fracture electron micrography of the same preparation. The bar indicates 200 nm.

number of particles with diameters of around 16 nm was detected, probably representing large micellae. Freeze-fracture electron micrography of proteoliposomes, presented in Fig. 3B, shows large and small vesicles, confirming the findings from Dynamic Light Scattering analysis. Similar vesicle sizes were obtained both with liposomes containing the enzyme and with liposomes prepared in the absence of proteins.

4.4. Properties of the Reconstituted Enzyme

The fundamental characteristics necessary to obtain functionally active Ca^{2+}-translocating ATPase in a reconstituted system are: (1) incorporation of the enzyme in an orientation spanning the liposomal membrane, and (2) exposure of the ATP catalytic site to the outside of the vesicles. If both objectives are met, it will be possible to obtain enzyme able to translocate Ca^{2+} across the membrane and also to assay the catalytic enzyme activity. In addition, it is necessary to obtain nonleaky liposomes in order to attain a Ca^{2+} concentration gradient across the membrane and to observe net Ca^{2+} uptake. We will discuss each of the criteria in the next subsections.

4.4.1. Coupling Efficiency

When ATP is hydrolyzed by the reconstituted Ca^{2+}-translocating ATPase, part of the free energy released during ATP hydrolysis is utilized in some way to translocate Ca^{2+} across the liposomal membrane, and therefore to form an electrochemical Ca^{2+} gradient. Both processes are mechanistically coupled; therefore, the free energy of the formed electrochemical Ca^{2+} gradient, $\Delta\bar{\mu}_{Ca^{2+}}$, is proportional to the free energy of hydrolysis of ATP (ΔG_{ATP}) according to equation (1).

$$\Delta G_{ATP} = m\Delta\bar{\mu}_{Ca^{2+}} \quad (1)$$

where m is the stoichiometric ratio of moles of Ca^{2+} translocated/mol of ATP hydrolysed.

When the values of both sides of equation (1) are equal, the Ca^{2+}-translocating ATPase stops both its Ca^{2+}-translocating and ATP hydrolytic activities. Conversely, if the electrochemical Ca^{2+} gradient is destroyed or prevented from being generated, the Ca^{2+}-translocation and ATPase will continue working at maximum rate. This simple relation provides a tool to analyze the coupling efficiency of the system numerically. The most popular method consists of the determination of the ATP hydrolysis control ratio. The hydrolysis control ratio is defined as the ratio of the rate of ATP hydrolysis in a system where $\Delta\bar{\mu}_{Ca^{2+}}$ is prevented from developing (for example, in the presence of a Ca^{2+} ionophore able to collapse an electrical gradient as well) over the rate of ATP hydrolysis in a system where $\Delta\bar{\mu}_{Ca^{2+}}$ is allowed to develop. When that ratio equals 1.0, the system is said to be uncoupled. When that ratio reaches infinity, the system is said to be perfectly coupled. To attain perfect coupling, it is necessary that the system meet two criteria: (1) all the active ATPase molecules should be able to translocate Ca^{2+} during ATP hydrolysis and (2) the membrane should be totally impermeable to Ca^{2+}. In practice, both criteria are never met; consequently, the coupling is invariably lower than perfect.

Table 2 summarizes the ATP hydrolysis control ratio obtained with reconstituted enzyme of various origins by different procedures, using the $Ca^{2+}/2H^{+}$ electroneutral exchanger A23187 to collapse or prevent the development of a chemical calcium gradient across the membrane. The use of A23187 should not prevent the development of an electrical gradient across the membrane if the net Ca^{2+} translocation is electrogenic. However, the combined use of A23187 plus other ionophores able to collapse electrical gradients does not result in a further increase of the ATP hydrolysis control ratio (Niggli et al., 1982; Villalobo and Roufogalis, 1986), indicating that the electrical component in that system

Table 2
ATP Hydrolysis Control Ratio of Plasma Membrane Ca^{2+}-Translocating ATPase of Different Origins Reconstituted in Proteoliposomes by Various Procedures

Reconstitution method	Phospholipids	Enzyme origin	ATP hydrolysis control ratio[a]	Reference
Freeze-thaw/sonication	Asolectin (soybean)	Pig erythrocyte	3.0	Haaker and Racker (1979)
Trinton X-100 removal (Bio-Beads)	Asolectin (soybean)	Human erythrocyte	2.0–4.0	Niggli et al. (1981b)
Triton X-100 removal (Bio-Beads)	Phosphatidylinositol	Human erythrocyte	3.0	Niggli et al. (1981a)
Cholate-dialysis	Phosphatidylcholine	Human erythrocyte	4.0	Niggli et al. (1981a)
Cholate-dialysis	Asolectin (soybean)	Human erythrocyte	9.8–11.0	Carafoli and Zurini (1982); Niggli et al. (1982)
Cholate-dialysis	Asolectin (soybean)	Human erythrocyte	8.6–10.0	Villalobo and Roufogalis (1986)
Cholate-dialysis	Phosphatidylcholine (soybean)	Human erythrocyte	8.5–10.7	Villalobo and Roufogalis (1986)
Cholate-dialysis	Phosphatidylcholine (egg yolk)	Human erythrocyte	5.8	Villalobo and Roufogalis (1986)
Cholate-dialysis	Asolectin (soybean)	Bovine heart	2.6–4.0	Caroni et al. (1983) Caroni et al. (1982)
Cholate-dialysis	Asolectin (soybean)	Pig antrum smooth muscle	8.0	Verbist et al. (1984)
Cholate-dialysis	Phosphatidylcholine	Pig antrum smooth muscle	3.6	Verbist et al (1984)

[a] The ATP hydrolysis control ratio was calculated using A23187 to collapse the developed Ca^{2+} concentration gradient.

appears to be of little significance, if any, in the overall process. The general data from Table 2 indicate that the average ATP hydrolysis control ratio obtained with the enzyme reconstituted by the cholate/dialysis method (Niggli et al., 1981a; Carafoli and Zurini, 1982; Niggli et al., 1982; Villalobo and Roufogalis, 1986; Caroni et al., 1983; Verbist et al., 1984) is approximately threefold better than when reconstituted by a freeze-thaw/sonication method (Haaker and Racker 1979) or a Triton X-100 depletion procedure (Niggli et al., 1981a,b).

4.4.2. Calcium Uptake

The ATP-dependent uptake of Ca^{2+} in reconstituted proteoliposomes has been measured by three different procedures. The first method used was the measurement of $^{45}Ca^{2+}$ entrapped in the proteoliposome inner space by liquid scintillation, after removal of the $^{45}Ca^{2+}$ from the outside of the vesicles by ion-exchange chromatography with a Dowex 50 column (Haaker and Racker, 1979) or by a rapid filtration system (pore size of the Millipore filters, 0.1 μm) (Niggli et al., 1981b). The second method consists of the measurement of the optical absorption of the metallochromic dye Arsenazo III, using a dual-wavelength spectrophotometer at the wavelength pair of 660–685 nm (Niggli et al., 1981b) or 675–685 nm (Verbist et al., 1984). The third procedure to measure Ca^{2+} uptake is by the use of a Ca^{2+}-selective electrode (Niggli et al., 1981b; Carafoli and Zurini, 1982; Niggli et al., 1982; Clark and Carafoli 1983; Villalobo and Roufogalis 1986; Caroni et al.,1982,1983). Quantitative determination of Ca^{2+} uptake by this last method requires careful calibration of the Ca^{2+} electrode using known amounts of $CaCl_2$, since the response of the electrode is not linear, but exponential. The total capacity of Ca^{2+} uptake in proteoliposomes could be increased, as expected, by preparing oxalate-loaded vesicle (Villalobo and Roufogalis, 1986).

4.4.3. (Ca^{2+} + Mg^{2+})-ATPase Activity

The ATP hydrolytic activity of the reconstituted human erythrocyte Ca^{2+}-translocating ATPase in a series of different phospholipid vesicles was found to vary depending on the phospholipid used. In a first attempt to reconstitute the enzyme in noncoupled vesicles, the maximum activity was found in phosphatidylserine vesicles (37–149 $nmol{\cdot}min^{-1}{\cdot}mg$ $prot.^{-1}$, depending on the purity of the enzyme employed), whereas no activity was found in phosphatidylethanolamine vesicles (Peterson et al., 1978). The activity was also very significant in vesicles of pure oleic acid (15–63 $nmol{\cdot}min^{-1}{\cdot}mg$ $prot.^{-1}$, depending on the purity of the enzyme employed) (Peterson et al., 1978).

When a more purified enzyme preparation from human erythrocyte was used in the reconstitution procedure, A23187 was included in the assay system to express all the ATP hydrolytic activity (Niggli et al., 1981a,b; Carafoli and Zurini, 1982; Villalobo and Roufogalis, 1986). The activity in the presence of calmodulin was highest in asolectin vesicles (2.1 $\mu mol \cdot min^{-1} \cdot mg\ prot.^{-1}$), whereas it was decreased in crude soybean L-α-phosphatidylcholine (type II-S from Sigma) and to a greater extent in purer preparations of L-α-phosphatidylcholine from soybean (type IV-S from Sigma) and egg yolk (type X-E from Sigma) (Villalobo and Roufogalis, 1986). Reconstituted enzyme from pig and human erythrocyte was found to be stimulated by calmodulin in asolectin and phosphatidylcholine vesicles (Haaker and Racker, 1979; Niggli et al., 1981a,b; Carafoli and Zurini, 1982), but not in the acidic phospholipids, phosphatidylinositol (Niggli et al., 1981a) or phosphatidylserine (Niggli et al., 1981a,b; Carafoli and Zurini, 1982). Furthermore, when the vesicles were prepared with a mixture of phosphatidylcholine and phosphatidylserine, the calmodulin stimulation decreased as the phosphatidylserine was progressively increased in the vesicles (Niggli et al., 1981a; Carafoli and Zurini, 1982).

When A23187 was not included in the assay medium, it was found that the initial high rate of ATP hydrolysis progressively decreased with time, because of the generation of a Ca^{2+} concentration gradient. This was found with the reconstituted enzyme from human erythrocyte (Niggli et al., 1981b; Carafoli and Zurini, 1982; Niggli et al., 1982; Clark and Carafoli, 1983; Villalobo and Roufogalis, 1986), heart sarcolemma (Caroni et al., 1983), and smooth muscle plasmalemma (Verbist et al., 1984).

4.5. Mechanism of Ca^{2+} Transport

The molecular events in the mechanism of Ca^{2+} transport by the Ca^{2+}-translocating ATPase are not yet understood in great detail. However, the reconstitution of the enzyme in liposomes has greatly advanced knowledge in this regard. Reconstitution of human erythrocyte enzyme has given strong evidence that the Ca^{2+}-dependent K^+ efflux channel is an independent entity from the Ca^{2+}-translocating ATPase, since Ca^{2+}-activated Rb^+ influx (Rb^+ is considered an analog of K^+) was absent in those proteoliposomes (Verma and Penniston, 1985). Moreover, studies in a reconstituted system also have given more information on the involvement of H^+ countertransport by the enzyme, the electrical properties of Ca^{2+} translocation, and the value of the mechanistic Ca^{2+}/ATP stoichiometric ratio. In the next subsections, we will discuss in some detail the present understanding of these points.

4.5.1. Proton Countertransport

The first evidence for Ca^{2+}/H^+ exchange carried out by the reconstituted Ca^{2+}-translocating ATPase from human erythrocyte come from the observation that, during ATP hydrolysis, there was an acidification of the external medium larger than the expected dissociation of H^+ during ATP hydrolysis at a given pH. Moreover, the extra H^+ released to the medium was avoided by the addition of the $Ca^{2+}/2H^+$ exchanger A23187 (Niggli et al., 1982). The calculated rate of vectorial H^+ ejection (after correction for the scaler H^+ production) was twice the rate of ATP hydrolysis. Further evidence for the involvement of H^+ countertransport by the enzyme has been provided by the observation that electroneutral H^+/monovalent cation ionophores, such as nigericin or monensin, stimulate the rate of ATP hydrolysis of the reconstituted enzyme by 2.3- or 3.7-fold, respectively (Villalobo and Roufogalis, 1986). In addition, acetate and ammonium ions, both able to collapse H^+ concentration gradients across the proteoliposome membrane, also stimulated the rate of ATP hydrolysis by the reconstituted enzyme. When a Ca^{2+} ionophore unable to translocate H^+, namely CYCLEX-2E, was used to stimulate the rate of ATP hydrolysis by the reconstituted enzyme, it was found to be less efficient that the $Ca^{2+}/2H^+$ exchanger A23187. However, when a series of proton-conducting agents were tested together with CYCLEX-2E for their ability to stimulate the rate of ATP hydrolysis by the reconstituted enzyme, a significant increase was found (Villalobo and Roufogalis, 1986). None of the ionophores employed in this study stimulated the rate of ATP hydrolysis by the nonreconstituted enzyme, demonstrating that their actions in the reconstituted system were indeed the result of the collapse of a H^+ concentration gradient (alkaline inside) developed during ATP hydrolysis (Villalobo and Roufogalis, 1986).

4.5.2. Electrical Properties

The generation of an electrical gradient by the enzyme translocating Ca^{2+} electrogenically has been proposed in the past when studies were done in inside-out plasma membrane vesicles (*see* Carafoli and Zurini, 1982; Villalobo and Roufogalis, 1986 for discussion). However, a typical mammalian cell holds a resting electrical potential across the plasma membrane of -70 to -90 mV (negative inside). Therefore, this could represent a large energetic barrier considering that the concentration difference of Ca^{2+} across plasma membranes could be as high as four or five orders of magnitude (10^{-8}–$10^{-7}M$ inside and approximately $10^{-3}M$ outside). Consequently, to attain an efficient Ca^{2+} extrusion from the cell in those conditions, an electroneutral exchange of Ca^{2+} for other cations could be of great energetic advantage.

In a reconstituted system, the proposed electroneutral exchange of $1Ca^{2+}/2H^{+}$ was based on the assumption that the Ca^{2+}/ATP stoichiometric ratio was equal to one, since only the stoichiometry of vectorial H^{+} translocated to ATP hydrolyzed was measured and found to approach a value of two (Carafoli and Zurini, 1982; Niggli et al., 1982). The suggested electroneutral translocation of Ca^{2+} was further supported by the observation that no uptake of the lipophilic anion tetraphenylborate (TPB) was observed during net Ca^{2+} uptake in this reconstituted system. In contrast, it was found that the uniport K^{+} carrier valinomycin, in the presence of potassium, significantly stimulated (1.4–1.9-fold) the rate of ATP hydrolysis by the reconstituted enzyme (Carafoli and Zurini, 1982; Niggli et al., 1982; Villalobo and Roufogalis, 1986), without any stimulatory effect on the nonreconstituted enzyme. However, this stimulatory effect of valinomycin in the reconstituted enzyme was considered by some authors to be trivial (Carafoli and Zurini, 1982; Niggli et al., 1982), since the initial rate of Ca^{2+} translocation appeared not to have been affected by this ionophore. Unfortunately, the lack of effect of valinomycin on Ca^{2+} transport was shown only in the first 15 s of ATP hydrolysis, whereas its effect on Ca^{2+} translocation over longer periods of time, during which the progressive decline of Ca^{2+} transport takes place, were not presented (Carafoli and Zurini, 1982; Niggli et al., 1982). The moderate stimulation of ATP hydrolysis induced by valinomycin on the reconstituted enzyme could be interpreted, indeed, as evidence of the generation of a small electrical gradient associated with Ca^{2+} transport, perhaps by the translocation of $1Ca^{2+}$ for $1H^{+}$.

We have found that the stimulatory effect of valinomycin was more pronounced at low buffer capacity than at high buffer capacity of the assay medium (Villalobo and Roufogalis, unpublished results). A large ΔpH (alkaline inside) across the proteoliposome membrane is more likely to develop in low buffer capacity media. Therefore, it is possible that total or partial uncoupling of H^{+} translocation from Ca^{2+} transport could happen in these conditions, rendering the system more electrogenic, and consequently explaining the larger stimulation observed by the addition of valinomycin. Conversely, at higher buffer capacity, the generation of ΔpH is minimized, and the enzyme could exchange H^{+} for Ca^{2+}, rendering the system more electroneutral and explaining the decreased stimulation by valinomycin in these conditions. *In vivo*, the proposed coupling/uncoupling mechanism of H^{+} could play an important role in adapting Ca^{2+} translocation to the physiological fluctuations not only of the ΔpH, but of the membrane potential as well. However, further work should be done to ascertain whether or not the enzyme is able to function alternately in an electroneutral and electrogenic mode depending upon the physiological conditions of the cell.

4.5.3. Stoichiometry

The reported stoichiometry of Ca^{2+} translocated/ATP hydrolyzed measured in inside-out plasma membrane vesicles or resealed ghosts varied from 1–2, apparently depending upon the net rate of ATP hydrolysis (*see* Roufogalis, 1979). However, the Ca^{2+}/ATP stoichiometry when measured in a reconstituted system appears to have a value approaching 1.0 (*see* Table 3). This value was found not only with the erythrocyte enzyme (Haaker and Racker, 1979; Niggli et al., 1981a,b; Clark and Carafoli, 1983), but with the plasma membrane enzyme isolated from heart (Carafoli and Zurini, 1982; Caroni et al., 1983, 1982) and smooth muscle (Verbist et al., 1984). The majority of the results on Ca^{2+}/ATP stoichiometry presented in Table 3 were obtained from experiments performed in the absence of calmodulin. However, the results using the pig erythrocyte enzyme indicate that no difference was found in the absence or presence of calmodulin (Haaker and Racker, 1979). Unfortunately, the results with the pig erythrocyte enzyme were obtained in liposomes exhibiting a low ATP hydrolysis control ratio.

The results in the absence of calmodulin could reflect the mechanistic Ca^{2+}/ATP stoichiometry, because of the fact that most proteoliposome preparations were very impermeable to ions. However, the question of whether or not the mechanistic Ca^{2+}/ATP stoichiometry could vary in the presence of calmodulin or depending on different physiological conditions should be considered open until this point is explored further.

5. Summary and Conclusions

In recent years, significant progress has been made in the purification of the erythrocyte plasma membrane Ca^{2+}-translocating ATPase to near homogeneity and in its reconstitution into defined liposomes able to transport calcium actively against an electrochemical gradient. The progress was the consequence of the use of calmodulin affinity chromatography for rapid purification of the enzyme, the maintenance of an active enzyme conformation in detergent or mixed detergent–lipid micelles, and development of detergent-depletion methods for the production of liposomes with sufficient internal space and integrity to allow calcium

Table 3
Stoichiometric Ratio of Ca^{2+} Translocated/ATP Hydrolyzed by the Plasma Membrane Ca^{2+}-Translocating ATPase of Different Origins Reconstituted in Proteoliposomes

Enzyme Origin	Reconstitution method	Ca^{2+}/ATP ratio	Reference
Pig erythrocyte	Freeze-thaw/ sonication	0.79[a] 0.77[b]	Haaker and Racker (1979)
Human erythrocyte	Triton X-100 (Bio-Beads)	1.08 ± 0.1[a]	Niggli et al. (1981b)
Human erythrocyte	Cholate-dialysis	0.94 ± 0.2[a]	Niggli et al (1981a); Clark and Carafoli (1983)
Bovine heart	Cholate-dialysis	≈1.00	Carafoli and Zurini (1982); Caroni et al. (1983)
Pig antrum smooth muscle	Cholate-dialysis	0.96 ± 0.18[a]	Verbist et al. (1984)

[a] Absence of calmodulin.
[b] Presence of calmodulin.

accumulation over measurable time periods. Both the activity and the calmodulin sensitivity of the purified Ca^{2+}-transport ATPase is highly dependent on the lipid environment. A similar dependency is shown for the reconstituted active Ca^{2+}-transport activity, emphasizing the close coupling between the ATP hydrolytic and translocation activities. Details such as the stoichiometric ratios of the transport of other ions during Ca^{2+} translocation, and the influence of electrical potential across the membrane, remain to be elucidated before a clearer understanding of the precise mechanisms of active calcium translocation can be obtained.

Acknowledgments

Experimental work from the authors' laboratory were supported by the Medical Research Council of Canada and the Canadian Heart Foundation (B.C.). We thank Drs. Pieter Cullis and Mick Hope for the analysis of liposome size, and Jim Harris for excellent technical assistance in the original experiments.

References

Allen, B. G., Katz, S., and Roufogalis, B. D. (1987) Effects of calcium, magnesium and calmodulin on formation and decomposition of the phosphorylated intermediate of the erythrocyte calcium transport ATPase. *Biochem. J.*, **244**, 617–623.

Cable, M. B., Fetter, J. J., and Briggs, F. N. (1985) Mechanism of allosteric regulation of the Ca,Mg-ATPase of sarcoplasmic reticulum: Studies with 5′-adenylyl methylenediphosphate. *Biochemistry* **24,** 5612–5619.

Carafoli, E. and Zurini, M. (1982) The Ca^{2+}/pumping ATPase of the plasma membranes. Purification, reconstitution and properties. *Biochim. Biophys. Acta* **683,** 279–301.

Carafoli, E., Zurini, M., Niggli, V., and Krebs, J. (1982) The calcium-transporting ATPase of erythrocytes. *Ann. N.Y. Acad. Sci.* (USA) **402,** 304–328.

Caroni, P., Zurini, M., and Clark, A. (1982) The calcium-pumping ATPase of heart sarcolemma. *Ann. N.Y. Acad. Sci. (USA)* **402,** 403–421.

Caroni, P., Zurini, M., Clark, A., and Carafoli, E. (1983) Further characterization and reconstitution of purified Ca^{2+}-pumping ATPase of heart sarcolemma. *J. Biol. Chem.* **258,** 7305–7310.

Clark, A. and Carafoli, E. (1983) The stoichiometry of the Ca^{2+}-pumping ATPase of erythrocytes. *Cell Calcium* **4,** 83–88.

Downes, P. and Michell, R. H. (1981) Human erythrocyte membranes exhibit a cooperative calmodulin-dependent Ca^{2+}-ATPase of high calcium sensitivity. *Nature (London)* **290,** 270–271.

Gietzen, K. and Kolandt, J. (1982) Large-scale isolation of human erythrocyte Ca^{2+}-transport ATPase. *Biochem. J.* **207,** 155–159.

Gietzen, K., Seiler, S., Fleischer, S., and Wolf, H. U. (1980a) Reconstitution of the Ca^{2+}-transport system of human erythrocytes. *Biochem. J.* **188,** 47–54.

Gietzen, K., Tejka, M., and Wolf, H. U. (1980b) Calmodulin affinity chromatography yields a functional purified erythrocyte (Ca^{2+} + Mg^{2+})-dependent adenosine triphosphatase. *Biochem. J.* **189,** 81–88.

Gopinath, R. M. and Vincenzi, F. F. (1977) Phosphodiesterase protein activator mimics red blood cell cytoplasmic activator of (Ca^{2+}-Mg^{2+})-ATPase. *Biochem. Biophys. Res. Commun.* **77,** 1203–1209.

Graf, E., Verma, A. K., Gorski, J. P., Lopaschuk, G., Niggli, V., Zurini, M., Carafoli, E., and Penniston, J. T. (1982) Molecular properties of calcium-pumping ATPase from human erythrocytes. *Biochemistry* **21,** 4511–4516.

Haaker, H. and Racker, E. (1979) Purification and reconstitution of the Ca^{2+}-ATPase from plasma membranes of pig erythrocytes. *J. Biol. Chem.* **254,** 6598–6602.

Holloway, P. W. (1973) A simple procedure for removal of Triton X-100 from protein samples. *Anal. Biochem.* **53,** 304–308.

Jarrett, H. W. and Penniston, J. T. (1977) Partial purification of the Ca^{2+}-Mg^{2+}-ATPase activator from human erythrocytes: Its similarity to the activator of 3′:5′-cyclic nucleotide phosphodiesterase. *Biochem. Biophys. Res. Commun.* **77,** 1210–1216.

Katz, S., Roufogalis, B. D., Landman, A. D., and Ho, L. (1979) Properties of (Mg^{2+} + Ca^{2+})-ATPase of erythrocyte membranes prepared by different procedures: Influence of Mg^{2+}, Ca^{2+}, ATP, and protein activator. *J. Supramol. Struct.* **10,** 215–225.

Kosk-Kosicka, D. and Inesi, G. (1985) Co-operative calcium binding and calmodulin regulation in the calcium-dependent adenosine triphosphatase purified from the erythrocyte membrane. *FEBS Lett.* **189,** 67–71.

Kosk-Kosicka, D., Scaillet, S., and Inesi, G. (1986) The partial reactions in the catalytic cycle of the calcium-dependent adenosine triphosphatase purified from erythrocyte membranes. *J. Biol. Chem.* **261,** 3333–3338.

Krug, A. and Wolf, H. U. (1982). Isolation and characterization of a calmodulin-insensitive calcium-dependent ATPase activity of human erythrocyte membranes. *Acta Physiol. Latino Americ.* **32,** 87–89.

Lynch, T. J. and Cheung, W. Y. (1979) Human erythrocyte Ca^{2+}-Mg^{2+}-ATPase: Mechanism of stimulation by Ca^{2+}. *Arch. Biochem. Biophys.* **194,** 165–170.

Minocherhomjee, A. M., Beauregard, G., Potier, M., and Roufogalis, B. D. (1983) The molecular weight of the calcium-transport-ATPase of the human red blood cell determined by radiation inactivation. *Biochem. Biophys. Res. Commun.* **116,** 895–900.

Niggli, V., Adunyah, E. S., and Carafoli, E. (1981a) Acidic phospholipids, unsaturated fatty acids, and limited proteolysis mimic the effect of calmodulin on the purified erythrocyte Ca^{2+}-ATPase. *J. Biol. Chem.* **256,** 8588–8592.

Niggli, V., Adunyah, E. S., Penniston, J. T., and Carafoli, E. (1981b) Purified (Ca^{2+}-Mg^{2+})-ATPase of the erythrocyte membrane. Reconstitution and effect of calmodulin and phospholipids. *J. Biol. Chem.* **256,** 395–401.

Niggli, V., Penniston, J. T., and Carafoli, E. (1979b) Purification of the (Ca^{2+}-Mg^{2+})-ATPase from human erythrocyte membranes using a calmodulin affinity column. *J. Biol. Chem.* **254,** 9955–9958.

Niggli, V., Ronner, P., Carafoli, E., and Penniston, J. T. (1979a) Effects of calmodulin on the (Ca^{2+}-Mg^{2+})ATPase partially purified from erythrocyte membranes. *Arch. Biochem. Biophys.* **198,** 124–130.

Niggli, V., Sigel, E., and Carafoli, E. (1982) The purified Ca^{2+} pump of human erythrocyte membranes catalyzes an electroneutral Ca^{2+}-H^{+} exchange in reconstituted liposomal systems. *J. Biol. Chem.* **257,** 2350–2356.

Penniston, J. T. (1982) Substrate specificity of the erythrocyte Ca^{2+}-ATPase. *Biochim. Biophys. Acta* **688,** 735–739.

Perevucnik, G., Schurtenberger, P., Lasic, D. D., and Hauser, H. (1985) Size analysis of biological membrane vesicles by gel filtration, dynamic light scattering and electron microscopy. *Biochim. Biophys. Acta* **821,** 169–173.

Peterson, S. W., Ronner, P., and Carafoli, E. (1978) Partial purification and reconstitution of the (Ca^{2+}-Mg^{2+})-ATPase of erythrocyte membranes. *Arch. Biochem. Biophys.* **186,** 202–210.

Quist, E. E. and Roufogalis, B. D. (1975) Calcium transport in human erythrocytes: separation and reconstitution of high and low Ca affinity (Mg^{2+} + Ca^{2+})-ATPase activities in membranes prepared at low ionic strength. *Arch. Biochem. Biophys.* **168,** 240–251.

Quist, E. E. and Roufogalis, B. D. (1977) Association of (Ca + Mg)-ATPase activity with ATP-dependent Ca uptake in vesicles prepared from human erythrocytes. *J. Supramol. Struct.* **6,** 375–381.

Roelofsen, B. and Schatzmann, H. J. (1977) The lipid requirement of the (Ca^{2+}-Mg^{2+})-ATPase in the human erythrocyte membrane, as studied by various highly purified phospholipases. *Biochim. Biophys. Acta* **464,** 17–36.

Ronner, P., Gazzotti, P., and Carafoli, E. (1977) A lipid requirement for the (Ca^{2+} + Mg^{2+})-activated ATPase of erythrocyte membranes. *Arch. Biochem. Biophys.* **179,** 578–583.

Roufogalis, B. D. (1979) Regulation of calcium translocation across the red blood cell membranes. *Can. J. Physiol. Pharmacol.* **57,** 1331–1349.

Scharff, O. (1972) The influence of calcium ions on the preparation of the (Ca^{2+}-Mg^{2+})-activated membrane ATPase in human red cells. *Scand. J. Clin. Lab. Invest.* **30,** 313–320.

Scharff, O. (1976) Ca^{2+} activation of membrane-bound (Ca^{2+}-Mg^{2+})-dependent ATPase from human erythrocytes prepared in the presence or absence of Ca^{2+}. *Biochim. Biophys. Acta.* **443,** 206–218.

Scharff, O. (1981) Kinetics of calmodulin-dependent [Ca^{2+}-Mg^{2+}]-ATPase in plasma membranes and solubilized membranes from erythrocytes. *Arch. Biochem. Biophys.* **209,** 72–80.

Schatzmann, H. J. (1966) ATP-dependent Ca^{2+} extrusion from human red cells. *Experientia* **22,** 364–365.

Schatzmann, H. J. and Rossi, G. L. (1971) (Ca^{2+}-Mg^{2+})-activated membrane ATPase in human red cells and their possible relations to cation transport. *Biochim. Biophys. Acta.* **241,** 379–392.

Sharma, R. K. and Wang, J. H. (1981) Inhibition of calmodulin-activated cyclic nucleotide phosphodiesterase by Triton X-100. *Biochem. Biophys. Res. Commun.* **100,** 710–715.

Stieger, J. and Luterbacher, S. (1981) Some properties of the purified (Ca^{2+} + Mg^{2+})-ATPase from human red cell membranes. *Biochim. Biophys. Acta.* **100,** 270–275.

Verbist, J., Wuytack, F., DeSchutter, G., Raeymaekers, L., and Casteels, R. (1984) Reconstitution of the purified calmodulin-dependent (Ca^{2+} + Mg^{2+})-ATPase from smooth muscle. *Cell Calcium* **5,** 253–263.

Verma, A. K. and Penniston, J. T. (1985) Evidence against involvement of the human erythrocyte plasma membrane Ca^{2+}-ATPase in Ca^{2+}-dependent K^+ transport. *Biochim. Biophys. Acta* **815,** 135–138.

Villalobo, A. and Roufogalis, B. D. (1986) Proton countertransport by the reconstituted erythrocyte Ca^{2+}-translocating ATPase: Evidence using ionophoretic compounds. *J. Membrane Biol.* **93,** 249–258.

Villalobo, A., Brown, L., and Roufogalis, B. D. (1986) Kinetic properties of the purified Ca^{2+}-translocating ATPase from human erythrocyte plasma membrane. *Biochim. Biophys. Acta* **854,** 9–20.

Wolf, H. U., Dieckvoss, G., and Lichtner, R. (1976) Partial purification of soluble high-affinity Ca^{2+}-ATPase of human erythrocyte membranes. *Experientia* **32,** 776.

Wolf, H. U., Dieckvoss, G., and Lichtner, R. (1977) Purification and properties of high-affinity Ca^{2+}-ATPase of human erythrocyte membranes. *Acta Biol. Med. Germ.* **36,** 847–858.
Wolf, H. U. and Gietzen, K. (1974) The solubilization of high-affinity Ca^{2+} -ATPase of human erythrocyte membranes. *Hoppe-Seyler's Z. Physiol.* **355,** 1272.

Catalytic Mechanisms of the Ca-Pump ATPase

Alcides F. Rega and Patricio J. Garrahan

1. Introduction

Under physiological conditions, the activity of calcium, that is, the concentration of free ionized calcium (Ca^{2+}) in the cytosol of human red cells, is less than $10^{-7}M$, which is about 10^4 times lower than that present in the blood plasma (*see* Rega and Garrahan, 1986). Although the passive permeability of the red-cell membrane to Ca^{2+} is very low, there is a small but permanent leak of Ca^{2+} into the cell. Hence, active transport of Ca^{2+} to the extracellular medium is required to maintain the low concentration of Ca^{2+} in the cytosol.

Active transport of Ca^{2+} across the plasma membrane was first described by H. J. Schatzmann (1966) in human red blood cells. Since that time, it has been demonstrated in cells from so many tissues and organisms (*see* Rega and Garrahan, 1986) that it can be concluded that the Ca^{2+}-pump system is present in most (if not all) plasma membranes from eukaryotic cells.

The extrusion of Ca^{2+} is strictly coupled to the hydrolysis of ATP. Accordingly, the Ca^{2+}-pump is a Ca^{2+}-activated ATPase or Ca-pump ATPase nomenclature that will be used throughout this chapter. The hydrolysis of ATP takes place through a phosphoenzyme intermediate with properties resembling those of the intermediates of the (Na^+ + K^+)-ATPase from plasma membranes and the Ca-pump ATPase from sarcoplasmic reticulum. It seems, therefore, that the Ca-pump ATPase from red cells belongs to the class of cation-motive E_1E_2 ATPases as defined by Mitchell and Koppenol (1982).

This chapter describes the characteristics of the transport of Ca^{2+} and the hydrolysis of ATP catalyzed by the Ca-pump ATPase of red blood cells. It also includes the description of partial reactions of the hydrolysis of ATP, which represent part of the transport cycle, as well as the hydrolysis of other organic phosphates by the Ca-pump ATPase.

2. The Main Properties of the Transport of Ca^{2+} and the Hydrolysis of ATP by the Ca-Pump ATPase

2.1. Transport of Ca^{2+}

Active Ca^{2+} transport by the Ca-pump ATPase from red cells has been studied using intact cells, resealed ghosts, inside-out vesicles, and purified Ca-pump ATPase reconstituted in lyposomes (*see* Rega and Garrahan, 1986).

2.1.1. Dependence on Ca^{2+}

Resealed ghosts resemble intact cells and allow the control of the composition of the intracellular medium during transport experiments. In calmodulin containing resealed ghosts from human red cells, active transport of Ca^{2+} increases with intracellular Ca^{2+} along a rectangular hyperbola with V_m of about 20 mmol/l cell/h (or 0.035 μmol/mg membrane protein/min) and is half maximal (K_{Ca}) at about 3 μ*M* Ca^{2+} (Kratje et al., 1983). Sr^{2+} can be transported instead of Ca^{2+} (Olson and Cazort, 1969). Alkali metal (Na^+, K^+, Rb^+) ions interact with the Ca^{2+}-ATPase from the intracellular medium and increase the V_m of Ca^{2+} extrusion by about 30% without changing K_{Ca} (Kratje et al., 1983). During activation, alkali metal ions are not transported by the Ca-pump ATPase (Kratje et al., 1983).

From studies made in preparations other than resealed ghosts, it is known that calmodulin, mild proteolysis, acidic lipid environment, and EGTA-Ca buffers lower the value of K_{Ca} from about 20 to about 3 μ*M* and increase about five times the maximum rate of the active transport (*see* Rega and Garrahan, 1986, and Chapter 7 in this book). The effects of these treatments are not additive and can be made apparent if the preparation used has been deprived of calmodulin.

Ca^{2+} applied on the external surface of the cell membrane inhibits active transport of Ca^{2+} from resealed ghosts with low apparent affinity ($K_i \simeq 9$ m*M*) along a rectangular hyperbola that tends to zero. Inhibition is independent of intracellular Ca^{2+}, and calmodulin is stimulated by alkaline pH and can be reproduced by Mg^{2+}. It is probably caused by the combination of Ca^{2+} at the site from which it is released after being transported from the cell interior (Kratje et al., 1985).

Intracellular Ca^{2+} inhibits Ca^{2+} transport from human red-cell ghosts in a way that is independent of external Ca^{2+}. The K_i is around 1–

1.5 mM, that is, about 7 times less than that for extracellular Ca^{2+} (Kratje et al., 1985). Nevertheless, inhibition by intracellular Ca^{2+} is probably of less physiological significance than that by external Ca^{2+}, since its K_i is more than 10^4 times higher than the physiological intracellular Ca^{2+} concentration, whereas the K_i for external Ca^{2+} is only about five times higher than the physiological extracellular Ca^{2+} concentration. Cytosolic free Ca^{2+} and CaATP may act as competitors of Mg^{2+} and MgATP, respectively, to inhibit Ca^{2+} transport (Kratje et al., 1985).

2.1.2. Electrical Balance

If the transport of Ca^{2+} by the Ca-pump ATPase takes place without compensating the two positive charges, the process will be electrogenic. Using fluorescent probes in inside-out vesicles of red-cell membranes, it has been shown that a member potential develops during Ca^{2+} transport by the Ca-pump ATPase (Gimble et al., 1982). In agreement with this, results obtained by measuring volume changes during Ca^{2+} transport in inside-out vesicles suspended in media of different ionic composition suggest that the Ca-pump ATPase is electrogenic, and that according to the condition chosen, either Cl^- or K^+ movements through passive channels provide electroneutrality to the overall ionic shift (Rossi and Schatzmann, 1982).

If Ca^{2+} transport is coupled to the transport of other ions, then the two positive charges of Ca^{2+} could be compensated, the transport process being electroneutral. It has been shown in a purified Ca-pump ATPase from human red cells reconstituted into isolectin liposomes that transported Ca^{2+} is exchanged by two H^+ and, hence, the pump is electroneutral (Niggli et al., 1982a). This conclusion, based on experiments made in a reconstituted system, is at variance with that based on experiments performed with intact membranes. The discrepancy is difficult to reconcile.

2.1.3. Coupling Between Ca^{2+} Transport and ATP Hydrolysis

It is now accepted that the energy-donating substrate of the Ca-pump ATPase is ATP, which is hydrolyzed into ADP and P_i, both of which are released in the cytosol. The transport reaction can, therefore, be written

$$nCa^{2+}_i + ATP_i \rightleftharpoons nCa^{2+}_o + ADP_i + P_i \tag{1}$$

where the subscript "i" denotes intracellular and "o" extracellular. Values of the stoichiometric coefficient number (n) ranging from 0.86–2.02 have been reported (Quist and Roufogalis, 1975; Larsen et al., 1978), making it uncertain whether the value of "n" is 1 or 2, or whether it shifts between 1 and 2 depending on the experimental conditions. Hence, whether the Ca-pump ATPase transports 1 of 2 Ca^{2+}/mol of ATP hydrolyzed still remains an open question.

According to reaction (1), for active Ca^{2+} outflow to take place, the Gibbs energy change of the hydrolysis of ATP plus the transport of Ca^{2+} must be negative. This can be calculated by means of the equation

$$\Delta G_t = \Delta G^{\circ\prime} + RT\ ln \frac{[ADP]\,[P_i]}{[ATP]} + nRT\ ln \frac{[Ca^{2+}_o]}{[Ca^{2+}_i]} + n2FE \quad (2)$$

where $\Delta G^{\circ\prime} = 30.55$ kj mol^{-1}, R = 8.31 J mol^{-1} K^{-1}, F = 96,500 C eq^{-1}, T = 310 K, and n = the stoichiometric coefficient. For a normal red cell, cytosolic ATP, ADP and P_i are near $1.5 \times 10^{-3}M$, $0.3 \times 10^{-3}M$ and $0.3 \times 10^{-3}M$, respectively. The concentration of cytosolic calcium ion $[Ca^{2+}_i] \simeq 10^{-7}M$, and the concentration of extracellular calcium ion $[Ca^{2+}_o]$ can be taken as $10^{-3}M$. E represents the membrane potential that, in the human red cell, is 0.01 V, negative inside. The value of ΔG_t will be negative even if the stoichiometric coefficient were 2 and, hence, there will be no thermodynamic restrictions for reaction (1) to take place from left to right.

2.1.4. Reversal of the Ca-Pump ATPase

Equation 2 predicts that, when the Gibbs energy change necessary for the extrusion of nCa^{2+} is made higher than that of the hydrolysis of ATP, the overall transport reaction will reverse. Consequently, a net synthesis of ATP will occur with the energy coming from the Ca^{2+} concentration difference across the membrane. It has been demonstrated in intact red cells (Ferreira and Lew, 1975), resealed ghosts (Rossi et al., 1978), and inside-out vesicles (Wüthrich et al., 1979) that incorporation of P_i into ADP takes place when the Ca^{2+} concentration at both sides of the membrane and the concentrations of ATP, ADP, and Pi are manipulated to make the value of ΔG_t positive. This was taken as evidence that the Ca-pump ATPase can be forced to run in a reverse fashion utilizing the energy derived from the Ca^{2+} concentration gradient to synthetize ATP.

2.2. Hydrolysis of ATP

2.2.1. Dependence on Ca^{2+}

ATP hydrolysis by the Ca-pump ATPase is activated by Ca^{2+} along a rectangular hyperbola (Schatzmann and Roelofsen, 1977). The kinetic parameters of the activation are highly dependent on calmodulin, acidic phospholipids, partial proteolysis, EGTA, and other factors, so that different experimental results are obtained depending on the procedure used for the isolation of the enzymatic preparation (Rega and Garrahan, 1986, and Chapter 7 in this book). When either purified Ca-pump ATPase or isolated red-cell membranes poor in acidic phospholipids treated with Ca^{2+} chelators to remove endogenous calmodulin and protected against endogenous proteolysis are assayed in the absence of EGTA, activation by Ca^{2+} takes place with a K_{Ca} of about 30 μM (Rega and Garrahan, 1986). Addition of calmodulin to these enzyme preparations lowers K_{Ca} to about 1 μM (*see* Chapter 7 in this book). Since Mg^{2+} competes with Ca^{2+}, the value of K_{Ca} will depend on the concentration of Mg^{2+} in the assay medium and, hence, should be calculated by extrapolation at zero Mg^{2+} concentration (Caride et al., 1986). Under the usual assay conditions (1–3 mM ATP), calmodulin increases 3–7 times the maximum effect of Ca^{2+}. As discussed below, this effect is probably caused by the action of calmodulin on the apparent affinity for ATP.

If activation by Ca^{2+} is extended to Ca^{2+} concentration in the mM range, the curve relating Ca-pump ATPase activity to Ca^{2+} concentration is biphasic. The activity reaches a maximum at about $10^{-4}M$ Ca^{2+} and then drops to almost zero with a K_i of 1.3 mM in medium with 5 mM $MgCl_2$. If calmodulin is present, the shape of the curve remains the same, except that the maximum is shifted towards lower Ca^{2+} concentration (Scharff, 1981; Muallem and Karlish, 1981; Kratje et al., 1985). Since the K_i value for the inhibition of Ca^{2+} transport by external Ca^{2+} is about 10 mM and the K_i for inhibition of the Ca^{2+}-ATPase is 1.3 mM (just about the value of the K_i for inhibition of Ca^{2+} transport by intracellular Ca^{2+} given above), it seems that inhibition of the Ca-pump ATPase by excess Ca^{2+} takes place at the internal surface of the cell membrane. The inhibitory species should be either Ca^{2+} or CaATP. It has been shown recently (Caride et al., 1986) that Ca^{2+} apparently lowers the binding of Mg^{2+} to the Ca-pump ATPase acting as a dead-end inhibitor of the activation of the Ca-pump ATPase by Mg^{2+}. Furthermore, as it is mentioned below, under adequate conditions, it can be demonstrated that Ca^{2+} inhibits the dephosphorylation reaction of the Ca-pump ATPase (Caride et al., 1986), an effect that could cause the inhibition of the overall reaction by Ca^{2+}. Apart from this evidence on the inhibitory effects of Ca^{2+}, it has been reported that CaATP inhibits the Ca-pump ATPase activity in a concentration-dependent fashion by displacing MgATP from the low-affinity site for ATP (*see below*) (Muallem and Karlish, 1981). Regardless of the underlying mechanism, it should be remembered that the concentration of Ca^{2+} necessary for the inhibition to become apparent is much higher than that in the cytosol of a normal red blood cell; hence, it is unlikely that inhibition by Ca^{2+} has any physiological meaning.

2.2.2. Dependence on ATP

In isolated human red blood cell membranes, CTP, GTP, ITP, or UTP does not substitute for ATP as the substrate (Rega et al., 1973). Similar results have been reported for a purified Ca-pump ATPase from human red cells (Graf et al., 1982). It seems that, in terms of substrate specificity, the Ca-pump ATPase is close to the (Na^+ + K^+)-ATPase, which hydrolyzes ATP only, than to the Ca-pump ATPase of sarcoplasmic reticulum that hydrolyzes the other nucleoside triphosphates as well (de Meis, 1980).

The curve relating Ca-pump ATPase activity to ATP concentration (Fig. 1) is biphasic with a component of low-velocity and high-apparent affinity and a component of high-velocity and low-apparent affinity (Richards et al., 1978; Muallem and Karlish, 1979). It is usual to represent the curve as the sum of two Michaelis-Menten equations, i.e.,

$$v = \frac{Vm_1\,[ATP]}{Km_1 + [ATP]} + \frac{Vm_2\,[ATP]}{Km_2 + [ATP]} \qquad (3)$$

where subscripts 1 and 2 denote the high and low affinity components, respectively. The second component is highly dependent on calmodulin (Muallem and Karlish, 1979; Rossi et al., 1985). This dependence seems to result from the ability of calmodulin to decrease the value of Km_2 without modifying Vm_2. In fact, it has been reported that in medium with optimal concentrations of calmodulin, Mg^{2+} and Ca^{2+}, $Vm_1 = 0.0178$ µmol/mg/min, $Vm_2 = 0.048$ µmol/mg/min, $Km_1 = 3.6$ *µM*, and $Km_2 = 500$ *µM*. In the same medium but without calmodulin, the value of Vm_1 decreases about 30%, the values of Km_1 and Vm_2 remain the same, and the value of Km_2 increases at least to 13,000 *µM* (Rossi et al., 1985).

Many different reaction mechanisms predict biphasic response of an enzyme to its substrate. In most of them, eq. (3) is a mathematically valid description of the substrate curve (Rossi and Garrahan, 1985). However, it is important to stress that mathematical validity may not necessarily bear a relationship with physical meaning. In fact, the parameters of eq. (3) will have their usual meanings only in the case that the substrate curve is generated by an enzyme having two independent and different active sites, which may not be the case for the Ca-pump ATPase.

As it has been mentioned above, it is now accepted that the Ca-pump ATPase hydrolyzes ATP. However, it is difficult to draw a definite conclusion on whether MgATP, CaATP, or free-ATP is the actual substrate. Detailed analysis of this problem can be made only if the require-

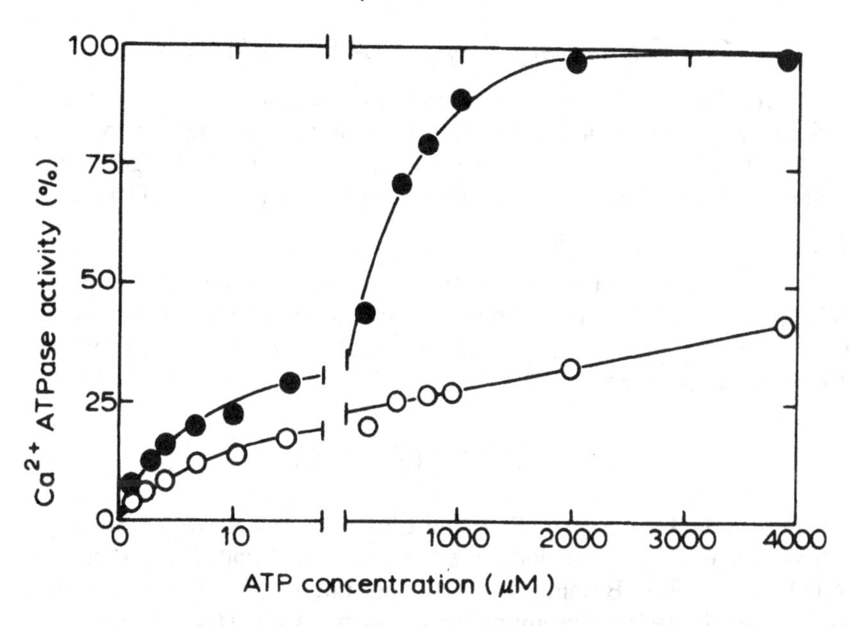

Fig. 1. The substrate curve of the Ca-pump ATPase of human red-cell membranes stripped of their calmodulin and incubated in media with 30 μ*M* calmodulin (●) and without (○). ATPase activity was measured at 37°C in media containing 120 m*M* KCl, 4 m*M* $MgCl_2$, 1 m*M* EDTA, 1 m*M* ouabain, 1.2 m*M* $CaCl_2$, 30 m*M* TrisHCl (pH 7.4 at 37°C), and various concentrations of ATP. Ca-pump ATPase activity was the difference between the activity in these media and the activity in the same media except that $CaCl_2$ was omitted. The activity corresponding to 100% was 0.066 μmol P_i/mg protein/min. The continuous curves are solutions of equation (3) for the values given in the text. (From Rossi et al., 1985, with permission.)

ments of ATP are considered separately for the high and low affinity components. From data available now, it seems that the Ca-pump ATPase does not discriminate between MgATP, CaATP, and free-ATP at the high-affinity component (Muallem and Karlish, 1981; Penniston, 1982). MgATP seems to be the effective ligand at the low affinity component with CaATP acting as a competitive inhibitor (Muallem and Karlish, 1981). This view is consistent with the observations that, in the absence of Mg^{2+}, only the high affinity component is apparent, and that, in the presence of excess Ca^{2+} to convert MgATP into CaATP, the low affinity component is inhibited.

2.3. Elementary Steps of ATP Hydrolysis

Hydrolysis of ATP by the Ca-pump ATPase takes place throughout a series of reactions involving the phosphorylation and dephosphorylation of the phosphoenzyme associated with conformational changes. This mechanism is essentially similar to that of the other E_1E_2 ATPases.

2.3.1. Phosphorylation

The first chemical reaction in the hydrolysis of ATP by the Ca-pump ATPase is the transference of the γ-phosphate of ATP to the enzyme to form a covalent phosphoenzyme (Knauf et al., 1974; Katz and Blostein, 1975; Rega and Garrahan, 1975)

$$ATP + E \overset{Ca^{2+}}{\rightleftharpoons} EP + ADP \quad (4)$$

where E is the Ca-pump ATPase and EP the phosphoenzyme. Although not shown, reaction (4) includes the binding of ATP and Ca^{2+} to form an ATPECa complex. Binding of ATP is not rate-limiting for the reaction (Adamo et al., 1988). The amino-acid sequence around the binding site is (Filoteo et al., 1987): -Met-Tyr-Ser-Lys-Gly-Ala-Ser-Glu-Ile-Ile-Leu-Arg-. The initial rate of phosphorylation at 37°C is 4–8 pmol P/mg prot./s. Mg and Ca^{2+} added together beforehand accelerate phosphorylation to 70 pmol P/mg prot./s. Under these conditions, calmodulin accelerated further the reaction to 280 pmol P/mg prot./s (Adamo et al., 1988).

The following are relevant properties of the reaction:

1. the level of EP increases with the concentration of Ca^{2+} with the same K_{Ca} as that of the Ca-pump ATPase
2. the steady-state level of EP as a function of the ATP concentration increases along a hyperbola with a value of $K_{0.5}$ that is similar to the value of Km_1 (eq. 3) for the Ca-pump ATPase
3. EP undergoes rapid turnover
4. The phosphate bond of EP is an acylphosphate formed with the β-carboxylate of an aspartyl residue. The amino-acid sequence around the phosphorylation site is (James et al., 1987): -Ile-Cys-Ser-Asp-Lys-Thr-Gly-Thr-Leu-Thr-. The residues are the same as those in a sequence common to the (Na^+, K^+)-ATPase, the Ca^{2+}-ATPase of sarcoplasmic reticulum, and the H^+,K^+-ATPase of gastric mucosa (James et al., 1987).

Out of the known elementary steps of the ATPase reaction, phosphorylation is the only one that is absolutely dependent on Ca^{2+}. Hence, phosphorylation seems to be responsible for the dependence of the Ca-pump ATPase on this cation. The steady-state level of phosphoenzyme in red-cell membranes is about 3 pmol P/mg protein. The level of phosphoenzyme can be substantially increased by La^{3+} (Luterbacher and Schatzmann, 1983), and high (m*M*) concentration of Ca^{2+} (Lichtner and Wolf, 1980a). This indicates that, under the usual assay conditions, only a fraction of the ATPase molecules are in the phosphorylated state, a fact that has to be considered when the level of Ca^{2+}-dependent phosphorylation is used to estimate the number of Ca-pump ATPase units in a given membrane. On this basis, it can be calculated that a human red blood cell possesses about 1500 Ca-pump ATPase molecules. Phosphorylation is not rate-limiting for the overall reaction, since its initial rate at 37°C is higher than ATPase activity (Adamo et al., 1988).

The decay of EP is accelerated by ADP in a Ca^{2+}-dependent fashion (Rega and Garrahan, 1978), suggesting that under these conditions EP transfers its phosphate to ADP and regenerates ATP. Hence, reaction (4) seems to proceed with energy conservation. This is in keeping with the fact that the phosphate is associated with the Ca-pump ATPase through an acylphosphate bond (Rega and Garrahan, 1975; Rega et al., 1979; Lichtner and Wolf, 1980b; James et al., 1987). An estimate of the Gibbs energy change based on the values of the rate constants k_1 and k_{-1} of reaction (4) measured in the presence of La^{3+} gives positive values up to 14 kj/mol (Schatzmann, 1983).

2.3.2. Dephosphorylation

The following are the salient features of the reaction of hydrolysis of EP (Rega and Garrahan, 1975; Garrahan and Rega, 1978; Caride et al., 1986):

1. it is enzymatically catalyzed
2. neither Mg^{2+} nor ATP alone modify the rate of hydrolysis
3. provided EP was made in media with Mg^{2+}, ATP increased 5- to 10-fold its rate of hydrolysis, even if Mg^{2+} was sequestered by chelators
4. Mg^{2+} and ATP added together during dephosphorylation increase 5- to 10-fold the rate of hydrolysis of EP made in media with 10 μM Ca^{2+} and without Mg^{2+}.

The Mg^{2+}-dependent effect of ATP is exerted at concentrations much higher than those needed for full occupation of the high affinity site for ATP. Most of the effects of Mg^{2+} and ATP on dephosphorylation can be explained by assuming that, as it has been proposed for other cation-transport ATPases, there is a conformational transition of the phosphoenzyme from a state E_1P of low reactivity towards water into a state E_2P (Rega and Garrahan, 1975; Garrahan and Rega, 1978).

$$E_1P \overset{Mg^{2+}}{\rightleftharpoons} E_2P \tag{5}$$

Reaction (5) is activated by Mg^{2+} ($K_{0.5} \simeq 80\ \mu M$) (Garrahan and Rega, 1978). Also, it does not imply any change in composition of the phosphoenzyme and is inhibited by La^{3+} (Luterbacher and Schatzmann, 1983).

Hydrolysis of the phosphoenzyme follows reaction (5) according to

$$E_2P + H_2O \overset{ATP}{\rightleftharpoons} E_2 + P_i \tag{6}$$

Hydrolysis of E_2P is largely accelerated by combination of ATP with E_2P at a site of low apparent affinity (Garrahan and Rega, 1978). It is thought that reaction (6) is associated with an energy change equivalent to that of the ATP hydrolysis (Schatzmann, 1983). It is tempting to think that acceleration of E_2P hydrolysis and Ca-pump ATPase activity at relatively high ATP concentration are expressions of the same phenomenon, a fact that is complicated by the finding that, although ATP accelerates the former reaction in media in which all Mg^{2+} is sequestered by chelators, the low affinity component of the substrate curve of the Ca-pump ATPase depends on MgATP and calmodulin (Muallem and Karlish, 1981). For the reaction cycle to be completed, E_2 in eq. (6) has to return to E_1 according to:

$$E_2 \rightleftharpoons E_1 \tag{7}$$

Information on the characteristics of reaction (7) is scarce. Indirect evidence has been used to suggest that E_2 possesses a noncatalytic site of low affinity for ATP that changes into a catalytic site of high affinity for ATP when E_2 is converted into E_1 (Muallem and Karlish, 1983). It has been suggested that ATP accelerates the rate of conversion of E_2 into E_1 by combination of ATP at the site in E_2, a mechanism that has been proposed for the Ca^{2+}-pump from sarcoplasmic reticulum and the Na^+-pump (Muallem and Karlish, 1983).

The requirements and effects of the physiological ligands Ca^{2+}, Mg^{2+}, and ATP on the partial reactions of the Ca-pump ATPase are summarized in Table 1. Figure 2 shows a scheme of the overall reaction of ATP hydrolysis by the Ca-pump ATPase based on the properties of the system and also what is known of the other E_1E_2 ATPases. The requirements of ligands other than ATP are indicated in Table 1. After phosphorylation up to E_2P, the reaction pathway is independent of the concentra-

Table 1
The Effects of Physiological Ligands on the Partial Reactions

Reaction	Ca^{2+}	Mg^{2+}	ATP
E_1 + ATP$\rightleftharpoons E_1P$ + ADP	Needed, Mg^{2+} displaces but does not replace	Needed, Ca^{2+} substitutes	Needed
$E_1P \rightarrow E_2P$	Not needed, prevents activation by Mg^{2+}	Activates	Not needed
$E_2P \rightarrow E_2 + P_i$	Not needed	Not needed	Activates
$E_2 \rightarrow E_1$	Not needed	Not needed	Activates?

tion of ATP. At low ATP concentrations (<10 μ*M*), the reaction continues via $E_2 \rightarrow E_1 \rightarrow ATPE_1$. As the concentration of ATP rises, the reaction proceeds via ATP $\rightarrow E_2P \rightarrow ATPE_2 \rightarrow ATPE_1$, which is faster than the other because E_2P hydrolysis is accelerated by ATP. This effect depends on Mg^{2+} to form the effective MgATP complex and on calmodulin. Full activation of the Ca-pump ATPase requires Mg^{2+} for acceleration of the $ATPE_1 \rightarrow E_1P$ and the $E_1P \rightarrow E_2P$ reactions.

2.4. Phosphatase Activity of the Ca-Pump ATPase

The Ca-pump ATPase, like the other known E_1E_2 ATPases, shows a phosphatase activity towards organic phosphates like *p*-nitrophenylphosphate (Rega et al., 1973; Caride et al., 1982; Verma and Penniston, 1984; Rossi et al., 1986). The physiological meaning of this activity, if any, is obscure. In contrast with what has been demonstrated in the Ca-pump ATPase from sarcoplasmic reticulum, phosphatase activity of human red cells does not drive Ca^{2+} transport (Rega et al., 1973; Caride et al., 1983; Rossi et al., 1986). Phosphatase activity from human red cells requires Mg^{2+} at concentrations higher than the optimal for the Ca-pump ATPase and is strongly activated by monovalent cations.

In the presence of calmodulin, activation of phosphatase activity by Ca^{2+} takes place with high affinity (K_{Ca} = 1.6 μ*M*) and is followed by high affinity inhibition by Ca^{2+} (K_i = 3.4 μ*M*), so that the activity becomes negligible at concentrations of Ca^{2+} above 30 μ*M* (Verma and Penniston, 1984). In membranes submitted to partial proteolysis with trypsin to mimic the effects of calmodulin, calmodulin and Ca^{2+} are no longer needed. Also, the activity is maximal in media without Ca^{2+} and

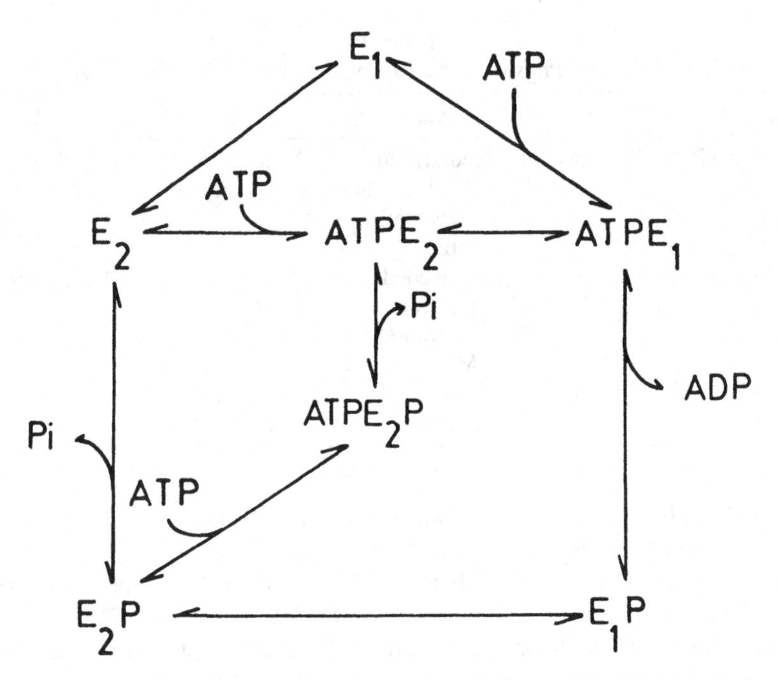

Fig. 2. A reaction scheme for phosphorylation and dephosphorylation of the Ca-pump ATPase. The requirements and effects of Ca^{2+} and Mg^{2+} are indicated in Table 1.

is inhibited by this cation with $K_i = 2.9\ \mu M$ (Rossi et al., 1986). These results have been taken as evidence that the dependence of the phosphatase activity on Ca^{2+} in media with calmodulin expresses the need of the cation to form the calmodulin–Ca complex, which is the effective cofactor. Under these conditions, binding of Ca^{2+} to the ATPase leads to inhibition (Rossi et al., 1986). In membranes with calmodulin or membranes submitted to partial proteolysis, ATP protects phosphatase activity against inhibition by Ca^{2+}, because it increases the K_i for Ca^{2+}. For this reason, the highest rate of phosphatase activity is attained in media with calmodulin and ATP. Protection by ATP also explains why, at Ca^{2+} concentrations above 30 μM, phosphatase activity becomes completely dependent on the nucleotide (Rossi et al., 1986).

In intact membranes without calmodulin, phosphatase activity can be elicited by ATP plus Ca^{2+} (Rega et al., 1973). This effect seems to be exerted by binding of ATP at the catalytic site and Ca^{2+} at the transport site (Caride et al., 1982).

2.5. Activators of the Ca-Pump ATPase

Activation by calmodulin and by substances and treatments that mimic the effect of calmodulin is treated in Chapters 5 and 7.

2.5.1. Magnesium

Mg^{2+} at the intracellular surface of the red-cell membrane activates the Ca^{2+}-ATPase. Activation is not associated with the transport of Mg^{2+} (Schatzmann, 1975). The only effect of extracellular Mg^{2+} is the low affinity inhibition of the transport of Ca^{2+} (Kratje et al., 1985).

There is hydrolysis of ATP by the Ca-pump ATPase in the absence of Mg^{2+} (Caride et al., 1986) at a rate lower than 10% of the optimum. Under these conditions, a low affinity component of the substrate curve is not apparent (Muallem and Karlish, 1981). Ca-pump ATPase activity shows a biphasic response to Mg^{2+} concentration: activity first rises, reaches a maximum at about 1.5 m*M* Mg^{2+}, and then declines progressively. Activation takes place along a rectangular hyperbola with a K_{Mg}, which varies linearly with the concentration of Ca^{2+} indicating that Ca^{2+} is a competitive inhibitor of the activation by Mg^{2+}. The value of K_{Mg} for activation extrapolated to zero Ca^{2+} concentration is 20 μM. K_i for Ca^{2+} as a competitor of Mg^{2+} during activation of the Ca-pump ATPase is 18 μM, suggesting that the site is not selective for Mg^{2+} (Caride et al., 1986).

Previous sections in this chapter have dealt with the effects of Mg^{2+} on particular aspects of the ATPase reaction. Taken together, these results allow the proposal that activation of the Ca-pump ATPase by Mg^{2+} is exerted by combination with high affinity at sites in the enzyme and by formation of the MgATP complex. Occupation of the site or sites in the enzyme leads to stimulation of Ca-pump ATPase activity by acceleration of the phosphorylation reaction and the $E_1P \rightarrow E_2P$ transition. In the first but not in the second effect, Mg^{2+} can be replaced by Ca^{2+} (Caride et al., 1986). In addition, as mentioned before, the MgATP complex seems to be the effective substrate for the low affinity site of the Ca-pump ATPase for ATP.

At concentrations higher than 2 m*M*, inhibition of Ca-pump ATPase activity by Mg^{2+} becomes apparent. This process takes place along a rectangular hyperbola with $K_i \simeq 10$ m*M* (Caride et al., 1986). Since maximum inhibition represents about 60% of the optimal activity and the kinetic parameters of inhibition are independent of calmodulin and Ca^{2+}, inhibition has to be attributed to a direct effect of Mg^{2+}, rather than to displacement of Ca^{2+} by the inhibitor (Caride et al., 1986).

2.5.2. Alkali Metal Ions

K^+, Na^+, and Rb^+ interact directly with the Ca^{2+}-pump to increase both Ca^{2+} transport and ATP hydrolysis, Cs^{2+} and Li^+ having little or no effect (Schatzmann and Rossi, 1971; Bond and Green, 1971; Sakardi et al., 1978; Wierichs and Bader, 1980; Romero, 1981; Kratje et al., 1983). The more salient features of the effect of these cations are:

1. stimulation is not additive
2. stimulation is insensitive to cardiac glycosides
3. the effects are exerted by combination of the alkali metal ions at site(s) on the inner surface of the membrane and are not associated to the transport of the cation
4. activation follows simple hyperbolic kinetics with $K_{0.5}$ = 6 m*M* for K^+ or Rb^+ and about 33 m*M* for Na^+, so that under physiological conditions the effects of K^+ are almost fully exerted
5. maximum activation is the same for all the cations and represents from 30–100% of the basal activity depending on the experimental conditions
6. activation is not associated to changes in the apparent affinities for Ca^{2+} and ATP
7. the ratio Ca-pump ATPase activity/phosphoenzyme level in the presence of K^+ is higher than in its absence, suggesting that alkali metal ions increase the turnover of the enzyme (Larocca et al., 1981).

2.6. Inhibitors of the Ca-Pump ATPase

The effects of drugs that inhibit the Ca-pump ATPase are described in detail in Chapter 14. No selective and specific inhibitors of the Ca-pump ATPase of red cells is yet available. Ca-pump ATPase is not affected by cardiac glycosides, so that these glycosides become a convenient tool to distinguish between the activities of the (Na^+ + K^+)-ATPase and of the Ca-pump ATPase in the red-cell membrane.

Lanthanum, holmium, praseodymium, gadolinium, and samarium ions inhibit Ca-pump ATPase activity and Ca^{2+} transport (Sarkadi et al., 1977; Schatzmann and Tschabold, 1971). Inhibition by La^{2+} is exerted with K_i in the range 2–3 μM and is associated with a large increase in the steady-state level of phosphoenzyme. Since neither the phosphorylation reaction nor the hydrolysis of E_2P are affected by La^{2+}, it has been proposed that it inhibits the $E_1P \rightarrow E_2P$ transition (Luterbacher and Schatzmann, 1983). Vanadate (VO_3^-) inhibits with high affinity Ca-pump ATPase (Bond and Hudgins, 1980; Barrabin et al., 1980) and Ca^{2+}-transport acting from the inner surface of the cell membrane (Rossi et al.,

1981). Inhibition is almost absent in Mg^{2+}-free reaction media. If Mg^{2+} is present, K^+ increases inhibition and Na^+, but not Li^+, replaces K^+. The effects of Mg^{2+} on the apparent affinity for vanadate during inhibition are exerted from sites at the intracellular surface of the membrane. The K_i value drops from 10^3–2 μ*M* VO_3^- in going from medium without to medium with Mg^{2+} plus K^+. External Ca^{2+} at concentrations similar to those that inhibit Ca-pump ATPase protects against inhibition by VO_3^- (Rossi et al., 1981). Vanadate acts as an uncompetitive inhibitor of the high affinity component of the substrate curve. These results are consistent with the idea that vanadate binds to E_2 and prevents the conversion of E_2 into E_1.

N-ethylmaleimide (NEM) is an irreversible inhibitor of Ca-pump ATPase activity. ATP, Na^+, and K^+ protect against inhibition, whereas Ca^{2+} increases the sensitivity of the pump to NEM (Richards et al., 1977). Fluorescein derivatives like fluorescein 5′-isothiocyanate (FITC) and tetraiodofluorescein (erythrosin B) inhibit Ca-pump ATPase activity. FITC can be made to bind covalently to the Ca-pump ATPase at the ATP sites. Kinetic analysis of the inhibition by FITC is consistent with the idea that the high and the low affinity sites for ATP do not coexist in the same enzyme unit (Muallem and Karlish, 1983). Erythrosin B binds reversibly and inhibits the Ca-pump ATPase with K_i about 70 μ*M* displacing ATP from the regulatory site (Mugica et al., 1984).

3. Summary and Conclusions

The Ca-pump ATPase couples the hydrolysis of ATP to the outflow of Ca^{2+} across the plasma membrane of human red cells. Maximum rate of Ca^{2+} transport is about 3 mmol Ca^{2+}/l cells/h or 0.05 μmol Ca^{2+}/mg membrane prot./min. The concentration of Ca^{2+} for half-maximal velocity is near 3 μ*M* and is highly dependent on calmodulin. No clear-cut experimental evidence to decide between a stoichiometry factor of either one or two Ca^{2+} transported/ATP hydrolyzed is available yet. The curve relating ATP hydrolysis by the Ca-pump ATPase from human red cells to ATP concentration can be described as the sum of two Michaelis-Menten-like equations, one with K_m = 3.5 μ*M* and V = 0.02 μmol/mg protein/min and another with K_m = 500 μ*M* and V = 0.05 μmol/mg protein/min. The component of higher affinity uses either ATP, MgATP, or CaATP as the substrate and is mostly independent of calmodulin, whereas the component of lower affinity requires MgATP and calmodulin. As in the other E_1E_2 ATPases, hydrolysis of ATP by the Ca-pump ATPase takes place throughout a series of reactions involving the synthesis and hydrolysis of the phosphoenzyme associated with conformational changes. Mg^{2+} and alkali metal ions at the cytosolic side of the membrane activate the Ca-pump ATPase without being transported. Excess Mg^{2+} and Ca^{2+} lower the enzyme activity. Among the inhibitors are protein reagents, lanthanides, and vanadate, but none is specific for the Ca-pump ATPase. Provided that ATP or calmodulin are present, the Ca-pump ATPase catalyzes the hydrolysis of *p*-nitrophenylphosphate in a way that is not associated to Ca^{2+} transport.

References

Adamo, H. P., Rega, A. F., and Garrahan, P. J. (1988) Pre-steady-state phosphorylation of the human red cell Ca^{2+}-ATPase. *J. Biol. Chem.* **263,** 17548–17544.

Barrabin, H., Garrahan, P. J., and Rega A. F. (1980) Vanadate inhibition of the Ca^{2+}-ATPase from human red cell membranes. *Biochim. Biophys. Acta* **600,** 796–804.

Bond, G. H. and Green J. W. (1971) Effects of monovalent cations on the (Mg^{2+} + Ca^{2+})-dependent ATPase of the red cell membrane. *Biochim. Biophys. Acta.* **241,** 393–398.

Bond, G. H. and Hudgins, P. (1980) Inhibition of the red cell Ca^{2+}-ATPase by vanadate. *Biochim. Biophys. Acta.* **600,** 781–790.

Caride, A. J., Rega, A. F., and Garrahan, P. J. (1982) The role of the sites for ATP of the Ca^{2+}-ATPase from human red blood cells during Ca^{2+} phosphatase activity. *Biochim. Biophys. Acta* **689,** 421–428.

Caride, A. J., Rega, A. F., and Garrahan, P. J. (1983) Effects of *p*-nitrophenylphosphate on Ca^{2+} transport in inside-out vesicles from human red-cell membranes. *Biochim. Biophys. Acta* **743,** 363–367.

Caride, A. J., Rega, A. F., and Garrahan, P. J. (1986) The reaction of Mg^{2+} with the Ca^{2+}-ATPase from human red cell membranes and its modification by Ca^{2+}. *Biochim. Biophys. Acta* **863,** 165–177.

de Meis, L. (1980) The sarcoplasmic reticulum. Transport and energy transduction, in *Transport in the Life Sciences,* **vol. 2** (Bittar, E. E., ed.), John Wiley & Sons, New York.

Ferreira, H. and Lew, V. L. (1975) Ca transport and Ca pump reversal in human red blood cells, *J. Physiol.* **252,** 86P.

Filoteo, A. G., Gorski, J. P., and Penniston, J. T. (1987) The ATP-binding site of the erythrocyte, membrane Ca^{2+} pump. *J. Biol. Chem.* **262,** 6526–6530.

Garrahan, P. J. and Rega, A. F. (1978) Activation of the partial reactions of the Ca^{2+}-ATPase from human red cells by Mg^{2+} and ATP. *Biochim. Biophys. Acta* **513,** 59–65.

Gimble, J. M., Waisman, D. M., Gustin, J., Goodman, D. B. P., and Rasmussen, H. (1982) Studies of the Ca^{2+} transport mechanism of human erythrocyte inside out membrane vesicles. Evidence for the development of a positive interior membrane potential. *J. Biol. Chem.* **257,** 10781–10788.

Graf, E., Verma, A. K., Gorski, J. P., Lopaschuk, G., Niggli, V., Zurini, M., Carafoli, E., and Penniston, J. T. (1982) Molecular properties of calcium-pumping ATPase from human erythrocytes. *Biochemistry* **21,** 4511–4616.

James, P., Zvaritch, E. I., Shakhparonov, M. I., Penniston, J. T., and Carafoli, E. (1987) The amino acid sequence of the phosphorylation domain of the erythrocyte Ca^{2+} ATPase. *Biochem. Biophys. Res. Commun.* **149,** 7–12.

Katz, S. and Blostein, R. (1975) Ca^{2+}-stimulated membrane phosphorylation and ATPase activity of the human erythrocyte. *Biochim. Biophys. Acta* **389,** 314–324.

Knauf, P. A., Proverbio, F., and Hoffman, J. F. (1974) Electrophoretic separation of different phosphoproteins associated with Ca-ATPase and Na, K-ATPase in human red cell ghosts. *J. Gen. Physiol.* **69,** 324–336.

Kratje, R. B., Garrahan, P. J., and Rega, A. F. (1983) The effects of alkali-metal ions on active Ca^{2+} transport in reconstituted ghosts from human red cells. *Biochim. Biophys. Acta* **731,** 40–46.

Kratje, R. B., Garrahan, P. J., and Rega, A. f. (1985) Two modes of inhibition of the Ca^{2+}-pump in red cells by Ca^{2+}. *Biochim. Biophys. Acta* **816,** 365–378.

Larocca, J. N., Rega, A. F., and Garrahan, P. J. (1981) Phosphorylation and dephosphorylation of the Ca^{2+} pump of human red cells in the presence of monovalent cations. *Biochim. Biophys. Acta* **645,** 10–16.

Larsen, F. L., Hinds, T. R., and Vincenzi, F. F. (1978) On the red blood cells Ca^{2+} pump: an estimate of stoichiometry. *J. Membrane Biol.* **41,** 361–376.

Lichtner, R. and Wolf, H. U. (1980a) Phosphorylation of the isolated high-affinity (Ca^{2+} + Mg^{2+})-ATPase of the human erythrocyte membrane. *Biochim. Biophys. Acta* **598,** 472–485.

Lichtner, R. and Wolf, H. U. (1980b) Characterization of the phosphorylated intermediate of the isolated high-affinity (Ca^{2+} + Mg^{2+})-ATPase of the human erythrocyte membrane. *Biochim. Biophys. Acta* **598,** 486–493.

Luterbacher, S. and Schatzmann, H. J. (1983) The site of action of La^{3+} in the reaction cycle of the human red cell membrane Ca^{2+}-pump ATPase, *Experientia* **39,** 311.

Mitchell, P. and Koppenol, W. H. (1982) Chemiosmotic ATPase mechanisms. *Ann. N.Y. Acad. Sci.* **402,** 584–601.

Muallem, S. and Karlish, S. D. (1979) Is the red calcium pump regulated by ATP? *Nature* **277,** 238–240.

Muallem, S. and Karlish, S. J. D. (1981) Studies on the mechanism of regulation of the red cell Ca^{2+} pump by calmodulin and ATP. *Biochim. Biophys. Acta* **647,** 73–86.

Muallem, S. and Karlish S. J. D. (1983) Catalytic and regulatory ATP-binding sites of the red cell Ca^{2+} pump studied by irreversible modification with fluorescein isothiocyanate. *J. Biol. Chem.* **258,** 169–175.

Mugica, H., Rega, A. F., and Garrahan, P. J. (1984) The inhibition of the calcium dependent ATPase from human red cells by erythrosin B. *Acta Physiol. Pharmacol. Lat.* **34,** 163–173.

Niggli, V., Sigel, E., and Carafoli, E. (1982) The purified Ca^{2+} pump of human erythrocyte membranes catalyzes an electroneutral Ca^{2+}-H^+ exchange in reconstituted liposomal systems. *J. Biol. Chem.* **257,** 2350–2356.

Olson, E. J. and Cazort, R. J. (1969) Active calcium and strontium transport in human erythrocyte ghosts. *J. Gen. Physiol.* **53,** 311–322.

Penniston, J. T. (1982) Substrate specificity of the erythrocyte Ca^{2+} ATPase. *Biochim. Biophys. Acta* **688,** 735–739.

Quist, E. E. and Roufogalis, B. D. (1975) Determinations of the stoichiometry of the calcium pump in human erythrocytes using lanthanum as a selective inhibitor. *FEBS Lett.* **50,** 135–139.

Rega, A. F. and Garrahan, P. J. (1975) Calcium ion-dependent phosphorylation of human erthrocyte membranes. *J. Membr. Biol.* **22,** 313–327.

Rega, A. F. and Garrahan, P. J. (1978) Calcium ion-dependent dephosphorylation of the Ca^{2+}-ATPase of red cells by ADP. *Biochim. Biophys. Acta* **507,** 182–184.

Rega, A. F., Richards, D. E., and Garrahan, P. J. (1973) Calcium ion-dependent *p*-nitrophenylphosphate phosphatase activity and calcium ion-dependent adenosine triphosphatase activity from human erythrocyte membranes. *Biochem. J.* **136,** 185–194.

Rega, A. F. and Garrahan, P. J. (1986) *The Ca^{2+} Pump of the Plasma Membranes* CRC Press Inc., Boca Ratón, FL.

Rega, A. F., Garrahan, P. J., Barrabin, H., Horenstein, A., and Rossi, J. P. (1979) Reaction scheme for the Ca-ATPase from human red blood cells, in *Cation Flux across Biomembranes*, (Mukohata, Y. and Packer, L., eds.), Academic Press, New York, pp. 77–86.

Richards, D. E., Rega, A. F., and Garrahan, P. J. (1977) ATPase and phosphatase activities from human red cell membranes. I. The effects of *N*-ethylmaleimide. *J. Membr. Biol.* **35,** 113–124.

Richards, D. E., Rega, A. F., and Garrahan, P. J. (1978) Two classes of site for ATP in the Ca^{2+}-ATPase from human red cell membranes. *Biochim. Biophys. Acta* **511,** 194–204.

Romero, P. J. (1981) Active calcium transport in red cell ghosts resealed in dextran solutions. *Biochim. Biophys. Acta* **649,** 404–418.

Rossi, R. C. ,and Garrahan, P. J. (1985) The substrate curve of the Na,K-ATPase, in *The Sodium Pump* (Glynn, I. M. and Ellory, C., eds.), The Company of Biologists Ltd., UK, pp. 443–455.

Rossi, J. P. F. C., and Schatzmann, H. J. (1982) Is the red cell calcium pump electrogenic? *J. Physiol.* **327,** 1–15.

Rossi, J. P. F. C., Garrahan, P. J., and Rega, A. F. (1978) Reversal of the calcium pump in human red cell. *J. Membr. Biol.* **44,** 37–46.

Rossi, J. P. F. C., Garrahan, P. J., and Rega, A. F. (1981) Vanadate inhibition of active Ca^{2+} transport across human red cell membranes. *Biochim. Biophys. Acta* **648,** 145–150.

Rossi, J. P. F. C., Garrahan, P. J., and Rega, A. F. (1986) The activation of phosphatase activity of the Ca^{2+}-ATPase from human red cell membranes by calmodulin, ATP and partial proteolysis. *Biochim. Biophys. Acta* **858,** 21–30.

Rossi, J. P. F. C., Rega, A. F., and Garrahan, P. J. (1985) Compound 48/80 and calmodulin modify the interaction of ATP with the (Ca^{2+} + Mg^{2+})-ATPase of red cell membranes. *Biochim. Biophys. Acta* **819,** 379–386.

Sarkadi, B., MacIntyre, J. D., and Gárdos, G. (1978) Kinetics of active calcium transport in inside-out red cell membrane vesicles. *FEBS Lett.* **89,** 78–82.

Sarkadi, B., Szász, I., Gerloczy, A., and Gárdos, G. (1977) Transport parameters and stoichiometry of active calcium ion extrusion in intact human red cells. *Biochim. Biophys. Acta* **464,** 93–107.

Scharff, O. (1981) Regulation of (Ca^{2+}, Mg^{2+})-ATPase in human erythrocytes dependent on calcium and calmodulin. *Acta Biol. Med. Ger.* **40,** 457–465.

Schatzmann, H. J. (1966) ATP-dependent Ca^{++} extrusion from human red cells. *Experientia* **22,** 364–368.

Schatzmann, H. J. (1975) Active calcium transport and Ca^{2+}-activated ATPase in human red cells. *Curr. Topics Membr. Transp.* **6,** 125–168.

Schatzmann, H. J. (1983) The red cell calcium pump. *Ann. Rev. Physiol.* **45,** 303–312.

Schatzmann, H. J. and Roelofsen, B. (1977) Some aspects of the Ca pump in human red blood cells, in *Biochemistry of Membrane Transport*, (Semenza, G. and Carafoli, E., eds.), Springer-Verlag, Berlin, pp. 389–400.

Schatzmann, H. J. and Rossi, G. L. (1971) Ca^{2+} + Mg^{2+}-activated membrane ATPases in human red cells and their possible relations to cation transport. *Biochim. Biophys. Acta* **241,** 379–392.

Schatzmann, H. J. and Tschabold, M. (1971) The lanthanide Ho^{2+} and Pr^{2+} as inhibitors of calcium transporting human red cells. *Experientia* **27,** 59–61.

Verma, A. K. and Penniston, J. T. (1984) Two Ca^{2+}-requiring *p*-nitrophenylphosphatase activities of the highly purified Ca^{2+}-pumping adenosine triphosphatase of human erythrocyte membranes, one requiring calmodulin and the other ATP. *Biochemistry* **23,** 5010–5015.

Wierichs, R. and Bader, H. (1980) Influence of monovalent ions on the activity of the (Ca^{2+} + Mg^{2+})-ATPase and Ca^{2+} transport of human red blood cells. *Biochim. Biophys. Acta* **596,** 325–328.

Wüthrich, A., Schatzmann, H. J., and Romero, P. (1979) Net ATP synthesis by running the red cell calcium pump backwards. *Experientia* **35,** 1789–1790.

Regulation of the Plasma Membrane Ca^{2+}-Pump

Frank F. Vincenzi

1. Introduction

This chapter will deal with regulation of the plasma membrane (PM) Ca^{2+}-pump. Emphasis will be on the pump as expressed in the human red blood cell (RBC). There are a number of recent reviews in related areas, including calmodulin (CaM) and drug interaction (Vincenzi, 1981; Roufogalis et al., 1983; Roufogalis, 1985; Weiss et al., 1985), regulation of calcium in cells (Carafoli, 1985), plasma membrane Ca^{2+} extrusion (Carafoli, 1984a, 1984b, 1985; Schatzmann, 1985, 1986), the effects of drugs on plasma membrane Ca^{2+} transport (Vincenzi and Hinds, 1987), and an excellent monograph on the PM Ca^{2+}-pump (Rega and Garrahan, 1986).

At the outset, it should be pointed out that it is the author's position that the real "regulator" of the PM Ca^{2+}-pump is the inward flux or "leak" of Ca^{2+}. A number of modifiers of the PM Ca^{2+}-pump have been described and examined, and some will be considered here. However, it seems apparent to this author that the function of the PM Ca^{2+}-pump is to remove Ca^{2+} from the cell cytosol in sufficient amounts and rapidly enough to allow the maintenance of intracellular Ca^{2+} at extremely low levels and to prevent the accumulation of Ca^{2+}-induced cell injury (Farber, 1981; Starke et al., 1986).

2. Does the RBC Need a Regulated Ca^{2+}-Pump?

The RBC is not a cell that would appear to require "Ca^{2+} signals" as a trigger for the basis for muscular contraction, release of neurotransmitters, and so on. It is not really clear if there is a physiological response to a transient increase in intracellular Ca^{2+} in RBCs, as there is in essentially all other cells. It could even be imagined that the RBC is created with a low permeability to Ca^{2+} and remains so throughout its

120-d life span. Such a view was the conclusion of Passow, who was aware that RBCs maintain low intracellular Ca^{2+}, but found no evidence for a Ca^{2+}-pump (Passow, 1961, 1963). However, a Ca^{2+}-pump was found (Schatzmann, 1966; Schatzmann and Vincenzi, 1969; Lee and Shin, 1969; Olson and Cazort, 1969), and the capacity of the Ca^{2+}-pump, when compared to the passive leak, is rather enormous. So, like all cells, RBCs maintain low intracellular Ca^{2+}. Whether Ca^{2+} is also used as an intracellular second messenger in RBCs is less clear.

3. ATP as a Regulator of the Ca^{2+}-Pump

The apparent K_m of the ATPase for ATP is in the micromolar range, although the estimate depends on the exact conditions of Mg^{2+}, Ca^{2+}, and so on. Considering that the RBC normally contains approximately 1–2 m*M* ATP, it appears that the Ca^{2+}-pump never lacks for energy substrate. It may be noted that the K_m for ATP of the Na,K-pump is considerably higher. Thus, in cells in which ATP is depleted, the Na,K-pump would fail before the Ca^{2+}-pump, based on substrate starvation. This kind of observation supports the notion advanced by our laboratory, as well as others, that the removal of Ca^{2+} from the cytoplasm is absolutely essential for healthy cellular life, whereas the extrusion of Na^+ may be of less fundamental importance.

A low affinity (approximately 100 μM K_m) ATP site has been postulated (Rega et al., 1974; Muallem and Karlish, 1983) to regulate turnover of the Ca^{2+}-pump ATPase. As pointed out by Schatzmann (1986), the "additional site" may represent the "same site" expressing a different affinity as the enzyme cycles through its E1 and E2 forms. Raess et al. (1985) investigated phenylglyoxal, an inhibitor that probably acts at the low affinity ATP site. Phenylglyoxal decreased the V_{max} and K_m of the Ca^{2+}-pump ATPase with no change in Ca^{2+} affinity. Phenylglyoxal reacts with arginyl residues, and Raess et al. (1985) concluded that the binding of ATP to the low affinity site involves an arginyl residue.

4. Calmodulin as a Regulator of the Ca^{2+}-Pump

Approximately 10 years ago, it was demonstrated that the cytoplasmic Ca^{2+}-pump activator protein (Bond and Clough, 1973) of the RBC is CaM (Jarrett and Penniston, 1977), and that CaM activates both the Ca^{2+}-pump ATPase (Gopinath and Vincenzi, 1977) and transport (Larsen and Vincenzi, 1979). CaM increases the V_{max} of the enzyme and increases its apparent affinity for Ca^{2+} (Rega and Garrahan, 1986). However, the Ca^{2+}-pump ATPase can operate without CaM (Hinds and Vincenzi, 1983).

Based on the binding to and activation of the Ca^{2+}-pump enzyme by CaM, it has been widely suggested that the Ca^{2+}-pump is "regulated" by CaM. Although the Ca^{2+}-pump ATPase is, indeed, a CaM-dependent enzyme, whether CaM is actually part of the pump in vivo under normal (or abnormal) conditions is not quite clear to this author. We have offered, admittedly approximate, calculations in support of the suggestion that, for all practical purposes, CaM could be considered a subunit of the pump in vivo (Vincenzi et al., 1980). On the other hand, convincing data and calculations from Scharff and Foder (1982) were interpreted to mean that CaM is not bound to the Ca^{2+}-pump ATPase under normal physiological conditions in the human RBC. Their data were compatible with the interpretation that a significant increase in intracellular Ca^{2+}, instead of causing an instantaneous signal to "begin pumping," would result in a relatively slow binding of the $CaM(Ca^{2+})_n$ complex to the enzyme. This results in hysteretic activation and deactivation of the pump, according to Scharff and Foder (1982). The idea is appealing, but seems only to displace the question of what "regulates" the Ca^{2+}-pump. Whether CaM buffers and/or delays the delivery of the message to accelerate the Ca^{2+}-pump, it is still Ca^{2+} that is the regulatory agent. Thus, the imposed Ca^{2+} load at the inner surface of the PM is what regulates the Ca^{2+}-pump, in the opinion of this author. If one accepts this position, then the question becomes: what are the determinants of Ca^{2+} influx in the human RBC? Although some binding sites for certain surface active agents may be found by binding assays, there is little evidence that RBCs contain active, receptor-coupled Ca^{2+} channels. Likewise, based upon the extremely low membrane potential of the RBC (about 9 mV negative inside, Jay and Burton, 1969), it is unlikely that the RBC contains (functional) voltage-operated Ca^{2+} channels. Varecka and Carafoli (1982) reported that verapamil inhibits Ca^{2+} influx in vanadate-treated RBCs. Since half maximal inhibition occurred only at about 100 μM, it seems unlikely that the effect proves the existence of Ca^{2+} channels in RBCs.

5. Anti-CaM Drugs as Modifiers of the Ca^{2+}-Pump

Because of the possible role of CaM in modulation of the Ca^{2+}-pump ATPase, a number of anti-CaM drugs have been examined for effects on the enzyme or transport. A variety of drugs bind to CaM, in the presence of Ca^{2+}. Wolff and Brostrom (1976) and Levin and Weiss (1977) demonstrated that phenothiazines, such as trifluoperazine (TFP), selectively antagonize certain CaM-dependent enzymes, including the RBC Ca^{2+}-pump ATPase (Raess and Vincenzi, 1980). Roufogalis et al. (1983) suggested that phenothiazines and other antipsychotics bind to CaM, probably at two hydrophobic sites that are Ca^{2+}-dependent. These sites are apparently not stereospecific. These insights have been expanded to show that a range of amphipathic cations, including calcium entry blockers (Bostrom et al., 1981; Schlondorff and Satriano, 1985) and tetracyclines (Schlondorff and Satriano, 1985) can bind to and antagonize the effects of CaM in many systems, including the RBC Ca^{2+}-pump ATPase (Vincenzi, 1981, 1982a; Weiss et al., 1985). Also, as has been known for some time, there is no simple relationship between therapeutic applications and anti-CaM activities of such drugs (Norman et al., 1979; Vincenzi, 1982a).

Anti-CaM agents are generally thought to prevent the binding of the $CaM(Ca^{2+})_n$ complex to its various effectors. It is usually assumed that most such agents interact directly with the $CaM(Ca^{2+})_n$ complex and not the enzyme. Because of the direct inhibition of the enzyme by anti-CaM agents in the absence of CaM, it is also apparent that these agents can bind directly to the enzyme, or closely associated lipids, or the detergent phase (Vincenzi et al., 1982). An important point to be made is that supposed anti-CaM drugs exert effects independent of their anti-CaM actions. A possible exception appears to be compound 48/80 (a mixture of polycationic polymers with average mol wt of about 1300), which Gietzen et al. (1983) reported to be a selective and potent CaM antagonist. That is, compound 48/80 appears to not inhibit the basal (CaM-independent) Ca^{2+}-pump ATPase activity. Presumably, compound 48/80 is a more selective inhibitor of CaM activation of the RBC Ca^{2+}-pump ATPase of isolated RBC membranes than drugs such as TFP, because it enters the lipid environment of the membrane to a lesser extent. The same property probably also limits the usefulness of compound 48/80 as an inhibitor of CaM in intact cells.

It should be noted that the antagonism of CaM by compound 48/80 may not be entirely simple and straightforward. We recently found that a fraction of compound 48/80 separated by CaM affinity chromatography was a potent inhibitor of both the CaM-activated and basal Ca^{2+}-pump ATPase activities of RBC membranes (Hinds et al., 1987). The fact that this fraction was a potent inhibitor of the basal ATPase shows that its inhibition cannot be accounted for on the basis of CaM antagonism alone.

6. Lipids as Regulators of the Ca^{2+}-Pump

It is interesting to compare results of "regulators" that presumably bind directly to the enzyme with those of changes in lipid composition of the membrane and associated Ca^{2+} leak. It would be predicted that such changes in the passive leak of the membrane would exert changes in the load presented to the membrane pump(s). This is what Rosier et al. (1986) found when they reported that depletion of cholesterol from RBC membranes decreases Ca^{2+} influx and intracellular Ca^{2+}. Cholesterol depletion also caused decreased Ca^{2+} efflux via the Ca^{2+}-pump. It was concluded that the effect on the pump was entirely indirect. This is a relatively straightforward kind of "regulation" to understand. Within the limits of the pump to respond to an increased load, increased passive leak would increase the activity of the pump by increasing the amount of Ca^{2+} available at the inner membrane surface. A different kind of impression of lipid regulation might be gained from experiments with isolated membranes. Thus, although Rosier et al. (1986) found that depletion of cholesterol reduced Ca^{2+} leak and Ca^{2+}-pump activity in intact cells, we found that cholesterol depletion increased the maximal activity of the Ca^{2+}-pump ATPase of isolated RBC membranes (Vincenzi and Hinds, 1987). This latter observation is apparently the result of an effect of the lipid on the maximal activity of the ATPase. A qualitatively opposite effect on the controlled pump mediated by Ca^{2+} flux in intact cells is almost certainly the more important effect in vivo.

It has been known for many years that the activity of the Ca^{2+}-pump ATPase in vitro may be increased by acidic phospholipids (Ronner et al., 1977) or even fatty acids (Vincenzi, 1982b; Pine et al., 1983). It appears that molecules containing hydrophobicity and negative charge may mimic the effect on the Ca^{2+}-pump ATPase of $CaM(Ca^{2+})n$. These results suggest that, in addition to altering the passive ion permeability properties of the membrane, the phospholipid composition of the membrane may exert effects on the enzymatic activity of the pump more directly. As noted above, we have observed that the cholesterol/phospholipid ratio of RBC membranes exerts a significant effect on the activity of the Ca^{2+}-pump ATPase, especially in its CaM-activated state (Vincenzi and Hinds, 1987). Under the same conditions, the cholesterol/phospholipid composition of the membrane exerts only minor effects on the basal Ca^{2+}-pump ATPase or on the Na,K-pump ATPase activity. It would appear that the activated status of the enzyme can be achieved in a number of ways, and that the status is also especially sensitive to potential "regulation." Thus, it appears that both the pump and leak aspect of PM

Ca^{2+} transport can be significantly affected by the lipid composition of the membrane.

Tokumura et al. (1985) performed a study of the structural requirements of lipid-related agents to activate the Ca^{2+}-pump ATPase. Although the results were in general compatible with the simple interpretation that hydrophobicity and a negative charge are necessary and sufficient features of a "calmodulinomimetic" (Vincenzi, 1981) agent, Tokumura et al. (1985) also found significant specific effects of structure. In particular, they concluded that the acidic moiety of lysophosphatidic acid is not an important structural determinant for activation of the Ca^{2+}-pump ATPase. They concluded that the nature of the hydrocarbon chain as well as the lyso structure were more critical.

7. Mechanically Induced Ca^{2+} Permeation as a Regulator of the Ca^{2+}-Pump

The notion of pump-leak arose in the RBC field around the Na,K-pump, and appears equally applicable to the question of PM Ca^{2+} transport. Thus, in vivo it would appear that the dominant "regulator" of the activity of the pump is the "leak." The magnitude of the maximal activity of the Ca^{2+}-pump ATPase is an issue we have considered in earlier reviews (Vincenzi and Hinds, 1980, 1987). We have been impressed for years that the capacity of the pump, as expressed in the ATPase activity of isolated membranes or inside-out vesicles (IOVs), is rather enormous compared to the passive transport of Ca^{2+} (Schatzmann and Vincenzi, 1969; Lew and Ferreira, 1978). Based on figures from a number of laboratories, it appears that the passive flux of Ca^{2+} across the human RBC membrane in vitro is approximately 17–170 nmole/min/L RBC. The capacity of the active Ca^{2+}-pump is in the order of 83,000–600,000 nmole/min/LRBC (Lew and Ferreira, 1978). From numbers summarized by Rega and Garrahan (1986), it can be estimated that the Ca^{2+}-pump of human RBCs runs at less than 1% of its capacity, at least in vitro.

For a number of years, we discussed this issue and developed the notion that RBC passive flux may be elevated in vivo. A corollary would be that in vitro estimates of passive Ca^{2+} permeability are underestimates of the real load on the Ca^{2+}-pump as it exists in vivo. We concluded that the most likely mechanism for increased passive permeability in vivo is the mechanical deformation that RBCs undergo repeatedly during their approximately 120-d life span. It should be noted that this idea was not without precedent. Lubowitz et al. (1974) provided evidence for increased Na^+ permeability of RBCs exposed to shear stress.

We suggested that RBC deformation in vitro might be employed to test the idea of mechanically induced increases in membrane permeability. A similar idea was tested by Larsen et al. (1981). The capacity of the Ca^{2+}-pump is great compared with the "resting" leak (Vincenzi and Hinds, 1980). Thus, it seemed unlikely that we could measure much accumulation of Ca^{2+} in RBCs in such experiments, although some Ca^{2+} accumulation had been reported by Larsen et al. (1981). We reasoned that the pump, if operating as it should, would "keep up" with the leak. We, therefore, chose to measure consumption of ATP in RBCs exposed to various levels of shear stress. To apply shear stress and to simultaneously monitor RBC deformation, we employed the technique of ektacytometry (Bessis and Mohandas, 1975; Mohandas et al., 1980). In order to prevent resynthesis of ATP, we included iodoacetic acid in the medium. The results showed that human RBCs exposed to shear stress increase their rate of ATP consumption some 5–6-fold (Vincenzi and Cambareri, 1985a). Of the increase in ATP splitting induced by shear stress, approximately 65% was prevented by EGTA (to prevent Ca^{2+} entry) and ouabain (to prevent ATP utilization by the Na^+,K^+-pump). Also, of the EGTA- and ouabain-sensitive increase in ATP utilization, some 62% was related to apparent activation of the Ca^{2+}-pump and the remainder to the Na^+,K^+-pump (Vincenzi and Cambareri, 1985b). These results support the notion that regulation of the Ca^{2+}-pump is just what would be expected of a well-designed pump-leak system. That is, the amount of ATP, Mg, CaM, and other so-called regulators is in great sufficiency. What controls the activity of the pump is the amount of Ca^{2+} present at the inner surface of the membrane, and this amount is normally the limiting factor (i.e., regulator).

Hysteretic delay of CaM activation (Scharff and Foder, 1982) probably does not play much of a role in the ektacytometry experiments, because we measured ATP utilization over the course of up to 40 min of constant shear stress and the disappearance followed first-order kinetics, with no apparent delay. Under physiological conditions, both the shear stress-induced increased in ion permeabilities and the recovery from these are presumably more transient.

The level of shear stress used in our experiments was sufficient (~2000 dyn/cm^2) to produce maximal RBC deformation, and resulted in up to 20% hemolysis at the end of 40 min of constant shear stress. It should be noted that the hematocrit of the system was very low. Any cells that lysed therefore lost their ATP into a large volume and then no longer participated in ATP breakdown. It should also be noted that there was no evidence of any ecto-ATPase activity in our system. Thus, to the extent that some RBCs lysed and lost their ATP to a much diluted extracellular medium, they no longer contributed to the estimate of mechanically induced ATPase activity.

Based upon these considerations, a speculative model of physiological regulation of the RBC Ca^{2+}-pump is presented in Fig. 1. It is suggested that the most significant inward leak of Ca^{2+} into RBCs is stimulated by deformation of RBCs as they pass through the microcirculation. It is suggested that the major effect of membrane deformation is on cation permeability. Anion permeability is normally so high, based upon the anion channel (band 3) that the mechanically induced increase (even if it did apply to anions) would be unlikely to increase significantly upon RBC deformation.

In contrast with this model of Ca^{2+} entry in RBCs, it is generally accepted that the major regulatory of Ca^{2+} transport in excitable cells is membrane depolarization. In other words, the activity of voltage-operated Ca^{2+} channels of the PM in heart cells (for example) is probably the major determinant of the rate of PM Ca^{2+} entry and extrusion. This is not to deny the importance, especially in skeletal muscle, of intracellular Ca^{2+} release and re-uptake. Rather, the important overview is based on the simplifying assumption that, on average, the amount of Ca^{2+} that enters a cell is the same amount that leaves a cell. Considering that most cells do not become calcified, in spite of a tremendous inward gradient for Ca^{2+}, this simpifying assumption seems reasonable.

Thus, it is suggested that a "physiological signal" of Ca^{2+} is part of the natural life of the human RBC. It is suggested that each pass through the microcirculation results in a transient, but significant increase in passive Ca^{2+} influx. Whether this "physiological signal" is needed for normal function of the RBC or is merely a side effect of the circulatory status of the RBC remains to be determined. Also to be determined is the magnitude of the terms "transient" and "significant increase." It can be estimated that human RBCs spend about 10% of the circulation time in the microcirculation. Just how much of the microcirculation (in diameter, distance, or time) represents a region of sufficiently high shear stress to produce increased membrane permeability is unknown to this author. Chien (1975) has considered shear stress in the circulation, but did not consider mechanically induced changes in Ca^{2+} permeability.

How much of an increase in membrane permeability can the pump respond to? It is suggested, based on our results with ektacytometry, that a 5–6-fold increase in the rate of passive permeation can be easily handled over relatively long periods of time (e.g., 20–40 min) (Vincenzi and

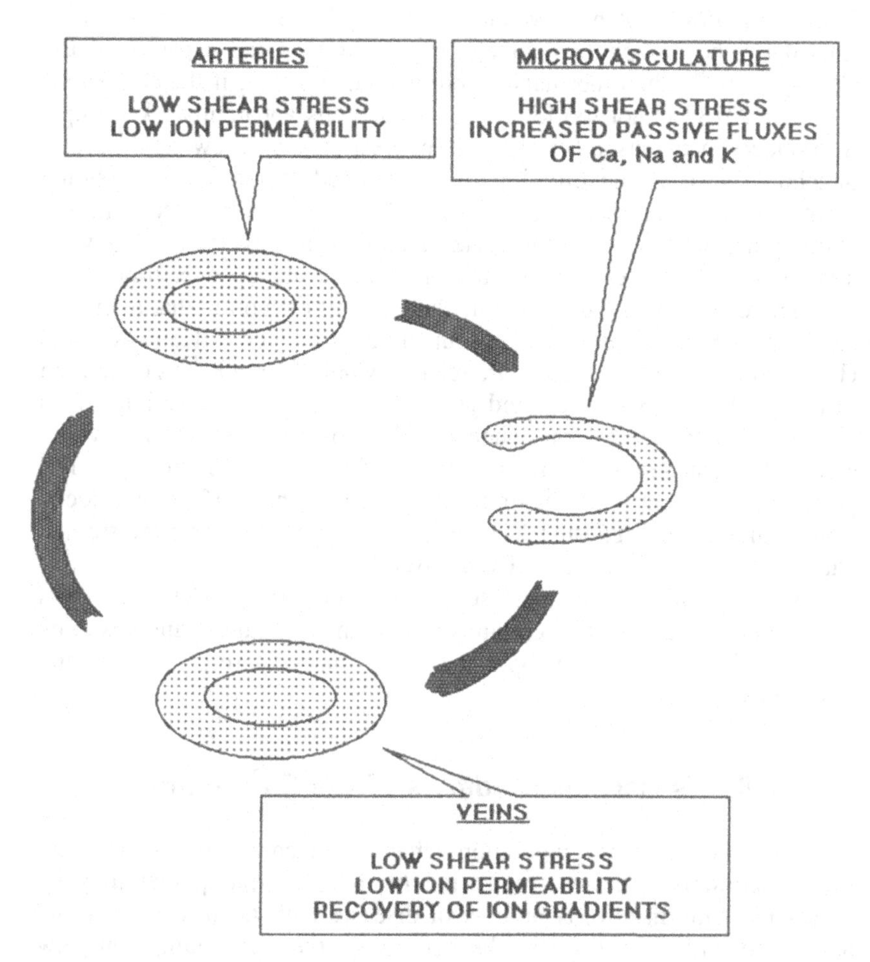

Fig. 1. Schematic diagram of the speculative model of regulation of the Ca^{2+}-pump ATPase of human RBCs in vivo. In this model, a significant increase in passive ion permeabilities occurs during RBC deformation in the microcirculation. It is suggested that shear stress-induced deformation increases Ca^{2+} (and Na^+) permeability, thereby imposing a transient but significantly increased load on the RBC Ca^{2+} (and Na^+,K^+) pump. It is further suggested that the relatively high capacity of the Ca^{2+}-pump ensures that recovery from the transient increase in Ca^{2+} influx is rapid. It is suggested that, if recovery were not rapid, then a variety of cellular damage processes would be initiated. In this speculative model, mechanically induced RBC deformation and the associated increase in Ca^{2+} permeability of the membrane are thus viewed as the major "physiological" regulators of the PM Ca^{2+}-pump.

Cambareri, 1985a). A massive and continuing increase in permeation, as caused by the ionophore A23187, may overcome the capacity of the Ca^{2+}-pump. On the other hand, there is evidence that, if the concentration of A23187 is not too high, then some cells apparently "keep up" (responders) whereas other do not (nonresponders) (Lew and Garcia-Sancho, 1985). If the duration of an induced increase in membrane permeability is on the order of seconds (as probably occurs in the microcirculation), then even greater increases in passive flux may be imagined without a deficit in transport when averaged over time.

There are already data availble that support the notion that microcirculatory changes in ion balance occur in humans in vivo. Kalsheker and Hales found that venous RBCs are relatively deficient in K^+ as compared to arterial RBCs (Kalsheker and Hales, personal communication). This presumably occurs as a consequence of K^+ loss from the cells down its electrochemical gradient; either because of a "direct" increase in K^+ permeability or via the Ca^{2+}-induced K^+ permeability (Gardos effect). The phenomenon suggested here is almost certainly of broader significance than in the life cycle of the RBC.

For example, in an interesting series of papers, Bevan and his coworkers have shown that certain vascular smooth muscle appears to be regulated by mechanically operated Ca^{2+} channels (e.g., Bevan and Hwa, 1985).

8. Diseases as Modifiers of the Ca^{2+}-Pump

Another aspect of "regulation" that can be considered is the question of possibly abnormal regulation of the Ca^{2+}-pump that may be associated with disease. A number of diseases have been reported to be associated with alterations in the activity of the Ca^{2+}-pump, but few definitive cause–effect relationships have been demonstrated. Nevertheless, the possibility that certain variations in the activity of the PM Ca^{2+}-pump might be critical in the expression of disease is so compelling as to demand attention.

Johnsson et al. (1983) found a significant increase in Ca^{2+} efflux from RBCs of patients with Duchenne muscular dystrophy (DMD). They found increased Na^+,K^+- and Ca^{2+}-pump ATPase activities in RBC membranes from patients with myotonic dystrophy, but no change compared to normals in DMD. There was increased rigidity (by filtration assay) of RBCs from patients with congenital myotonia, and no change in RBC deformability in patients with DMD or myotonic dystrophy. Such results are, in the author's experience, typical of those that one may obtain in the study of specific disease(s). Typically, the number of patients involved is very small and the significance of the findings is unclear. We have made similar and probably equally nonspecific observation in several diseases, including sickle cell anemia (Gopinath and Vincenzi, 1979; Vincenzi et al., 1986), and suggest that much restraint is needed in interpretation of such results.

In another example of attempts to link diseases and RBC Ca^{2+}, Kluge and Kuhne (1985) found altered kinetic properties of the Ca^{2+}-pump ATPase of RBC membranes in some subgroups of affective disorders and schizophrenia. These data more or less complement and extend the findings of Linnolia et al. (1983). Based on evidence from several laboratories, it is suggested that the above findings are related to changes in the lipid environment of the membrane. Meuleman and Odenheimer (1985) found decreased Ca^{2+} uptake in fibroblasts cultured from patients with Alzheimer's disease. More Ca was surface associated and less was taken up by these presumptively abnormal cells. Such differences from normal may be more related to Ca^{2+} channels than to the Ca^{2+}-pump. On the other hand, the impact of altered leak on the activity of the pump may well be significant.

David-Dufilho and Devynck (1985) described altered Ca^{2+} binding and transport in heart PMs from spontaneously hypertensive rats (SHRs) compared to the normal Wistar Kyoto strain (WKY) rats. The differences were subtle and not strictly concentration-dependent, but it appears that SHR membranes transported more Ca^{2+} under some conditions. This is opposite to the decreased (compared to normal) basal activity of the Ca^{2+}-pump ATPase of RBC membranes from hypertensive humans (Vincenzi et al., 1986). Our data showing decreased basal activity of the Ca^{2+}-pump ATPase were obtained using saponin lysates of RBCs rather than isolated membranes. Thus, in order to define the basal activity, it was necessary to add $10^{-4}M$ TFP (Raess and Vincenzi, 1980). The result may be related to the use of that anti-CaM drug, because in a subsequent study using compound 48/80 to define the basal activity, no difference was found (Vincenzi et al., 1987). The more recent result may be interpreted to mean that there is no inherent defect in the PM Ca^{2+}-pump associated with hypertension. On the other hand, because TFP, which is a drug with significant membrane perturbing activity, resulted in a lower basal activity in RBC lysates of hypertensives than of normotensives, whereas compound 48/80, which has little or no membrane perturbing activity (Gietzen et al., 1983), resulted in equal activities, it is suggested that a subtle difference exists in the lipid part of the PMs of hypertensives and normotensives. In any event, major methodological differences make comparison of these data difficult.

David-Dufilho and Devynck (1985) prepared heart membranes from prehypertensive SHRs. They measured Ca^{2+} accumulation at 2×10^{-8} M and $4 \times 10^{-7}M$ free Ca^{2+}. ATP-dependent Ca^{2+} accumulation was higher in SR vesicles at these concentrations. Neither CaM nor calmidazolium (R 24 571) (Van Belle, 1981) altered Ca^{2+} binding, or Ca^{2+} accumulation at the lower Ca^{2+} concentration. On the other hand, CaM led to increased uptake of Ca^{2+} at $4 \times 10^{-7}M$, and the increase was greater in vesicles derived from WKY than from SHR hearts. Alterations in the lipid compositions of SHR membranes were suggested as a possible basis for the results. It was not clear whether these vesicles were derived exclusively or mainly from SR, SL, or both, although an SL source was claimed. Preparation of the vesicles did involve a $100{,}000 \times g$ step, a procedure that we have found to be detrimental to Ca^{2+}-pump ATPase activity; at least in IOVs from human RBCs (Halvorsen, Hinds and Vincenzi, unpublished).

David-Dufilho et al. (1984) in a similar study found that nifedipine led to increased Ca^{2+} transport at low (2×10^{-8}), but not high ($4 \times 10^{-7}M$) Ca^{2+}. The stimulation was more sensitive in vesicles derived from SHR than from WKY rats. There was no change in Ca^{2+} binding produced by nifedipine. The authors suggested that the results "raise the question of the mechanism of nifedipine on the calcium pump." We suggest, alternatively, that unless shown otherwise, it seems reasonable to assume that nifedipine would act like any other anti-CaM agent (Vincenzi, 1981), i.e., as an amphipathic cation. More than one action was apparently present. The authors obtained a biphasic dose–response curve to nifedipine, with maximal stimulation of uptake occurring at about $10^{-6}M$. Higher concentrations resulted in less or no stimulation. The results of these workers point to some subtle differences in, probably, the lipid environment of membranes of SHRs compared to WKYs, but the differences that may impact on both pump and leak components of Ca^{2+} transport remain to be elucidated.

Postnov et al. (1984a), examined both affinity and maximal activity of the Ca^{2+}-pump of human RBCs. These properties were the same in EGTA-treated RBC membranes derived from normals and hypertensive subjects, when assayed in the absence of added CaM. On the other hand, if CaM was added, a smaller increase in Ca^{2+}-pump ATPase activity and in Ca^{2+} affinity occurred in HT compared to NL membranes. Using IOVs, these authors concluded that differences were the result of an alteration of the interaction between CaM and the Ca^{2+}-pump ATPase in the hypertensive group. In my view, an altered lipid pattern is also compatible with the human data and could be anticipated based on earlier studies with other chronic diseases. Postnov et al. (1984b) also found altered Ca^{2+} handling in RBC membranes of SHRs.

Dearborn et al. (1984) found that cystic fibrosis was associated with decreased RBC ATPase activities. These included the Ca^{2+}-pump ATPase, both in its basal and CaM-activated states, as well as the Na^+, K^+-pump ATPase. The changes were significant in patients with pancreatic insufficiency, but not in patients without pancreatic insufficiency. There was no correlation between ATPase activities and clinical scores in these patients. Such a correlation would not necessarily be expected, partly because clinical scores in CF depend very much on lung infections, and partly because the ATPase may be responding to altered lipid metabolism in the patients rather than being causative to the clinical manifestations of the disease. A different interpretation, more directly related to changes in the Ca^{2+}-pump ATPase itself, is more consistent with the results of Katz et al. (1984). These authors found decreased Ca^{2+}-pump ATPase activity in RBC membranes of CF patients and decreased activity, compared to normal, in fibroblasts grown from patients. The data in fibroblasts represent a strong argument against altered lipid metabolism in vivo as (the only) basis for the Ca^{2+}-pump changes observed in CF. By contrast, Clough and Hubbard (1984) found no changes in CF vs normals when the Ca^{2+}-pump ATPase was monitored. On the other hand, they found that Na^+ activated the Ca^{2+}-pump ATPase of CF membranes 40% more than of normal RBC membranes. The matter of the Ca^{2+}-pump ATPase and its possibly altered regulation in CF remains to be resolved.

9. RBCs as Models for Ca^{2+} Transport

Human RBCs have been extremely useful for the development of the PM Ca^{2+}-pump model (Hinds and Vincenzi, 1983), but human RBCs have certain important limitations that must be appreciated. One such limitation is that human RBCs lack some of the features of cells for which they are often used as models. For example, lacking functional Ca^{2+} channels represents a drawback. Lack of a Na^+/Ca^{2+} exchanger also represents a limitation. Heart muscle is thought to transport most of the Ca^{2+} out of the cell via a Na^+/Ca^{2+} exchanger. The heart may represent a somewhat unique example of cells with a large membrane potential (at rest) and with a high capacity Na^+/Ca^{2+} exchanger, which thus extrudes substantial amounts of Ca^{2+} across the plasma membrane as part of its beat-by-beat physiology. A similar suggestion has been made for nerve terminals, based upon work with synaptosomes (Nachshen et al., 1986).

Since dog RBCs exhibit Na^+/Ca^{2+} exchange, we sought to explore them as a possible model of cardiac and nerve cells. However, there are two limitations with the use of dog RBCs, one of which we were aware at the outset, and one of which was unexpected. Thus, as has been known for a number of years, mature dog RBCs lack a Na^+/K^+-pump (Parker, 1977, 1979), but as Parker has shown so well, dog RBCs actively pump Ca^{2+} and then exchange Ca^{2+} for Na^+. They thus maintain intracellular Na^+ at somewhat less than extracellular at the expense of running the Ca^{2+}-pump. It is a familiar theme, but inverted from that which is advanced as the mechanism for *removal* of intracellular Ca^{2+} in a wide variety of excitable cells. On the other hand, it is a phenomenon not without precedent. In fact, Borle and his coworkers have shown convincingly that the Na^+/Ca^{2+} exchanger can serve as a Ca^{2+} *influx* pathway in kidney cells, for example (Snowdowne and Borle, 1985). Under the usual physiological conditions of ion gradients and membrane potential that these cells exhibit, the exchanger is more or less "balanced," and carries little or no net flux of either Na^+ or Ca^{2+}. However, in the face of significant depolarization or changes in ion gradients (especially sensitive to altered internal Na^+), the exchanger can carry amounts of Ca^{2+} into the cell that are sufficient to significantly, if transiently, increase intracellular-free Ca^{2+} (Snowdowne and Borle, 1985).

A second limitation of dog RBCs as models of other cells is that the PM Ca^{2+}-pump ATPase appears to act anomalously. That is, although in essentially (but not) all cells examined, the Ca^{2+} pump ATPase behaves as elucidated in human RBCs, there may be something different about dog RBCs. In most cells studied to date, the Ca^{2+}-pump ATPase has a mol wt of approximately 138,000 dalton (Carafoli et al., 1984b), and it binds and is activated by low concentrations of CaM. By contrast, the Ca^{2+}-pump ATPase of membranes isolated from dog RBCs does not bind and is not activated by CaM (Schmidt et al., 1985). Because of a lack of cross-linking with CaM, no good estimate of the ATPase mol wt is available. It seems unlikely that the explanation is that dog RBCs lack a Ca^{2+}-pump ATPase, because Parker has shown that dog RBCs do actively extrude Ca^{2+} (which they may exchange for Na^+). By using a rather different approach than measurement of ATPase activities of isolated RBC membranes, we have convinced ourselves that the Ca^{2+}-pump ATPase of dog RBCs is not inherently anomalous. Thus, we assessed the activity of the Ca^{2+}-pump in intact dog RBCs by measuring the loss of ATP in cells exposed to Ca^{2+} and the ionophore A23187. The loss of ATP is relatively rapid (half-life = approximately 9 min) (Hinds and

Vincenzi, 1986), but is prolonged in an inherited hemolytic anemia in beagles (Maggio-Price et al., 1987) and is sensitive to the anti-CaM agent TFP (Hinds and Vincenzi, 1986). For these reasons (Hinds and Vincenzi, 1986), we have suggested that dog RBCs contain a CaM-sensitive Ca^{2+}-pump ATPase, and that, by mechanisms yet to be elucidated, the Ca^{2+}-pump ATPase loses its CaM dependence during membrane isolation. Apparently, the human enzyme does not lose its CaM dependence upon membrane isolation (or does so much less completely than dog).

10. Summary

A PM Ca^{2+}-pump appears to be universally present in cells, including RBCs. In RBCs, a number of potential modifiers of pump activity are present, but the second-to-second regulation of the rate of Ca^{2+} efflux appears to dependent in turn on the rate of Ca^{2+} influx. Factors determining Ca^{2+} influx in RBCs are not well understood. It is suggested that mechanical deformation of RBCs in the microcirculation may transiently and significantly increase Ca^{2+} influx under normal physiological conditions in vivo.

References

Bessis, M., and Mohandas, N. (1975) A diffractometric method for the measurement of cellular deformability. *Blood Cells,* **1,** 307–313.

Bevan, J. A. and Hwa J. J. (1985) Myogenic tone and cerebral vascular autoregulation: The role of a stretch-dependent mechanism. *Ann. Biomed. Eng.* **13,** 281–286.

Bond, G. H. and Clough, D. L. (1973) A soluble protein activator of ($Mg^{2+} + Ca^{2+}$)-dependent ATPase of human red blood cells. *Biochim. Biophys. Acta* **323,** 592–599.

Bostrom, S.-L., Ljung, B., Mardh, S., Forsen, S., and Thulin E. (1981) Interaction of the antihypertensive drug felodipine with calmodulin. *Nature* **292,** 777–778.

Carafoli, E. (1984a) Molecular, mechanistic, and functional aspects of the plasma membrane calcium pump in: *Epithelial Calcium and Phosphate Transport: Molecular and Cellular Aspects,* **Vol 168** (Bronner, F., Peterlik, M., eds.) Alan R. Liss, Inc., New York, pp. 13–17.

Carafoli, E. (1984b) Calmodulin-sensitive calcium-pumping ATPase of plasma membranes: Isolation, reconstitution, and regulation. *Fed. Proc.* **43,** 3005–3010.

Carafoli, E. (1985) The homeostasis of calcium in heart cells. *J. Mol. Cell Cardiol.* **17,** 203–212.

Chien, S. (1975) Biophysical behavior of red cells in suspensions, in: *The Red Blood Cell,* **Volume II** (MacN. Surgenor, D., ed.) Academic Press, New York, pp. 1031–1133.

Clough, D. L. and Hubbard, V. S. (1984) Red cell membrane Ca-ATPase in cystic fibrosis: Increased activation by Na. *Clin. Chim. Acta* **138,** 259–265.

David-Dufilho, M. and Devynck, M. A. (1985) Calmodulin abolishes the changes in Ca^{2+} binding and transport by heart sarcolemmal membranes of spontaneously hypertensive rats. *Life Sci.* **36,** 2367–2373.

David-Dufilho, M., Devynck, M.-A., Kazda, S., and Meyer, P. (1984) Stimulation by nifedipine of calcium transport by cardiac sarcolemmal vesicles from spontaneously hypertensive rats. *Eur. J. Pharmacol.* **97,** 121–127.

Dearborn, D. G., Wityk, R. J., Johnson, L. R., Poncz, L., and Stern, R. C. (1984) Calcium-ATPase activity in cystic fibrosis erythrocyte membranes: Decreased activity in patients with pancreatic insufficiency. *Ped. Res.* **18,** 890–895.

Farber, J. L. (1981) The role of calcium in cell death. *Life Sci.* **29,** 1289–1295.

Gietzen K., Adamczyk-Engelmann P., Wuthrich, A., Konstantinova, A., and Bader, H. (1983) Compound 48/80 is a selective and powerful inhibitor of calmodulin-regulated functions. *Biochim. Biophys. Acta* **736,** 102–118.

Gopinath, R. M. and Vincenzi, F. F. (1977) Phosphodiesterase protein activator mimics red blood cell cytoplasmic activator of (Ca^{2+} + Mg^{2+})-ATPase. *Biochem. Biophys. Res. Comm.* **77,** 1203–1209.

Gopinath, R. M. and Vincenzi, F. F. (1979) (Ca^{2+} + Mg^{2+})-ATPase of sickle cell membranes: decreased activation by red blood cell cytoplasmic activator. *Amer. J. Hematol.* **7,** 303–312.

Hinds, T. R. and Vincenzi, F. F. (1983) The red blood cell as a model for calmodulin-dependent Ca^{2+} transport, in *Methods in Enzymology, Volume 102, Hormone Action, Part G, Calmodulin and Calcium-Binding Proteins* (Means, A. R., and O'Malley, B. W., eds.), Academic Press, New York, pp. 47–62.

Hinds. T. R., and Vincenzi, F. F. (1986) Evidence for a calmodulin activated Ca^{2+} pump ATPase in dog erythrocytes. *Proc. Soc. Exper. Biol. Med.* **181,** 542–549.

Hinds, T. R., Di Julio, and Vincenzi, F. F. (1987) Red blood cell Ca^{2+} Pump ATPase: Inhibition by compound 48/80. *Proc. West. Pharmacol. Soc.*, **30,** 93–95.

Jarrett, H. W. and Penniston, J. T. (1977) Partial purification of the Ca^{2+} + Mg^{2+} ATPase activator from human erythrocytes: its similarity to the activator of 3′5′-cyclic nucleotide phosphodiesterase. *Biochem. Biophys. Res. Commun.* **40,** 1210–1216.

Jay, A. W. L. and Burton, A. C. (1969) Direct measurement of potential difference across the human red blood cell membrane. *Biophys. J.* **9,** 115–121.

Johnsson, R., Somer, H., Karli, P., and Saris, N.-E. (1983) Erythrocyte flexibility, ATPase activities and Ca efflux in patients with Duchenne Muscular

Dystrophy, Myotonic Muscular Dystrophy and congenital Myotonia. *J. Neurol. Sci.* **58,** 399–407.

Katz S., Schoni, M. H., and Bridges, M. A. (1984) The calcium hypothesis of cystic fibrosis. *Cell Calcium* **5,** 421–440.

Kluge, H. and Kuhne, G.-E. (1985) Preliminary findings on calmodulin-stimulated Ca^{2+}-ATPase of erythrocyte ghosts in psychotic patients. *Eur. Arch. Psychiatr. Neurol. Sci.* **235,** 57–59.

Larsen, F. L. and Vincenzi, F. F. (1979) Ca^{2+} Transport across the plasma membrane: Stimulation by calmodulin. *Science* **204,** 306–309.

Larsen, F. L., Katz, S., Roufogalis, B. D., and Brooks, D. E. (1981) Physiological shear stresses enhance the Ca^{2+} permeability of human erythrocytes. *Nature* **294,** 667–668.

Lee, K. S., and Shin, B. C. (1969) Studies on the active transport of calcium in human red cells. *J. Gen. Physiol.* **54,** 713–729.

Levin, R. M. and Weiss, B. (1977) Binding of trifluoperazine to the calcium-dependent activator of cyclic nucleotide phosphodiesterase. *Molec. Pharmacol.* **13,** 690–697.

Lew, V. L. and Ferreira, H. G. (1978) Calcium transport and the properties of a calcium-activated potassium channel in red cell membranes. *Curr. Top. Membr. Transp.* **10,** 217–277.

Lew, V. L. and Garcia-Sancho, J. (1985) Use of the ionophore A23187 to measure and control cytoplasmic Ca^{2+} levels in intact red cells. *Cell Calcium* **6,** 15–23.

Linnolia, M., MacDonald, E., Reinila, M., Leroy, A. Rubinow, D. R., and Goodwin, F. K. (1983) RBC membrane adenosine triphosphatase activities in patients with major affective disorders. *Arch. Gen. Psychiatry* **40,** 1021–1026.

Lubowitz, H., Harris, F., Mehrjardi, M. H., and Sutera, S. P. (1974) Shear-induced changes in permeability of human RBC to sodium. *Trans. Amer. Soc. Artif. Int. Organs.* **20,** 470–473.

Maggio-Price, L., Emerson, C. L., Hinds, T. R., Vincenzi, F. F., and Hammond, W. P. (1987) Inherited nonspherocytic hemolytic anemia in beagle dogs., *Amer. J. Vet. Res.* **49,** 1020–1025.

Meuleman, J. R., and Odenheimer, G. L. (1985) Altered calcium uptake in cultured skin fibroblasts from patients with Alzheimer's disease. *New Eng. J. Med.* **312,** 1063–1065.

Mohandas, N., Clark, M. R., Health, B. P., Rossi, M., Wolfe, L. C., and Lux, S. E. (1980) Analysis of factors regulating erythrocyte deformability. *J. Clin. Invest.* **66,** 563–573.

Muallem, S., and Karlish, S. J. D. (1983) Catalytic and regulatory ATP-binding sites of the red cell Ca^{2+} pump studied by irreversible modification with fluorescein isothiocyanate. *J. Biol. Chem.* **258,** 169–175.

Nachshen, D. A., Sanchez-Armass, S., and Weinstein, A. M. (1986) The regulation of cytosolic calcium in rat brain synaptosomes by sodium-dependent calcium efflux. *J. Physiol (Lond.)* **381,** 17–28.

Norman, J. A., Drummond, A. H., and Moser, P. (1979) Inhibition of calcium-dependent regulator-stimulated phosphodiesterase activity by neuroleptic drugs is unrelated to their clinical efficacy. *Molec. Pharmacol.* **16,** 1089–1094.

Olson, E. J., and Cazort, R. J. (1969) Active calcium and strontium transport in human erythrocyte ghosts. *J. Gen. Physiol.* **63,** 590–600.

Parker, J. C. (1977) Solute and water transport in dog and cat red blood cells, in *Membrane Transport in Red Cells* (Ellory, J. C. and Lew, V. L., eds.) MIC Press, New York, pp. 427–465.

Parker, J. C. (1979) Active and passive Ca movements in dog red blood cells and resealed ghosts. *Amer. J. Physiol.* **237,** C10–C16.

Passow, H. (1961) Zusammenwirken von Membranstruktur und Zellstoffwechsel bei der Regulierung der Ionenpermeabilitaet roter Blutkoerperchen. 12. Colloq. Ges. physiol. Chem., Springer-Verlag, Berlin, pp. 54–55.

Passow, H. (1963) Metabolic control of passive cation permeability in human red blood cells, in *Cell Interphase Reactions* (Brown, H. D., ed.), Scholars Library, New York, pp. 57–107.

Pine, R. W., Vincenzi, F. F., and Carrico, C. J. (1983) Apparent inhibition of the plasma membrane Ca^{2+} pump by oleic acid. *J. Trauma* **23,** 366–371.

Postnov, Y. V., Orlov, S. N., Reznikova, M. B., Rjazhsky, G. G., and Pokudin, N. I. (1984a) Calmodulin distribution and Ca^{2+} transport in the erythrocytes of patients with essential hypertension. *Clin. Sci.* **66,** 459–463.

Postnov, Y. V., Orlov, S. N., Kravtsov, G. M., and Gulak, P. V. (1984b) Calcium transport and protein content in cell plasma membranes of spontaneously hypertensive rats. *J. Cardiovas. Pharmacol.* **6,** S21–S27.

Raess, B. U. and Vincenzi, F. F. (1980) Calmodulin activation of red blood cell ($Ca^{2+}+Mg^{2+}$)-ATPase and its antagonism by phenothiazines. *Molec. Pharmacol.* **18,** 253–258.

Raess, B. U., Record, D. M., and Tunnicliff, G. (1985) Interaction of phenylglyoxal with the human erythrocyte ($Ca^{2+}+Mg^{2+}$)-ATPase. *Molec. Pharmacol.* **27,** 444–450.

Rega, A. F. and Garrahan, P. J. (1986) *The Ca^{2+} Pump of Plasma Membranes* (CRC Press, Boca Raton, FL).

Rega, A. F., Richards, D. E., and Garrahan, P. J. (1974) The effects of Ca^{2+} on ATPase and phosphatase activities of erythrocyte membranes. *Ann. N.Y. Acad. Sci.* **242,** 317–323.

Ronner, P., Gazzotti, P., and Carafoli, E. (1977) A lipid requirement for the ($Ca^{2+}+Mg^{2+}$)-activated ATPase of erythrocyte membranes. *Arch. Biochem. Biophys.* **179,** 578–583.

Rosier, F., M'Zali, H., and Giraus, F. (1986) Cholesterol depletion affects the Ca^{2+} influx but not the Ca^{2+} pump in human erythrocytes. *Biochim. Biophys. Acta* **863,** 253–263.

Roufogalis, B. D. (1985) Calmodulin antagonism, in *Calcium and Cell Physiology* (Marme, D., ed.), Springer, Verlag, New York, pp. 148–169.

Roufogalis, B. D., Minocherhomjee, A.-E.-V. M., and Al-Jobore, A. (1983) Pharmacological antagonism of calmodulin. *Can. J. Biochem. Cell Biol.* **61,** 927–933.

Scharff, O. and Foder, B. (1982) Rate constants for calmodulin binding to Ca^{2+}-ATPases in erythrocyte membranes. *Biochim. Biophys. Acta* **691,** 133–143.

Schatzmann, H. J. (1966) ATP-dependent Ca^{++} extrusion from human red cells. *Experientia* **22,** 364–365.

Schatzmann, H. J. (1985) Calcium extrusion across the plasma membrane by the calcium-pump and the Ca^{2+}-Na^{+} exchange system, in *Calcium and Cell Physiology* (Marme, D., ed.) Springer-Verlag, New York, pp. 18–52.

Schatzmann, H. J. (1986) The plasma membrane calcium pump, in *Intracellular Calcium Regulation* (Bader, H., Gietzen, K., Rosenthal, J., Ruedel, R., and Wolf, H. U., eds.) Manchester University Press, Manchester, pp. 47–56.

Schatzmann, H. J. and Vincenzi, F. F. (1969) Calcium movements across the membrane of human red cells. *J. Physiol. (Lond.)* **201,** 369–395.

Schlondorff, D. and Satriano, J. (1985) Interactions with calmodulin: Potential mechanism for some inhibitory actions of tetracyclines and calcium channel blockers. *Biochem. Pharmacol.* **34,** 3391–3393.

Schmidt, J. W., Hinds, T. R., and Vincenzi, F. F. (1985) On the Failure of calmodulin to activate the Ca^{2+} pump ATPase of dog red blood cells. *Comp. Biochem. Physiol.* **82A,** 601–607.

Snowdowne, K. W., and Borle, A. B. (1985) Effects of low extracellular sodium on cytosolic ionized calcium. *J. Biol. Chem.* **260,** 14998–15007.

Starke, P. E., Hoek, J. B., and Farber, J. L. (1986) Calcium-dependent and calcium-independent mechanisms of irreversible cell injury in cultured hepatocytes. *J. Biol. Chem.* **261,** 3006–3012.

Tokumura, A., Mostafa, M. H., Nelson, D. R., and Hanahan, D. J. (1985) Stimulation of (Ca^{2+} + Mg^{2+})-ATPase activity in human erythrocyte membranes by synthetic lysophosphatidic acids and lysophosphatidylcholines. Effects of chain length and degree of unsaturation of the fatty acid groups. *Biochim. Biophys. Acta* **812,** 568–574.

Van Belle, H. (1981) R 24 571: A potent inhibitor of calmodulin-activated enzymes. *Cell Calcium* **2,** 483–494.

Varecka, L. and Carafoli, E. (1982) Vanadate-induced movements of Ca^{2+} and K^{+} in human red blood cells. *J. Biol. Chem.* **257,** 7414–7421.

Vincenzi, F. F. (1981) Calmodulin pharmacology. *Cell Calcium* **2,** 387–409.

Vincenzi, F. F. (1982a) The pharmacology of calmodulin antagonism: A reappraisal, in *Calmodulin and Intracellular Ca^{++} Receptors* (Kakiuchi, S., Hidaka, H., and Means, A. R., eds.), Plenum Press, New York, pp. 1–17.

Vincenzi, F. F. (1982b) Pharmacological modification of the Ca^{2+}-pump ATPase activity of human erythrocytes. *Ann. N.Y. Acad. Sci.* **402,** 368–380.

Vincenzi, F. F. and Cambareri, J. J. (1985a) Apparent ionophoric effects of red

blood cell deformation, in *Cellular and Molecular Aspects of Aging: The Red Cell as a Model* (Eaton, J. W., Konzen, D. L., and White, J. G., eds.) Alan R. Liss, New York, pp. 213–222.
Vincenzi, F. F., and Cambareri, J. J. (1985b) Apparent ionophoric effects of red blood cell deformation. *Proc. West. Pharmacol. Soc.* **28,** 263–264.
Vincenzi, F. F. and Hinds, T. R. (1980) Calmodulin and plasma membrane calcium transport, in *Calcium and Cell Function,* **Vol 1,** (Cheung, W. Y., ed.), Academic Press, New York, pp. 127–165.
Vincenzi, F. F., and Hinds, T. R. (1987) Drug effects on plasma membrane calcium transport, in *Calcium and Drug Actions* (Baher, P. F., ed.) Springer-Verlag, New York, pp. 147–162.
Vincenzi, F. F., Hinds, T. R., and Raess, B. U. (1980) Calmodulin and the plasma membrane calcium pump. *Ann. N.Y. Acad. Sci.* **356,** 232–244.
Vincenzi, F. F., Adunyah, E. S., Niggli, V., and Carafoli, E. (1982) Purified red blood cell Ca^{2+}-pump ATPase: Evidence for direct inhibition of by presumed anti-calmodulin drugs in the absence of calmodulin. *Cell Calcium* **3,** 545–559.
Vincenzi, F. F., Di Julio, D., Morris, C. D., and McCarron, D. A. (1987) Measurements on the activity of the plasma membrane Ca pump ATPase in human hypertension, in *Cellular Calcium and Phosphate Transport in Health and Disease* (Bronner, F. and Peterlik, M., eds.) Alan R. Liss, New York, pp. 379–383.
Vincenzi, F. F., Morris, C. D., Kinsel, L. B., Kenny, M., and McCarron, D. A. (1986) Decreased *in vitro* Ca^{2+} pump ATPase activity in red blood cells of hypertensive subjects. *Hypertension* **8,** 1058–1066.
Weiss, B., Sellinger-Barnette, M., Winkler, J. D., Schechter, L. E., and Prozialeck, W. C. (1985) Calmodulin antagonists: structure-activity relationships, in *Calmodulin Antagonists and Cellular Physiology* (Hidaka, H. and Hartshorne, D. J., eds.), Academic Press, Inc., New York, pp. 45–62.
Wolff, D. J. and Brostrom, C. O. (1976) Calcium-dependent cyclic nucleotide phosphodiesterase from brain: Identification of phospholipids as calcium-independent activators. *Arch. Biochem. Biophys.* **173,** 720–731.

The Anion Transport Protein

Structure and Function

Michael L. Jennings

1. Introduction

The major 95 kdalton transmembrane protein of the red-cell membrane, denoted band 3 (Fairbanks et al., 1971) or capnophorin (Wieth and Bjerrum, 1983), catalyzes a tightly coupled exchange of anions. The purpose of this chapter is to summarize the current state of knowledge regarding the structure of this protein, with emphasis on those aspects of the structure that relate to anion transport. The amount of information currently available about band 3 is quite large, and this chapter will deal only superficially with several important topics that are somewhat related to the transport mechanism. These include lateral and rotational mobility of the protein (Beth et al., 1986; Mühlebach and Cherry, 1985); carbohydrate composition and structure (Tsuji et al., 1981; Fukuda et al., 1984); band 3 biosynthesis (Braell and Lodish, 1981); role of band 3 in red-cell senescence (Kay, 1984; Low et al., 1985); interaction with lipids (Köhne et al., 1983); and the physiology of CO_2 exchange in the pulmonary and systemic capillaries (*see* Wieth et al., 1982). Certain other topics that relate centrally to the transport mechanism, including transport kinetics, reversible inhibitors, pH dependence, and chemical modification of arginine residues and carboxyl groups, are discussed in detail in Chapters 9 and 15. For more comprehensive recent reviews of molecular aspects of band 3-mediated anion transport, the reader is referred to Passow (1986) and to Jay and Cantley (1986).

2. Structural Domains

Band 3 consists of two structurally distinct domains. Exposure of the cytoplasmic side of red-cell membranes to proteolytic enzymes under very mild conditions causes the cleavage of band 3 at a site that is located about 43 kdalton from the *N*-terminus (Steck et al., 1976). The resultant *N*-terminal fragment represents a water-soluble cytoplasmic domain. The remainder of the protein, the membrane domain, has an M_r of 52 kdalton and is hydrophobically associated with the lipid bilayer (Steck et al., 1976, 1978).

The amino acid sequence of mouse band 3 was recently deduced from the sequence of the cloned cDNA (Kopito and Lodish, 1985). The membrane domain of mouse band 3 is extremely homologous with those regions of human band 3 membrane domain that have been sequenced (Kopito and Lodish, 1985; Mawby and Findlay, 1982; Brock et al., 1983; Brock and Tanner, 1986). There are also regions of homology in the cytoplasmic domains of human and mouse band 3, but the homology is weaker (Kopito and Lodish, 1985; Kaul et al., 1983).

2.1. Cytoplasmic Domain

2.1.1. Isolation and Characterization

The cytoplasmic domain can be isolated from unsealed membranes or inside-out vesicles that have been treated with low concentrations of chymotrypsin or trypsin (Appell and Low, 1981). The isolated cytoplasmic domain assumes a highly elongated conformation and can form stable dimers (Appell and Low, 1981). There is a proline-rich segment roughly in the middle of the sequence of the cytoplasmic domain (residues 176–191 of human band 3) that appears to act as a hinge connecting the two halves of the domain (Low et al., 1984).

2.1.2. Intracellular Ligands

The cytoplasmic domain contains the binding site for ankyrin, the 200,000 dalton protein that is responsible for anchoring the assembly of structural proteins collectively known as the membrane skeleton (Bennett and Stenbuck, 1980; Bennett, 1985). Low et al. (1984) have suggested that the ankyrin binding site is located just toward the *N*-terminus from the hinge region in the middle of the sequence of the cytoplasmic domain; the precise location of the site is not yet known. In addition to containing the attachment site for ankyrin, the cytoplasmic domain of band 3 interacts directly with another component of the membrane skeleton, band 4.1 (Pasternack et al., 1985).

Steck and coworkers (Kliman and Steck, 1980; Murthy et al., 1981; Tsai et al., 1982) have demonstrated that the cytoplasmic domain of band 3 contains binding sites for the glycolytic enzymes aldolase and glyceraldehyde-3-phosphate dehydrogenase. The enzymatic activities of the bound forms are much lower than the soluble forms. Another glycolytic enzyme, phosphofructokinase, can also be bound to the cytoplasmic domain of band 3 (Higashi et al., 1979). The bound form of this enzyme is active, but has altered allosteric properties (Karadsheh and Uyeda, 1977).

Work in several laboratories has established that hemoglobin can bind to the cytoplasmic domain of band 3. Walder et al. (1984) showed that the first 11 amino acid residues at the *N*-terminus of human band 3 can occupy a site that overlaps the 2,3-diphosphoglycerate binding site of hemoglobin. The binding is much stronger with the deoxyhemoglobin tetramer than to the oxyhemoglobin tetramer. The binding of hemoglobin to the membrane does not have a significant effect on the total oxygen-carrying capacity of the cells, but it is possible that hemoglobin binding to band 3 could have important effects on red-cell metabolism by virtue of the displacement of glycolytic enzymes (*see* Walder et al., 1984).

2.1.3. Independence Between Cytoplasmic and Membrane Domains

The thermal denaturation of the membrane domain of band 3 is independent of the cytoplasmic domain (Appell and Low, 1982). There are, however, indications that the two domains can influence each other. Salhany et al. (1980) found that DIDS* (bound to a site accessible only from the extracellular medium) increases the amount of hemoglobin retained by isolated membranes. This indicates that DIDS bound to the membrane domain can influence the conformation of the cytoplasmic domain. Hsu and Morrison (1983) showed that covalent DIDS attachment causes increased retention of the cytoskeleton. This effect does not imply a direct interaction between the membrane domain and cytoplasmic domain, because the DIDS effect is apparently mediated by another protein, which can be cleaved by extracellular trypsin under conditions in which sialoglycoproteins, but not band 3, are cleaved.

2.1.4. Is There a Role for the Cytoplasmic Domain in Anion Transport?

In a study that was completed when little was known about the domain structure of band 3, Lepke and Passow (1976) showed that trypsin sealed inside red-cell ghosts does not prevent stilbenedisulfonate-

*Abbreviations used: DNDS: 4,4′-dinitrostilbene-2,2′-disulfonate; DIDS: 4,4′-diisothiocyanatostilbene-2,2′-disulfonate; H_2DIDS: 4,4′-diisothiocyanatodihydrostilbene-2, 2′-disulfonate; APMB: 2-(4′-aminophenyl)-6-methylbenzenethiazol-3′, 7-disulfonate; pCMBS: parachloromercuribenzenesulfonate; DTNB: 5,5′-dithiobis(2-nitrobenzioc acid); BSSS: bis)sulfosuccinimidyl)suberate; CNBr: cyanogen bromide; NBD-Cl: 7-chloro-4-nitrobenzo-2-oxa-1,3-diazole.

sensitive anion transport. Under the conditions of these experiments, the cytoplasmic domain (and membrane skeleton) are proteolytically degraded. Grinstein et al. (1978) subsequently demonstrated that mild proteolysis of inside-out vesicles does not inhibit anion transport under conditions in which the cytoplasmic domain is completely removed from the membrane. Taken together, these studies are strong evidence that the cytoplasmic domain is not absolutely necessary for anion exchange. Another argument against a major role for the cytoplasmic domain in anion transport is that the Cl exchange flux is independent of intracellular pH (Funder and Wieth, 1976) in a range (pH 7–10) over which substantial structural transitions in the cytoplasmic domain take place (Low et al., 1984).

It is premature, however, to conclude that the cytoplasmic domain has absolutely no influence on anion transport, because there may be subtle interactions between the two domains that were not revealed in the above studies. A very recent preliminary report by Garcia et al. (1987) indicates that the whole band 3 gene product induces a higher Cl flux in frog oocytes that does the polypeptide coded by the DNA sequence for the membrane domain alone.

2.2. Membrane Domain

The abundance of band 3 makes it possible to obtain direct information regarding the portions of the primary sequence that are exposed on the extracellular side of the membrane. In this context, it is important to define carefully what is meant by *extracellular*. Sites identified by the use of macromolecules such as enzymes or antibodies are probably at or near the true outer surface of the membrane. Sites that are accessible to small, impermeant (or slowly permeant) hydrophilic chemical modifiers are on the extracellular side of the permeability barrier, but such sites may reside in a hydrophilic cavity some distance below the plane of the lipid head groups of the outer leaflet.

Table 1 is a list of most of the extracellular sites in human band 3 that have been localized (or partly localized) in the primary structure. The residue numbers refer to mouse band 3 (Kopito and Lodish, 1985), because the sequence of human band 3 is not yet completely known. In addition to the extracellular sites, a few intracellular sites have been localized. These are listed in Table 2.

From the markers known thus far, it is possible to make some limited statements about the numbers and positions of the membrane-spanning segments of the protein. The first membrane crossing segment must be on the *N*-terminal side of Lys 449, and the polypeptide must

Table 1
Extracellular Markers in the Primary Structure of Band 3

Method of detection	Location	Reference
Reductive methylation	Lys 449	Jennings and Nicknish, 1984
Extracellular chymotrypsin or thermolysin	Tyr 572	Drickamer, 1976 Steck et al., 1978 Williams et al., 1979
Extracellular papain	Gln 569	Jennings and Adams, 1981
	Gln 582 Gln 648	Jennings et al., 1984
Extracellular trypsin, at low ionic strength	Lys 580	Brock et al., 1983
Carbohydrate attachment	Asn 660	Kopito and Lodish, 1985 Jay, 1986
Covalent reaction with H_2DIDS, other negatively charged arylisothiocyanates	Lys 558 or Lys 561	See Kopito and Lodish, 1985 Mawby and Findlay, 1982
Covalent reaction with other end of H_2DIDS	C-terminal 168 residues	Jennings et al., 1986
Pyridoxal phosphate	Lys 869	Kawano et al., 1988

cross the membrane an even number of times between Lys 449 and Tyr 572. The residues between the extracellular sites Gln 569 and Gln 648 have a sequence that is very strongly suggestive of a U-shaped loop that crosses the membrane (Brock et al., 1983). The polypeptide must cross the membrane an odd number of times between the glycosylation site (Asn 660) and the intracellular trypsin cleavage site at Lys 761. Because one of the (extracellular) H_2DIDS-binding lysines appears to be between (intracellular) Lys 761 and the (intracellular) C-terminus, the C-terminal 168 residues must contain either two or four membrane-spanning segments. In all, there is direct evidence for eight membrane-spanning segments of polypeptide.

The above discussion has emphasized the notion of membrane-spanning helical segments of polypeptide, because the limits of current methods are to define sites as "intracellular" or "extracellular." Moreover, there is evidence that the membrane domain of band 3 is very high in helix content (Oikawa et al., 1985). It should be recognized, though, that it is unlikely that the membrane domain consists entirely of transmembrane helices and short hydrophilic connectors at each surface.

Table 2
Intracellular Markers in the Primary Structure of Band 3 Membrane Domain

Method of detection	Location	Reference
N-terminus of cytoplasmic domain	Near Residue 375	Mawby and Findlay, 1982 Markowitz and Marchesi, 1981
Intracellular trypsin	Lys 761	Jennings et al., 1986
Intracellular carboxypeptidase	C-terminal Val	Lieberman, et al., 1987
Strongly inferred from sequence and from lack of labeling in intact cells	Tyr 614	Brock et al., 1983
Sulfhydryls labeled by *N*-ethylmaleimide	C-terminal 70 residues (mouse)	Kopito and Lodish, 1985 Rao, 1979 Roa and Reithmeier, 1979
Phosphorylation	Not detected in membrane domain, main site is Tyr 8 of cytoplasmic domain	Waxman, 1979 Dekowski et al., 1983
Chymotrypsin	Site about 20 residues from *N*-terminus of membrane domain	Ramjeesingh and Rothstein, 1982

3. The Anion Transport Pathway

3.1. Minimum Structure Required for Transport

As mentioned above, the cytoplasmic domain is not required for anion transport (Lepke and Passow, 1976). Vigorous proteolysis by extracellular papain (Jennings and Passow, 1979) or bilateral chymotrypsin (duPre and Rothstein, 1981) causes inhibition of transport. In the case of papain, the inhibition is not complete and is caused by inhibition of the efflux portion of the catalytic cycle; the substrate binding site is not detectably altered (Jennings and Adams, 1981). Falke et al. (1985) used ^{35}Cl nuclear magnetic resonance to show that extensive bilateral papain digestion damages, but does not destroy, the line broadening attributed to Cl binding to a transport site on band 3. The papain digestion was sufficient to reduce the protein to small fragments of 3–7 kdalton; the fragments apparently remain associated in the membrane after the proteolysis. In summary, it is probable that the transport pathway consists of several portions of the sequence of the membrane domain. Other than the immediate vicinity of the extracellular chymotrypsin cleavage site (Tyr 572), no portion of the membrane domain has been eliminated as a possible constituent of the transport pathway.

3.2. Extracellular Aspect: The Stilbenedisulfonate Site

The work of Cabantchik et al. (reviewed in 1978) originally showed that stilbenedisulfonates are potent inhibitors of band 3-mediated anion exchange. Both reversibly acting and covalently binding stilbenedisulfonates inhibit the transport, and they do so by binding to a site that is accessible only from the outer surface of the membrane (Kaplan et al., 1976; Barzilay and Cabantchik, 1979). Although the site cannot be reached from inside the cells, the bound stilbenedisulfonate appears to be a considerable distance below the outer surface of the membrane, as indicated by fluorescence energy transfer and collisional quenching experiments (Rao et al., 1979; Macara et al., 1983).

Stilbenedisulfonates, bound either reversibly or covalently, inhibit band 3-mediated anion transport by over 99% (Funder et al., 1978; Passow, 1986). Inhibition of Cl–Cl exchange by reversibly bound H_2 DIDS or DNDS is competitive (Shami et al., 1978; Fröhlich, 1982), although there is evidence that Cl and stilbenedisulfonate can be bound simultaneously (Dix et al., 1986). The simultaneous binding of Cl and stilbenedisulfonate would be possible if the extracellular substrate anion binding site were contained in the stilbenedisulfonate site such that bound substrate anion does not completely prevent stilbenedisulfonate binding.

The most rapid covalent reaction between band 3 and H_2DIDS is with a lysine residue on the C-terminal side of the major CNBr cleavage site (Met 454) of the membrane-bound portion of the 60-kdalton chymotryptic fragment (Ramjeesingh et al., 1980). The same lysine residue probably reacts with other negatively charged arylisothiocyanate inhibitors of anion transport (Drickamer, 1977; Kempf et al., 1981; Mawby and Findlay, 1982). In mouse band 3, there are only two lysine residues, at positions 558 and 561, between Met 454 and the chymotrypsin cleavage site, and it is likely that one of these lysines is the site of rapid covalent H_2DIDS reaction in human band 3.

Although the most rapid covalent reaction is with a lysine residue in the 60-kdalton chymotryptic fragment, H_2DIDS also reacts quantitatively with the complementary 35-kdalton fragment; this results in a covalent intramolecular cross-link (Jennings and Passow, 1979; Kampmann et al., 1982). There are many candidates for the H_2DIDS-reactive lysine residue in the 35-kdalton chymotryptic fragment (Kopito and Lodish, 1985). Papain proteolysis experiments (Jennings et al., 1984) showed that the reactive residue must be between Gln 648 and the C-terminus. Proteolysis experiments with internal trypsin have recently provided evidence that H_2DIDS must react with a lysine residue that is between Lys 761 and the C-terminus (Jennings et al., 1986).

3.3. Role of Amino Groups in Anion Transport

Current evidence indicates that the amino groups with which H_2DIDS reacts covalently are not absolutely necessary for anion transport. Passow and coworkers (Passow, 1986; Rudloff et al., 1983) have performed detailed studies on the function of the most reactive lysine residue, i.e., the H_2DIDS-binding lysine in the 60-kdalton chymotryptic fragment. The rate of modification of this lysine residue by dinitrophenylation (with 1-fluoro-2,4-dinitrobenzene) is sensitive to the conformation of the protein and is modulated by various ligands. This lysine is not the transport site, but is allosterically linked to the transport site and to two other lysine residue that are not at the stilbenedisulfonate site (*see* Passow, 1986).

Hamasaki and coworkers (Nanri et al., 1983; Matsuyama et al., 1983; Kawano and Hamasaki, 1986, Kawano et al., 1988) have shown that modification of red cells with pyridoxal phosphate/B^3H_4 causes labeling of a lysine residue (Lys 869) in the 28-kdalton C-terminal papain fragment. The modification of this lysine residue correlates with the inhibition of anion transport. This finding further suggests that this lysine residue either lies near the transport pathway or is allosterically linked to it. Because of the bulk and negative charge of pyridoxal phosphate, the transport inhibition does not imply that the lysine residue itself has a role in the anion exchange.

Jennings et al. (1985) have used the active ester bis(sulfosuccinimidyl) suberate (BSSS) to investigate the functional role of lysine residues at the H_2DIDS binding site. As first shown by Staros and coworkers (Staros, 1982; Staros and Kakkad, 1983), BSSS forms both intramolecular and intermolecular cross-links at the extracellular side of band 3. The intramolecular cross-link blocks (and is blocked by) stilbenedisulfonate binding, and the cross-link is between the same two papain fragments as are cross-linked by H_2DIDS (Jennings and Nicknish, 1985; Jennings et al., 1985). It is quite possible, therefore, that the same two lysine side chains are involved in the intramolecular cross-link by each of the two agents. A major difference between BSSS and H_2DIDS is that BSSS introduces only an aliphatic chain (with amides at each end) between the lysine residues. When anion transport (Cl–Br exchange) is measured in cells pretreated with BSSS, the transport is strongly inhibited when assayed at pH 8 and above. However, lowering the extracellular pH in the pretreated cells activates the anion exchange; the transport assayed at extracellular pH 6 is only slightly slower than that of control cells. The detailed interpretation of this transport activation in the treated cells by extracellular protons is not yet clear, but the result does demonstrate that neither of the lysines that bind BSSS covalently (at the stilbenedisulfonate site) is absolutely necessary for anion exchange.

In addition to causing this major change in the pH dependence of transport, modification of the lysines by BSSS abolishes substrate inhibition by extracellular bromide and iodide (Jennings et al., 1985). The extracellular modifier site at which these halides inhibit transport noncompetitively may therefore contain one of the lysines that can be modified by BSSS. Wieth and Bjerrum (1982), on the basis of the pH dependence of Cl–Cl exchange at various Cl concentrations, concluded that noncompetitive inhibition by extracellular halides involves an arginine residue. However, the pH dependence of the modifier effect is equally consistent with the involvement of a lysine residue.

Another approach to the study of the functional role of the H_2DIDS-reactive lysines is chemical modification by reductive methylation. Reductive methylation does not alter the charge on the side chain (at neutral pH, the native and the methylated side chains are positively charged). Nonetheless, methylation of the H_2DIDS-binding lysine residues causes about 75% inhibition of band 3-mediated anion exchange. The extent of inhibition correlates with the modification of the H_2DIDS-reactive lysine on the 35-kdalton chymotryptic fragment (Jennings, 1982). The substantial inhibition of transport by such a minor modification of the lysine residue is perhaps surprising in light of the subsequent finding that neither of the lysines at the stilbenedisulfonate site is necessary for transport. The cause of the inhibition by the methylated (and positively charged) lysine residue could relate to the involvement of the residue in extracellular substrate inhibition. The methylation could enhance substrate inhibition (or noncompetitive inhibition by gluconate, which was used as the spectator anion).

There is a lysine (Lys 608) residue in the pepsin fragment sequenced by Brock et al. (1983) that reacts with phenylisothiocyanate under conditions in which this agent inhibits transport irreversibly (Sigrist et al., 1980; Kempf et al., 1981). The sequence of the protein suggests that this lysine lies in a region of positive charge that may be important for intracellular substrate binding. This lysine is the only functionally important residue identified thus far that is not in contact with extracellular water.

3.4. Histidine Modification

Chiba et al. (1986) have recently shown that the histidine modification reagent diethylpyrocarbonate, at pH 6.3, inhibits sulfate transport. Moreover, extracellular pH affects the fluorescence of covalently bound eosinisothiocyanate in a manner consistent with the titration of a histidine side chain. Matsuyama et al. (1986), studying the pH dependence of phosphate and phosphoenolpyruvate transport, have presented evidence that protonation of an intracellular histidine residue causes inhibition of transport. The sequence of mouse band 3 contains no histidine in the membrane domain between residues 315 and 669 (Kopito and Lodish, 1985). Therefore, the histidines that may be important for transport are very likely in the C-terminal 260 residues.

3.5. Arginine Residues and Carboxyl Groups

As reviewed by Bjerrum in Chapter 15, there is very good evidence for the involvement of arginine residues in band 3-mediated anion exchange. Bjerrum also reviews recent work on the chemical modification of carboxyl groups by carbodiimides. Jennings and Anderson (1987) have used the negatively charged reagent, Woodward's reagent K, to modify extracellular carboxyl groups in band 3. At least two glutamate residues are modified, and both residues can be protected by stilbenedisulfonate. These glutamate residues are probably distinct from those that are modified by carbodiimides. Jennings and Al-Rhaiyel (1988) recently showed that one of the carboxyl groups modified by Woodward's reagent K and BH_4^- is very likely the group that binds protons during the proton–sulfate cotransport associated with chloride–sulfate exchange. Moreover, the reactive carboxyl group has characteristics that suggest that, when the protein is in the outward-facing state, the group is in contact with extracellular water, but when the protein is in the inward-facing configuration, the group has access to the intracellular water. The location of this carboxyl group in the primary structure is not yet known.

3.6. Other Amino Acid Residues

Passow and coworkers (Lepke and Passow, 1982; Berghout et al., 1985; Raida and Passow, 1985; Passow, 1986) have shown that dansylation of red-cell membranes causes dramatic effects on anion transport.

Sulfate transport is enhanced at pH above 6.3 and Cl transport is slightly inhibited. The degree of enhancement of sulfate transport can be modulated by several agents that are allosterically linked to the stilbenedisulfonate site. For example, APMB potentiates the enhancement and DNDS prevents the enhancement. The effects of dansylation do not require the modification of lysine residues at the stilbenedisulfonate site, and it is not yet known what residues are modified.

Craik et al., (1986) have demonstrated that chemical modification with NBD-Cl, tetranitromethane, or *p*-nitrobenzene sulfonyl fluoride causes inhibition of band 3-mediated anion transport. This is consistent with the involvement of a tyrosine residue in the inhibition; the location of the modified tyrosine(s) in the sequence is not yet known.

Sulfhydryl modification does not inhibit band 3-mediated anion transport (Knauf and Rothstein, 1971). There is a cysteine residue (Cys 498 in mouse band 3) that reacts with pCMBS and alters the interaction between stilbenedisulfonates and band 3 (Solomon et al., 1983). The suggestion that this sulfhydryl group has a role in water permeability is discussed below.

3.7. Transport-Related Conformational Changes

Although the kinetics of anion exchange in red cells are complex, most evidence indicates that band 3-catalyzed anion exchange takes place by way of a ping-pong mechanism, in which the anions take turns crossing the membrane (*see* Knauf, Chapter 9). This mechanism implies that there must be at least two distinct conformations of the protein: inward-facing and outward-facing. A key to the understanding of the translocation event will be to determine the structural differences between these two conformations. There is now considerable information regarding the influence of various ligands on the conformation of band 3 and on the allosteric relations in the chemical reactivity of particular amino-acid residues (*see* Passow, 1986). From the large activation volume of sulfate transport, it is probable that the transport-related conformational change is physically quite substantial (Canfield and Macey, 1984). It is not yet possible, however, to specify any amino-acid residue that definitely participates in the conformational change that accompanies ion translocation.

4. Oligomeric Structure

Beginning with the work of Steck (1972), evidence from several laboratories has indicated that band 3 exists as a dimer or higher oligomer in the membrane. This evidence has been reviewed recently (Jennings, 1984; Passow, 1986) and will be summarized very briefly. For references to the original literature before 1985, *see* the above review articles.

Chemical cross-linking, ultracentrifugation, and a variety of spectroscopic data all indicate that band 3 is not monomeric in the native membrane. The evidence is consistent with the idea that the protein is tetrameric, but that the tetramer is a dimer of dimers, rather than a tetramer with fourfold rotational symmetry. There appear to be subunit interactions in both the membrane domain and cytoplasmic domain. Although the protein is oligomeric, there is no evidence that the catalytic mechanism of transport requires concerted interactions between subunits. Certain large stilbenedisulfonates exhibit negative cooperativity in binding, suggesting that the stilbenedisulfonate site may be at the interface between subunits. The exact location of the stilbenedisulfonate site relative to the subunit interface is not yet known.

Two recent radiation inactivation studies have investigated the oligomeric structure of band 3. Cuppoletti et al. (1985) showed that the target size for the anion transport function of band 3 is about 200 kdalton; this is consistent with the idea that there is efficient energy transfer between subunits, and that a radiation hit on one subunit (in either the membrane domain or cytoplasmic domain) is sufficient to inhibit transport by both halves of the dimer. Verkman et al. (1986) found that the target size of stilbenedisulfonate binding is only about 59 kdalton. This suggests that a single hit on the dimer, though sufficient to destroy anion transport, is not sufficient to destroy stilbenedisulfonate binding. This is not the first example of different target sizes for different aspects of the function of the same protein (*see* Verkman et al., 1986).

5. Transport Activities of Band 3 Other than Cl-HCO_3 Exchange

5.1. Anion Selectivity

Although the physiological function of band 3 is Cl-HCO_3 exchange, many other monovalent and divalent anions are also transported. Band 3-mediated transport of a given anion can be detected even if the rate of that transport is 10^{-6} that of Cl, and essentially all inorganic anions thus far tested are transported at a measurable rate by band 3. Band 3 also is a major transport pathway for anionic metal complexes (Funder et al., 1978; Simons, 1986). It should be noted that there is a separate transport pathway for monocarboxylates, even though many monocarboxylates are transported at detectable rates by band 3 (*see* Deuticke, 1982).

5.2. Proton Transport

The extracellular pH dependence of chloride transport in the acid range is consistent with the requirement that a titratable group of pK_a 5 be unprotonated in order for transport to take place (Wieth et al., 1982; Milanick and Gunn, 1982). Conversely, a protonation with similar pK_a is required for sulfate influx into chloride-loaded cells (Milanick and Gunn, 1984). These two characteristics of the chloride and sulfate transport suggest that, as originally postulated in the titratable carrier model of Gunn (1973), the protonation of a single titratable group converts the transport system from a form that prefers monovalent anions to one that transports divalent anions at an appreciable rate. As predicted by this model, the net influx of sulfate into Cl-loaded cells is accompanied by a stoichiometric influx of protons (Jennings, 1976). The proton influx presumably involves the same titratable group that is responsible for the pH dependence of the sulfate influx.

5.3. Anion Conductance

The chloride conductance of the red-cell membrane is several orders of magnitude smaller than would be expected if the ^{36}Cl exchange flux took place through a conductive pathway. The conductive flux, though much smaller than the exchange, is mediated largely by band 3 (Knauf et al., 1977, 1983). The mechanism of the conductive flux and its relationship to the exchange mechanism are discussed in Chapter 9. Another interesting band 3-mediated transport process is the cation permeability that is stimulated at low ionic strength (Jones and Knauf, 1985). This mode of transport is discussed in Chapter 9.

5.4. Postulated Water Transport

In 1975, Brown et al. labeled intact red cells with (^{14}C) DTNB [5,5′-dithiobis(2-nitrobenzoic acid)], an inhibitor of hydraulic water conductivity in red cells, and showed that band 3 is the major labeled species. This finding suggested that band 3 may function as a transport pathway for water as well as for anions. Solomon and coworkers (Solomon et al., 1983; Lukacovic et al., 1984; Toon et al., 1985) have investigated this question in considerable detail and have shown that pCMBS, an inhibitor of water permeability, interacts with band 3 in ways that are consistent with the idea that band 3 is a water permeation pathway. It should be noted in this context that most stilbenedisulfonates do not by themselves inhibit water permeability (Solomon et al., 1983), although a mercurial derivative of stilbenedisulfonate does inhibit water transport (Yoon et al., 1984).

Benz et al. (1984) showed that the conductance induced in lipid bilayers by solubilized band 3 is inhibitable by mercurials and is proportional to the fourth power of the protein concentration. This finding suggests that a channel formed by the tetramer could be the transport pathway for water. It is not yet clear how closely this system resembles native band 3, because the incorporated protein did not catalyze anion exchange. Recent radiation inactivation studies have shown that the target sizes for anion transport and water transport are very different (Dix et al., 1985), suggesting that a band 3 dimer or tetramer is unlikely to constitute the water transport pathway. The water transport protein has not been identified, and it remains possible that band 3 mediates water transport. If so, the transport pathway through the protein is distinct from that for anions.

6. Band 3-Related Proteins in Nonerythroid Cells

In keeping with the theme of this volume, it is appropriate to discuss the red-cell Cl-HCO_3 exchanger in the context of nonerythroid cells. There are several major unanswered questions about the relationship between red-cell band 3 and anion transport in general: How many kinds of Cl-HCO_3 exchangers are there? What is the difference between Cl-HCO_3 exchange proteins that are posttranslationally regulated and those that are not? Is a protein identical to red-cell band 3 expressed in other tissues? What are the similarities between Cl-HCO_3 exchangers and other (e.g., SO_4, PO_4) anion transport proteins? The state of understanding of these issues is primitive, but there have been some recent advances that are discussed briefly below.

One point that should be emphasized at the outset is that the red cell, because of its specialized role in CO_2 transport, has a capacity for Cl transport that far exceeds that of other cells. The half-time for ^{36}Cl–Cl exchange in red cells at 37°C is about 50 ms (Brahm, 1977). The half-time for the same process in human neutrophils is 30 min (Simchowitz and DeWeer, 1986; Simchowitz et al., 1986), that in erythroleukemic cells is about 10 min at 25°C (Law et al., 1983), and that in Ehrlich ascites tumor cells is about 10 min at 22°C (Levinson, 1978). This does not imply that band 3 is absent from nonerythroid cells, but rather that, if it is present, the number of copies/cell must be far less than in red cells (or less likely, the turnover number of the protein is much lower than in red cells).

6.1. Stilbenedisulfonate-Inhibitable Anion Transport Processes

Reports of stilbenedisulfonate inhibition (and noninhibition) of anion transport in nonerythroid cells are too numerous to review here (*see* Hoffmann, 1986). In several systems, stilbenedisulfonate-binding proteins have been labeled (e.g., Levinson et al., 1979; Cheng and Levy, 1980; Zoccoli et al., 1982; Jessen et al., 1986). Too little is presently known about these proteins to comment on their relationships with band 3. It should be emphasized that stilbenedisulfonates are potent inhibitors of red-cell anion transport, but they should not be thought of as specific. DIDS, for example is known to inhibit both the Na,K-ATPase (Teisinger et al., 1984) and the plasma membrane calcium pump (Niggli et al., 1982).

One anion that has been examined in several systems is sulfate, and it is useful to comment on the relationship between red-cell band 3 and sulfate transport proteins in other cells. Sulfate is a commonly used substrate anion for red-cell band 3, and band 3 represents the only known mediated pathway for sulfate transport in red cells. However, SO_4 is transported by red-cell band 3 at a rate that (at neutral pH) is about 100,000 times slower than the rate of Cl–HCO_3 exchange. Therefore, although many interesting things have been learned about anion transport pathway from the study of sulfate (e.g., Milanick and Gunn, 1982, 1984), red-cell band 3 should not be thought of as a sulfate transport protein. Cells in which the transport rates for Cl and SO_4 are of the same order of magnitude, such as the Ehrlich ascites tumor cell (Levinson, 1978), must have an anion transport protein that is functionally quite different from red-cell band 3. Another cell in which anion exchange differs markedly from that in red cells is the neutrophil; in these cells, sulfate transport is inhibitable by DIDS (Korchak et al., 1980), whereas chloride exchange is not inhibited by stilbenedisulfonate (Simchowitz et al., 1986). It appears, then, that there exists a stilbenedisulfonate-sensitive SO_4 transport pathway that has a SO_4/Cl selectivity very different from that of red-cell band 3.

Another stilbenedisulfonate-inhibitable transport process that is of considerable interest is the Na-dependent Cl-HCO_3 exchanger that has been studied most extensively in invertebrates (*see* Thomas, 1977; Russell et al., 1983). The function of the transport is to extrude acid from the cytoplasm and thereby help regulate intracellular pH, which in the steady state is known to be more alkaline than would be expected on the basis of Donnan equilibrium (Roos and Boron, 1981). Recently, work in two laboratories has provided evidence that a similar system is operative in mammalian cells as well (Jentsch et al., 1985; L'Allemain et al., 1985). The protein responsible for this transport has not been identified in either the vertebrate or the invertebrate systems.

6.2. Na-Independent Cl-HCO_3 Exchange in Nonerythroid Cells

The remainder of this discussion will be limited to the anion exchange process that is catalyzed by red-cell band 3, i.e., Na-independent Cl-HCO_3 exchange. There is no known general cellular function for Cl-HCO_3 exchange. With the exception of cells in tissues responsible for net HCO_3 transport (e.g., red cells, certain epithelia), there is little reason to expect a cell to have a Na-independent Cl-HCO_3 exchange protein in its plasma membrane. As noted above, animal cells regulate cytoplasmic pH at a value that is above what would be expected on the basis of electrochemical equilibrium for protons (Roos and Boron, 1981). In order to do this, there must be acid-extruding mechanisms, e.g., the amiloride-sensitive Na–H exchanger and the Na-dependent Cl–HCO_3 exchanger. Under the conditions that prevail in most cells, an Na-independent Cl-HCO_3 exchanger would act as an *alkali*-extruding system, because the inward Cl gradient exceeds the inward HCO_3 gradient. Accordingly, red-cell band 3, if present in most cells, would not contribute to the extrusion of acid under physiological conditions. Instead, the purpose of a Cl–HCO_3 exchange would be expected to be protection of cells from an alkali load (Olsnes et al., 1986; Simchowitz and Roos, 1985).

6.3. Acid-Secreting Epithelia

Although there is no compelling reason to expect a Cl-HCO_3 exchanger to be present in most cells, cells that are involved in acid secretion would be expected to have Cl–HCO_3 exchange capability that would allow the intracellular pH to remain neutral in the presence of an acid efflux. For example, there is good electrophysiological and kinetic evidence that the urinary bladder of the turtle (*see* Husted et al., 1979) and the mammalian renal collecting duct (*see* Stone et al., 1983) contain acid-secreting cells that have a proton pump in the apical membrane and a stilbenedisulfonate-sensitive Cl–HCO_3 exchanger in the basolateral membrane. There is also good evidence that the basolateral membranes of gastric acid-secreting cells contain a stilbenedisulfonate-sensitive Cl–HCO_3 exchanger (*see* Paradiso et al., 1987).

In 1985, Cox et al. presented immunocytochemical evidence that certain cells in the chicken kidney express a protein that is cross-reactive with chicken red-cell band 3. The state of knowledge of the renal physiology of the chicken is not sufficient to make it possible to propose a definite function for the band 3-like protein. Drenckhahn et al. (1985) showed that a subset of cells in the cortical and medullary collecting ducts of the rat kidney stain with affinity-purified polyclonal antibody against rat band 3. The staining was confined to the basolateral membrane of intercalated cells. Most of the immunoglobulins in the antiserum were directed against the cytoplasmic domain, but the authors reported similar staining with antibodies against the membrane domain. The staining with anti-band 3 colocalized with that with antibodies against ankyrin, suggesting that ankyrin and band 3 form the same association in the kidney as in the red cell. Schuster et al. (1986), using a monoclonal antibody against red-cell band 3 membrane domain, demonstrated a staining pattern in rabbit kidney that is very similar to that found by Drenckhahn et al. (1985). No convincing staining of apical membranes was observed in either the cortical or medullary collecting duct. The number of band 3-positive cells increased in parallel with the number of acid-secreting cells in the progression from the cortical to medullary collecting duct. Therefore, it is very likely that the acid-secreting intercalated cells contain a band 3-like Cl–HCO_3 exchange protein in the basolateral membrane; this protein presumably serves as an exit pathway for HCO_3.

The actual protein in the kidney that cross-reacts with antibody against red-cell band 3 has not yet been characterized. In the rat, the protein has an apparent mol wt that is slightly higher than that of red-cell band 3 (Drenckhahn et al., 1985), but the details of its similarities with and differences from red-cell band 3 are not known. Molecular biological techniques have shown that the mouse erythroid band 3 gene is expressed in the kidney (Kopito et al., 1987); the initiation sites for the expression of the gene differ between renal and erythroid tissue. In addition to erythroid band 3, a nonerythroid band 3 homolog is also expressed in renal tissue (Alper et al., 1987). The sequence of this protein is indistinguishable from that of a nonerythroid band 3 homolog that is expressed in nonepithelial cells (*see below*).

6.4. Cells Not Involved in Net Acid or Base Secretion

There is immunocytochemical evidence that proteins related to band 3 are present in nonerythroid cells that have no role in net acid or base secretion (Kay et al., 1983; Drenckhahn et al., 1984). As discussed above, the function of these cross-reactive proteins is unclear. In one cell, the neutrophil, there appears to be a large amount of a plasma membrane protein that is immunologically cross-reactive with band 3 (Kay et al., 1983). It is very unlikely that this protein carries out a transport function that is similar to red-cell band 3, because chloride transport in neutrophils is 10,000-fold slower than in red cells and is not inhibited by stilbenedisulfonate (Simchowitz and DeWeer, 1986; Simchowitz et al., 1986).

Hazen-Martin et al. (1986) used antibody against the band 3 cytoplasmic domain to stain sections of human skin. Sweat duct basolateral membranes were stained equally in normal samples and in those from individuals with cystic fibrosis. LM55 glial cells also have a membrane protein that is immunologically cross-reactive with red-cell band 3 (Wolpaw and Martin, 1986). The functions of these cross-reactive proteins are unknown.

Demuth et al. (1986) recently published the cDNA sequence for a human protein that is very similar to, but not identical with, red-cell band 3. The protein has very extensive homology with red-cell band 3 membrane domain, except for an insertion of 29 amino-acid residues in the vicinity of the site that in red-cell band 3 is cleaved by extracellular chymotrypsin. The protein is expressed in a variety of nonerythroid cells, but its expression is turned off when erythroleukemia cells are induced to differentiate. It appears, then, that the protein coded by this cDNA represents a nonerythroid band 3 homolog. It is not known what the precise function of the protein is, and it is not known which of the immunologically cross-reactive proteins mentioned above are coded by this DNA sequence.

In summary, it is clear that proteins with immunological cross-reactivity and/or major sequence homology with red-cell band 3 exist in nonerythroid cells. The structural and functional relationships between these anion transporters and red-cell band 3 remains to be determined.

References

Alper, S., Kopito, R. R., and Lodish, H. F. (1987) A molecular biological approach to the study of anion transport. *Kidney International* **32, Suppl. 23,** S117–S128.

Appell, K. C. and Low, P. S. (1981) Partial structural characterization of the cytoplasmic domain of the erythrocyte membrane protein, band 3. *J. Biol. Chem.* **256,** 11104–11111.

Appell, K. E. and Low, P. S. (1982) Evaluation of structural independence of membrane spanning and cytoplasmic domains of band 3. *Biochemistry* **21,** 2151–2157.

Barzilay, M. and Cabantchik, Z. I. (1979) Anion transport in red blood cells. III. Sites and sideness of inhibition by high affinity reversibly binding probes. *Membr. Biochem.* **2,** 297–322.

Bennett, V. (1985) The membrane skeleton of human erythrocytes and its implications for more complex cells. *Ann. Rev. Biochem.* **54,** 273–304.

Bennett, V. and Stenbuck, P. J. (1980) Association between ankyrin and the cytoplasmic domain of band 3 isolated from the human erythrocyte membrane. *J. Biol. Chem.* **255,** 6424–6432.

Benz, R., Tosteson, M. T., and Schubert, D. (1984) Formation and properties of tetramers of band 3 protein from human erythrocyte membranes in planar lipid bilayers. *Biochim. Biophys. Acta* **775,** 347–355.

Berghout, A., Raida, M., Romano, L., and Passow, H. (1985) pH dependence of phosphate transport across the red blood cell membrane after modification by dansyl chloride. *Biochim. Biophys. Acta* **815,** 281–286.

Beth, A. H., Conturo, T. E., Venkataramu, S. D., and Staros, J. D. (1986) Dynamics and interactions of the anion channel in intact human erythrocytes: An electron paramagnetic resonance spectroscopic study employing a new membrane-impermeant bifunctional spin-label. *Biochemistry* **25,** 3824–3832.

Braell, W. A. and Lodish, H. F. (1981) Biosynthesis of the erythrocyte anion transport protein. *J. Biol. Chem.* **256,** 11337–11344.

Brahm, J. (1977) Temperature-dependent changes of chloride transport kinetics in human red blood cells. *J. Gen. Physiol.* **70,** 283–306.

Brock, C. J. and Tanner, M. J. A. (1986) The human erythrocyte anion-transport protein. Further amino acid sequence from the integral membrane domain homologous with the murine protein. *Biochem. J.* **235,** 899–901.

Brock, C. J., Tanner, M. J. A., and Kempf, C. (1983) The human erythrocyte anion-transport protein. Partial amino acid sequence, conformation, and a possible molecular mechanism for anion exchange. *Biochem. J.* **213,** 577–586.

Brown, P. A., Feinstein, M. B., and Sha'afi, R. I. (1975) Membrane proteins related to water transport in human erythrocytes. *Nature* **254,** 523–525.

Cabantchik, Z. I., Knauf, P. A., and Rothstein, A. (1978) The anion transport system of the red blood cell. The role of membrane protein evaluated by use of "probes". *Biochim. Biophys. Acta* **515,** 239–302.

Canfield, V. A. and Macey, R. I. (1984) Anion exchange in human erythrocytes has a large activation volume. *Biochim. Biophys. Acta* **778,** 379–384.

Cheng, S. and Levy, D. (1980) Characterization of the anion transport system in hepatocyte plasma membranes. *J. Biol. Chem.* **255,** 2637–2640.

Chiba, T., Sato, Y., and Suzuki, Y. (1986) Amino acid residues complexed with eosin 5-isothiocyanate in band 3 protein of the human erythrocyte. *Biochim. Biophys. Acta* **858,** 107–117.

Cox, J. V., Moon, R. T., and Lazarides, E. (1985) Anion transporter: highly cell-type-specific expression of distinct polypeptides and transcripts in erythroid and non-erythroid cells. *J. Cell. Biol.* **100,** 1548–1557.

Craik, J. D., Gouaden, K., and Reithmeier, R. A. F. (1986) Inhibition of phosphate transport in human erythrocytes by 7-chloro-4-nitrobenzo-2-oxa-1,3-diazole (NBD-Cl). *Biochim. Biophys. Acta* **856,** 602–609.

Cuppoletti, J., Goldinger, J., Kang, B., Jo, I., Berenski, C., and Jung, C. Y. (1985) Anion carrier in the human erythrocyte exists as a dimer. *J. Biol. Chem.* **260,** 15714–15717.

Dekowski, S. A., Rybicki, A., and Drickamer, K. (1983) A tyrosine kinase associated with the red cell membrane phosphorylates band 3. *J. Biol. Chem.* **258,** 2750–2753.

Demuth, D. R., Showe, L. C., Ballantine, M., Palumbo, A., Fraser, P. J., Cioe, L., Rovera, G., and Curtis, P. J. (1986) Cloning and structural characterization of a human non-erythroid band 3-like protein. *EMBO J.* **5,** 1205–1214.

Deuticke, B. (1982) Monocarboxylate transport in erythrocytes. *J. Membr. Biol.* **70,** 89–103.

Dix, J. A., Verkman, A. S., and Solomon, A. K. (1986) Binding of chloride and a disulfonic stilbene transport inhibitor to red cell band 3. *J. Membr. Biol.* **89,** 211–223.

Dix, J. A., Ausiello, D. A., Jung, C. Y., and Verkman, A. S. (1985) Target analysis studies of red cell water and urea transport. *Biochim. Biophys. Acta* **821,** 243–252.

Drenckhahn, D., Schlüter, K., Allen, D. P., and Bennett, V. (1985) Colocalization of band 3 with ankyrin and spectrin at the basal membrane of intercalated cells in the rat kidney. *Science* **230,** 1287–1290.

Drenckhahn, D., Zinke, K., Schauer, U., Appell, K. C., and Low, P. S. (1984) Identification of immunoreactive forms of human erythrocyte band 3 in nonerythroid cells. *Eur. J. Cell Biol.* **34,** 144–150.

Drickamer, L. K. (1976) Fragmentation of the 95,000-Dalton transmembrane polypeptide in human erythrocyte membranes. Arrangement of the fragments in the lipid bilayer. *J. Biol. Chem.* **251,** 5115–5123.

Drickamer, L. K. (1977) Fragmentation of the band 3 polypeptide from human erythrocyte membranes. Identification of regions likely to interact with the lipid bilayer. *J. Biol. Chem.* **252,** 6906–6917.

DuPre, A. M. and Rothstein, A. (1981) Inhibition of anion transport associated with chymotryptic cleavages of red blood cell band 3 protein. *Biochim. Biophys. Acta* **646,** 471–478.

Fairbanks, G., Steck, T. L., and Wallach, D. F. H. (1971) Electrophoretic analysis of the major polypeptides of the human erythrocyte membrane. *Biochemistry* **10,** 2606–2617.

Falke, J. J., Kanes, K. J., and Chan, S. I. (1985) Thc minimal structure containing the band 3 anion transport site. *J. Biol. Chem.* **260,** 13294–13303.

Frölich, O. (1982) The external anion binding site of the human erythrocyte anion transporter: DNDS binding and competition with chloride. *J. Membr. Biol.* **56,** 111–123.

Fukuda, M., Dell, A., Oates, J. E., and Fukuda, M. N. (1984) Structure of branched lactoseamino-glycan, the carbohydrate moiety of band 3 isolated from adult human erythrocytes. *J. Biol. Chem.* **259,** 8260–8273.

Funder, J. and Wieth, J. O. (1976) Chloride transport in human erythroctes and ghosts: a quantitative comparison *J. Physiol.* **262,** 679–698.

Funder, J., Tosteson, D. C., and Wieth, J. O. (1978) Effects of bicarbonate on lithium transport in human red cells. *J. Gen. Physiol.* **71,** 721–746.

Garcia, A. M., Kopito, R., and Lodish, H. V. (1987) Expression of band 3 in Xenopus laevis oocytes microinjected with in vitro prepared mRNA. *Biophys. J.* **51,** 567a.

Grinstein, S., Ship, S., and Rothstein, A. (1978) Anion transport in relation to proteolytic dissection of band 3 protein. *Biochim. Biophys. Acta* **507,** 294–304.

Gunn, R. B. (1973) A titratable carrier model for monovalent and divalent inorganic anions in red blood cells, in *Erythrocytes, Thrombocytes, Leucocytes* (Gerlach, E., Moser, K., Deutsch, E., and Wilmanns, W., eds.), Thieme, Stuttgart, pp. 77–79.

Hazen-Martin, D. J., Pasternack, G., Spicer, S. S., and Sens, D. A. (1986) Immunolocalization of band 3 protein in normal and cystic fibrosis skin. *J. Histochem. Cytochem.* **34,** 823–826.

Higashi, T., Richards, C. S., and Uyeda, K. (1979) The interaction of phosphofructokinase with erythrocyte membranes. *J. Biol. Chem.* **254,** 9542–9550.

Hoffmann, E. K. (1986) Anion transport systems in the plasma membrane of vertebrate cells. *Biochim. Biophys. Acta* **864,** 1–31.

Hsu, L. and Morrison, M. (1983) The interaction of human erythrocyte band 3 with cytoskeletal components. *Arch Biochem. Biophys.* **227,** 31–38.

Husted, R. F., Cohen, L. H., and Steinmetz, P. R. (1979) Pathways for bicarbonate transfer across the serosal membrane of turtle urinary bladder: studies with a disulfonic stilbene. *J. Membr. Biol.* **47,** 27–37.

Jay, D. G. (1986) Glycosylation site of band 3, the human erythrocyte anion-exchange protein. *Biochemistry* **25,** 554–556.

Jay, D. and Cantley, L. (1986) Structural aspects of the red cell anion exchange protein. *Ann. Rev. Biochem.* **55,** 511–538.

Jennings, M. L. (1976) Proton fluxes associated with erythrocyte membrane anion exchange. *J. Membr. Biol.* **28,** 187–205.

Jennings, M. L. (1982) Reductive methylation of the two H_2DIDS-binding lysine residues of band 3, the human erythrocyte anion transport protein. *J. Biol. Chem.* **257,** 7554–7559.

Jennings, M. L. (1984) Oligomeric structure and the anion transport function of human erythrocyte band 3 protein. *J. Membr. Biol.* **80,** 105–117.

Jennings, M. L. and Adams, M. F. (1981) Modification by papain of the structure and function of band 3, the erythrocyte anion transport protein. *Biochemistry* **20,** 7118–7122.

Jennings, M. L. and Nicknish, J. S. (1984) Erythrocyte band 3 protein. Evidence for multiple membrane-crossing segments in the 17000-dalton chymotryptic fragment. *Biochemistry* **23,** 6432–6436.

Jennings, M. L. and Nicknish, J. S. (1985) Localization of a site of intermolecular cross-linking in human red blood cell band 3 protein. *J. Biol. Chem.* **260,** 5472–5479.

Jennings, M. L. and Passow, H. (1979) Anion transport across the erythrocyte membrane, in situ proteolysis of band 3 protein, and cross-linking of proteolytic fragments by 4,4′-diisothiocyano-dihydrostilbene-2,2′-disulfonate. *Biochim. Biophys. Acta* **554,** 498–519.

Jennings, M. L. and Al-Rhaiyel, S. (1988) Modification of a carboxyl group that appears to cross the permeability barrier in the red blood cell anion transporter. *J. Gen. Physiol.* **92,** 161–178.

Jennings, M. L., Adams-Lackey, M., and Denney, G. H. (1984) Peptides of human erythrocyte band 3 protein produced by extracellular papain cleavage. *J. Biol. Chem.* **259,** 4652–4660.

Jennings, M. L., Anderson, M. P., and Monaghan, R. (1986) Monoclonal antibodies against human erythrocyte band 3 protein. *J. Biol. Chem.* **261,** 9002–9010.

Jennings, M. L. and Anderson, M. P. (1987) Chemical modification and labeling of glutamate residues at the stilbenedisulfonate site of human red blood cell band 3 protein. *J. Biol. Chem.* **262,** 1691–1697.

Jennings, M. L., Monaghan, R., Douglas, S. M., and Nicknish, J. S. (1985) Functions of extracellular lysine residues in the human erythrocyte anion transport protein. *J. Gen. Physiol.* **86,** 653–669,

Jentsch, T. J., Stahlknecht, T. R., Hollwede, H., Fischer, D. G., Keller, S. K., and Wiederhold, M. (1985) A bicarbonate-dependent process inhibitable by disulfonic stilbenes and a Na^+/H^+ exchange mediate $^{22}Na^+$ uptake into cultured bovine corneal endothelium. *J. Biol. Chem.* **260,** 795–801.

Jessen, F., Sjøholm, C., and Hoffman, E. K. (1986) Identification of the anion exchange protein of Ehrlich cells: A kinetic analysis of the inhibitory effects of 4,4′-diisothiocyano-2,2′-stilbene disulfonic acid (DIDS) and labeling of membrane proteins with ^{3}H-DIDS. *J. Membr. Biol.* **92,** 195–205.

Jones, G. S. and Knauf, P. A. (1985) Mechanism of the increase in cation permeability of human erythrocytes in low-chloride media. Involvement of the anion transport protein capnophorin. *J. Gen. Physiol.* **86,** 721–738.

Kampmann, L., Lepke, S., Fasold, H., Fritzch, G., and Passow, H. (1982) The kinetics of intramolecular cross-linking of the band 3 protein in the red blood cell membrane by 4,4′-diisothiocyano dihydrostilbene-2,2′disulfonic acid (H_2DIDS). *J. Membr. Biol.* **70,** 199–216.

Kaplan, J. H., Skorah, K., Fasold, H., and Passow, H. (1976) Sidedness of the inhibitory action of disulfonic acids on chloride equilibrium exchange and net transport across the human erythrocyte membrane. *FEBS Lett.* **62,** 182–185.

Karadsheh, N. S. and Uyeda, K. (1977) Changes in allosteric properties of phosphofructokinase bound to erythrocyte membranes. *J. Biol. Chem.* **252,** 7418–7420.

Kaul, R. K., Murthy, S. N. P., Reddy, A. G., Stech, T., and Köhler, H. (1983) Amino acid sequence of the *N*-terminal 201 residues of the human erythrocyte membrane band 3. *J. Biol. Chem.* **258,** 7981–7990.

Kawano, Y. and Hamasaki, N. (1986) Isolation of a 5,300 dalton peptide containing a pyridoxal phosphate binding site from the 38,000-dalton domain of band 3 of human erythrocyte membranes. *J. Biochem.* **100,** 191-199.

Kawano, Y., Okubo, K., Tokunaga, F., Miyata, T., Iwanaga, S., and Hamasaki, N. (1988) Localization of the pyridoxal phosphate binding site at the COOH- terminal region of erythrocyte band 3 protein. *J. Biol. Chem.* **263,** 8232–8238.

Kay, M. M. B. (1984) Localization of senescent cell antigen on band 3. *Proc. Natl. Acad. Sci. USA* **81,** 5753–5757.

Kay, M. M. B., Tracey, C. M., Goodman, J. R., Cone, J. C., and Bassel, P. A. (1983) Polypeptides immunologically related to band 3 are present in nucleated somatic cells. *Proc. Natl. Acad. Sci. USA* **80,** 6882–6886.

Kempf, C., Brock, C., Sigrist, H., Tanner, M. J. A., and Zahler, P. (1981) Interaction of phenylisothiocyanate with human erythrocyte band 3 protein. II. Topology of phenylisothiocyanate binding sites and influence of p-sulfophenylisothiocyanate on phenylisothiocyanate modification. *Biochim. Biophys. Acta* **641,** 88–98.

Kliman, H. J. and Steck, T. L. (1980) Association of glyceraldehyde-3-phosphate dehydrogenase with the human red cell membrane. A kinetic analysis. *J. Biol. Chem.* **255,** 6314–6321.

Knauf, P. A. and Rothstein, A. (1971) Chemical modification of membranes. I. Effects of sulfhydryl and amino reactive reagents on anion and cation permeability of the human red blood cell. *J. Gen. Physiol.* **58,** 190–210.

Knauf, P. A. Law, F. -Y., and Marchant, P. J. (1983) Relationship of net chloride flow across the human erythrocyte membrane to the anion exchange mechanism. *J. Gen. Physiol.* **81,** 95–126.

Knauf, P. A., Fuhrmann, G. F., Rothstein, S., and Rothstein, A. (1977) The relationship between anion exchange and net anion flow across the human red blood cell membrane. *J. Gen. Physiol.* **69,** 363–386.

Köhne, W., Deuticke, B., and Haest, C. W. M. (1983) Phospholipid dependence of the anion transport system of the human erythrocyte membrane. *Biochim. Biophys. Acta* **730,** 139–150.

Kopito, R. R. and Lodish, H. F. (1985) Primary structure and transmembrane orientation of the murine anion exchange protein. *Nature* **316,** 234–238.

Kopito, R. R., Anderson, M. A., and Lodish, F. H. (1987) Multiple tissue-specific sites of transcriptional initiation of the mouse anion transport gene in erythroid and renal cells. *Proc. Natl. Acad. Sci. USA* **84,** 7149–7153.

Korchak, H. M., Eisenstat, B. A., Hoffstein, S. T., Dunham, P. B., and Weissmann, G. (1980) Anion channel blockers inhibit lysosomal enzyme secretion from human neutrophils without affecting generation of superoxide anion. *Proc. Nat. Acad. Sci. USA* **77,** 2721–2725.

L'Allemain, G., Paris, S., and Pouysségur, J. (1985) Role of a Na^+-dependent Cl^-/HCO_3^+ exchange in regulation of intracellular pH in fibroblasts. *J. Biol. Chem.* **260,** 4877–4883.

Law, F. -Y. Steinfeld, R., and Knauf, P. A. (1983) K562 cell anion exchange differs markedly from that of mature red blood cells. *Am J. Physiol.* **244,** C68–C74.

Lepke, S. and Passow, H. (1976) Effects of incorporated trypsin on anion exchange and membrane proteins in human red blood cell ghosts. *Biochim. Biophys. Acta* **455,** 353–370.

Lepke, S. and Passow, H. (1982) Inverse effects of dansylation of red blood cell membrane on band 3 protein-mediated transport of sulphate and chloride. *J. Physiol.* **328,** 27–48.

Levinson, C. (1978) Chloride and sulfate transport in Ehrlich ascites tumor cells: Evidence for a common mechanism. *J. Cell Physiol.* **95,** 23–32.

Levinson, C., Corcoran, R. J., and Edwards, E. H. (1979) Interaction of tritium-labeled H_2DIDS (4,4′-diisothiocyano-1,2,diphenyl ethane-2,2′disulfonic acid) with the Ehrlich mouse ascites tumor cell. *J. Membr. Biol.* **45,** 61–79.

Lieberman, D. M., Natriss, M., and Reithmeier, R. A. F. (1987) Carboxypeptidase Y digestion of band 3, the anion transport protein of human erythrocyte membranes. *Biochim. Biophys. Acta* **903,** 37–47.

Low, P. S., Waugh, S. M., Zinke, K., and Drenckhahn, D. (1985) The role of hemoglobin denaturation and band 3 clustering in red blood cell aging. *Science* **227,** 531–533.

Low, P. S., Westfall, M. A., Allen, D. P., and Appell, K. C. (1984) Characterization of the reversible conformational equilibrium of the cytoplasmic domain of erythrocyte membrane band 3. *J. Biol. Chem.* **259,** 13070–13076.

Lukacovic, M. F., Verkman, A. S., Dix, J. A., and Solomon, A. K. (1984) Specific interactions of the water transport inhibitor, pCMBS, with band 3 in red blood cell membranes. *Biochim. Biophys. Acta* **778,** 253–259.

Macara, I. G., Kuo, S., and Cantley, L. C. (1983) Evidence that inhibitors of anion exchange induce a transmembrane conformational change in band 3. *J. Biol. Chem.* **258,** 1785–1792.

Markowitz, S. and Marchesi, V. T. (1981) The carboxyl-terminal domain of human erythrocyte band 3. Description, isolation and location in the bilayer. *J. Biol. Chem.* **256,** 6463–6468.

Matsuyama, H., Kawano, Y., and Hamasaki, N. (1983) Anion transport activity in the human erythrocyte membrane modulated by proteolytic digestion of the 38,000-dalton fragment in band 3. *J. Biol. Chem.* **258,** 15376–15381.

Matsuyama, H., Kawano, Y., and Hamasaki, N. (1986) Involvement of a histidine residue in inorganic phosphate and phosphoenolpyruvate transport across the human erythrocyte membrane. *J. Biochem.* **99,** 495–501.

Mawby, W. J. and Findlay, J. B. C. (1982) Characterization and partial sequence of di-iodosulphophenyl isothiocyanate-binding peptide from human erythrocyte anion-transport protein. *Biochem. J.* **205,** 465–475.

Milanick, M. A. and Gunn, R. B. (1982) Proton-sulfate co-transport: Mechanism of H^+ and sulfate addition to the chloride transporter of human red blood cells. *J. Gen. Physiol.* **78,** 547–568.

Milanick, M. A. and Gunn, R. B. (1984) Proton-sulfate cotransport: External proton activation of sulfate influx into human red blood cells. *Am. J. Physiol.* **247,** C247–C259.

Mühlebach, T. and Cherry, R. J. (1985) Rotational diffusion and self-association of band 3 in reconstituted lipid vesicles. *Biochemistry* **24,** 1975–1983.

Murthy, P. S. N., Liu, T., Kaul, R. K., Köhler, H., and Steck, T. L. (1981) The aldolase-binding site of the human erythrocyte membrane is at the NH_2 terminus of band 3. *J. Biol. Chem.* **256,** 11203–11208.

Nanri, H., Hamasaki, N., and Minakami, S. (1983) Affinity labeling of erythrocyte band 3 protein with pyridoxal 5-phosphate. *J. Biol. Chem.* **258,** 5985–5989.

Niggli, V., Sigel, E., and Carafoli, E. (1982) Inhibition of the purified and reconstituted calcium pump of erythrocytes by uM levels of DIDS and NAP-taurine. *FEBS Lett.* **13,** 164–166.

Oikawa, K., Lieberman, D. M., and Reithmeier, R. A. F. (1985) Conformation and stability of the anion transport protein of human erythrocyte membranes. *Biochemistry* **24,** 2843–2848.

Olsnes, S., Tønnessen, T. I., and Sandvig, K. (1986) pH-regulated anion antiport in nucleated mammalian cells. *J. Cell Biol.* **102,** 967–971.

Paradiso, A. M., Tsien, R. Y., and Machen, T. E. (1987) Digital image processing of intracellular pH in gastric oxyntic and chief cells. *Nature* **325,** 447–450.

Passow, H. (1986) Molecular aspects of band 3 protein-mediated anion transport across the red blood cell membrane. *Rev. Physiol. Biochem. Pharmacol.* **103,** 61–203.

Pasternack, G. R., Anderson, R. A., Leto, T. L., and Marchesi, V. T. (1985) Interactions between protein 4.1 and band 3. *J. Biol. Chem.* **260,** 3676–3683.

Raida, M. and Passow, H. (1985) Enhancement of divalent anion transport across the human red blood cell membrane by the water-soluble dansyl chloride derivative 2-(*N*-piperidine)ethylamine-1-naphthyl-5-sulfonyl-chloride (PENS-Cl). *Biochim. Biophys. Acta* **812,** 624–632.

Ramjeesingh, M. and Rothstein, A. (1982) The location of a chymotrypsin cleavage site and of other sites in the primary structure of the 17,000-dalton transmembrane segment of band 3, the anion transport protein of red cell. *Membr. Biochem.* **4,** 259–269.

Ramjeesingh, M., Gaarn, A., and Rothstein, A. (1980) The location of a disulfonic stilbene binding site in band 3, the anion transport protein of the red blood cell membrane. *Biochim. Biophys. Acta* **599,** 127–139.

Rao, A. (1979) Disposition of the Band 3 polypeptide in the human erythrocyte membrane. The reactive sulfhydryl groups. *J. Biol. Chem.* **254,** 3503–3511.

Rao, A. and Reithmeier, R. A. F. (1979) Reactive sulfhydryl groups of the Band 3 polypeptide from human erythrocyte membranes. Locations in the primary structure. *J. Biol. Chem.* **254,** 6144–6150.

Rao, A., Martin, P., Reithmeier, R. A. F., and Cantley, L. C. (1979) Location of the stilbene disulfonate binding site of the human erythrocyte anion-exchange system by resonance energy transfer. *Biochemistry* **18,** 4505–4516.

Roos, A. and Boron, W. F. (1981) Intracellular pH. *Physiol. Rev.* **61,** 296–434.

Rudloff, V., Lepke, S., and Passow, H. (1983) Inhibition of anion transport across the red cell membrane by dinitrophenylation of a specific lysine residue at the H_2DIDS binding site of the band 3 protein. *FEBS Lett.* **163,** 14–21.

Russell, J. M., Boron, W. F., and Brodwick, M. S. (1983) Intracellular pH and Na fluxes in barnacle muscle with evidence for reversal of the ionic mechamism of intracellular pH regulation. *J. Gen. Physiol.* **82,** 47–78.

Salhany, J. M., Cordes, K. A., and Gaines, E. D. (1980) Light-scattering measurements of hemoglobin binding to the erythrocyte membrane. Evidence for transmembrane effects related to a disulfonic stilbene binding to band 3. *Biochemistry* **19,** 1447–1454.

Schuster, V. L., Bonsib, S. M., and Jennings, M. L. (1986) Two types of collecting duct mitochondria-rich (intercalated) cells: Lectin and band 3 cytochemistry. *Am. J. Physiol.* **251,** C347–C355.

Shami, Y., Rothstein, A., and Knauf, P. A. (1978) Identification of the Cl transport site of human red blood cells by a kinetic analysis of the inhibitory effects of a chemical probe. *Biochim. Biophys. Acta* **508,** 357–363.

Sigrist, H., Kempf, C., and Zahler, P. (1980) Interaction of phenylisothiocyanate with human erythrocyte Band 3. I. Covalent modification and inhibition of phosphate transport. *Biochim. Biophys. Acta* **597,** 137–144.

Simchowitz, L. and DeWeer, P. (1986) Chloride movement in human neutrophils. Diffusion, exchange, and active transport. *J. Gen. Physiol.* **88,** 195–217.

Simchowitz, L. and Roos, A. (1985) Regulation of intracellular pH in human neutrophils. *J. Gen. Physiol.* **85,** 443–470.

Simchowitz, L., Ratzlaff, R., and DeWeer, P. (1986) Anion/anion exchange in human neutrophils. *J. Gen. Physiol.* **88,** 219–236.

Simons, T. J. B. (1986) The role of anion transport in the passive movement of lead across the human red cell membrane. *J. Physiol.* **378,** 287–312.

Solomon, A. K., Chasan, B., Dix, J. A., Lukacovic, M. F., Toon, M. R., and Verkman, A. (1983) The aqueous pore in the red cell membrane: Band 3 as a channel for anions, cations, non-electrolytes, and water. *Ann. N.Y. Acad. Sci.* **414,** 97–124.

Staros, J. V. (1982) *N*-hydroxysulfosuccinimide active esters: Bis(*N*-hydroxysulfosuccinimide esters of two dicarboxylic acids are hydrophilic, membrane-impermeant, protein cross-linkers. *Biochemistry* **21,** 3950–3955.

Staros, J. V. and Kakkad, B. P. (1983) Cross-linking and chymotryptic digestion of the extracytoplasmic domain of the anion exchange channel in intact human erythrocytes. *J. Membr. Biol.* **74,** 247–254.

Steck, T. L. (1972) Cross-linking the major proteins of the isolated erythrocyte membrane. *J. Mol. Biol.* **66,** 295–305.

Steck, T. L., Ramos, B., and Strapazon, E. (1976) Proteolytic dissection of band 3, the predominant transmembrane polypeptide of the human erythrocyte membrane. *Biochemistry* **15,** 1154–1161.

Steck, T. L., Koziarz, J. J., Singh, M. K., Reddy, G., and Kohler, H. (1978) Preparation and analysis of seven major, topographically defined fragments of band 3, the predominant transmembrane polypeptide of the human erythrocyte membrane. *Biochemistry* **17,** 1216–1222.

Stone, D. K., Seldin, D. W., Kokko, J. P., and Jacobson, H. R. (1983) Anion dependence of medullary collecting duct acidification. *J. Clin. Invest.* **71,** 1505–1505.

Teisinger, J., Zemkova, H., and Vyskocil, F. (1984) The effect of anion channel blockers on enzymatic activity of Na^+/K^+-ATPase and the electrogenic Na^+ $/K^+$ pump. *FEBS Lett.* **175,** 275–278.

Thomas, R. C. (1977) The role of bicarbonate, chloride and sodium ions in the regulation of intracellular pH in snail neurones. *J. Physiol. (Lond.)* **273,** 317–338.

Toon, M. R., Dorogi, P. L., Lukacovic, M. F., and Solomon, A. K. (1985) Binding of DTNB to band 3 in the human red cell membrane. *Biochim. Biophys. Acta* **818,** 158–170.

Tsai, I. H., Murphy, S. N. P., and Steck, T. L. (1982) Effect of red cell membrane binding on the catalytic activity of glyceraldehyde-3-phosphate dehydrogenase. *J. Biol. Chem.* **257,** 1438–1442.

Tsuji, T., Irimura, T., and Osawa, T. (1981) The carbohydrate moiety of band 3 glycoprotein of human erythrocyte membranes. *J. Biol. Chem.* **256,** 10497–10502.

Verkman, A. S., Skorecki, K. L., Jung, C. Y., and Ausiello, D. A. (1986) Target molecular weights for red cell band 3 stilbene and mercurial binding sites. *Am. J. Physiol.* **251,** C541–C548.

Walder, J. A., Chatterjee, R., Steck, T. L., Low, P. S., Musso, G. F., Kaiser, E. T., Rogers, P. H., and Arnone, A. (1984) The interaction of hemoglobin with the cytoplasmic domain of band 3 of the human erythrocyte membrane. *J. Biol. Chem.* **259,** 10238–10246.

Waxman, L. (1979) The phosphorylation of the major proteins of the human erythrocyte membrane. *Arch. Biochem. Biophys.* **195,** 300–314.

Wieth, J. O. and Bjerrum, P. J. (1982) Titration of transport and modifier sites in the red cell anion transport system. *J. Gen. Physiol.* **79,** 253–282.

Wieth, J. O. and Bjerrum, P. J. (1983) Transport and modifier sites in capnophorin, the anion transport protein of the erythrocyte membrane, in *Structure and Function of Membrane Proteins* (Quagliariello, E. and Palmieri, F., eds.), Elsevier Science Pub., Amsterdam, pp. 95–106.

Wieth, J. O., Andersen, O. S., Brahm, J., Bjerrum, P. J. and Borders, C. L. (1982) Chloride-bicarbonate exchange in red blood cells: Physiology of transport and chemical modification of binding sites. *Phil. Trans. R. Soc. Lond. B* **299,** 383–399.

Williams, D. G., Jenkins, R. E., and Tanner, M. J. A. (1979) Structure of the anion-transport protein of the human erythrocyte membrane. Further studies on the fragments produced by proteolytic digestion. *Biochem. J.* **181,** 477–493.

Wolpaw, E. W. and Martin, D. L. (1986) A membrane protein in LRM55 glial cells cross-reacts with antibody to the anion exchange carrier of human erythrocytes. *Neurosci. Lett.* **67,** 42–47.

Yoon, S. C., Toon, M. R., and Solomon, A. K. (1984) Relation between red cell anion exchange and water transport. *Biochim. Biophys. Acta* **778,** 385–389.

Zoccoli, M. A., Hoopes, R. R., and Karnovsky, M. L. (1982) Rat liver microsomal glucose-6-P translocase. *J. Biol. Chem.* **257,** 11296–11300.

Kinetics of Anion Transport

Philip A. Knauf

1. Introduction

This chapter will deal primarily with the question of what we can learn about the red-cell anion transport system from kinetic studies, that is, measurements of ion fluxes under different conditions and in the presence of various inhibitors. It is important to recognize, however, that one of the chief advantages of the red-cell anion transport system has been the availability of structural data on the transport protein as well as kinetic data. Largely for reasons of brevity, this chapter will focus on only a few subjects that, in my opinion, will be important in linking the kinetic information with structural studies (*see* Chapter 8) to develop a molecular model for the transport process. For a broader overview of the kinetics of transport, the reader is referred to several comprehensive reviews (Macara and Cantley, 1983; Knauf, 1979, 1986; Passow, 1986; Fröhlich and Gunn, 1986).

1.1. Functions of Anion Exchange

The primary function of the anion exchange system in the red blood cell is to enhance the ability of the blood to carry CO_2. In the tissues, CO_2 produced by metabolism rapidly diffuses into the red cells, where it is hydrated by the aid of carbonic anhydrase to produce H^+ and HCO_3^-. The protons are buffered by hemoglobin and the other product of the reaction, HCO_3^-, leaves the cell in exchange for plasma Cl^-. The net result is that CO_2 can be stored in the form of plasma HCO_3^-, which roughly doubles the CO_2-carrying capacity of the blood (Gunn, 1979; Wieth and Brahm, 1985). In the lungs, the process is reversed, and plasma HCO_3^- enters the cells to be converted to CO_2 and exhaled. Because the anion exchange system is intimately involved in the removal of CO_2, which may be thought of as the "smoke" resulting from the fires of cellular metabolism, Wieth and Bjerrum (1983) suggested that an appropriate name for the 95,000-dalton band 3 protein that catalyzes the transport would be capnophorin, from the Greek roots for "smoke" and "carrier." To equilibrate HCO_3^- during the less than 1 s that the red cell may spend in a capillary, the cell is highly specialized for this transport function, such that capnophorin accounts for approximately 30% of the total red-cell membrane protein (Fairbanks et al., 1971). This specialization has made the red cell an excellent system in which to relate the kinetics of the transport system to the structural biochemistry of the transport protein, discussed in Chapters 8 and 15.

Although CO_2 transport is the primary function of the red-cell anion transporter, it also has the effect of bringing the intracellular pH into equilibrium with the Cl^- gradient. Because the anion exchange system involves no external energy input, and because it exchanges one anion inside for one anion outside, it will bring Cl^- and HCO_3^- to equilibrium such that $\tilde{\mu}_{Cl} = \tilde{\mu}_{HCO_3^-}$. At equilibrium, therefore,

$$[HCO_3^-]_i/[HCO_3^-]_o = [Cl^-]_i/[Cl^-]_o \tag{1}$$

where i and o refer to the concentrations inside and outside the cell, respectively. Since protons are in equilibrium with HCO_3^- through the reactions of CO_2 hydration and carbonic acid dissociation, if CO_2 is constant inside and outside the cell:

$$[H^+]_i[HCO_3^-]_i = [H^+]_o[HCO_3^-]_o \tag{2}$$

From equations (1) and (2):

$$[H^+]_o/[H^+]_i = [Cl^-]_i/[Cl^-]_o \tag{3}$$

Thus, even if the membrane is relatively impermeable to H^+ and OH^-, in the presence of CO_2, the anion exchange system will effectively equilibrate protons with Cl^- ions.

In the red blood cell, the permeability for net Cl^- flow (although smaller than that for exchange, *see below*) is sufficiently large relative to the permeabilities for other ions, so that Cl^- is virtually at equilibrium with the membrane potential, about -9 mV (Hoffman and Laris, 1974). Thus, the anion exchanger simply brings protons to electrochemical equilibrium with the membrane potential. In cells where Cl^- is not at equilibrium, however, the anion exchanger may play an important role in intracellular pH regulation (Simchowitz and Roos, 1985; Jans et al., 1987; Restrepo et al., 1988). Such a transport system may also be

Table 1A
Dependence of Cl^- Efflux on External Cl^- at 37–38°C

Cl_o (mM)	$t_{1/2}$ (s)	J (ions/b3-s)	% Inhibition
150	0.052[a]	48,400[b]	0
0[c]	175.9	14.3	99.97[d]

[a]pH 7.2 at 38°C, Brahm, 1977.

[b]Calculated from Cl^- flux, J, assuming 1 x 10^6 band 3 (b3) monomers/red blood cell.

[c]150mM Cl_o was replaced by 25 mM K_3-citrate and 200 mM sucrose; Si-qiong Liu, unpublished data.

[d]Inhibition relative to flux with 150 mM Cl_o

important for epithelia whose function involves transport of either protons or bicarbonate. Indeed, such epithelia have been found to have analogous transport systems, some of which are structurally related to the red-cell anion transport protein (*see* Chapter 8).

In addition to the functions of CO_2 transport and H^+ transport, the anion exchange system is also capable of transporting a wide variety of other anions, including such diverse anions as superoxide, phosphate, sulfate, and cAMP. In doing so, it may play a role in other phenomena involving transport of these substances, as well as in the transport of anionic drugs.

1.2. Basic Transport Characteristics

As described in detail in Chapter 8, capnophorin probably forms a transmembrane structure that spans the lipid bilayer and that superficially resembles the structure proposed for channel-forming proteins and peptides. Despite this, two kinds of evidence show that the vast majority of Cl^- transport does not involve free diffusion through this "channel." First, the efflux of radioactive $^{36}Cl^-$ is strongly dependent on the presence of a suitable exchange partner, such as Cl^-, HCO_3^-, or another transportable anion, at the outside of the cell (Table 1A). When there is no exchangeable anion in the external medium, the efflux is inhibited by more than 99.97%, and even the small efflux remaining may be the result of contaminating exchangeable anions (e.g., HCO_3^-) in the medium. Second, if one measures net flow of Cl^- driven by a membrane potential (set up by use of an ionophore such as valinomycin or gramicidin), this net flow (Table 1B) is less than 1/10,000th of the rate of Cl^- exchange (Hunter, 1971; Knauf et al., 1977). Thus, it is a gross misnomer to refer to capnophorin as the "red-cell anion channel," since in fact the transport process involves a very tightly coupled one-for-one exchange of anions, completely unlike diffusion through a channel.

Table 1B
Comparison of Net and Exchange Cl^- Fluxes at 37–38°C

Type of flux	P_{Cl}[a] (cm/s) x 10^8	% of exchange
^{36}Cl efflux	50,000[b]	100
Net Cl efflux with valinomycin	2.8[c]	0.0056

[a]Calculated from flux data, using Goldman (1943) equation, assuming free diffusion of ions.
[b]pH 7.2 at 38°C, Brahm, 1977.
[c]pH 7.1 at 37°C, Knauf et al., 1977.

2. Models for Tightly Coupled Anion Exchange

2.1. Comparison of Ping-Pong and Simultaneous Models

Two kinds of models could account for the tight one-for-one coupling of anion efflux and influx. In the ping-pong (Cleland, 1963) model (Fig. 1A), there are two different conformations of capnophorin, one with the transport site facing in (E_i) and another with it facing out (E_o). According to this model, one of the products (e.g., Cl_i^-) is released before the second substrate (e.g., $HCO_3{}^-{}_i$) is bound. That is, only one anion is transported across the membrane at a time. The tight coupling of efflux and influx is explained by proposing that the change from E_i to E_o or vice versa can only occur if a suitable substrate (such as Cl^-) is bound to the transport site, to form the corresponding ECl_i or ECl_o form. Thus, capnophorin would transport one anion at a time alternately out or in, just as in a game of ping-pong the ball moves alternately in opposite directions. According to the simultaneous model (Fig. 1B), on the other hand, anions are bound at both internal and external sites, after which the two anions are translocated simultaneously. Thus, the two substrates (e.g., Cl_o^- and HCO_{3i}^-) are bound before either of the products (Cl_o^- and $HCO_3{}^-{}_i$) are released. This mechanism would provide a direct mechanical coupling of anion flows in opposite directions.

2.2. Evidence for the Ping-Pong Model

2.2.1. Single Turnover Experiment

Perhaps the most direct evidence for the ping-pong model is the demonstration by Jennings (1982) that, if red-cell ghosts are loaded with a very low Cl^- concentration (about 70 μM) and then suspended at 0°C in

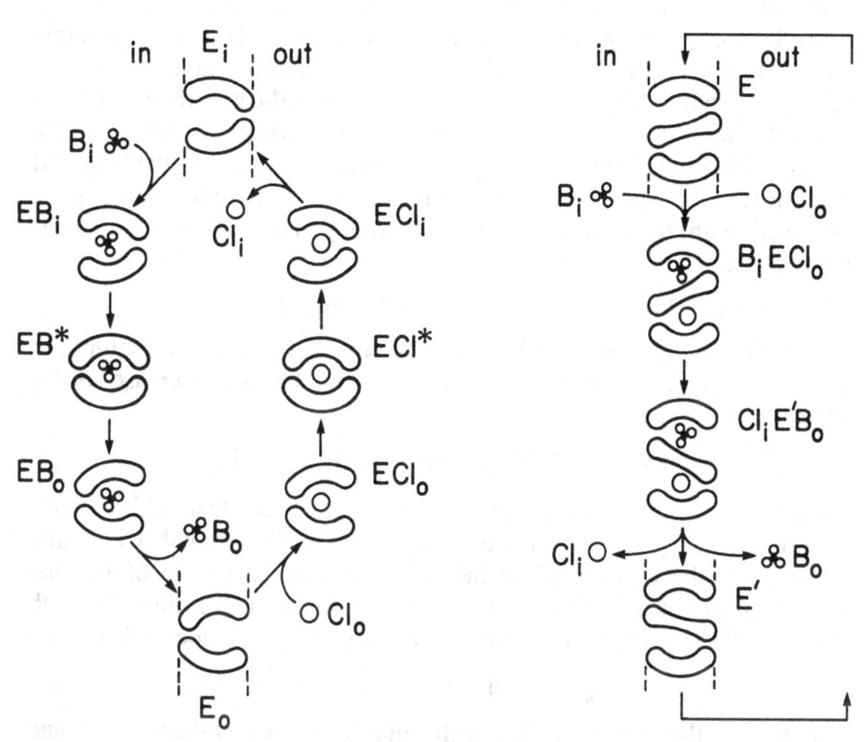

Fig. 1. Comparison of the ping-pong and simultaneous models for Cl^--HCO_3^- exchange. In each case, the steps required for exchange of a bicarbonate (B) inside for a chloride outside are shown, with the reactions proceeding in the direction of the arrows. All reactions are, in fact, reversible. A. Ping-pong model. The bicarbonate and chloride cross the membrane in *separate* steps, in each case proceeding through a transition state (EB* or ECl*) in which the anion is probably occluded, that is, not accessible from either side of the membrane. The model is shown as a lock-carrier, where the anion itself does not move during the I/O transition, but rather access to the two solutions changes because of the opening and closing of gates on either side of the anion binding site. This model is kinetically equivalent to a diffusible carrier model, such as depicted in Fig. 2 (Patlak, 1957). B. Simultaneous model. In this case, the movement of bicarbonate out and chloride in takes place in the *same* step, after both anions are bound. After the anion exchange, the system is in a new state, E′, with the accessibility of the binding sites reversed. It returns to the original state (E) by a series of steps that are the reverse of those involved in the E to E′ conversion.

a medium with a very slowly transported anion, sulfate, there is a rapid efflux of Cl^- ions, which corresponds to the transport of a single Cl^- ion by each capnophorin molecule from inside to outside. The reason that this can be observed is that the number of capnophorin molecules is so large (about 10^6/cell or 20 μmol/L cells) that a significant fraction of the intracellular Cl^- is transported by this process. This demonstration of the "ping without the pong" strongly supports the idea that the outward translocation of anions takes place independently of (rather than simultaneously with) the inward translocation step, in agreement with the ping-pong model.

2.2.2. Dependence of Cl^- Flux on Cl_i and Cl_o

A ping-pong model predicts that the concentration of Cl_o required to give half-maximal flux ($K1/2_o$) with constant Cl_i will increase as Cl_i increases:

$$K1/2_o = K1/2_{o,max}/[1 + (K1/2_{i,max}/Cl_i)] \quad (4)$$

[where $K1/2_{o,max}$ is the concentration of external Cl^- that half-saturates the transport system with saturating (maximal) internal Cl^-, and $K1/2_{i,max}$ is the corresponding half-saturation concentration of internal Cl^- with saturating external Cl^-]. The maximum flux at high Cl_o with constant Cl_i (V_{m_o}) has the same dependence on Cl_i as does $K1/2_o$:

$$V_{m_o} = V_m/[1 + (K1/2_{i,max}/Cl_i)] \quad (5)$$

where V_m is the maximum flux with saturating concentrations of Cl_i and Cl_o. From the above, it is clear that the ratio $K1/2_o/V_{m_o}$ will be constant as a function of Cl_i for a ping-pong system. For simultaneous transport systems, $K1/2_o$ may also be a function of Cl_i, but the functional dependencies of V_{m_o} and $K1/2_o$ are different, so their ratio is not constant as a function of Cl_i (Gunn and Fröhlich, 1982). The arguments above also apply to the case where Cl_o is kept constant and Cl_i is varied to determine $K1/2_i$ and V_{m_i}.

The pioneering work of Gunn and Fröhlich (1979, 1980, 1982) clearly demonstrated that $K1/2_o$ and V_{m_o} are functions of Cl_i. The ratio $K1/2_o/V_{m_o}$ is nearly constant, but there is considerable scatter in the data for very low Cl_i, and a tendency for the ratio to increase at low Cl_i. Jennings (1982) has shown that $K1/2_i$ becomes very small if Cl_o is low, as predicted. A study of Cl^- half-saturation at each side of the membrane, using resealed red-cell ghosts (Hautmann and Schnell, 1985), indicated that the slope of a Lineweaver-Burk plot of 1/flux vs $1/Cl_o$ (which is equivalent to $K1/2_o/V_{m_o}$) increases from 14.1 at 103 mM Cl_i to 33.2 at 1.3

mM Cl_i. Similarly, the slope of a plot of 1/flux vs 1/Cl_i ($K1/2_i/V_{m_i}$) increases from 44.8 at 100 mM Cl_o to 64.4 at 1.2 mM Cl_o. In this study, however, as in the earlier work (Cass and Dalmark, 1973; Gunn et al., 1973; Dalmark, 1975; Brazy and Gunn, 1976), no indication of sigmoidicity was found when Cl_i and Cl_o were varied, keeping $Cl_i = Cl_o$, as would be expected for the simultaneous model where two Cl^- ions must bind to band 3 before transport can occur. Also, the Cl^- concentration which half-saturates the system with $Cl_i = Cl_o$, K1/2, is approximately equal to the sum of $K1/2_{o,max}$ and $K1/2_{i,max}$, as predicted by the ping-pong model (Fröhlich and Gunn, 1986). Thus, the results in general support the ping-pong model, but there are some discrepancies that require further study. At present, it seems likely that these may result from experimental problems at low Cl^- concentrations. Among these is the possibility that the ionic strength may affect Cl^- binding. Even though ionic strength was maintained by nonpenetrating ions, such as citrate, there is some evidence that citrate may not be an effective counterion in the restricted area near the transport sites (Wieth and Bjerrum, 1982; Wieth et al., 1982a).

2.2.3. Transmembrane Effects

A fundamental prediction of the ping-pong model is that substrates or transport-site inhibitors at one side of the membrane will have an effect on interactions of inhibitors or substrates with the transport site at the opposite side of the membrane. This is expected because the transport site can be accessible to one side of the membrane or the other, but not both, so recruitment of sites to one side necessarily affects the number of available sites at the other side. In contrast, for a simple simultaneous model with one site at each side of the membrane, interactions with the transport site at one side will take place independently of whether or not substrates or inhibitors are bound at the other side.

One of the most direct demonstrations of such an effect, using the substrate Cl^- itself, involves use of nuclear magnetic resonance (NMR) of ^{35}Cl to measure Cl^- binding to the transport site. Binding of Cl^- to the transport site causes a broadening of the line-width of the NMR signal that is proportional to the fraction of total Cl^- ions in the system that are bound (Shami et al., 1977; Falke et al., 1984a). For a saturable site, the fraction bound decreases with increasing Cl^- concentration. In this way, Falke et al. (1984a) have measured a Cl^- binding component that, on the basis of affinity and competition with inhibitors, represents Cl^- binding to the transport site. By using appropriate systems, it is possible to measure either total Cl^- binding to both E_i and E_o in leaky ghosts or to E_o

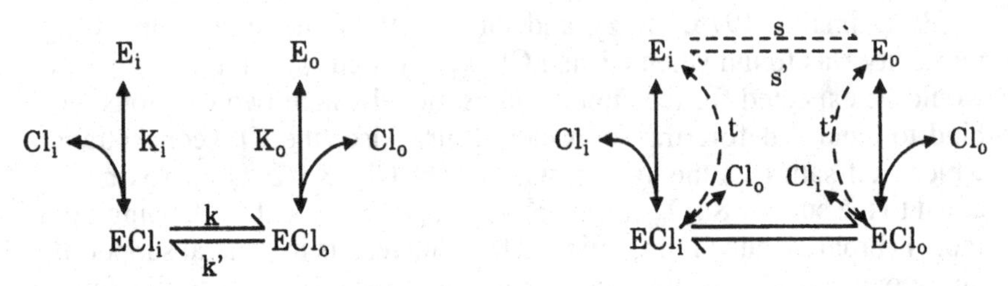

Fig. 2. Kinetic description of the ping-pong model. A. The simple ping-pong model for Cl^- exchange. Although depicted as a diffusible carrier for convenience, this model is kinetically equivalent to the lock-carrier model shown in Fig. 1A (Patlak, 1957). K_i and K_o are the dissociation constants for Cl^- binding to E_i and E_o, and k and k′ are the rate-constants for the I/O and O/I transitions, respectively. B. Ping-pong model with slippage and tunneling. The rates of I/O and O/I conversion (slippage) of the unloaded transport sites are designated s and s′. The rate constant t represents the conversion of ECl_i to E_i and Cl_o, which involves a tunneling process in which the Cl^- ion bound to E_i crosses the diffusion barrier (*see* Fig. 1A) to reach the external solution, resulting in net transport of a Cl^- from inside to outside. Net efflux of Cl^- can also occur if an inside Cl^- crosses the diffusion barrier to bind to E_o (t′). Net influx could occur by the reverse of these processes.

alone (e.g., in intact cells). Falke et al. (1984b) showed that the external inhibitors DNDS (4,4′-dinitro-stilbene-2,2′-disulfonate) or pNBS (*p*-nitrobenzenesulfonate), which bind only to E_o, are able to prevent binding of Cl^- to E_i, presumably by recruiting all of the band 3 molecules to the E_o conformation with inhibitor bound, thereby depleting E_i, exactly as predicted by the ping-pong model.

Similar demonstrations have made use of reversible competitive inhibitors, such as DNDS and H_2DIDS (4,4′-diisothiocyano-1,2-diphenylethane-2,2′-disulfonate) (at 0°C where the covalent reaction of H_2DIDS is slow). Fröhlich (1982) found that the binding affinity for DNDS is affected by variations in Cl_i and Cl_o, or Cl_o alone, as predicted by the ping-pong model. Furuya et al. (1984) examined the effects of changes in Cl_i on the ability of H_2DIDS to inhibit Cl^- exchange. If Cl_i is increased, this will tend to push the reaction cycle in Fig. 2 toward ECl_i, which then is translocated to ECl_o, which dissociates to E_o. E_o can only return to the inward-facing conformation after binding an external Cl^-

Thus, the balance of E_i and E_o is affected by the Cl^- ratio, such that E_o/E_i increases if $Cl_i > Cl_o$. If E_o increases, by the law of mass action, this will favor the formation of the complex of E_o with H_2DIDS, and will therefore increase the apparent affinity for H_2DIDS. In fact, the inhibitory potency of *external* H_2DIDS does increase dramatically with increasing *internal* Cl^- concentration, providing evidence for a transmembrane effect as predicted.

Several experiments have demonstrated that one reagent that reacts covalently near the transport site on one side of the membrane can prevent reaction of another transport site label at the opposite side (*see*, e.g., Knauf, 1986; Passow, 1986). Perhaps the clearest example of this is the demonstration by Grinstein et al. (1979) that, if cells are first reacted with external DIDS (4,4′-diisothiocyano-stilbene-2,2′-disulfonate), the subsequent labeling of band 3 from the inside surface with NAP-taurine [*N*-(4-azido-2-nitrophenyl)-2-aminoethylsulfonate] (which is a competitive inhibitor from the inside, Knauf et al., 1978) is inhibited. Labeling of proteins other than band 3 is also affected, however, indicating allosteric effects of DIDS that complicate the interpretation of the results.

2.2.4. Evidence against the Ping-Pong Model

Although most evidence with monovalent anion substrates favors the ping-pong model, some observations, particularly with divalent anions as substrates, are difficult to reconcile with it. There is some evidence for sigmoidicity in the saturation kinetics for divalent anions, such as phosphate and sulfate (Schnell et al., 1981; Schnell and Besl, 1984). This suggests that more than one anion binds to capnophorin at a time, contrary to the simple ping-pong model. Salhany and Rauenbuehler (1983) have found that, for sulfate–dithionite exchange, $K1/2_o$ and V_{m_o} for dithionite have different dependencies on the internal sulfate concentration. This led them to construct a hybrid of the ping-pong and simultaneous models, in which only one anion crosses the membrane at a time, but where binding of a second anion occurs before release of the first anion. This model is very complex and, hence, not very useful as a predictive model. Also, Falke and Chan (1985) found that the dissociation constant for Cl^- release from ECl_o was not dependent on the external Cl^- concentration, contrary to the expected result if Salhany and Rauenbuehler's model were applicable to chloride exchange. Thus, it seems so far that the ping-pong model is sufficient to explain monovalent anion transport, although there may be some complexities with divalent anions that require a more involved model.

2.2.5. Extensions of the Ping-Pong Model: Modifier Sites

One of the greatest advantages of the ping-pong model is its simplicity and predictive power. With only two anion-binding conformations (E_i and E_o) and four constants (Fig. 2), it can explain most observations of Cl^- flux as a function of Cl_i and Cl_o, and can even explain such apparently complex phenomena as the different dependencies of flux on Cl^- when $Cl_i = Cl_o$ or when only Cl_o is varied, as well as the effects of Cl^- gradients on the potency of competitive inhibitors, such as H_2DIDS and DNDS (*see* sections 2.2.2. and 2.2.3).

Unfortunately, there are other phenomena that require additions to the ping-pong model. From the earliest work (Cass and Dalmark, 1973; Dalmark, 1975, 1976), it was recognized that Cl^- flux decreases at high Cl^- concentrations. This was explained by postulating a separate "modifier" site at which Cl^- noncompetitively inhibits Cl^- exchange (self-inhibition), with a dissociation constant of about 335 m*M*. Chloride binding to a site with this affinity is not seen by NMR (Falke et al., 1984a), but there is some binding that is not saturable at $[Cl^-] < 500$ m*M* and that might be responsible for the inhibition. Tanford (1985) has suggested that self-inhibition might be explained if there are low-affinity binding sites for Cl^- in series with the transport site. When these sites are occupied by Cl^-, release of Cl^- from the transport site is inhibited. Structural data suggests that there are positive charges near the entrances of the "channel" structure formed by capnophorin (Jay and Cantley, 1986; Chapter 8) that might correspond to such "approach" sites. NMR evidence, however, indicates that the association-dissociation of Cl^- with E_o or E_i is *not* rate-limiting for transport (Falke et al., 1985), at least at 0°C, which would argue against this model.

Recent evidence (Knauf and Mann, 1986) suggests that the principal Cl^- self-inhibitory effect is the result of binding to *internal* sites. In addition to Cl^-, halides such as Br^- and I^- inhibit anion exchange by binding to internal sites (Gunn and Fröhlich, 1979). Internal Ca also inhibits transport at μ*M* concentrations, whereas Mg stimulates transport (Low, 1978, Gunn et al., 1979). On the external surface are sites where I^- and Br^- (but not Cl^-, except perhaps at low pH, Milanick and Gunn, 1986) inhibit transport. There is also a site with a low Cl^- affinity where the inhibitors NAP-taurine and NIP-taurine [*N*-(4-isothiocyano-2-nitrophenyl)-2-aminoethylsulfonate] inhibit (Knauf et al., 1978; 1987a), which may be identical to the external site at which I^- inhibits Cl^- exchange (Knauf and Spinelli, 1988). These modifier sites add unwelcome complexity to the ping-pong model, but further investigation may reveal that such effects are intimately connected with and indicative of the molecular structure of the anion exchange protein.

3. Nature of the Transporting Conformational Change

3.1. Kinds of Conformational Change

In considering how capnophorin mediates anion exchange, two kinds of protein conformational change are important. Perhaps most central is the question of how the protein changes from the ECl_i conformation, in which the transport site faces inward, to the ECl_o form, thereby transporting a Cl^- ion from inside to outside, the I/O (or the reverse O/I) change. A related question concerns the change in capnophorin structure caused by binding of a substrate anion such as Cl^-, that is, the changes from E_i to ECl_i and from E_o to ECl_o. It is clear that these changes are also substantial and important, since the I/O conformational change *cannot* occur (or is highly disfavored, *see* section 6) unless an appropriate substrate anion is bound. The question here is: what does the substrate anion do to reduce the activation-free energy for the in-to-out conformational change?

3.2. Conformational Distribution and Asymmetry

Before one can obtain information about the differences between conformations, it is necessary to know what fraction of the capnophorin molecules is in each conformation, and to have ways of altering this distribution. The first question one can ask is whether or not the system is symmetric, that is, are there equal numbers of transport sites facing each side of the membrane, or are most of the sites normally at one side, with few at the other. The answer to this question has implications for the energy differences between the various conformations, as well as for experiments designed to probe the structural differences among the various conformations of capnophorin.

3.2.1. Asymmetry of Unloaded Transport Sites

For a ping-pong model, the ratio of E_o to E_i is given by (Knauf et al., 1984):

$$E_o/E_i = (Cl_i/Cl_o)(kK_o/k'K_i) \quad (6)$$

where k and k′ are the rate-constants for the in-to-out or out-to-in conformational change, and K_o and K_i are the Cl^- dissociation constants of E_o and E_i, respectively, as shown in Fig. 2. This assumes that the association–dissociation reactions are fast relative to k and k′, for which there is NMR evidence, at least at 0°C (Falke et al., 1985). The intrinsic asymmetry of the unloaded sites, as distinct from any extrinsic asymmetry imposed by the Cl^- gradient across the membrane, may be defined as the ratio of E_o/E_i with $Cl_i = Cl_o$. This asymmetry factor has been named *A* and from equation (6) is:

$$A = kK_o/k'K_i \quad (7)$$

A can be determined in a number of ways, for example, by measuring the dependence of the Cl^- exchange flux on Cl_i and Cl_o, as described in section 2.2.2. From the values of $K1/2_{o,max}$ and $K1/2_{i,max}$, A is given simply as $K1/2_{o,max}/K1/2_{i,max}$ (Fröhlich et al., 1983; Knauf et al., 1984). Values of *A* by this method range from .060–.324 (Gunn and Fröhlich, 1979; Hautmann and Schnell, 1985; Knauf and Brahm, 1988). That is, 1/A, which represents the ratio E_i/E_o, ranges from 3.1–16.7. Much of the variation is the result of differences in the values of $K1/2_{i,max}$ (often calculated from the difference between K1/2, the Cl^- concentration that half-saturates with $Cl_i = Cl_o$, and $K1/2_{o,max}$). The best estimate for 1/A is probably about 15 (Gunn and Fröhlich, 1979), since this is based on a flux K1/2 value (Brazy and Gunn, 1976; Dalmark, 1976) that best agrees with independent NMR measurements (Falke et al., 1984a; Falke and Chan 1985, 1986a,b,c). In spite of the experimental difficulties in measuring A, all data agree that the system is asymmetric at 0°C, such that $E_i > E_o$. The asymmetry may be even larger at 38°C, with 1/A almost 30 (Knauf and Brahm, 1986).

3.2.2. Asymmetry of Substrate-Loaded Transport Sites

By analogy with A, it is possible to define a second asymmetry factor, L, which is equal to ECl_o/ECl_i. If Cl^- is at equilibrium, influx must equal efflux. Since influx equals $k'ECl_o$ and efflux equals $kECl_i$, it follows that:

$$L = k/k' \quad (8)$$

Note that ECl_o/ECl_i is not affected by the Cl^- gradient across the membrane, unlike the ratio E_o/E_i [Eq. (6)]. One way of determining L is to measure the Cl^- dissociation constants for E_o and E_i, K_o and K_i. If these are known, L is given simply as $A(K_i/K_o)$.

Unfortunately, no practical method is yet known for measuring K_i and K_o for a rapidly transported substrate such as Cl^-. The reason for this is that Cl^- at one side of the membrane rapidly equilibrates with *all* forms of capnophorin, so the rate-constants k and k' strongly affect the apparent dissociation constants (Knauf et al., 1984; Passow, 1986; Fröhlich and Gunn, 1986). For substrates that are transported more slowly, however, this is possible. For example, if one measures exchange of Cl^- on the inside for a slowly transported anion (X^-) outside, nearly all of the capnophorin molecules will be recruited to the outside (Jennings and

Adams, 1981), and the concentration of the slow anion that gives half-maximum flux is equal to K_o for that anion (Passow, 1986; Fröhlich and Gunn, 1986). Another method, which does not require total recruitment [with the associated problems of maintaining zero (Cl^-) at one side of the membrane], is to measure the inhibitory effect of the slow anion at one side (e.g., the outside) of the membrane on Cl^- exchange. If this is done at two (or more) different external Cl^- concentrations, and the data are plotted as 1/flux vs X_o (Dixon plot), the lines for the different Cl_o values intersect at a point whose x-value is $-K_o(X)$. For this method to work, all of the inhibition by X_o must be competitive, there must be no appreciable modifier site effects, and the flux of X^- must be small (<10%) compared to that of Cl^-. Measurements of this sort for iodide at 0°C indicate a value of $K_o(I)$ of 2.5–3 m*M* (Milanick and Gunn, 1986; Knauf et al., 1986a,b), while $K_i(I)$ is much larger, around 22 m*M* (Knauf et al., 1986a,b). This implies that the structure of the transport site is different for the E_i and E_o conformations, such that iodide binds more favorably to the E_o form. Thus, for iodide the asymmetry in E_o and E_i is primarily the result of differences in $K_o(I)$ and $K_i(I)$, and the iodide-loaded forms are much more symmetrically distributed, with $L(I) = 0.5$. The ratio K_i/K_o will depend on the nature of the substrate anion, however, so the result for iodide does not apply to other anions, such as Cl^-.

4. Noncompetitive Inhibitors as Probes of Conformation

By definition, competitive inhibitors will bind only to the unloaded forms of capnophorin, E_i and E_o, and nonpenetrating competitive inhibitors will bind only to the form with the transport side facing the side where the inhibitor is present. This feature has, in fact, been used to demonstrate recruiting effects as predicted by the ping-pong model (*see* section 2.2.3.) as well as to measure asymmetry. For noncompetitive inhibitors, the situation is more complex, but potentially rich in information. In theory, a noncompetitive inhibitor, designated N, can bind to all forms of capnophorin, but with different affinities defined by the dissociation constants K_e, K_f, K_g, and K_h (Fig. 3). The inhibitor can thus be used to report changes in conformation upon binding of substrate or I/O conformational changes. Also, such inhibitors will tend to recruit capnophorin to the form with the highest affinity for inhibitor.

Evidence that different capnophorin conformations have different affinities for certain inhibitors was first obtained by experiments in which the Cl^- concentration was varied on one side of the membrane, with the inhibitor at the opposite side. Transmembrane effects on inhibitor poten-

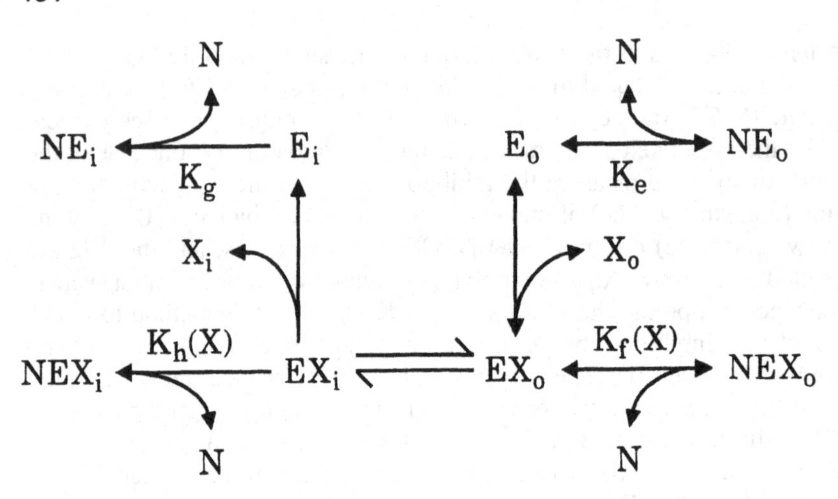

Fig. 3. Interaction of a noncompetitive inhibitor (N) with the various forms of capnophorin. X_i and X_o represent substrate anions, such as Cl^- or I^-. K_e and K_g are the dissociation constants for binding of N to E_o and E_i, respectively, while $K_f(X)$ and $K_h(X)$ are the dissociation constants for binding of N to the substrate-loaded forms, EX_o and EX_i.

cy were interpreted as meaning that the inhibitor has a preference for some forms of capnophorin over others (Knauf and Mann, 1984; Knauf et al., 1984). The actual dissociation constants for N binding to E_o and E_i can be determined by experiments similar to those used to measure the iodide affinity of E_o and E_i (section 3.2.2.; Fröhlich and Gunn, 1986; Knauf et al., 1987a). The affinity of N for EX_o and EX_i (with a slowly transported anion bound) can be measured from Dixon plots (1/flux vs N_o) with different concentrations of X_o, for example. The lines intersect at a point whose x-value is $-K_f(X)$, the dissociation constant for N binding to EX_o (Knauf et al., 1986b; 1987b,c). A similar technique yields $K_h(X)$ for binding to EX_i. Effects of binding of rapidly transported anions cannot as yet be quantitated precisely.

4.1. Effects of Transport Site Orientation

For several noncompetitive inhibitors, it has been found that the affinity for the E_o form is greater than that for the E_i form. For example, NIP-taurine (and probably NAP-taurine as well), which binds to an external inhibitory site, prefers E_o by a factor of about 4 relative to E_i (Knauf et al., 1987a). For niflumic acid (NA), the preference is about 3.5-fold (Knauf et al., 1987b), and for flufenamic acid (FA) about 3.9-

fold (Knauf et al., 1987c). Eosin-5-maleimide (EM) also seems to exhibit a preference for the E_o conformation (Si-qiong Liu, personal communication). These results show that the I/O conformational change affects regions of capnophorin in addition to the transport site itself, including those responsible for tight binding of certain noncompetitive inhibitors.

The similarity in the extent of preference for E_o relative to E_i might suggest that similar interactions between capnophorin and the inhibitors are involved in this selectivity. Although FA, NA, and EM compete with each other, NIP-taurine does not prevent binding of NA (Knauf and Mann, unpublished data) or FA (Knauf and Severski, unpublished data), making this explanation unlikely. Also, phloretin, which does compete with FA (Soland and Knauf, unpublished data), shows little or no conformational preference (Fröhlich et al., 1988). All of these inhibitors compete with external H_2DIDS and DNDS, for which (as nonpenetrating competitive inhibitors) the preference for E_o relative to E_i is absolute. This reemphasizes that inhibitors that compete with each other may have very different specific interactions with capnophorin, only overlapping to a slight degree (or else interacting allosterically) (Passow, 1986).

The existence of conformational preferences for E_o or E_i complicates the kinetics of substrate–inhibitor and inhibitor–inhibitor interactions. If an inhibitor prefers E_o, increases in external substrate (which decrease E_o) will decrease inhibitory potency, an effect that may be confused with competitive inhibition. Increases in internal substrate will increase inhibitory potency (Knauf et al., 1984; Knauf and Mann, 1984; Fröhlich and Gunn, 1986). Also, if two inhibitors prefer different conformations, they will seem to "compete" with each other, even though no direct competition exists.

4.2. Effects of Substrate Binding

Binding of iodide to E_o increases the dissociation constant for flufenamic acid or niflumic acid binding by a factor of 1.6–1.7 (Knauf et al., 1987b,c); there is little or no effect on binding of NIP-taurine (Spinelli and Knauf, unpublished data). Binding of sulfate increases the NA or FA dissociation constant by more than sevenfold, suggesting that larger and/or more highly charged ions have a larger effect.

Surprisingly, binding of iodide to E_i has a stronger effect on NA or FA affinity than does I^- binding to E_o, despite the fact that the binding site for these inhibitors is very likely located at the outside (Brahm and Knauf, unpublished data; Cousin and Motais, 1979). The dissociation constant is increased by about 2.4-fold for NA and fivefold for FA, indicating a strong transmembrane effect of internal iodide binding on the binding of external NA or FA (Knauf et al., 1987b,c). This result demonstrates that there is a very substantial conformational change in capnophorin caused by substrate binding to E_i, sufficient to affect binding of a noncompetitive inhibitor to an external site markedly.

Further evidence suggestive of protein conformational changes upon substrate binding comes from Passow and co-workers (Passow, 1986; Passow et al., 1980), who showed that dinitrophenylation of a particular lysine residue in band 3 (lys a) by fluorodinitrobenzene is strongly affected by the nature of the substrate anion present. Cl^- enhances reactivity, whereas sulfate decreases it. The effects of Cl^- on the reactive lysine, confirmed by Falke and Chan (1986c), are opposite to those seen for reaction of phenylglyoxal (PG) with an arginine residue near the transport site (Passow, 1986). Changes in capnophorin conformation upon substrate binding have also been reported using an EPR probe (Ginsburg et al., 1981).

4.3. Asymmetry of Chloride-Loaded Sites

If the dissociation constants for binding of a noncompetitive inhibitor such as FA to all forms of capnophorin are known, it is possible to predict the inhibitory potency (measured as concentration for half-inhibition, the IC50) for any given substrate concentrations in and out and for any values of A and L. If one further assumes that binding of Cl^- to E_i or E_o has effects on FA binding intermediate between no effect and the effect produced by the larger substrate, iodide, then predictions can be made for the FA IC50 as a function of Cl^- concentration, keeping Cl_i = Cl_o. For $1/A = 15$, the predictions fit well only if L is nearly equal to A and if Cl^- has nearly as large an effect on FA affinity as does iodide. If the effect of Cl^- is the same as that of iodide, a minimum value for $1/L$ ($= ECl_i/ECl_o$) of 7 is obtained (Knauf and Spinelli, 1987). Thus, it seems likely that for Cl^-, K_i and K_o are similar, A and L are similar, and most of the asymmetry is the result of differences between k and k′, suggesting that ECl_i has a lower free energy (is more stable) than ECl_o by about 6.2 kj/mol. This is a very small energy difference, less than that (about 10 kj/mol) for formation of a single hydrogen bond in a nonpolar environment (Guidotti, 1986). The situation with Cl^- contrasts with that for iodide as substrate, where $K_i > K_o$, k and k′ are similar, and EI_i and EI_o have similar free energies (differing by about 1.6 kj/mol). Thus, the proportion of inward-facing vs outward-facing transport sites may change, depending on the substrate present.

5. Transport Mechanism: Thin Barrier Vs Zipper

Two kinds of models have been proposed to explain the actual anion translocation event at the molecular level. In the first of these, termed a single-site, thin barrier, or lock-in mechanism (Macara and Cantley, 1983; Passow, 1986; Knauf, 1986), external anions can bind to a site within the transport protein to which internal anions do not have access. After binding, a conformational change occurs such that the site now becomes accessible from the inside, but not the outside. Thus, there is a single "site" that is open to the inside in one form and to the outside in another. The second type of model, known as the "zipper" model (Wieth et al., 1982b; Brock et al., 1983), involves a chain of anion binding sites. The substrate anion binds to a site at one side of the membrane, and then through a series of changes in conformation is shuttled from one intramembrane site to another until it finally reaches a site at the opposite side of the membrane, from which it can be released into the medium. In the single-site model, anions diffuse through aqueous regions in capnophorin to the transport site, and in principle it should be possible to label a single amino-acid residue in the transport site from either side of the membrane. In the zipper model, substrate anions bind to different sites at the two sides of the membrane, and there is no residue that can be labeled from both sides.

Structural models of the band 3 protein indicate that there may be clusters of positive charges near the entrance of a pore-like structure at either side of the membrane (Jay and Cantley, 1986), and kinetic evidence supports this concept, at least for the external-facing transport site (Passow, 1986). There seem to be overlapping binding sites for a number of different competitive and noncompetitive inhibitors, involving several positively charged ligands, and it is possible that these charges establish a surface potential that can affect inhibitor reactivity (Wieth et al., 1982a). Other evidence suggests that these charges may be sensitive to the I/O conformation of the transport site (Passow et al., 1980; Passow, 1986) or to substrate binding (Passow, 1986). These elements could fit with either model. Fluorescence energy transfer experiments indicate that the binding site for the stilbene disulfonates (such as DIDS), competitive inhibitors that may overlap the transport site, is located within the lipid bilayer, some distance from the external surface (Rao et al., 1979). This would fit with the single-site model, although it would not preclude a chain of sites shorter than the bilayer thickness. Differences in iodide binding (*see* section 3.2.2.) and affinity for competitive inhibitors (Passow, 1986) between the external-facing and internal-facing transport sites indicate that these two conformations differ in structure, which fits with the zipper concept, but can also be accommodated within the single-site model, since the site may have a slightly different structure in the I and O conformations.

The single-site concept was supported by evidence that eosin maleimide (EM), when bound covalently from the outside, becomes accessible to nonpenetrating fluorescence quenching agents present at the inside (Macara et al., 1983). Since EM competes with stilbene disulfonates, this was taken as evidence that an agent bound to the transport site from the outside can become accessible to the inside, while remaining covalently attached, an unlikely event if the O/I conversion requires movement along a series of binding sites. More recently, however, it has been shown that EM does not compete with Cl^- for binding to the transport site (Knauf et al., 1985). Thus, the apparent translocation of EM is not clearly relevant to the behavior of the *transport* site, but rather to a noncompetitive inhibitory site. Very recently, Jennings and Al-Rhaiyel (1988) found that external Woodward's reagent K (WRK) plus borohydride, which converts glutamic acid carboxyl groups in capnophorin to the corresponding alcohols (Jennings and Anderson, 1987), changes the dependence of sulfate transport on *both* internal and external pH in the same way. Sulfate transport normally requires cotransport of a proton that, when bound to its site, facilitates the binding of sulfate to form a transportable complex (Milanick and Gunn, 1982). For this reason, sulfate transport increases as pH is lowered from 7.5 to 6.5. Under conditions where WRK plus borohydride primarily modifies an externally accessible glutamic-acid residue in the C-terminal third of capnophorin (Jennings and Al-Rhaiyel, 1988), sulfate flux is no longer dependent on external *or internal* pH over this range, and the proton flux that normally accompanies sulfate flux is inhibited. The simplest interpretation for this is that *external* WRK has modified a carboxyl group in the transport site that can interact with either external or *internal* protons. That is, one residue is accessible alternately from either side of the membrane, as predicted by the single-site model.

At present, therefore, the evidence seems to favor the single-site model, but much further work is needed to establish this and to relate the kinetics to the protein structure. One consequence of such a model is that, if there is free diffusion of anions to the transport site, and if the site occupies only a small fraction of the membrane thickness, there will be a very high voltage field at this point if a potential is applied across the membrane. One might, therefore, expect that, if there is any charge reorientation in capnophorin associated with the I/O conformational change, this will be reflected in a potential-dependence of the rate-constants k and k′ shown in Fig. 2. Alternatively, if the pathways for ion diffusion to the transport sites are highly restricted, the potential at the transport site may be affected by the transmembrane voltage, resulting in a potential dependence of K_i or K_o (Läuger and Jauch, 1986; Läuger, 1987). Either effect would be reflected as changes in transport site asymmetry in response to membrane potential, and/or changes in transport rate, despite the fact that the exchange itself is electroneutral and thus not potential-dependent in principle. Grygorczyk et al. (1987) have observed an increase in Cl^- transport in response to membrane hyperpolarization for mouse band 3 inserted into *Xenopus* oocytes, and have interpreted this in terms of changes in the orientation of transport sites in response to potential (corresponding to movement of -0.1 elementary charge during the ECl_i to ECl_o transition). Experiments in which membrane potential has been altered in human erythrocytes by means of ionophores, however, have shown no significant effects on Cl^- transport or on iodide binding to the external transport site (Gunn and Fröhlich, 1979; Wieth et al., 1980; Fröhlich et al., 1983; Furuya et al., 1984; Spinelli, Restrepo, and Knauf, unpublished data).

6. Net Anion Transport

6.1. Involvement of Capnophorin

Although net flow of Cl^- takes place much more slowly than Cl^- exchange (Table 1), a major fraction of net Cl^- flux is inhibited by DIDS strictly in parallel with anion exchange (Knauf et al., 1977). At highly negative membrane potentials, there is also a DIDS-insensitive component of net Cl^- flux (Knauf et al., 1977, 1983a), but this may be smaller or nonexistent at membrane potentials nearer to zero (Freedman and Novak, 1987). Further evidence that the major component of net flux is mediated by band 3 comes from the fact that other competitive inhibitors of anion exchange also inhibit net chloride flow (Fröhlich et al., 1983; Kaplan et al., 1976). In addition, Woodward's reagent K plus borohydride causes a >10-fold increase in net Cl^- flow, suggesting that a glutamate in the transport pathway forms part of the barrier to net Cl^- flux (Jennings, 1987).

6.2. Mechanism of Net Anion Transport

In terms of the ping-pong model (Fig. 2), net Cl^- transport might occur by either of two major mechanisms. The first of these involves "slippage" of the unloaded E_o or E_i forms into the opposite conformation. For example, if the E_o form slips back into the E_i form, when this E_i binds a Cl^- ion, returns to the ECl_o form, and releases the Cl^- ion into the outside medium to form E_o again, the result of this cycle is net transport of a Cl^- ion from inside to outside. The second mechanism, termed "tunneling" (Fröhlich et al., 1983) or simply "diffusion" (Kaplan et al., 1983), depends upon the fact that the barriers that prevent internal anions from binding to the transport site in the E_o form, for example, may not be perfectly tight. If there is a small diffusive leakage of ions through these barriers, the result will be a net flow of anions without a concomitant translocation of transport sites (such as occurs during the anion exchange process).

There is strong evidence that slippage plays at best a very minor role in net Cl^- transport. If the Cl^- concentrations inside and outside are increased in parallel (Kaplan et al., 1983; Knauf et al., 1983a), the net Cl^- efflux increases with little evidence of saturation even at $[Cl^-] = 300$ m*M*. If slippage were the mechanism, net efflux should decrease with increasing Cl_o, as the fraction of capnophorin in the E_o form decreases. Fröhlich et al. (1983) discovered that net Cl^- efflux does indeed decrease if Cl_o is raised, keeping Cl_i constant, in parallel with the decrease in E_o. This could either indicate slippage under these conditions or else that the E_o form has a higher diffusive permeability than other forms of capnophorin. Evidence for the latter interpretation comes from experiments (Fröhlich, 1984) showing that the magnitude of the net anion flux depends on the kind of anion inside the cell; this would not be expected for slippage where the rate-limiting step is return of the empty E_o to the inside, which should not be affected by the nature of the inside anion. Phloretin and other noncompetitive inhibitors, which inhibit anion exchange by preventing the ECl_i to ECl_o transition, might be expected to inhibit slippage, since it involves a similar change from E_o to E_i. Phloretin, however (Bennekou, 1985; Fröhlich and King, 1987; Fröhlich et al., 1988), has no effect on net Cl^- flux. For other inhibitors, such as niflumic acid (Knauf et al., 1983b) and NAP-taurine (Knauf et al., 1983a), the situation is more complex because of recruiting effects, but the data still argue against the slippage model.

These data strongly suggest that net Cl^- flux involves a diffusive process, which does not require the same kind of conformational change as does Cl^- exchange. The precise relationship of this process to the Cl^- exchange mechanism is still unclear. One problem is that the net flux does not saturate appreciably at Cl^- concentrations where exchange does saturate (Knauf et al., 1983a; Kaplan et al., 1983; Fröhlich, 1984; Bennekou, 1985; Passow, 1986). Although Passow (1986) argues that this can be accounted for, a quantitative analysis has not yet been presented. One possibility is that the net Cl^- flux does not involve binding to the same transport sites as does exchange. The ability of competitive inhibitors to affect net flux does not preclude this, since such bulky inhibitors could alter regions of capnophorin distant from the exchange transport site, either sterically or allosterically. In fact, Schwarz and Passow (*see* Passow, 1986) have observed anion-conductive channels with a single-channel conductance of about 5 pS, which can be inhibited by blockers of anion exchange. This raises the interesting possibility that net Cl^- flow might be the result of a low-frequency, sporadic, high-conductance state of capnophorin, so that the entire Cl^- flux would result from the presence on average of less than one open channel (e.g., ~0.4) with a very high ion flux (2.5×10^6 ions/s).

6.3. Rates of Slippage, Diffusion, and Exchange

Net flow of sulfate is much slower than that of chloride (Knauf et al., 1977; Fröhlich and King, 1988). Recently, Fröhlich and King (1988) have attempted to measure slippage in experiments using sulfate, where the rates of diffusion and slippage might be of comparable magnitude, using phloretin sensitivity as a criterion for slippage. Expressed in terms of ions/capnophorin molecule/s at 20°C, the rate of slippage from E_o to E_i is about 0.3 s^{-1}. This compares with a maximal Cl^- exchange rate of about 12,300 s^{-1} at 20°C (Brahm, 1977). From these rates, it is possible to construct an approximate thermodynamic profile for the transport-related conformational changes as shown in Fig. 4. If we assume that K_i and K_o are equal and similar to their values at 0°C, and if k/k' is similar at 0°C and 20°C, there will be about a 6.2 kj/mol difference in free energy (ΔG^0) between ECl_i and ECl_o and between E_i and E_o. The binding free energy for Cl^-, corresponding to a dissociation constant of 65 m*M* (Dalmark, 1976; Brazy and Gunn, 1976), is about 6.7 kj/mol. The ΔG^* for the ECl_i to ECl^* transition is 48.7 kj/mol and for ECl_o to ECl^* is 42.5 kj/mol (Glasstone and Lewis, 1960). From the rate-constant for slippage, the transition from E_o to E^* has a ΔG^* of about 75.7 kj/mol. Thus, the dramatic difference in rates of exchange and slippage can be related to only a 33.2 kj/mol difference in activation free energies, corresponding to the free energy of formation of just over 3 hydrogen bonds in a hydrophobic environment. Seen in these terms, it is entirely plausible that the low-energy interaction with substrates changes the structure of capnophorin sufficiently to cause this lowering of the free energy of activation. From the temperature dependence of Cl^- exchange, the enthalpy of activation is about 82.4 kj/mol at 15–38°C and 126.4 kj/mol below 15°C (Brahm, 1977). The calculated entropy change, $T\Delta S^*$, is about 33.7 kj/

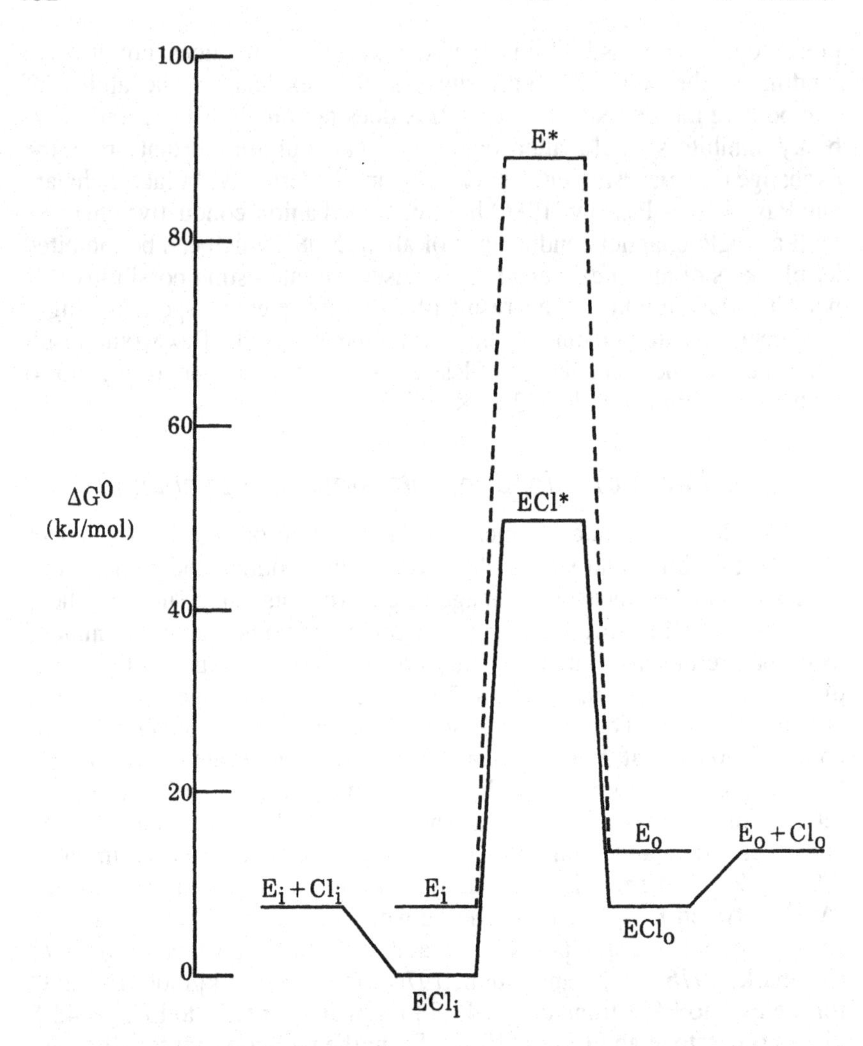

Fig. 4. Approximate thermodynamics of anion transport at 20°C. Free energies (ΔG^0) for formation of the substrate-loaded transition state (ECl*) are calculated from the Cl^- exchange rate at 20°C (Brahm, 1977), whereas those for the unloaded site transition state (E*) are calculated from the slippage rates reported by Fröhlich and King (1988), also at 20°C. Free energy differences for the other forms are calculated from data at 0°C.

mol at 20°C, indicating that the transition state for exchange (ECl*) is associated with a very large increase in enthalpy, which is partially compensated by a rather large increase in entropy. This, together with the large activation volume of 150 cm^3/mol (Canfield and Macey, 1984), suggests that there is a substantial change in protein conformation involved in the ECl_i to ECl_o transition.

In view of the uncertainties regarding the mechanism of tunneling, it would be unwise to interpret the magnitude of the tunneling rates in thermodynamic terms. The fact that the Cl^- tunneling through the E_o form (30 s^{-1}) is larger than that through the other forms of capnophorin (3 s^{-1}, Fröhlich and King, 1988), however, and the small magnitude of the tunneling rates indicates that the ECl* transition state form probably involves at least a transient occluded form, in which the Cl^- ion bound to the transport site is inaccessible to either solution (Fig. 1A). The alternative, that there is a transition state with the gates open on both sides, would imply a higher tunneling rate and also that the Cl^--loaded forms (with the transiently open gates) should have a larger tunneling rate than the unloaded forms.

7. Relation to Other Transport Systems

As mentioned in Chapter 8, there is increasing evidence that proteins exist in other tissues that have structural homology to capnophorin, as revealed by antibodies or by similarities in nucleic-acid sequence, and some of these proteins may play an important role in anion transport, e.g. in the cortical collecting duct in the kidney. It is also probable that many if not all transport proteins have an essentially similar structure, with several α-helical segments crossing the membrane and self-associating to form a channel-like structure (Guidotti, 1986). On this basis, one might speculate that the wide variety of transport systems observed in nature might have a common structural basis, with variations that have evolved to suit the particular needs of the system.

If this is true, it follows that studies of the red-cell anion exchange protein may provide valuable insights into the operation of other, more complex transport systems that are less amenable to detailed analysis. It is already clear that capnophorin can perform a variety of functions, including a very tightly coupled one-for-one exchange of anions, cotransport of sulfate ions with protons (Milanick and Gunn, 1982), and net flow of anions, as well as perhaps some net cation flow under special circumstances (Jones and Knauf, 1985). Although band 3 does not catalyze any primary active transport, it would be easy to imagine how a protein with two conformations of the transport site, inward- and outward-facing, could mediate active flow of ions if the conformational change were coupled to metabolic energy. Analysis of the details of the transporting conformational changes in capnophorin may thus provide insights into the operation of many other systems. Capnophorin already has been instructive in illustrating how dependence of the I/O conformational change upon substrate binding can create a system that mediates very tightly coupled exchange and that is far more selective for anions with respect to cations (Knauf, 1979; Wieth and Brahm, 1985) than a positively charged channel might be (Solomon, 1960). The concept of a gated channel operating as an exchange carrier (Patlak, 1957) is a very versatile one, which bridges the channel-carrier dichotomy and which should prove to be a useful framework for constructing transport models at the molecular level.

Acknowledgments

Assistance of Drs. Otto Fröhlich, Michael Jennings, and Hermann Passow in providing preprints prior to publication is gratefully acknowledged, as well as the comments of Drs. H. Passow, M. Jennings, O. Fröhlich, J. Brahm, and D. Restrepo on the manuscript. This work was supported by NIH grant DK 27495 and by Fogarty Senior International Fellowship TW 00975.

References

Bennekou, P. (1985) Chloride net permeability of the human red cell—a comparison with the exchange permeability. *Acta Physiol. Scand.* **124 (Suppl. 542),** 154.

Brahm, J. (1977) Temperature-dependent changes of chloride transport kinetics in human red cells. *J. Gen. Physiol.* **70,** 283–306.

Brazy, P. C. and Gunn, R. B. (1976) Furosemide inhibition of chloride transport in human red blood cells. *J. Gen. Physiol.* **68,** 583–599.

Brock, C. J., Tanner, M. J. A., and Kempf, C. (1983) The human erythrocyte anion transport protein: Partial amino acid sequence, conformation and a possible molecular mechanism for anion exchange. *Biochem. J.* **213,** 577–586.

Canfield, V. A. and Macey, R. I. (1984) Anion exchange in human erythrocytes has a large activation volume. *Biochim. Biophys. Acta* **778,** 379–384.

Cass, A. and Dalmark, M. (1973) Equilibrium dialysis of ions in nystatin-treated red cells. *Nature New Biology* **244,** 47–49.

Cleland, W. W. (1963) The kinetics of enzyme-catalysed reactions with two or more substrates or products. I. Nomenclature and rate equations. *Biochim. Biophys. Acta* **67,** 104–137.

Cousin, J. L. and Motais, R. (1979) Inhibition of anion permeability by amphiphilic compounds in human red cell: Evidence for an interaction of niflumic acid with the band 3 protein. *J. Membr. Biol.* **47,** 125–153.

Dalmark, M. (1975) Chloride transport in human red cells. *J. Physiol. (Lond.)* **250,** 39–64.

Dalmark, M. (1976) Effects of halides and bicarbonate on chloride transport in human red blood cells. *J. Gen. Physiol.* **67,** 223–234.

Fairbanks, G. L., Steck, T. L., and Wallach, D. F. H. (1971) Electrophoretic analysis of the major polypeptides of the human erythrocyte membrane. *Biochemistry* **10,** 2606–2617.

Falke, J. J. and Chan, S. I. (1985) Evidence that anion transport by band 3 proceeds via a ping-pong mechanism involving a single transport site. *J. Biol. Chem.* **260,** 9537–9544.

Falke, J. J. and Chan, S. I. (1986a) Molecular mechanisms of band 3 inhibitors. 1. Transport site inhibitors. *Biochemistry* **25,** 7888–7894.

Falke, J. J. and Chan, S. I. (1986b) Molecular mechanisms of band 3 inhibitors. 2. Channel blockers. *Biochemistry* **25,** 7895–7898.

Falke, J. J. and Chan, S. I. (1986c) Molecular mechanisms of band 3 inhibitors. 3. Translocation inhibitors. *Biochemistry* **25,** 7899–7906.

Falke, J. J., Kanes, K. J., and Chan, S. I. (1985) The kinetic equation for the chloride transport cycle of band 3. A^{35}Cl and ^{37}Cl NMR study. *J. Biol. Chem.* **260,** 9545–9551.

Falke, J. J., Pace, R. J., and Chan, S. I. (1984a) Chloride binding to the anion transport binding sites of band 3. A^{35}Cl NMR study. *J. Biol. Chem.* **259,** 6472–6480.

Falke, J. J., Pace, R. J., and Chan, S. I. (1984b) Direct observation of the transmembrane recruitment of band 3 transport sites by competitive inhibitors. A^{35}Cl NMR study. *J. Biol. Chem.* **259,** 6481–6491.

Freedman, J. C. and Novak, T. S. (1987) Chloride conductance of human red blood cells at varied E_K. *Biophys. J.* **51,** 565a.

Fröhlich, O. (1982) The external anion binding site of the human erythrocyte anion transporter: DNDS binding and competition with chloride. *J. Membr. Biol.* **65,** 111–123.

Fröhlich, O. (1984) Relative contributions of the slippage and tunneling mechanisms to anion net efflux from human erythrocytes. *J. Gen. Physiol.* **84,** 877–893.

Fröhlich, O. and Gunn, R. B. (1986) Erythrocyte anion transport: the kinetics of a single-site obligatory exchange system. *Biochim. Biophys. Acta* **864,** 169–194.

Fröhlich, O. and King, P. A. (1987) Mechanisms of anion net transport in the human erythrocyte. *J. Gen. Physiol.* **90,** 6a.

Fröhlich, O. and King, P. A. (1988) Mechanism of net anion transport in the human erythrocyte, in *Cell Physiology of Blood* (Gunn, R. B. and Parker, J. C., eds.), Rockefeller University Press, New York, pp. 181–192.

Fröhlich, O., Bain, D., and Weimer, L. (1989) The effect of phloretin and DNDS on chloride net transport in erythrocytes. Submitted for publication.

Fröhlich, O., Leibson, C., and Gunn, R. B. (1983) Chloride net efflux from intact erythrocytes under slippage conditions. Evidence for a positive charge on the anion binding/transport site. *J. Gen. Physiol.* **81,** 127–152.

Furuya, W., Tarshis, T., Law, F.-Y., and Knauf, P. A. (1984) Transmembrane effects of intracellular chloride on the inhibitory potency of extracellular H_2DIDS. Evidence for two conformations of the transport site of the human erythrocyte anion exchange protein. *J. Gen. Physiol.* **83,** 657–681.

Ginsburg, H., O'Connor, S. E., and Grisham, C. M. (1981) Evidence from electron paramagnetic resonance for function-related conformation changes in the anion-transport protein of human erythrocytes. *Eur. J. Biochem.* **114,** 533–538.

Glasstone, S. and Lewis, D. (1960) *Elements of Physical Chemistry, Second Edition* (Van Nostrand, Princeton).

Goldman, D. E. (1943) Potential, impedance and rectification in membranes. *J. Gen. Physiol.* **27,** 37–60.

Grinstein, S., McCulloch, L., and Rothstein, A. (1979) Transmembrane effects of irreversible inhibitors of anion transport in red blood cells. Evidence for mobile transport sites. *J. Gen. Physiol.* **73,** 493–514.

Grygorczyk, R., Schwarz, W., and Passow, H. (1987) Potential dependence of the "electrically silent" anion exchange across the plasma membrane of *Xenopus* oocytes mediated by the band-3 protein of mouse red blood cells. *J. Membr. Biol.* **99,** 127–136.

Guidotti, G. (1986) Membrane proteins: structure, arrangement, and disposition in the membrane, in *Physiology of Membrane Disorders, Second Edition* (Andreoli, T. E., Hoffman, J. F., Fanestil, D. D., and Schultz, S. G., eds.), Plenum, New York, pp. 45–55.

Gunn, R. B. (1979) Transport of anions across red cell membranes, in *Transport Across Biological Membranes, Vol. II* (Giebisch, G., Tosteson, D., and Ussing, H. H., eds.), Springer-Verlag, Heidelberg, pp. 59–80.

Gunn, R. B. and Fröhlich, O. (1979) Asymmetry in the mechanism for anion exchange in human red blood cell membranes: Evidence for reciprocating sites that react with one transported anion at a time. *J. Gen. Physiol.* **74,** 351–374.

Gunn, R. B., and Fröhlich, O. (1980) The kinetics of the titratable carrier for anion exchange in erythrocytes. *Ann. N.Y. Acad. Sci.* **341,** 384–393.

Gunn, R. B. and Fröhlich, O. (1982) Arguments in support of a single transport site on each anion transporter in human red cells, in *Chloride Transport in Biological Membranes* (Zadunaisky, J., ed.), Academic Press, New York, pp. 33–59.

Gunn, R. B., Dalmark, M., Tosteson, D. C., and Wieth, J. O. (1973) Characteristics of chloride transport in human red blood cells. *J Gen. Physiol.* **61,** 185–206.

Gunn, R. B., Fröhlich, O., Macintyre, J. D., and Low, P. S. (1979) Calcium modification of the anion transport mechanism in red blood cells. *Biophys. J.* **25,** 106a.

Hautmann, M. and Schnell, K. F. (1985) Concentration dependence of the chloride selfexchange and homoexchange fluxes in human red cell ghosts. *Pflügers Arch.* **405,** 193–201.

Hoffman, J. F., and Laris, P. C. (1974) Determination of membrane potentials in human and *Amphiuma* red blood cells by means of a fluorescent probe. *J. Physiol. (Lond.)* **239,** 519–552.

Hunter, M. J. (1971) A quantitative estimate of the non-exchange restricted chloride permeability of the human red cell. *J. Physiol. (Lond.)* **218,** 49P–50P.

Jans, A. W. H., Krijnen, E. S., Luig, J., and Kinne, R. K. H. (1987) A ^{31}P-NMR study on the recovery of intracellular pH in LLC-PK_1/Cl_4 cells from intracellular alkalinization. *Biochim. Biophys. Acta* **931,** 326–334.

Jay, D. and Cantley, L. (1986) Structural aspects of the red cell anion exchange protein. *Ann. Rev. Biochem.* **55,** 511–538.

Jennings, M. L. (1982) Stoichiometry of a half-turnover of band 3, the chloride-transport protein of human erythrocytes. *J. Gen. Physiol.* **79,** 169–185.

Jennings, M. L. (1987) Functional roles of carboxyl groups in human red blood cell band 3. *J. Gen. Physiol.* **90,** 5a.

Jennings, M. L. and Adams, M. F. (1981) Modification by papain of the structure and function of band 3, the erythrocyte anion transport protein. *Biochemistry* **20,** 7118–7123.

Jennings, M. L. and Al-Rhaiyel, S. (1988) Modification of a carboxyl group that appears to cross the permeability barrier in the red blood cell anion transporter. *J. Gen. Physiol.* **92,** 161–178.

Jennings, M. L. and Anderson, M. P. (1987) Chemical modification and labeling of glutamate residues at the stilbenedisulfonate site of human red blood cell band 3 protein. *J. Biol. Chem.* **262,** 1691–1697.

Jones, G. S. and Knauf, P. A. (1985) Mechanism of the increase in cation permeability of human erythrocytes in low-chloride media. Involvement of the anion transport protein capnophorin. *J. Gen. Physiol.* **86,** 721–738.

Kaplan, J. H., Pring, M., and Passow, H. (1983) Band 3 protein-mediated anion conductance of the red cell membrane: Slippage versus ionic diffusion. *FEBS Lett.* **156,** 175–179.

Kaplan, J. H., Scorah, K., Fasold, H., and Passow, H. (1976) Sidedness of the inhibitory action of disulfonic acids on chloride equilibrium exchange and net transport across the human erythrocyte membrane. *FEBS Lett.* **62,** 182–185.

Knauf, P. A. (1979) Erythrocyte anion exchange and the band 3 protein: transport kinetics and molecular structure. *Curr. Topics Membr. Trans.* **12,** 249–363.

Knauf, P. A. (1986) Anion transport in erythrocytes, in *Membrane Transport Disorders, 2nd Edition* (Andreoli, T. E., Schultz, S. G., Hoffman, J. F., and Fanestil, D. D., eds.), Plenum, New York, pp. 191–220.

Knauf, P. A. and Brahm, J. (1986) Asymmetry of the human red blood cell anion transport system at 38°C. *Biophys. J.* **49,** 579a.

Knauf, P. A. and Brahm, J. (1989) Functional asymmetry of the anion exchange protein, capnophorin: Effects on substrate and inhibitor binding. *Methods Enzymol.* **173,** 432–453.

Knauf, P. A. and Mann, N. (1984) Use of niflumic acid to determine the nature of the asymmetry of the human erythrocyte anion exchange system. *J. Gen. Physiol.* **83,** 703–725.

Knauf, P. A. and Mann, N. A. (1986) Location of the chloride self-inhibitory site of the human erythrocyte anion exchange system. *Am. J. Physiol.* **251** (*Cell Physiol.* **20**), C1–C9.

Knauf, P. A. and Spinelli, L. J. (1987) Asymmetry of the Cl-loaded forms of the human erythrocyte anion exchange protein, band 3. *J. Gen. Physiol.* **90,** 24a.

Knauf, P. A. and Spinelli, L. J. (1988) Evidence that external NIP-taurine, NAP-taurine and iodide inhibit red blood cell anion exchange by binding to the same site on band 3. *Biophys. J.* **53,** 532a.

Knauf, P. A., Fuhrmann, G. F., Rothstein, S., and Rothstein, A. (1977) The relationship between anion exchange and net anion flow across the human red blood cell membrane. *J. Gen. Physiol.* **69,** 363–386.

Knauf, P. A., Law, F.-Y., and Marchant, P. J. (1983a) Relationship of net chloride flow across the human erythrocyte membrane to the anion exchange mechanism. *J. Gen. Physiol.* **81,** 95–126.

Knauf, P. A., Mann, N., and Kalwas, J. E. (1983b) Net chloride transport across the human erythrocyte membrane into low chloride media: Evidence against a slippage mechanism. *Biophys. J.* **41,** 164a.

Knauf, P. A., Mann, N. A., and Penikas, J. (1985) Noncompetitive partial inhibition of human red cell chloride exchange by eosin (E) and eosin maleimide (EM). *The Physiologist* **28,** 294.

Knauf, P. A., Law, F.-Y., Tarshis, T., and Furuya, W. (1984) Effects of the transport site conformation on the binding of external NAP-taurine to the human erythrocyte anion exchange system: Evidence for intrinsic asymmetry. *J. Gen. Physiol.* **83,** 683–701.

Knauf, P. A., Mann, N., Brahm, J., and Bjerrum, P. (1986a) Asymmetry in iodide affinities of external and internal-facing red cell anion transport sites. *Fed. Proc.* **45,** 1005.

Knauf, P. A., Brahm, J., Bjerrum, P., and Mann, N. (1986b) Kinetic asymmetry of the human erythrocyte anion exchange system, in *Proceedings 8th School on Biophysics of Membrane Transport Agricultural University of Wroclaw, Wroclaw, Poland* (Kuczera, J. and Przestalski, S., eds.), vol. 1, pp. 157–169.

Knauf, P. A., Mann, N. A., Kalwas, J. E., Spinelli, L. J., and Ramjeesingh, M. (1987a) Interactions of NIP-taurine, NAP-taurine, and Cl^- with the human erythrocyte anion exchange system. *Am. J. Physiol.* **253** (*Cell Physiol.* **22**), C652–C661.

Knauf, P. A., Mann, N. A., and Spinelli, L. J. (1987b) Effects of transport site conformation and anion binding on the affinity of the human red blood cell anion transport protein for niflumic acid (NA). *Biophys. J.* **51,** 566a.

Knauf, P. A., Spinelli, L. J., and Mann, N. A. (1987c) Affinities of flufenamic acid (FA) for different conformations of the human erythrocyte anion transport protein. *Fed. Proc.* **46,** 534.

Knauf, P. A., Ship, S., Breuer, W., McCulloch, L., and Rothstein, A. (1978) Asymmetry of the red cell anion exchange system: different mechanisms of reversible inhibition by *N*-(4-azido-2-nitrophenyl)-2-aminoethylsulfonate (NAP-taurine) at the inside and outside of the membrane. *J. Gen. Physiol.* **72,** 607–630.

Läuger, P. (1987) Voltage dependence of sodium-calcium exchange: predictions from kinetic models. *J. Membr. Biol.* **99,** 1–11.

Läuger, P. and Jauch, P. (1986) Microscopic description of voltage effects on ion-driven cotransport systems. *J. Membr. Biol.* **91,** 275–284.

Low, P. S. (1978) Specific cation modulation of anion transport across the human erythrocyte membrane. *Biochim. Biophys. Acta* **514,** 264–273.

Macara, I. G. and Cantley, L. C. (1983) The structure and function of band 3, in *Cell Membranes: Methods and Reviews* (Elson, E., Frazier, W., and Glaser, L., eds.), Plenum Press, New York, pp. 41–87.

Macara, I. G., Kuo, S., and Cantley, L. C. (1983) Evidence that inhibitors of anion exchange induce a transmembrane conformational change in band 3. *J. Biol. Chem.* **258,** 1785–1792.

Milanick, M. A. and Gunn, R. B. (1982) Proton-sulfate co-transport: Mechanism of H^+ and sulfate addition to the chloride transporter of human red blood cells. *J. Gen. Physiol.* **79,** 87–113.

Milanick, M. A. and Gunn, R. B. (1986) Proton inhibition of chloride exchange: asynchrony of band 3 proton and anion transport sites? *Am. J. Physiol.* **250** (*Cell Physiol.* **19**), C955–C969.

Passow, H. (1986) Molecular aspects of the band 3 protein-mediated anion transport across the red blood cell membrane. *Rev. Physiol. Biochem. Pharmacol.* **103,** 61–203.

Passow, H., Fasold, H., Gärtner, E. M., Legrum, B., Ruffing, W., and Zaki, L. (1980) Anion transport across the red blood cell membrane and the conformation of the protein in band 3. *Ann. N.Y. Acad. Sci.* **341,** 361–383.

Patlak, C. S. (1957) Contributions to the theory of active transport: II. The gate type non-carrier mechanism and generalizations concerning tracer flow, efficiency, and measurement of energy expenditure. *Bull. Math. Biophys.* **19,** 209–235.

Rao, A., Martin, P., Reithmeier, R. A. F., and Cantley, L. C. (1979) Location of the stilbenedisulfonate binding site of the human erythrocyte anion-exchange system by resonance energy transfer. *Biochemistry* **18,** 4505–4516.

Restrepo, D., Kozody, D. J., and Knauf, P. A. (1988) pH homeostasis in promyelocytic leukemic HL60 cells. *J. Gen. Physiol.* **92,** 489–507.

Salhany, J. M. and Rauenbuehler, P. B. (1983) Kinetics and mechanism of erythrocyte anion exchange. *J. Biol. Chem.* **258,** 245–249.

Schnell, K. F. and Besl, E. (1984) Concentration dependence of the unidirectional sulfate and phosphate flux in human red cell ghosts under selfexchange and under homoexchange conditions. *Pflügers Arch.* **402,** 197–206.

Schnell, K. F., Besl, E., and V. der Mosel, R. (1981) Phosphate transport in human RBC: Concentration dependence and pH dependence of the unidirectional phosphate flux at equilibrium conditions. *J. Membr. Biol.* **61,** 173–192.

Shami, Y., Carver, J., Ship, S., and Rothstein, A. (1977) Inhibition of Cl^- binding to anion transport protein of the red blood cell by DIDS (4,4′-diisothiocyano-2,2′-stilbene disulfonic acid) measured by (^{35}Cl)NMR. *Biochim. Biophys. Res. Comm.* **76,** 429–436.

Simchowitz, L. and Roos, A. (1985) Regulation of intracellular pH in human neutrophils. *J. Gen. Physiol.* **85,** 443–470.

Solomon, A. K. (1960) Red cell membrane structure and ion transport. *J. Gen. Physiol.* **43 (Suppl 2),** 1–15.

Tanford, C. (1985) Simple model can explain self-inhibition of red cell anion exchange. *Biophys. J.* **47,** 15–20.

Wieth, J. O. and Bjerrum, P. J. (1983) Transport and modifier sites in capnophorin, the anion transport protein of the erythrocyte membrane, in *Structure and Function of Membrane Proteins* (Quagliariello, E., and Palmieri, F., eds.), Elsevier, Amsterdam, pp. 95–106.

Wieth, J. O. and Bjerrum, P. J. (1982) Titration of transport and modifier sites in the red cell anion transport system. *J. Gen. Physiol.* **79,** 253–282.

Wieth, J. O. and Brahm, J. (1985) Cellular anion transport, in *The Kidney: Physiology and Pathophysiology, Chapter 4* (Seldin, D. W. and Giebisch, G., eds.), Raven Press, New York, pp. 49–89.

Wieth, J. O., Brahm, J., and Funder, J. (1980) Transport and interactions of anions and protons in the red blood cell membrane. *Ann. N.Y. Acad. Sci.* **341,** 394–418.

Wieth, J. O., Bjerrum, P. J., and Borders, Jr., C. L. (1982a) Irreversible inactivation of red cell chloride exchange with phenylglyoxal, an arginine-specific reagent. *J. Gen. Physiol.* **79,** 283–312.

Wieth, J. O., Bjerrum, P. J., Brahm, J., and Andersen, O. S. (1982b) The anion transport protein of the red cell membrane. A zipper mechanism of anion exchange. *Tokai J. Exp. Clin. Med.* **7 (Suppl),** 91–101.

Ca^{2+}-Activated Potassium Channels

Javier Alvarez and Javier García-Sancho

1. Introduction

It was 50 years ago that the presence in the human red-cell membrane of a K^+-selective pathway activatable by Pb (Ørskov, 1935) or fluoride (Wilbrandt, 1937) was reported. Gárdos described a similar pathway that was activated by incubation with iodoacetate and adenosine, and pointed out the absolute requirement of calcium for activation ("Gárdos effect," Gárdos, 1956, 1958). The site for the Ca^{2+} effect was later documented to face towards the cytoplasmic side of the membrane (Whittam, 1968; Lew, 1970; Blum and Hoffman, 1972). During the last 15 yr, K^+-selective channels that are activated by the increase of ionized calcium levels of the cytoplasm have also been discovered in a large variety of cells. The participation of these channels in important physiological functions, such as the control of membrane potential of excitable cells, the secretion of hormones or neurotransmitters, and the control of cell volume and transephithelial ion transport, is now widely acknowledged (for reviews, *see* Meech, 1976, 1978; Putney, 1979; Grinstein et al., 1982; Schwarz and Passow, 1983; Petersen and Maruyama, 1984; Sarkadi and Gárdos, 1985; Hoffman, 1985).

The actual molecular substrate of the Ca^{2+}-dependent K^+ channels is still unknown. The degree of similarity among the channels present in different cell lines is also unclear. Important differences in terms of Ca^{2+}-sensitivity, voltage-dependence, unitary conductance, and sensitivity to several inhibitors have been reported, leading to the classification of Ca^{2+}-dependent K^+ channels in at least three different groups (*see below*). However, common characteristics still keep open the possibility of a basic structural similarity among these channels and red cells continue to be a popular model for the study of Ca^{2+}-dependent K^+ transport.

Reviews on different aspects of this subject have appeared with certain periodicity (Riordan and Passow, 1973; Lew, 1974; Lew and Ferreira, 1977; Lew and Ferreira, 1978; Lew and Beaugé, 1979; Hoffman et al., 1980; Parker, 1981, Passow, 1981; Lew, 1983; Schwarz and Passow, 1983; Sarkadi and Gárdos, 1985).

2. Some Experimental Points

Red cells offer a rich variety of experimental preparations to study Ca^{2+}-dependent K^+ transport. Intact cells can be submitted to one of the ATP-depleting treatments known to activate the K^+ channels (*see below*). If the intracellular Ca^{2+} level is to be controlled, the methods making use of the divalent cation ionophore A23187 (Ferreira and Lew, 1976; Lew and Ferreira, 1976) are preferable. It must be taken into account that it is not possible with the above procedure to control conveniently intracellular Ca^{2+} at the submicromolar level when fed cells are used. This impossibility arises from the fact that, at extracellular Ca^{2+} concentrations below 1 μM, even the highest ionophore concentrations (ca. 100–200 μmol/l cell) are not enough to overcome the powerful Ca-pump in all the cells. Since equilibrium distribution of Ca^{2+} cannot be reached and the maximum capacity for Ca-pumping differs among cells, the treatment conduces to heterogeneity of Ca distribution among cells (Lew and García-Sancho, 1985). For these reasons, ATP-depleted cells, in which homogeneity of Ca distribution on treatment with ionophore A23187 has been documented (Simonsen et al., 1982), are to be preferred for the above purposes. The treatments with iodoacetate or iodoacetamide plus adenosine or inosine (Lew, 1971a) are the most widely employed ATP-depleting procedures. Minor differences regarding the activation of Ca^{2+}-dependent K^+ transport by these treatments have been reported (Gárdos, 1967; Plishker, 1985). Hydrolysis of 2,3-diphosphoglycerate can sustain some ATP synthesis (about 200 μmol/l cell/h, *see* Glynn and Lew, 1970) and Ca-pumping capacity (about 10–15 μmol/l cell/h, Alvarez et al., 1988) in cells incubated with iodoacetate plus inosine. This can be avoided by previous depletion of 2,3-diphosphoglycerate stores by treatment with bisulphite (Szász et al., 1974) or by inhibiting its hydrolysis with tetrathionate (Alvarez et al., 1988). Resealed ghosts offer the opportunity to control the composition of internal medium by loading them with Ca buffers (Simons, 1976a) or other "intracellular" components. Flux measurements are always technically worse than in intact cells because of the presence of a contaminating population of highly permeable ghosts (*see* Bodemann and Passow, 1972). On the other hand, some of the K^+ channels could be inactivated during the preparation procedure, and this factor is, in our hands, difficult to control (Alvarez et al. 1985).

A good option for the study of the effects of intracellular constituents, including Ca^{2+} concentrations, is the everted vesicles preparation. It should be kept in mind that Steck-type inside-out vesicles (Steck and Kant, 1974) are not suitable for such studies, since K^+ channels are altered or inactivated during the preparation procedure (Grinstein and Rothstein, 1978; Sze and Solomon, 1979; Szebeni, 1981). Fortunately, a new preparation procedure has been described that gives a functional preparation of inside-out vesicles (Lew et al., 1982a). It should be remarked that, because of the small size of the vesicles, the mean number of channels per vesicle could be close to 1. The response is then easier to evaluate as the fraction of activated vesicles (García-Sancho et al., 1982, *see* section 3.5.) than as the first-order rate constant for uptake or efflux of tracer, since the interpretation of the last data is complicated by the size heterogeneity of the vesicles (Alvarez et al., 1986c). Possible modifications of the response by proteolytic attack of membrane proteins during preparation and manipulation of ghosts or vesicles should be kept in mind. Both selectivity and Ca^{2+}-sensitivity of the K^+ channels have been reported to be affected by low levels of trypsin incorporated into resealed ghosts (Wood and Mueller, 1984; but *see also* Passow et al., 1986), and the increase of cytoplasmic Ca^{2+} concentration is known to activate some intracellular proteases (Grasso et al., 1986).

Recently, the advent of single-channel conductance techniques has made possible the study of the individual behavior and properties of Ca^{2+}-activated K^+ channels in a variety of cells (Marty, 1983; Moczydlowski and Latorre, 1983; Latorre and Miller, 1983; Miller, 1983; Petersen and Maruyama, 1984; Blatz and Magleby, 1986). Patch-clamp studies in red cells are complicated by the need for "cleaning" the cell coat in order to obtain a successful seal. However, the application of this technique to the red-cell channel (Hamill et al., 1981; Hamill, 1981 and 1983; Grygorczyk and Schwarz, 1983, 1985; Grygorczyk et al., 1984) has also thrown new light on some of the characteristics of this complex system, and the correlation between single-channel conductance and flux measurements promises to be very fruitful in the next few years.

Under physiological conditions, with a K^+ gradient directed from the intracellular to the extracellular phase, the activation of the K^+ channels is conducive to membrane hyperpolarization, net loss of KCl, and shrinkage of the red cells. The activation process can be followed by measuring the changes of K^+ content of the cells or of the incubation medium. In the last case, the use of a K^+ electrode allows continuous recording of K^+ movements (Sanchez et al., 1980). For routine experiments, the continuous monitoring of volume changes by light scattering is an excellent experimental approach (Lew and García-Sancho, 1985; Shields et al., 1985; Alvarez et al., 1986d). The volume changes can be accelerated by adding to the medium a permeant anion such as SCN^- or NO_3^-(Lew and García-Sancho, 1985). Changes in cell density subsequent to shrinking can also be used to separate cells whose K^+ channels have been activated from those that are silent, by density centrifugation over phthalate oils. This allows separate analysis of the contents of both subpopulations of cells (Lew and García-Sancho, 1985). A valid alternative to the last procedure is differential hypotonic hemolysis of shrunken and unshrunken cells (Riordan and Passow, 1973; Lew and García-Sancho, 1985). The use of radioactive isotopes, ^{42}K or ^{86}Rb (*see* 3.2.), allows measurements under equilibrium exchange conditions for K^+, thus avoiding changes in cell volume, pH, or membrane potential when these are unwanted. Usually ^{86}Rb, with a longer half-life, is preferred for technical reasons. Finally, changes in membrane potential or current can be used to follow activation of Ca^{2+}-dependent K^+ channels when cell size allows microelectrode impalement, as in the case of the *Amphiuma* red cell (Lassen et al., 1973; 1976; Gárdos et al., 1976), or by patch-clamp procedures (Hamill et al., 1981; Hamill, 1981, 1983; Grygorczyk and Schwarz, 1983, 1985; Grygorczyk et al., 1984).

3. Properties of Ca^{2+}-Dependent K^+ Transport

3.1. Mechanisms of Activation

A selective increase of K^+ transport has been reported in red cells in response to several experimental maneuvers: a variety of ATP-depleting treatments, treatment with divalent cation ionophores in the presence of Ca, treatment with propranolol, artificial electron donors, or lead. Similarities in the phenomenology of the fluxes and sensitivity to inhibitors suggest that the same transport system is involved in all the cases.

The mechanism responsible for the activation of the K^+ channel in the "Gárdos effect" includes several steps, as stated by Lew (1974):

1. The incubation of cells with iodoacetate and adenosine results in a very fast and complete ATP depletion
2. ATP depletion is a necessary requirement for the increase of the intracellular Ca^{2+} concentration to the μM range. The role of ATP depletion is to inhibit the potent red-cell calcium pump, able in normal conditions to keep levels of intracellular Ca^{2+} as low as

20–30 n*M* (Lew et al., 1982b)
3. The intracellular Ca^{2+} interacts then with the internal site of the channel, inducing its activation.

An increase of Ca^{2+} sensitivity because of the ATP depletion has also been proposed, but no specific mechanism for this effect has yet been reported (*see* Riordan and Passow, 1973; Lew and Ferreira, 1978). Modifications of Ca^{2+}-dependent K^+ transport that were initially attributed to changes of 2,3-diphosphoglycerate metabolism (Gárdos, 1966, 1967; *see also* Parker, 1969) can be explained by their influence on cell ATP levels (Gárdos et al., 1975). ATP depletion does not produce by itself an increase in the calcium permeability, since the rate of calcium uptake in ATP-depleted red cells is not greater than in chelator-loaded fresh cells (Lew et al., 1982b; Tiffert et al., 1984; Alvarez et al., 1988).

The effect of fluoride seems to occur by the same mechanism, i.e., ATP depletion. In the presence of fluoride, both calcium (0.05 m*M*) and magnesium (0.5 m*M*) are effective (Lepke and Passow, 1968), but the effect of magnesium could be owing to contamination of the magnesium salt used by calcium (Lew and Ferreira, 1978). Direct effects on the channel cannot be excluded, since the presence of fluoride in ATP-free ghosts decreases by two orders of magnitude the Ca^{2+} concentration needed to activate the channel (Riordan and Passow, 1973). Iodoacetate has the same effect in red-cell ghosts, although less intense (Riordan and Passow, 1973). Fluoride is known to activate some Ca^{2+}-dependent enzymes, such as adenylate cyclase (Cheung et al., 1978). Possible similarities with its effects on Ca^{2+}-dependent K^+ channels are still to be assessed.

The K^+ permeability can be activated in intact red cells without previous ATP depletion by using the Ca^{2+} ionophore A23187, which transports Ca^{2+} across the red-cell membrane in electroneutral exchange with Mg^{2+} and H^+ (Reed, 1976). Lew and Ferreira (1976) studied the Ca^{2+} sensitivity of the K^+ channel at two ionophore concentrations (ca. 10 and 100 μmol/l cell). The estimated Ca^{2+}-sensitivity was three orders of magnitude higher (μ*M* range) when measured using the high ionophore concentration than when studied with the low ionophore concentration (Ca^{2+} sensitivity in the m*M* range). They proposed that the "minimally perturbed" state of the low ionophore concentration could approach the physiological state of the cell. Different Ca^{2+}-sensitivities for the K^+ channel have been reported in the literature. Ca^{2+} sensitivity in the μ*M* range was reported in ATP- or Mg-depleted red cells. (Lew, 1970; Kregenow and Hoffman, 1972; Szász et al., 1980; Simons, 1982) and in red-cell ghosts loaded with Ca buffers (Simons, 1976a,b; Yingst and

Hoffman, 1978, 1981, 1984). On the other hand, 10^{-3}–$10^{-4}M$ Ca^{2+} was needed to activate K^+ channels in ghosts loaded with unbuffered Ca (Hoffman, 1962), 3–4-wk-old bank erythrocytes (Romero and Whittam, 1970), or red cells exposed shortly to ionophore A23187 (Sarkadi et al., 1976; Gárdos et al., 1977). Sickle cells have a high cellular level of calcium, but they do not show channel activity unless A23187 is added (Lew and Bookchin, 1980). However, this is not owing to the channel being in a "refractory state," as proposed initially, but to the accumulation of calcium in intracellular inside-out vesicles (Bookchin and Lew, 1983; Lew et al., 1985).

Although the existence of high and low affinity states of the channel cannot be excluded, the results of Lew and Ferreira (1976) can now be reinterpreted in terms of heterogeneity in the calcium contents among different cells. In the low ionophore condition most of the calcium is accumulated near equilibrium in a fraction of cells, whereas the other fraction shows barely a detectable increase in Ca content. Hoffman and coworkers were the first to propose this possibility (Hoffman et al., 1980; Yingst and Hoffman, 1984), although concluding evidence about this point has only been provided more recently (*see* Lew and García-Sancho, 1985). Under this new perspective, not only the original results of Lew and Ferreira (1976), but most of the experimental results obtained under conditions of out-of-equilibrium pump-leak steady-state in fed cells must be reevaluated. These limitations do not apply to ATP-depleted red cells, where the homogeneity of Ca distribution has been conveniently documented (Simonsen et al., 1982).

The artificial electron-donor system ascorbate plus phenazine methosulphate (asc-PMS) increases the K^+ permeability in intact red cells in a Ca^{2+}-dependent manner. This effect is much greater in ATP-depleted cells than in normal cells, and it has been attributed to an increase of the Ca^{2+}-sensitivity of the K^+ channels (*see* García-Sancho and Herreros, 1983). This effect will be discussed further in connection with the redox regulation of the channel.

Propranolol also induces a Ca^{2+}-dependent increase of K^+ permeability in fresh red cells (Ekman et al., 1969; Manninen, 1970). The effect is probably not related to its β-blocking activity, as the concentrations needed to activate the K^+ channel (0.5–1 mM) are two orders of magnitude higher than those known to have antiarrythmic effects and the sterospecificity is the opposite (*see* Glynn and Warner, 1972; Lew and Ferreira, 1978). The origin of this effect is not yet fully understood, although three possible mechanisms have been proposed:

1. Displacement of Ca from internal binding sites (Szász and Gárdos, 1974; Porzig, 1975)
2. Increase of the Ca^{2+} permeability (Szász and Gárdos, 1974) and
3. Increase of the Ca^{2+} sensitivity of the channel (Simons, 1980).

Complex and unexpected interactions between propranolol and electron donors have recently been described (Skulskii and Manninen, 1984; *see below*). High concentrations of propranolol (3–5 m*M*) inhibit Ca^{2+}-dependent K^+ transport (Szász et al., 1978). Practolol, pronethanol, tetracaine, and other local anesthetics have much the same effects as propranolol (Szász and Gárdos, 1974; Porzig, 1975). Ethanol and other alcohols have also been shown to stimulate the K^+ channel (Yamamoto and Harris, 1983), but the mechanism has not been studied. Possible nonspecific effects of some of these drugs by interaction with the lipid bilayer should be kept in mind.

Pb added to red-cell suspensions is rapidly taken up by the cells, probably through the anion transport system (Simons, 1986a,b), and seems to be able to substitute for Ca^{2+} in the activation of the K^+ channels (*see* Riordan and Passow, 1973; Passow, 1981). It has been shown in ghosts loaded with Pb buffers that the activation takes place at nanomolar concentrations of Pb^{2+} (Simons, 1985). In single-channel conductance studies, Pb^{2+} produced the same activation pattern as Ca^{2+} (Grygorczyk and Schwarz, 1983). However, the mechanisms of action of Pb^{2+} and Ca^{2+} present some differences (*see* Riordan and Passow, 1971; Passow, 1981). This is especially relevant in the ability of Pb^{2+} to inhibit activation of K^+ channels under certain conditions. Both in intact cells and isolated membrane patches, low concentrations of Pb (10 μ*M*) activated K^+ channels, whereas high concentrations (100 μ*M*) inhibited them (Shields et al., 1985). Using one-step inside-out vesicles was found to affect the apparent threshold sensitivity to Ca^{2+} with a biphasic time course, first decreasing and later increasing it. The inhibitory effect was not observed when the addition of Ca preceded the addition of Pb or when Mg^{2+} was absent (Alvarez et al., 1986b). Mg itself antagonizes the effect of Ca^{2+} (Porzig, 1975; Simons, 1976b; García-Sancho et al., 1982). These complex interactions suggest the existence of several divalent cation binding sites that can accept either Ca, Mg, or Pb. The effect of the occupation of these sites on channel opening or allosteric modification of other site(s) (occlusion, stabilization, change of affinity) would vary depending on the actual cation bound to it.

It has recently been reported that incubation of red cells at low ionic strength increases the membrane permeability to K. The effect was attributed to structural changes of cell membrane that made K^+ channels accessible to external Ca^{2+} (Adorante and Macey, 1986).

3.2. Selectivity and Kinetics

In ATP-depleted red cells treated with calcium and A23187, the permeability ratio K^+/Na^+ is probably greater than 10^4 (Lew, 1974). In resealed ghosts, the following transport ratios have been measured (Simons, 1976b): 1(K):1.5(Rb):0.05(Cs). Li, like Na, is not transported. A K^+/Na^+ selectivity as low as 15:1 has nevertheless been reported in single-channel conductance experiments, with permeability ratios of 1(K):0.77(Rb):O(Cs) (Grygorczyk and Schwarz, 1983, 1985). The reason for these discrepancies is not known yet.

The nature of the system responsible for the Ca^{2+}-dependent K^+ fluxes, either "channel" or "carrier," has been subject to discussion. Early observations of saturation by external K^+ and countertransport, with both transient cell accumulation of ^{42}K added to the medium at low external K^+ concentrations and stimulation of the exit of preloaded ^{42}K by external K^+ (Manninen, 1970; Blum and Hoffman, 1971), led Blum and Hoffman (1971) to postulate that the Ca^{2+}-dependent K^+ transport was carrier-mediated. However, these observations could alternatively be explained by electrical coupling of diffusional fluxes taking into account changes of the membrane potential when the K^+ permeability is made greater than the Cl^- permeability (Glynn and Warner, 1972). The fact that inhibitors of Cl^- permeability inhibit the net K^+ efflux induced by Ca^{2+} and, at the same time, increase the uptake of ^{42}K (Gárdos et al., 1975; Cotterrell, 1975) gives further support for this explanation. Direct microelectrode measurements in the giant red cells of *Amphiuma* have shown membrane hyperpolarization during activation of Ca^{2+}-dependent K^+ channels induced by insertion of the electrode in high-Ca medium (Lassen, et al., 1973, 1976) or treatment with A23187 + Ca (Gárdos et al., 1976). Direct estimations of the membrane potential are difficult in human red cells. Microelectrode measurements are unreliable in these small cells, carbocyanine dyes inhibit the transport (Simons, 1976c, 1979), and the oxonol dye WW 781 stimulates it (Freedman and Novak, 1983). In spite of these problems, studies carried out with these fluorescent indicators (Pape, 1982; Freedman and Novak, 1983) or with the lipophilic cation tetraphenylphosphonium (Valdeolmillos et al., 1986) showed a considerable hyperpolarization on activation of the K^+ channel. The same result has been reported by Macey et al. (1978) and

Vestergaard-Bogind and Bennekou (1982) using an indirect method. In their technique, cells are incubated in unbuffered medium in the presence of the protonophore CCCP. Under these conditions, changes in the extracellular pH can be related to changes in the membrane potential. More recently, single-channel conductance experiments have shown unequivocally that the red cell Ca^{2+}-dependent K^+ fluxes take place across a channel (*see below*).

The hypothesis of electrical coupling explains saturation by external K^+ and countertransport-like kinetics. It does not explain, however, the reported stimulation of tracer uptake, either ^{42}K (Blum and Hoffman, 1971) or ^{86}Rb (García-Sancho et al., 1979; Manninen and Skulskii, 1981), by the increase of extracellular K^+ concentration in the 0–1 m*M* range. The presence of small amounts of extracellular K^+ is an absolute requirement for the activation of the exit of K^+ from red blood cell ghosts induced by addition of Ca (Knauf et al., 1975). Similar results have been reported for intact cells treated with propranolol (Manninen and Skulskii, 1981). Along the same lines, it has been reported that incubation of red-cell ghosts at 37°C in the absence of K^+ produces an irreversible inactivation of the channel, which is half-maximal in 7–10 min. As little as 0.5 m*M* external K^+ is enough to prevent this inactivation. To explain these facts, the presence of a specific modulating site for K^+ on the external side of the membrane has been proposed (Heinz and Passow, 1980). An intracellular site for Na^+ has also been postulated to explain the inhibition produced by intracellular Na^+. This inhibition is counteracted by extracellular K^+. This observation led to the suggestion of transmembrane interactions between the sites for extracellular K^+ and intracellular Na^+ (Knauf et al., 1975; Heinz and Passow, 1980). Similar observations have been made in membrane patches by the single-channel conductance technique (Grygorczyk and Schwarz, 1983, 1985). In other preparations, these interactions have been interpreted as interactions of the ions inside a single-file pore with multiple sites (Yellen, 1984). Evidence for single-file diffusion with at least three sites in the human red-cell channel has been recently provided (Vestergaard-Bogind et al., 1985), although conflicting results favoring noninteracting diffusion have also been reported (Kregenow and Hoffman, 1972). It is not possible at present to decide if these ions interact directly inside the pore, via a transmembrane allosteric effect, or both ways.

The number of channels per cell is an object of controversy. Lew et al. (1982a), using one-step inside-out vesicles, compared the fraction of intravesicular space that became free of previously loaded ^{86}Rb on activation of either sodium pumps or Ca^{2+}-dependent K^+ channels. The fraction was similar in both cases, which was interpreted to mean, under the assumption that some vesicles do not release ^{86}Rb because they do not possess the tested transport system, that the number of K^+ channels and sodium pumps per unit of surface area was similar, i.e., about 100–200 channel/cell. Our own estimates, also in inside-out vesicles, calculated from the mean number of channels per vesicle (estimated from the fraction of nonresponding vesicles assuming a Poisson distribution of the channels) and mean vesicular size, are consistent with these figures (Alvarez and García-Sancho, 1987). On the other hand, by comparison of the single-channel conductance, measured in patch-clamp experiments, and the rate of ^{42}K exchange, measured in intact cells treated with A23187 + Ca, a mean number of about 10 channels/cell has been proposed (Grygorczyk and Schwarz, 1983; Grygorczyk et al., 1984). Flux measurements indicated an extreme heterogeneity in channel distribution, with 75% of the cells having only 1–5 channels and 25% between 11 and 55 channels/cell. The measurements were performed in intact cells loaded with Ca using a low concentration of the ionophore A23187. This condition is similar to that described previously by Lew and Ferreira (1976) in which heterogeneity of Ca distribution among cells has been reported (Lew and García-Sancho, 1985). If heterogeneity of Ca distribution among cells applied to the experiments of Grygorczyk and coworkers, then that would produce flux curves that do not fit single exponentials, simulating a heterogeneity of channel distribution among cells. On the other hand, this situation would also lead to an underestimation of the rate constant for ^{42}K exchange under conditions of maximal activation and of the number of channels per cell estimated by comparison with the unitary conductance of the channels. In our hands, similar experiments, performed with higher ionophore concentrations produced ^{86}Rb uptake curves that fitted single exponentials (Alvarez et al., 1986a). Differences in methodology could also contribute to discrepancies in the estimates of the number of channels per cell (*see* Passow et al., 1986).

3.3. Inhibitors

Among the inhibitors of the red cell Ca^{2+}-dependent K^+ channel, quinine and its optical isomer quinidine are the best studied. The effect of these drugs is reversible and not the result of modification of the calcium fluxes (Armando-Hardy et al., 1975). Since the inhibition is antagonized by external K^+ and potentiated by internal Na^+, a mechanism of action has been postulated involving competition for the external activating site for K^+ (Reichstein and Rothstein, 1981). These drugs also inhibit the Ca^{2+}-dependent K^+ channel in other cells (Grinstein et al., 1982; Valdeolmillos et al., 1982; Iwatsuki and Petersen, 1985; Sanchez et al., 1986). However, this effect is not general or specific. For example, in rat submandibular gland, quinidine inhibits the adrenaline-induced K^+ efflux more strongly than the A23187 plus Ca-induced K^+ efflux (Kurtzer and Roberts, 1982), and in insulin-secreting cells, quinine inhibits Ca^{2+}-independent K^+ channels without affecting the Ca^{2+}-dependent ones (Findlay et al., 1985).

Ouabain partially inhibits the Ca^{2+}-dependent K^+ transport in energy-depleted red cells. This effect, along with the complex kinetic interactions between Na^+ and K^+ mentioned above, was originally interpreted as evidence for the identity between the sodium pump and the Ca^{2+}-dependent K^+ channel (Blum and Hoffman, 1971). However, Lew (1971b, 1974) showed that the action of ouabain can be explained by changes in the intracellular level of ATP arising from its effect on the operation of the sodium pump. More evidence against an identical nature of both systems will be discussed later.

Oligomycin has been reported to inhibit K^+ channel activation in intact red cells and ghosts (Blum and Hoffman, 1972; García-Sancho et al., 1979; Sanchez et al., 1980). It is ineffective, however, in one-step inside-out vesicles (Alvarez et al., 1984), and it has a paradoxical stimulating effect in some ghost preparations in which the K^+ channels were partially inactivated (Alvarez et al., 1985). Oligomycin inhibits the effect of Pb^{2+}, but only at maximally activating concentrations, whereas it has a stimulating effect at intermediate Pb^{2+} concentrations (Riordan and Passow, 1971). Oligomycin was unable to inhibit Ca^{2+}-dependent K^+ transport in the Ehrlich ascites-tumor cell (Valdeolmillos et al., 1982). This extremely variable behavior suggests that the interaction may involve a loosely bound component of the channel. The fact the oligomycin is also an inhibitor of the sodium pump (Whittam et al., 1964) was one of the arguments in favor of both systems being identical.

Other inhibitors of the red-cell channel whose effect has been studied in less detail include lanthanum (Szász et al., 1978), furosemide (Blum and Hoffman, 1970), ethacrinic acid, uncoupling agents, and antimycin A (García-Sancho et al., 1979), the *in vitro* antisickling agent cetiedil (Berkowitz and Orringer, 1981), and the carbocyanine dyes used to measure membrane potential (Simons, 1976c, 1979). Tetraethylammonium ions exhibited a curious behavior, inhibiting much more strongly the uptake of ^{42}K than the net exodus of KCl (Szász et al., 1978). Inhibition by excess Ca^{2+} has also been reported (Simons, 1976b; Vestergaard-Bogind, 1983). Inhibition by other divalent cations such as Mg^{2+} and Pb^{2+} has already been discussed (*see* section 3.1.). The effect of oxidoreductase inhibitors and neuroleptic drugs will be treated later.

Apamin, a component of bee venom, is able to inhibit with high affinity Ca^{2+}-dependent K^+ channels present in smooth muscle (Banks et al., 1979), neuroblastoma cells (Hugues et al., 1982b), and hepatocytes (Burguess et al., 1981), although it is inactive against the channels of mollusk neurons (Hermann and Hartung, 1983), pancreatic β cells (Atwater et al., 1983), red cells (Burguess et al., 1981), ascites tumor cells, or thymocytes (Abia et al., 1986). It has been recently shown by "patch-clamp" experiments in rat skeletal muscle (Blatz and Magleby, 1986) that the channels that are sensitive to apamin are those showing low unitary conductance (10–14 pS) and little or no voltage-dependence. They are responsible for the afterhyperpolarization that follows the action potential in many excitable cells. The physiological importance of these channels has been emphasized by the discovery in the mammalian brain of a natural peptide with the same effect as apamin (Fosset et al., 1984). In contrast, other Ca^{2+}-dependent K^+ channels repeatedly observed in "patch-clamp" experiments (Marty, 1983; Moczydlowski and Latorre, 1983; Latorre and Miller, 1983; Miller, 1983; Petersen and Maruyama, 1984) present higher conductance (150–250 pS), are strongly voltage-dependent, and are inhibitable by external tetraethylammonium (TEA) and not by apamin. Both kinds of channels apparently coexist in some cells (Romey and Lazdunski, 1984; Blatz and Magleby, 1986; *see also* Lazdunski, 1983). The red-cell channel would be different from both because it has small conductance (10–40 pS) with inward rectification, is very little voltage-dependent (Hamill, 1981, 1983; Grygorczyk and Schwarz, 1983), is not sensitive to apamin (Burguess et al., 1981), and is partially sensitive to TEA (Gárdos et al., 1975; Szász et al., 1978).

A component of the scorpion *Leiurus quinquiestriatus* venom, charybdotoxin, has been shown recently to inhibit reversibly the large-conductance, apamin-insensitive channel of skeletal muscle T-tubules (Miller et al., 1985), cultured mammalian kidney cells (Guggino et al., 1987), and the low-conductance channels of *Aplysia* (Hermann, 1986). The toxin has been purified and shown to be a highly stable basic polypeptide with a mass of 10 kdalton, that inhibits the skeletal muscle channel with an apparent dissociation constant of 3.5 n*M* (Smith et al., 1986). On the other hand, crude *Leiurus quinquiestriatus* venom inhibits the apamin-insensitive channels of human red cells, Ehrlich ascites-tumor cells, and rat thymocytes, as well as the apamin-sensitive channel of guinea-pig hepatocytes (Abia et al., 1986). By fractionation of the venom, two polypeptide toxins could be separated: one blocks apamin-sensitive K^+ fluxes in hepatocytes and inhibits ^{125}I-monoiodoapamin binding, and the other, more basic, blocks apamin-insensitive channels in red cells (Castle and Strong, 1986) and large-conductance channels in vascular smooth muscle (Beech et al., 1987). Both toxins showed a mol wt of about 5 kdalton by SDS-PAGE, and the second one probably corresponds to charybdotoxin, which shows itself as a band at about 6 kdalton by SDS-PAGE (Smith et al., 1986) and displays very similar pharmacological properties. The high affinity ($IC_{50} < 40$ n*M*) shown by the toxin active against the red-cell channel suggests that it may be an ideal ligand for use in purifying the channel. Note that the recent results with charybdotoxin suggest structural similarities between some low- and high-conductance K^+ channels distinguished by patch-clamp measurements, since channels of both kinds were sensitive to that toxin.

Irreversible inhibition or inactivation of the K^+ channels by several experimental treatments has been described. The inactivation by the lack of external K^+ observed in red-cell ghosts has already been discussed (*see* section 3.2.). Slow irreversible inhibition has also been reported in intact cells by prolonged treatment with iodoacetate but not with iodoacetamide (Plishker, 1985). Vesiculation of ghosts at a pH value of 8 or higher has been reported to yield inside-out vesicles with inactivated K^+ channels (Lew et al., 1982a). We have observed a time-dependent inactivation of K^+ channels by prolonged incubation (24–48 h) of intact cells at 37°C with or without metabolic substrates. This kind of inactivation can be reversed by treatment with the electron donor system ascorbate plus phenazine methosulphate.

3.4. Regulation of the Activity

It has often been postulated that factors related to the metabolic state of the red cell could modify the Ca^{2+} sensitivity of the K^+ channels (*see* Riordan and Passow, 1973; Passow, 1981). This idea originated from the fact that some of the treatments known to activate Ca^{2+}-dependent K^+ channels involved alteration of the red-cell metabolism. Even though we know today that the main effect of these treatments is to facilitate calcium accumulation in the cytoplasm by inhibition of the Ca-pump resulting from ATP-depletion, there are some observations that cannot be explained on these grounds, such as the increase of the sensitivity to Ca^{2+} effected by fluoride and iodoacetate in ATP-depleted ghosts (*see* Riordan and Passow, 1973).

The potential importance of the NAD^+/NADH ratio as a possible control mechanism was first suggested by Passow (*see* Passow, 1981). We have already mentioned that the Ca^{2+}-dependent K^+ permeability of human red cells can be activated by treatment with the artificial electron donor system ascorbate plus phenazine methosulphate (asc-PMS). In resealed ghosts containing Ca^{2+} buffers, asc-PMS increased the sensitivity of the K^+ channel to cytoplasmic Ca^{2+} (García-Sancho et al., 1979). Using inside-out vesicles incubated with redox buffers, the sensitivity to Ca^{2+} of the K^+ channel has been shown to vary with redox potential in the 0–100 mV range, and the suggestion has been made that the affinity for Ca^{2+} depends on the redox state of a membrane component able to exchange two electrons and with an apparent standard redox potential (E'_0, pH 7.5) of about 50 mV. Reduced and oxidized states would show high and low affinity, respectively, the difference between both extremes corresponding to a variation of 15 times in the Ca^{2+} concentration needed to produce half-maximum activation. The presence of PMS is necessary to produce the reduced high affinity state, but the oxidized state can be obtained by adding oxidized cytochrome c alone (Alvarez et al., 1984). In addition to its effect on the Ca^{2+} sensitivity, asc-PMS also produces an increase in the transport rate at maximally activating Ca^{2+} concentrations. This effect has been interpreted as a redox modulation of the conductance of the channels (Alvarez et al., 1986c). Redox regulation has been reported recently in intact red cells, where variations of the NAD^+/NADH ratio modified the Ca^{2+} sensitivity of the K^+ channel (Alvarez et al., 1986a). Treatment with asc-PMS has also been shown to restore the sensitivity to Ca^{2+} of previously inactivated channels in red blood cells ghosts (Alvarez et al., 1985) and intact cells (unpublished results, *see above*). Propranolol has a paradoxical effect inhibiting the ^{86}Rb efflux induced by asc-PMS in intact cells. For the development of this inhibitory effect, the addition of propranolol must be done after the addition of asc-PMS (Skulskii and Manninen, 1984).

It has been suggested that plasma membrane NADH-dehydrogenase could be involved in the redox modulation of Ca^{2+}-dependent K^+ transport on the basis of the effects of several oxidoreductase inhibitors like atebrin, chlorpromazine, oligomycin, and antimycin A (García-Sancho et al., 1979; Sanchez et al., 1980). The cholesterol-lowering drug, probucol, inhibits both the channel and the oxidoreductase activity in the same concentration range (Howland et al., 1984). Pb, menadione, and menadione analogs also affect both activities, although not always in a parallel way (Fuhrmann et al., 1985). However, no correlation was found between the membrane-linked NADH-dehydrogenase and the Ca^{2+}-dependent K^+ transport activities in erythrocytes of several mammalian species (Miner et al., 1983). Although the mechanism is still unclear, the potential importance of this regulation system, able to link cell metabolism and cell permeability, deserves further study.

The involvement of cytoplasmic proteins in the modulation of Ca^{2+}-dependent K^+ transport activity is controversial at present. Sarkadi et al.

(1980) reported that a heat-labile cytoplasmic protein concentrate partially restored Ca^{2+}-induced K^+ transport in inside-out vesicles in which this transport system was inactivated. However, this factor could not be evidenced in other laboratories (Grinstein and Rothstein, 1978; Sze and Solomon, 1979; Lew et al., 1982a). More recently, Pape and Rothstein, 1978; Sze and Solomon, 1979; Lew et al., 1982a). More recently, Pape and Kristensen (1984) reported a stimulating effect of calmodulin on the K^+ channel activity of a similar preparation of inside-out vesicles. This result could not be reproduced in a preparation of one-step inside-out vesicles, which is known to preserve much better Ca^{2+}-dependent K^+ transport activity (*see above*), even though (Ca + Mg)-ATPase activity, measured simultaneously, was significantly stimulated by calmodulin (Alvarez et al., 1986d).

A second line of evidence favoring the involvement of calmodulin on Ca^{2+}-dependent K^+ transport modulation relies on the ability of some neuroleptic drugs, well-known calmodulin antagonists (Weiss and Levin, 1978), to inhibit the K^+ channel activity (Lackington and Orrego, 1981). However, in a recent study (Alvarez et al., 1986d):

1. no correlation was found between K_I for inhibition of calmodulin-dependent processes and K_I for inhibition of the K^+ channel
2. the value of K_I depended strongly on the hematocrit used, suggesting accumulation of the drugs in the cells, and presumably in the lipid bilayer because of their lipophilic character and
3. effects on the valinomycin-induced K^+ fluxes, either activation or inhibition, were detected for every drug.

The effects of neuroleptic drugs on Ca^{2+}-induced ^{86}Rb uptake by one-step inside-out vesicles were extremely variable among the different drugs, with some of them (trifluoperazine) increasing the sensitivity to Ca^{2+} (Alvarez et al., 1986d). Stimulation of Ca^{2+}-dependent K^+ transport by some neuroleptic drugs has been reported by Plishker (1984) in intact human red cells, but this action was explained as a side effect of the inhibition of the Ca-pump. Regarding the data obtained in other cell lines, the evidence is conflicting as well. In human lymphocytes (Grinstein et al., 1982) and in Ehrlich cells (*see* Hoffman, 1985), pimozide, trifluoperazine, and chlorpromazine, well-known calmodulin antagonists, were reported to inhibit regulatory volume decrease, a mechanism thought to involve Ca^{2+}-dependent K^+ channels. However, chlorpromazine and trifluoperazine were unable to inhibit KCl loss induced by treatment with A23187 + Ca in the Ehrlich cell (M. Valdeolmillos, unpublished results), and neither calmodulin nor trifluoperazine injected in snail neurons was able to modify Ca^{2+}-activated K^+ current (Levitan and Levitan, 1986).

The Ca^{2+}-dependent K^+ transport in human red cells is also dependent on the cell pH (Hoffman et al., 1980; Freedman and Novak, 1983; Stampe and Vestergaard-Bogind, 1985). Although the modulation takes place outside the physiological values of red-cell pH, the possible importance of this regulation system in other cell lines has been suggested (Stampe and Vestergaard-Bogind, 1985).

3.5. All or Nothing Behavior

Ca^{2+}-dependent K^+ transport in red cells, ghosts, or inside-out vesicles responds under certain conditions in an all or nothing manner to changes in the concentration of agonist. All or nothing behavior means that, at submaximal agonist concentrations, only a fraction of the intracellular or intravesicular space equilibrates its tracer (^{42}K or ^{86}Rb) with the external medium, and the increase of agonist concentration increases the fraction of the intracellular space that becomes accessible to the tracer. In a gradual response, on the other hand, variations in the concentration of agonist would produce variations in the rate of equilibration, the whole space accessible to the tracer being the same at all the agonist concentrations. As emphasized in a recent review on this subject (Lew et al., 1983), the meaning of all or nothing response reported in intact cells or ghosts is different from that reported in vesicle preparations, where the mean number of channels per vesicle could be close to 1. In the first case, the all or nothing response implies coordination of the thresholds for activation of all the channels in each cell, whereas in the second, it implies differences in the behavior of the individual channels.

Several artifacts could simulate all or nothing response in intact cells, especially a heterogeneous distribution of agonist among different cells. The heterogeneity of Ca distribution that appears in fed cells treated with low ionophore concentrations has already been mentioned. Heterogeneity of Ca distribution has also been observed in ATP-depleted cells incubated in calcium-containing medium (Alvarez et al.,1988), one of the situations in which all or nothing responses appear (Szász et al., 1974). In this case, heterogeneity seems to arise from differences in the mobilization or the utilization by the calcium pump of the reserve of ATP contained in the 2,3-diphosphoglycerate. Added lead enters red cells quickly without the need of divalent cation ionophores (Simons, 1983, 1986a,b) and activates K^+ channels with clear all or nothing cell responses (*see* Riordan and Passow, 1973). However, the recent demonstration that the calcium pump is also able to extrude lead (Simons, 1984) throws doubts on a possible heterogeneity in the Pb^{2+} distribution, since a pump-leak steady-state with individual variations from cell to cell could then be established. In fact, a small but significant heterogeneity of lead distribution has been recently detected (Alvarez et al., 1988). All or nothing responses were also reported in ghosts loaded with Ca^{2+} buffers (Heinz and Passow, 1980). In this case, heterogeneity of agonist distribution seems difficult to think of. Recent work, however, has shown a clear gradual response in conditions in which a highly homogeneous Ca^{2+} concentration in the submicromolar range was forced by the addition of an excess of ionophore A23187 (100 μmol/l cells) and different calcium concentrations to chelator-loaded red cells (Alvarez et al., 1986a).

Heterogeneity of agonist distribution is not a problem when using inside-out vesicles, because the concentration of agonist, present in the external medium, is the same for every channel. All or nothing responses reported in this case would then reflect an all or nothing behavior of the individual channels (García-Sancho et al., 1982; Lew et al., 1982a). The incongruity of this phenomenon with the graded response observed in patch-clamp experiments has been pointed out (Lew et al., 1983). Recently, the kinetics of the ^{86}Rb uptake in inside-out vesicles has been examined in detail, and it has been shown that the increase in the Ca^{2+} concentration produces an increase of both the accessible ^{86}Rb space and the rate of tracer equilibration (Alvarez et al., 1986c). The all or nothing response in the vesicles can then be compatible with the "patch-clamp" findings, if we assume that Ca^{2+}-sensitivity varies from channel to channel and that the activation curve for each channel as a function of Ca^{2+} concentration is very sharp. The last condition is necessary for the coexistence of completely activated channels with others having a very low open-state probability. In molecular terms, it means cooperativity in the mechanism of activation or activation by more than one Ca^{2+} ion. This has already been proposed for Ca^{2+}-dependent K^+ channels studied by patch-clamp in other cell preparations (Moczydlowski and Latorre, 1983; Latorre and Miller, 1983; Miller, 1983). For the red-cell channel, the activation curves for Ca^{2+} (Porzig, 1977; Simons, 1976a,b) and Pb^{2+} (Passow, 1981) are sigmoidal, suggesting also the binding of two or more ions/channel. The reason for the presence of variations of Ca^{2+} sensitivity from channel to channel in inside-out vesicles is not clear. This phenomenon has also been observed in patch-clamp experiments, even in channels from the same preparation (Moczydlowski and Latorre, 1983; Latorre and Miller, 1983). Possible reasons for this variability might include protein phosphorylation (De Peyer et al., 1982; Wen et al., 1984), differences in the lipidic environment (Moczydlowski and

Latorre, 1983; Miller, 1983) or redox state (García-Sancho and Herreros, 1983), or existence of dissociable subunits. Interactions with the residual cytoskeleton in the vesicles cannot be excluded.

3.6. Molecular Substrate

The identity of the molecular substrate of Ca^{2+}-activated K^+ transport in red cells has long been object of controversy. Both sodium pump (Blum and Hoffman, 1971) and calcium pump (Lassen et al., 1980) have been suggested as the structural basis for this transport system. However, considerable evidence exists against these hypotheses. Apart from the already mentioned absence of any direct effect of ouabain on the K^+ channel, additional evidence against the identity with the sodium pump is based on the lack of correlation between the presence of sodium pumps and Ca^{2+}-dependent K^+ channels in red cells of different animal species (Jenkins and Lew, 1973; Richhardt et al., 1979; Miner et al., 1983; Parker, 1983; Brown et al., 1978), the lack of effect on the K^+ channel of an antibody able to inhibit all the transport mediated by the pump (*see* Lew and Beaugé, 1979), the different kinetics shown by the Ca^{2+}-activation of K^+ transport and the Ca^{2+} inhibition of the sodium pump (Brown and Lew, 1983; Yingst and Hoffman, 1984), and the fact that proteoliposomes reconstituted from purified sodium pumps do not show Ca^{2+}-dependent K^+ transport (Karlish et al., 1981). Some evidence is also available against the involvement of the calcium pump. Antibodies against the purified calcium ATPase inhibited ATPase activity in resealed erythrocytes, but had no effect on the Ca^{2+}-dependent K^+ transport, and reconstituted liposomes containing purified active calcium pump did not show any channel activity (Verma and Penniston, 1985).

The very small number of channels (100–200/cell at most) makes very difficult any attempt of isolation and identification. The availability of a high affinity toxin may facilitate these tasks in the same way as apamin is being used to characterize its receptor. The mol wt and the subunit composition of the Ca^{2+}-dependent K^+ channel sensitive to apamin are now being studied by radiation inactivation and affinity labeling with apamin (Hugues et al., 1982a; Schmid-Antomarchi et al., 1984; Seagar et al., 1986).

4. Summary

The last 15 yr have been a parallel development in the characterization of the properties of the Ca^{2+}-dependent K^+ channel of the red cell and the discovery of many similar channels in other cell lines. Although the physiological role of the red-cell channel is still unknown, a variety of physiological functions have been proposed for the channels present in several excitable and nonexcitable cells. The degree of similarity among the channels present in different cell lines is unclear. Important differences have been reported in terms of unitary conductance, voltage-dependence, and sensitivity to several inhibitors, but basic structural similarities cannot be excluded and constitute a useful working hypothesis. Recent results showing inhibition of both high- and low-conductance K^+ channels by a new highly active inhibitor, charybdotoxin, support this view. The red cell can then still be considered a good general model for the study of Ca^{2+}-dependent K^+ channels.

The effect of most of the known mechanisms of activation is to facilitate the access of Ca^{2+} to the internal activatory site. More than one Ca ion seems necessary for activation. Pb^{2+} seems to be able to substitute for Ca^{2+}, although complex interactions among Ca^{2+}, Pb^{2+}, and Mg^{2+} seem to take place at the activation or regulatory site(s). The presence of external K^+ is also required, and the increase of cytoplasmic Ca^{2+} in the absence of extracellular K^+ leads to an irreversible inactivation of the channel. The mechanism of action of the electron donor system ascorbate plus phenazine methosulphate involves an increase of the Ca^{2+} sensitivity of the channel. Recent results suggest that redox modulation of the K^+ channel could be relevant under physiological conditions. A role for calmodulin in the regulation of the K^+ channel has been proposed, but the evidence is contradictory and no clear conclusion about its implication can yet be decided.

There is now very strong evidence in favor of the ''channel'' nature of this transport system, the most conclusive one being provided by the results from single-channel conductance experiments. The arguments for the existence of this channel as an independent entity, unrelated to the sodium or calcium pumps, are also very strong.

Estimates of the mean number of channels/cell oscillate between 10 and 200 depending on the methodology used. Coordination of the thresholds for Ca^{2+} of all the channels within each cell with differences of Ca^{2+}-sensitivity among different cells has been proposed on the basis of observed all or none cell responses. Possible heterogeneity of agonist distribution among cells obscures, however, the interpretation of this interesting phenomenon. The all or nothing response observed in one-step inside-out vesicles has a different meaning. It seems to reflect differences in Ca^{2+} threshold among individual channels. The origin and relevance of this behavior are unknown at present. The biochemical substrate of the Ca^{2+}-dependent K^+ channel is still unknown, but the recent availability of a high affinity inhibitor may help for purification and identification.

Acknowledgment

Support from the Spanish DGICYT (PB-86/0312) is gratefully acknowledged. We thank Prof. B. Herreros and Dr. V. L. Lew for critical reading of the manuscript.

References

Abia, A., Lobatón, C. D., Moreno, A., and García-Sancho, J. (1986) *Leiurus quinquestriatus* venom inhibits different kinds of Ca^{2+}-dependent K^+ channels. *Biochim. Biophys. Acta* **856,** 403–407.

Adorante, J. S. and Macey, R. I. (1986) Calcium-induced transient potassium efflux in human red blood cells. *Am. J. Physiol.* **250,** C55–C64.

Alvarez, J. and García-Sancho, J. (1987) An estimate of the number of Ca^{2+}-dependent K^+ channels in the human red cell. *Biochim. Biophys. Acta* **903,** 543–546.

Alvarez, J., García-Sancho, J., and Herreros, B. (1984) Effects of electron donors on Ca^{2+}-dependent K^+ transport in one-step inside-out vesicles from the human erythrocyte membrane. *Biochim. Biophys. Acta* **771,** 23–27.

Alvarez, J., García-Sancho, J., and Herreros, B. (1985) Paradoxical effects of oligomycin on Ca-dependent ^{86}Rb transport in human red cell ghosts. *J. Physiol.* **369,** 113P.

Alvarez, J., García-Sancho, J., and Herreros, B. (1986b) Inhibition of Ca^{2+}-dependent K^+ channels by lead in one-step inside-out vesicles from human red cell membranes. *Biochim. Biophys. Acta* **857,** 291–294.

Alvarez, J., García-Sancho, J., and Herreros, B. (1986c) Analysis of the all or nothing behavior of Ca-dependent K channels in one-step inside-out vesicles from human red cell membranes. *Biochim. Biophys. Acta* **859,** 56–60.

Alvarez, J., García-Sancho, J., and Herreros, B. (1986d) The role of calmodulin on Ca^{2+}-dependent K^+ transport regulation in the human red cell. *Biochim. Biophys. Acta* **860,** 25–34.

Alvarez, J., García-Sancho, J., and Herreros, B. (1988) All or none cell responses of Ca^{2+}-dependent K channels elicited by calcium or lead in human red cells can be explained by heterogeneity of agonist distribution. *J. Membr. Biol.* **104,** 129–138.

Alvarez, J., Camaleño, J. M., García-Sancho, J., and Herreros, B. (1986a) Modulation of Ca^{2+}-dependent K^+ transport by modifications of thc NAD^+/NADH ratio in intact human red cells. *Biochim. Biophys. Acta* **856,** 408–411.

Armando-Hardy, M., Ellory, J. C., Ferreira, H. G., Fleminger, S., and Lew, V. L. (1975) Inhibition of the calcium-induced increase in the potassium permeability of human red blood cells by quinine. *J. Physiol. (Lond.)* **250,** 32P–33P.

Atwater, I., Rosario, L., and Rojas, E. (1983) Properties of the Ca^{2+}-activated K^+ channel in pancreatic beta-cells. *Cell Calcium* **4,** 451–461.

Banks, B. E. C., Brown, C., Burgess, G. M., Burnstock, G., Claret, M., Cocks, T. M., and Jenkinson, D. H. (1979) Apamin blocks certain neurotransmitter-induced increases in potassium permeability. *Nature* **282,** 415–417.

Beech, D. J., Bolton, T. B., Castle, N. A., and Strong, P. N. (1987) Characterization of a toxin from scorpion (*Leiurus quinquestriatus*) venom that blocks in vitro both large (BK) K^+-channels in rabbit vascular smooth muscle and intermediate (IK) conductance Ca^{2+}-activated K^+ channels in human red cells. *J. Physiol. (Lond.)* **387,** 32P.

Berkowitz, L. R. and Orringer, E. P. (1981) Effect of cetiedil, an in vitro antisickling agent on erythrocyte membrane cation permeability. *J. Clin. Invest.* **68,** 1215–1220.

Blatz, A. L. and Magleby, K. L. (1986) Single apamin-blocked Ca-activated K^+ channels of small conductance in cultured rat skeletal muscle. *Nature* **323,** 718–720.

Blum, R. M. and Hoffman, J. F. (1970) Carrier mediation of Ca-induced K transport and its inhibition in red blood cells. *Fed. Proc.* **29,** 663a.

Blum, R. M. and Hoffman, J. F. (1971) The membrane locus of Ca-stimulated K transport in energy depleted human red blood cells. *J. Membr. Biol.* **6,** 315–328.

Blum, R. M. and Hoffman, J. F. (1972) Ca-induced K transport in human red cells: localization of the Ca-sensitive site to the inside of the membrane. *Biochem. Biophys. Res. Commun.* **46,** 1146–1152.

Bodemann, H. and Passow, H. (1972) Factors controlling the resealing of the membrane of human erythrocyte ghosts after hypotonic hemolysis. *J. Membr. Biol.* **8,** 1–26.

Bookchin, R. M., and Lew, V. L. (1983) Red cell membrane abnormalities in sickle cell anemia, in *Progress in Haematology*, Vol XIII (Brown, E. B., ed.), Grune and Stratton, New York, pp. 1–23.

Brown, A. M. and Lew, V. L. (1983) The effect of intracellular calcium on the sodium pump of human red cells. *J. Physiol. (Lond.)* **343,** 455–493.

Brown, A. M., Ellory, J. C., Young, J. D., and Lew, V. L. (1978) A calcium-activated potassium channel present in foetal red cells of sheep but absent from reticulocytes and mature red cells. *Biochim. Biophys. Acta* **511,** 163–175.

Burguess, G. M., Claret, M., and Jenkinson, D. H. (1981) Effects of quinine and apamin on the calcium-dependent potassium permeability of mammalian hepatocytes and red cells. *J. Physiol. (Lond.)* **317,** 67–90.

Castle, N. A. and Strong, P. N. (1986) Identification of two toxins from scorpion (*Leiurus quinquestriatus*) venom which block distinct classes of calcium-activated potassium channel *FEBS Lett.* **209,** 117–121.

Cheung, W. Y., Lynch, T. J., and Wallace, R. W. (1978) An endogenous Ca^{2+}-dependent activator protein of brain adenylate cyclase and cyclic nucleotide phosphodiesterase, in *Advances in Cyclic Nucleotide Research*, vol 9 (George, W. J. and Ignarro, L. J., eds.), Raven Press, New York, pp. 233–251.

Cotterrell, D. (1975) The action of inhibitors of anion transfer on potassium and calcium movements in metabolically depleted human red cells. *J. Physiol. (Lond.)* **246,** 51P–52P.

De Peyer, J. E., Cachelin, A. B., Levitan, I. B., and Reuter, H. (1982) Ca^{2+}-activated K^+ conductance in internally perfused snail neurones is enhanced by protein phosphorilation. *Proc. Natl. Acad. Sci. USA* **79,** 4207–4211.

Ekman, A., Manninen, V., and Salminen, S. (1969) Ion movements in red cells treated with propranolol, *Acta Physiol. Scand.* **75,** 333–344.

Ferreira, H. G. and Lew, V. L. (1976) Use of ionophore A23187 to measure cytoplasmic Ca buffering and activation of the Ca pump by internal Ca. *Nature* **259,** 47–49.

Findlay, I., Dunne, M. J., Ullrich, S., Wollheim, C. B., and Petersen, O. H. (1985) Quinine inhibits Ca^{2+}-independent K^+ channels whereas tetraethylammonium inhibits Ca^{2+}-activated K^+ channels in insulin-secreting cells. *FEBS Lett.* **185,** 4–8.

Fosset, M., Schmid-Antomarchi, H., Hugues, M., Romey, G., and Lazdunski, M. (1984) The presence in pig brain of an endogenous equivalent of apamin, the bee venom peptide that specifically blocks Ca^{2+}-dependent K^+ channels. *Proc. Natl. Acad. Sci. USA* **81,** 7228–7232.

Freedman, J. C. and Novak, T. S. (1983) Membrane potentials associated with Ca-induced conductance in human red blood cells: studies with a fluorescent oxonol dye, WW 781. *J. Membr. Biol.* **72,** 59–74.

Fuhrmann, G. F., Schwarz, W., Kersten, R., and Sdun, H. (1985) Effects of vanadate, menadione and menadione analogs on the Ca^{2+}-activated K^+ channels in human red cells. Possible relations to membrane-bound oxidoreductase activity. *Biochim. Biophys. Acta* **820,** 223–234.

García-Sancho, J. and Herreros, B. (1983) Effects of redox agents on the Ca^{2+}-activated K^+ channel. *Cell Calcium* **4,** 493–497.

García-Sancho, J., Sanchez, A., and Herreros, B. (1979) Stimulation of monovalent cation fluxes by electron donors in the human red cell membrane. *Biochim. Biophys. Acta* **556,** 118–130.

García-Sancho, J., Sanchez, A., and Herreros, B. (1982) All or nothing response of the Ca^{2+}-dependent K^+ channel in inside-out vesicles. *Nature* **296,** 744–746.

Gárdos, G. (1956) The permeability of human erythrocytes to potassium. *Acta Physiol. Acad. Sci. Hung.* **10,** 185–189.

Gárdos, G. (1958) The function of calcium in the potassium permeability of human erythrocytes. *Biochim. Biophys. Acta* **30,** 653–654.

Gárdos, G. (1966) The mechanism of ion transport in human erythrocytes. I. The role of 2,3-diphosphoglyceric acid in the regulation of potassium transport. *Acta Biochim. Biophys. Acad. Sci. Hung.* **1,** 139–148.

Gárdos, G. (1967) Studies on potassium permeability changes in human erythrocytes. *Experientia* **23,** 19–20.

Gárdos, G., Lassen, U. V., and Pape, L. (1976) Effects of antihistamines and chlorpromazine on the calcium-induced hyperpolarization of *Amphiuma* red cell membrane. *Biochim. Biophys. Acta* **448,** 599–606.

Gárdos, G., Szász, I., and Sarkadi, B. (1975) Mechanism of Ca-dependent K transport in human red cells, in *Biomembranes: Structure and Function, FEBS Proc.*, vol 35 (Gárdos, G., and Szász, I., eds.), North-Holland, Amsterdam, pp. 167–180.

Gárdos, G., Szász, I., and Sarkadi, B. (1977) Effect of intracellular calcium on the cation transport processes in human red cells. *Acta Biol. Med. Germ.* **36,** 823–829.

Glynn, I. M. and Lew, V. L. (1970) Synthesis of adenosine triphosphate at the expense of the downhill cation movements in intact human red cells. *J. Physiol. (Lond.)* **207,** 393–402.

Glynn, I. M. and Warner A. E. (1972) Nature of the calcium-dependent potassium leak induced by (+)-propranolol, and its possible relevance to the drug's antiarrhythmic effect. *Br. J. Pharmacol.* **44,** 271–278.

Grasso, M., Morelli, A. and De Flora, A. (1986) Calcium-induced alterations in the levels and subcellular distribution of proteolytic enzymes in human red blood cells. *Biochem. Biophys. Res. Commun.* **138,** 87–94.

Grinstein, S. and Rothstein, A. (1978) Chemically induced cation permeability in red cell membrane vesicles. The sidedness of the response and the proteins involved, *Biochim. Biophys. Acta* **508,** 236–245.

Grinstein, S., Dupre, A., and Rothstein, A. (1982) Volume regulation by human lymphocytes, Role of calcium. *J. Gen. Physiol.* **79,** 849–868.

Grygorczyk, R. and Schwarz, W. (1983) Properties of the Ca^{2+}-activated K^+ conductance of human red cells as revealed by the patch-clamp technique. *Cell Calcium* **4,** 499–510.

Grygorczyk, R. and Schwarz, W. (1985) Ca^{2+}-activated K^+ permeability in human erythrocytes: Modulation of single-channel events. *Eur. Biophys. J.* **12,** 57–65.

Grygorczyk, R., Schwarz, W., and Passow, H. (1984) Ca^{2+}-activated K^+ channels in human red cells: Comparison of single-channel currents with ion fluxes. *Biophys. J.* **45,** 693–698.

Guggino, S. E., Guggino, W. B., Green, N., and Sacktor, B. (1987) Blocking agents of the Ca^{2+}-activated K^+ channels in cultured medullary thick ascending limb cells. *Am. J. Physiol.* **252,** C128–C137.

Hamill, O. P. (1981) Potassium channel currents in human red blood cells. *J. Physiol. (Lond.)* **314,** 125P.

Hamill, O. P. (1983) Potassium and chloride channels in red blood cells, in *Single Channel Recording* (Sakmann, B., and Neher, E., eds.), Plenum Press, New York, pp. 451–471.

Hamill, O. P. Marty, A., Neher, E., Sakmann, B., and Sigworth, F. J. (1981)

Improved patch-clamp techniques for high-resolution current recording from cells and cell-free membrane patches. *Pflugers Archiv.* **391,** 85–100.

Heinz, A. and Passow, H. (1980) Role of external potassium in the calcium-induced potassium efflux from human red blood cell ghosts. *J. Membr. Biol.* **57,** 119–131.

Hermann, A. (1986) Selective blockade of Ca-activated K current in *aplysia* neurons by charybdotoxin. *Pflugers Archiv.* suppl. **406,** 204.

Hermann, A. and Hartung, K. (1983) Ca^{2+}-activated K^+ conductance in molluscan neurones. *Cell Calcium* **4,** 387–406.

Hoffman, E. K. (1985) Role of separate K^+ and Cl^- channels and of Na^+/Cl^- cotransport in volume regulation in Ehrlich cells. *Fed. Proc.* **44,** 2513–2519.

Hoffman, J. F. (1962) Cation transport and structure of the red cell plasma membrane. *Circulation* **26,** 1201–1213.

Hoffman, J. F., Yingst, D. R., Goldinger, J. M., Blum, R. M., and Knauf, P. A. (1980) On the mechanism of Ca-dependent K transport in human red blood cells, in *Membrane Transport in Erythrocytes* Alfred Benzon Symposium 14 (Lassen, U. V., Ussing, H. H., and Wieth, J. O., eds.), Munksgaard, Copenhagen, pp. 178–195.

Howland, J. L., Daughtey, J. N., Donatelli, M., and Theorfrastous, J. P. (1984) Inhibition of the erythrocyte calcium-sensitive potassium channel by probucol. *Pharmacol. Res. Commun.* **16,** 1057–1064.

Hugues, M., Schmid, H., and Lazdunski, M. (1982a) Identification of a protein component of the Ca^{2+}-dependent K^+ channel by affinity labelling with apamin. *Biochem. Biophys. Res. Commun.* **107,** 1577–1582.

Hugues, M., Romey, G., Duval, D., Vincent, J. P., and Lazdunski, M. (1982b) Apamin as a selective blocker of the calcium-dependent potassium channel in neuroblastoma cells: Voltage-clamp and biochemical characterization of the toxin receptor. *Proc. Natl. Acad. Sci. USA* **79,** 1308–1312.

Iwatsuki, N. and Petersen, O. H. (1985) Inhibition of Ca^{2+}-activated K^+ channels in pig pancreatic acinar cells by Ba^{2+}, Ca^{2+}, quinine and quinidine. *Biochim. Biophys. Acta* **819,** 249–257.

Jenkins, D. M. G. and Lew, V. L. (1973) Ca uptake by ATP-depleted red cells from different species with and without associated increase in K permeability. *J. Physiol. (Lond.)* **234,** 41P–42P.

Karlish, S. J. D., Ellory, J. C., and Lew, V. L. (1981) Evidence against Na^+-pump mediation of Ca^{2+}-activated K^+ transport and diuretic-sensitive (Na^+/K^+)-cotransport. *Biochim. Biophys. Acta* **646,** 353–355.

Knauf, P. A., Riordan, J. R., Schuhmann, B., Wood-Guth, I., and Passow, H. (1975) Calcium-potassium-stimulated net potassium efflux from human erythrocyte ghosts. *J. Membr. Biol.* **25,** 1–22.

Kregenow, F. M. and Hoffman, J. F. (1972) Some kinetic and metabolic characteristics of calcium-induced potassium transport in human red cells. *J. Gen. Physiol.* **60,** 406–429.

Kurtzer, R. J. and Roberts, M. L. (1982) Calcium-dependent K^+ efflux from rat submandibular gland. The effects of trifluoperazine and quinidine. *Biochim. Biophys. Acta* **693,** 479–484.

Lackington, I. and Orrego, F. (1981) Inhibition of calcium-activated potassium conductance of human erythrocytes by calmodulin inhibitory drugs. *FEBS Lett.* **133,** 103–106.

Lassen, U. F., Pape, L., and Vestergaard-Bogind, B. (1973) Membrane potential of *Amphiuma* red cells: Effect of calcium, in *Erythrocytes, Thrombocytes, Leukocytes* (Gerlach, E., Moser, K., Deutsch, E., and Wilmanns, W., eds.) Georg Thieme, Stuttgard, pp. 33–36.

Lassen, U. V. Pape, L., and Vestergaard-Bogind, B. (1976) Effect of calcium on the membrane potential of *Amphiuma* red cells. *J. Membr. Biol.* **26,** 51–70.

Lassen, U. V., Pape, L., and Vestergaard-Bogind, B. (1980) Calcium related transient changes in membrane potential of red cells, in *Membrane Transport in Erythrocytes* Alfred Benzon Symposium 14 (Lassen, U. V., Using, H. H., and Wieth, J. O., eds. Munksgaard, Copenhagen, pp. 255–273.

Latorre, R. and Miller, C. (1983) Conduction and selectivity in potassium channels. *J. Membr. Biol.* **71,** 11–30.

Lazdunski, M. (1983) Apamin, a neurotoxin specific for one class of Ca^{2+}-dependent K^+ channels. *Cell Calcium* **4,** 421–428.

Lepke, S. and Passow, H. (1968) Effects of fluoride on potassium and sodium permeability of the erythrocyte membrane. *J. Gen. Physiol.* **51,** 365s–372s.

Levitan, E. S. and Levitan, I. B. (1986) Apparent loss of calcium-activated potassium current in internally perfused snail neurons is due to accumulation of intracellular calcium. *J. Membr. Biol.* **90,** 59–65.

Lew, V. L. (1970) Effects of intracellular calcium on the potassium permeability of human red cells. *J. Physiol. (Lond.)* **206,** 35P–36P.

Lew, V. L. (1971a) On the ATP-dependence of the Ca^{2+}-induced increase in K^+ permeability observed in human red cells. *Biochim. Biophys. Acta* **233,** 827–830.

Lew, V. L. (1971b) Effect of ouabain on the Ca^{2+}-dependent increase in K^+ permeability in ATP depleted guinea-pig red cells. *Biochim. Biophys. Acta* **249,** 236–239.

Lew, V. L. (1974) On the mechanism of the Ca-induced increase in K permeability observed in human red cell membranes, in *Comparative Biochemistry and Physiology of Transport* (Bolis, L., Bloch, K., Luria, S. E., and Lyden, F., eds.), North-Holland, Amsterdam, pp. 310–316.

Lew, V. L., ed. (1983) Ca^{2+}-activated K^+ channels. *Cell Calcium*, vol 4, (Churchill-Livingstone, Edinburgh).

Lew, V. L. and Beaugé, L. (1979) Passive cation fluxes in red cell membranes, in *Membrane Transport in Biology*, vol II: *Transport Across Single Biological Membranes* (Giebich, G., Tosteson, D. C., and Ussing, H. H., eds.), Springer-Verlag, Berlin, pp. 81–115.

Lew, V. L. and Bookchin, R. M. (1980) Ca^{2+}-refractory state of the Ca-sensitive K^+ permeability mechanisms in sickle cell anemia red cells. *Biochim. Biophys. Acta* **602,** 196–200.

Lew, V. L. and Ferreira, H. G. (1976) Variable Ca sensitivity of a K-selective channel in intact red cell membranes. *Nature* **263,** 336–338.

Lew, V. L. and Ferreira, H. G. (1977) The effect of Ca on the K permeability of red cells, in *Membrane Transport in Red Cells* (Ellory, J. C., and Lew, V. L., eds.) Academic Press, London, pp. 93–100.

Lew, V. L. and Ferreira, H. G. (1978) Calcium transport and the properties of a calcium-activated potassium channel in red cell membranes, in *Current Topics in Membranes and Transport*, vol 10 (Bronner, F. and Kleinzeller, A., eds.), Academic Press, New York, pp. 217–277.

Lew, V. L. and García-Sancho, J. (1985) Use of the ionophore A23187 to measure and control cytoplasmic Ca^{2+} levels in intact red cells. *Cell Calcium* **6,** 15–23.

Lew, V. L., Muallem, S., and Seymour, C. A. (1982a) Properties of the Ca^{2+}-activated K^+ channel in one-step inside-out vesicles from human red cell membranes. *Nature* **296,** 742–744.

Lew, V. L., Muallem, S., and Seymour, C. A. (1983) The Ca^{2+}-activated K^+ channel of human red cells: All or none behavior of the Ca^{2+}-gating mechanism. *Cell Calcium* **4,** 511–517.

Lew, V. L., Tsien, R. Y., Miner, C., and Bookchin, R. M. (1982b) Physiological $(Ca^{2+})_i$ level and pump leak turnover in intact red cells measured using an incorporated Ca chelator. *Nature* **298,** 478–481.

Lew, V. L., Hockaday, A., Sepulveda, M. I., Somlyo, A. P., Somlyo, A. V., Ortiz, O. E., and Bookchin, R. M. (1985) Compartmentalization of sickle-cell calcium in endocytic inside-out vesicles. *Nature* **315,** 586–589.

Macey, R. I., Adorante, J. S., and Orme, F. W. (1978) Erythrocyte membrane potential determined by hydrogen ion distribution. *Biochim. Biophys. Acta* **512,** 284–295.

Manninen, V. (1970) Movements of sodium and potassium ions and their tracers in propranolol-treated red cells and diaphragm muscle. *Acta Physiol. Scand.* **Suppl. 355,** 1–76.

Manninen, V. and Skulskii, I. A. (1981) Effect of extracellular potassium on the loss of potassium from human red blood cells treated with propranolol. *Acta Physiol. Scand.* **111,** 361–365.

Marty, A. (1983) Ca^{2+}-dependent K^+ channels with large unitary conductance. *Trends in Neurosci.* **6,** 262–265.

Meech, R. W. (1976) Intracellular calcium and the control of membrane permeability, in *Calcium in Biological Systems* XXX Symp. Exptl. Biol. (Duncan, C. J., ed.), Cambridge University Press, London, pp. 161–191.

Meech, R. W. (1978) Calcium-dependent potassium activation in nervous tissues. *Ann. Rev. Biophys. Bioeng.* **7,** 1–18.

Miller, C. (1983) Integral membrane channels: studies in model membranes. *Physiol. Rev.* **63,** 1209–1242.

Miller, C., Moczydlowski, E., Latorre, R., and Phillips, M. (1985) Charybdotoxin, a protein inhibitor of single Ca^{2+}-activated K^+ channels from mammalian skeletal muscle. *Nature* **313,** 316–318

Miner, C., Lopez-Burillo, S., García-Sancho, J., and Herreros, B. (1983) Plasma membrane NADH dehydrogenase and Ca^{2+}-dependent potassium transport in erythrocytes of several animal species. *Biochim. Biophys. Acta* **727,** 266–272.

Moczydlowski, E. and Latorre, R. (1983) Gating kinetics of Ca^{2+}-activated K^+ channels from rat muscle incorporated into planar lipid bilayers. *J. Gen. Physiol.* **82,** 511–542.

Ørskov, S. L. (1935) Untersuchungen über des einfluss von kohlensaure und blei auf die permeabilität der blutkörperchen für kallium und rubidum. *Biochem. Z.* **279,** 250–261.

Pape, L. (1982) Effect of extracellular Ca^{2+}, K^+ and CH^- on erythrocyte membrane potential as monitored by the fluorescent probe 3,3-dipropylthiodicarbocyanine. *Biochim. Biophys. Acta* **686,** 225–232.

Pape, L. and Kristensen, B. I. (1984) A calmodulin-activated Ca^{2+}-dependent K^+ channel in human erythrocyte membrane inside-out vesicles. *Biochim. Biophys. Acta* **770,** 1–6.

Parker, J. C. (1969) Influences of 2,3-diphosphoglycerate metabolism on sodium-potassium permeability in human red blood cells: Studies with bisulfite and other redox agents. *J. Clin. Invest.* **48,** 117–125.

Parker, J. C. (1981) Effects of drugs on calcium related phenomena in red blood cells. *Fed. Proc.* **40,** 2872–2876.

Parker, J. C. (1983) Hemolytic action of potassium salts in dog red cells. *Am. J. Physiol.* **244,** C313–317.

Passow, H. (1981) Selective enhancement of potassium efflux from red blood cells by lead, in *The Function of Red Blood Cells: Erythrocyte Pathobiology* (Wallach, D. F. H., ed.), Alan R. Liss Inc., New York, pp. 79–104.

Passow, H., Shields, M., and La Celle, P. (1986) Effects of calcium on structure and function of the human red blood cell membrane, in *The Cytoskeleton* (Clarkson, T. W., Sager, P. R., and Syversen, T. L. M., eds.), Plenum Publishing Corporation, New York, pp. 177–186.

Petersen, O. H. and Maruyama, Y. (1984) Calcium-activated potassium channels and their role in secretion. *Nature* **307,** 693–696.

Plishker, G. A. (1984) Phenothiazine inhibition of calmodulin stimulated calcium-dependent potassium efflux in human red blood cells. *Cell Calcium* **5,** 177–185.

Plishker, G. A. (1985) Iodoacetic acid inhibition of calcium-dependent potassium efflux in red blood cells. *Am. J. Physiol.* **248,** C419–C424.

Porzig, H. (1975) Comparative study of the effects of propranolol and tetracaine on cation movements in resealed human red cell ghosts. *J. Physiol. (Lond.)* **249,** 27–49.

Porzig, H. (1977) Studies on the cation permeability of human red cell ghosts. *J. Membr. Biol.* **31,** 317–349.

Putney, J. W. (1979) Stimulus-permeability coupling: role of calcium in the receptor regulation of membrane permeability. *Pharmacol. Rev.* **30,** 209–245.

Reed, P. W. (1976) Effects of the divalent cation ionophore A23187 on potassium permeability of rat erythrocytes. *J. Biol. Chem.* **251,** 3489–3494.

Reichstein, E. and Rothstein, A. (1981) Effects of quinine on Ca^{2+}-induced K^+ efflux from human red blood cells. *J. Membr. Biol.* **59,** 57–63.

Richhardt, H. W., Fuhrmann, G. F., and Knauf, P. A. (1979) Dog red blood cells exhibit a Ca stimulated increase in permeability in the absence of (Na,K)-ATPase activity. *Nature* **279,** 248–250.

Riordan, J. R. and Passow, H. (1971) Effects of calcium and lead on potassium permeability of human erythrocyte ghosts. *Biochim. Biophys. Acta* **249,** 601–605.

Riordan, J. R. and Passow, H. (1973) The effects of calcium and lead on the potassium permeability of human erythrocytes and erythrocyte ghosts, in *Comparative Physiology* (Bolis, L., Schmidt-Nielsen, K., and Maddrell, S. H. P., eds.) North-Holland, Amsterdam, pp. 543–581.

Romero, P. J. and Whittam, R. (1970) The control by internal calcium of membrane permeability to sodium and potassium. *J. Physiol. (Lond.)* **214,** 481–507.

Romey, G. and Lazdunski, M. (1984) The coexistence in rat muscle cells of two distinct classes of Ca^{2+}-dependent K^+ channels with different pharmacological properties and different physiological functions. *Biochem. Biophys. Res. Commun.* **118,** 669–674.

Sanchez, A., García-Sancho, J., and Herreros, B. (1980) Effects of several inhibitors on the K^+ efflux induced by activation of the Ca^{2+}-dependent channel and by valinomycin in the human red cell. *FEBS Lett.* **110,** 65–68.

Sanchez, A., Valdeolmillos, M., García-Sancho, J., and Herreros, B. (1986) Ca^{2+}-dependent K^+ transport in lymphocytes. *Rev. Esp. Fisiol.* **42,** 459–464.

Sarkadi, B., and Gárdos, G. (1985) Calcium-induced potassium transport in cell membranes, in *The Enzymes of Biological Membranes*, vol 3 (Martonosi, A. N., ed.), Plenum Press, New York and London, pp. 193–234.

Sarkadi, B., Szász, I., and Gárdos, G. (1976) The use of ionophores for rapid loading of human red cells with radioactive cations for cation pump studies. *J. Membr. Biol.* **26,** 357–370.

Sarkadi, B., Szebeni, J., and Gárdos, G. (1980) Effects of calcium on cation transport processes in inside-out red cell membrane vesicles, in *Membrane Transport in Erythrocytes* Alfred Benzon Symposium 14 (Lassen, U. V., Ussing, H. H., and Wieth, J. O., eds.), Munksgaard, Copenhagen, pp. 220–235.

Schmid-Antomarchi, H., Huges, M., Norman, R., Ellory, C., Borsotto, M., and Lazdunski, M. (1984) Molecular properties of the apamin-binding component of the Ca^{2+}-dependent K^+ channel. Radiation-inactivation, affinity labelling and solubilization. *Eur. J. Biochem.* **142,** 1–6.

Schwarz, W. and Passow, H. (1983) Ca^{2+}-activated K^+ channels in erythrocytes and excitable cells. *Ann. Rev. Physiol.* **45,** 359–374.

Seagar, M. J., Labbe-Jullie, C., Granier, C., Goll, A., Glossman, H., Van Rietschoten, J., and Couraud, F. (1986) Molecular structure of rat brain apamin receptor: Differential photoaffinity labelling of putative K^+ channel subunits and target size analysis. *Biochemistry* **25,** 4051–4057.

Shields, M., Grygorczyk, R., Fuhrmann, G. F., Schwarz, W., and Passow, H. (1985) Lead-induced activation and inhibition of potassium-selective channels in the human red blood cell. *Biochim. Biophys. Acta* **815,** 223–232.

Simons, T. J. B. (1976a) The preparation of human red cell ghosts containing calcium buffers. *J. Physiol. (Lond.)* **256,** 209–225.

Simons, T. J. B. (1976b) Calcium-dependent potassium exchange in human red cell ghosts. *J. Physiol. (Lond.)* **256,** 227–244.

Simons, T. J. B. (1976c) Carbocyanine dyes inhibit Ca-dependent K efflux from human red cell ghosts. *Nature* **264,** 467–469.

Simons, T. J. B. (1979) Actions of a carbocyanine dye on calcium-dependent potassium transport in human red cell ghosts. *J. Physiol. (Lond.)* **288,** 481–507.

Simons, T. J. B. (1980) Effect of propranolol on calcium-dependent potassium permeability of red cell membranes. *J. Physiol. (Lond.)* **300,** 35P–36P.

Simons, T. J. B. (1982) A method for estimating free Ca within human red blood cells, with application to the study of their Ca-dependent permeability. *J. Membr. Biol.* **66,** 235–247.

Simons, T. J. B. (1983) The transport of lead ions across human red cell membranes. *J. Physiol. (Lond.)* **345,** 108P.

Simons, T. J. B. (1984) Active transport of lead by human red blood cells. *FEBS Lett.* **172,** 250–254.

Simons, T. J. B. (1985) Influence of lead ions on cation permeability in human red cell ghosts. *J. Membr. Biol.* **84,** 61–71.

Simons, T. J. B. (1986a) Passive transport and binding of lead by human red blood cells. *J. Physiol. (Lond.)* **378,** 267–286.

Simons, T. J. B. (1986b) The role of anion transport in the passive movement of lead across the human red cell membrane. *J. Physiol. (Lond.)* **378,** 287–312.

Simonsen, L. O., Gomme, J. and Lew, V. L. (1982) Uniform ionophore A23187 distribution and cytoplasmic calcium buffering in intact human red cells. *Biochim. Biophys. Acta* **692,** 431–440.

Skulskii, I. A. and Manninen, V. (1984) Interaction between propranolol and electron donors in altering the calcium ion-dependent potassium ion-permeability of the human red blood cell membrane. *Acta Physiol. Scand.* **120,** 329–332.

Smith, C., Phillips, M., and Miller C. (1986) Purification of charybdotoxin, a specific inhibitor of the high-conductance Ca^{2+}-activated channel. *J. Biol. Chem.* **261,** 14607–14613.

Stampe, P. and Vestergaard-Bogind, B. (1985) The Ca^{2+} sensitive K^+-

conductance of the human red cell membrane is strongly dependent on cellular pH. *Biochim. Biophys Acta.* **815,** 313–321.

Steck, T. L. and Kant, J. A. (1974) Preparation of impermeable ghosts and inside-out vesicles from human erythrocyte membranes. *Meth. Enzymol.* **31,** 172–180.

Szász, I. and Gárdos, G. (1974) Mechanism of various drug effects on the Ca^{2+}-dependent K^+-efflux from human red blood cells. *FEBS Lett.* **44,** 213–216.

Szász, I., Sarkadi, B., and Gárdos, G. (1974) Erythrocyte parameters during induced Ca^{2+}-dependent rapid K^+ efflux: Optimum conditions for kinetic analysis. *Haematologia* **8,** 143–151.

Szász, I., Sarkadi, B., and Gárdos, G. (1978) Effects of drugs on the calcium-dependent rapid potassium transport in calcium-loaded intact red cells. *Acta Biochim. Biophys. Acad. Sci. Hung.* **13,** 133–141.

Szász, I., Sarkadi, B., and Gárdos, G. (1980) Calcium sensitivity of calcium-dependent functions in human red blood cells, in *Advances in Physiological Sciences*, vol 6 (Eds. S. R. Hollan, G. Gárdos and B. Sarkadi), pp. 211–221, Pergamon Press, Akademiai Kiado, Budapest.

Szász, I., Sarkadi, B., Schubert, A., and Gárdos, G. (1978) Effects of lanthanum on calcium-dependent phenomena in human red cells. *Biochim. Biophys. Acta* **512,** 331–340.

Sze, H. and Solomon, A. K. (1979) Calcium-induced potassium pathway in sided erythrocyte membrane vesicles. *Biochim. Biophys. Acta* **554,** 180–194.

Szebeni, J. (1981) The Ca^{2+}-sensitive K^+ transport in inside-out red cell membrane vesicles. *Acta Biochim. Biophys. Acad. Sci. Hung.* **16,** 77–82.

Tiffert, T., García-Sancho, J., and Lew, V. L. (1984) Irreversible ATP depletion caused by low concentrations of formaldehyde and of calcium-chelator esters in intact human red cells. *Biochim. Biophys. Acta* **773,** 143–156.

Valdeolmillos, M., García-Sancho, J., and Herreros, B. (1982) Ca^{2+}-dependent K^+ transport in the Ehrlich ascites tumor cell. *Biochim. Biophys. Acta* **685,** 273–278.

Valdeolmillos, M., García-Sancho, J., and Herreros, B. (1986) Differential effects of transmembrane potential on two Na^+-dependent transport systems for neutral amino acids. *Biochim. Biophys. Acta* **858,** 181–187.

Verma, A. K. and Penniston, J. T. (1985) Evidence against involvement of the human erythrocyte plasma membrane Ca^{2+}-ATPase in Ca^{2+}-dependent K^+ transport. *Biochim. Biophys. Acta* **815,** 135–138.

Vestergaard-Bogind, B. (1983) Spontaneous inactivation of the Ca^{2+}-sensitive K^+ channels of human red cells at high intracellular Ca^{2+} activity. *Biochim. Biophys. Acta* **730,** 285–294.

Vestergaard-Bogind, B. and Bennekou, P. (1982) Calcium induced oscillations in the K^+ conductance and membrane potential of human erythrocytes mediated by the ionophore A23187. *Biochim. Biophys. Acta* **688,** 37–44.

Vestergaard-Bogind, B., Stampe, P. and Christophersen, P. (1985) Single-file diffusion through the Ca^{2+}-activated K^+ channels of human red cells. *J. Membr. Biol.* **88,** 67–75.

Weiss, B. and Levin, B. M. (1978) Mechanism for selectively inhibiting the activation of cyclic nucleotide phosphodiesterase and adenylate cyclase by antipsychotic agents, in *Advances in Cyclic Nucleotide Research*, vol 9 (George, W. J., and Ignarro, L. J., eds.), Raven Press, New York, pp. 285–303.

Wen, Y., Famulski, K. S., and Carafoli, E. (1984) Ca^{2+}-dependent K^+ permeability of heart sarcolemmal vesicles. Modulation by cAMP-dependent protein kinase activity and by calmodulin. *Biochem. Biophys. Res. Commun.* **122,** 237–243.

Whittam, R. (1968) Control of membrane permeability to potassium in red blood cells. *Nature* **219,** 610.

Whittam, R., Wheeler, K. P., and Blake, A. (1964) Oligomycin and active transport reactions in cell membranes. *Nature* **203,** 720–724.

Wilbrandt, W. (1937) A relation between the permeability of red cell and its metabolism. *Trans. Faraday Soc.* **33,** 956–959.

Wood, P. G. and Mueller, H. (1984) Modification of the cation selectivity filter and the calcium receptor of the Ca-stimulated K channel in resealed ghosts of human red blood cells by low levels of incorporated trypsin. *Eur. J. Biochem.* **141,** 91–95.

Yamamoto, H. and Harris, R. A. (1983) Calcium-dependent ^{86}Rb efflux and ethanol intoxication: studies of human red blood cells and rodent brain synaptosomes. *Eur. J. Pharmacol.* **88,** 357–363.

Yellen, G. (1984) Relief of Na^+ block of Ca^{2+}-activated K^+ channels by external cations, *J. Gen. Physiol.* **84,** 187–199.

Yingst, D. R. and Hoffman, J. F. (1978) Changes of intracellular Ca^{2+} as measured by arsenazo III in relation to the K permeability of human erythrocyte ghosts. *Biophys. J.* **23,** 463–471.

Yingst, D. R. and Hoffman, J. F. (1981) Effects of intracellular Ca on inhibiting Na-K pump and stimulating Ca-induced K transport in resealed human red cell ghosts. *Fed. Proc.* **40,** 543.

Yingst, D. R. and Hoffman, J. F. (1984) Ca-induced transport in human red blood cell ghosts containing arsenazo III: Transmembrane interactions of Na, K and Ca and the relationship to the functioning Na-K pump. *J. Gen. Physiol.* **83,** 19–45.

Amino Acid Flux Across the Human Red Cell Plasma Membrane

Godfrey Tunnicliff

1. Introduction

Certain criteria, if met, will point to the involvement of a specific process in the movement of solute across the plasma membrane. The primary criteria are the existence of specificity, saturation kinetics, and competitive inhibition by structural analogues. Transport of solute can be accomplished by either a facilitated diffusion or an active process, the latter of course requiring a source of free energy. Where the solute is an amino acid, both types of transport have been shown to occur in human red cells.

Relatively speaking, we have considerable knowledge of those processes governing sugar and anion permeability through human erythrocyte membranes. Notably, the transporters responsible for these solute fluxes have been isolated, purified, and characterized, including the elucidation of their primary structures (*see* Carruthers, Chapter 12; Jennings, Chapter 8).

Amino-acid transport, on the other hand, has received comparatively scant attention owing primarily, no doubt, to the widely held view that red cells carry out few functions that require amino acids. Furthermore, it has been only within the last few years that we have come to appreciate the extent and diversity of the processes involved in amino-acid flux. Only fairly recently, for instance, have we become fully aware that some amino-acid transport systems of human erythrocytes are Na^+-dependent (Young and Ellory, 1977).

The first indications that red-cell membranes were permeable to amino acids stemmed from the work of Constantino (1913a,b). He observed not only that red cells contained higher concentrations of amino-acid-type compounds than did serum, but that these cells were able to take up glycine and asparagine in vitro. Subsequent experiments by Abderhalden and Kurten (1921), however, led to the conclusion that the amino acids merely bound to the surface of the plasma membrane. Nevertheless, based on the observation that a portion of the alpha-amino nitrogen could be readily washed from red blood cells, Danielson (1933) argued that red-cell membranes were indeed permeable to amino acids. He also made the interesting suggestion that the alpha-amino nitrogen remaining after washing was chiefly glutathione. A decade later, Ussing (1943) monitored the uptake of leucine by erythrocytes and discovered that the amino acid readily crossed the plasma membrane. This was in sharp contrast to acidic amino acids, to which the membrane appeared impermeable. In 1947, Christensen and colleagues demonstrated that, when either plasma glycine or alanine was elevated in human subjects by ingestion of the appropriate pure amino acid, red cells showed a slow increase in the concentration of amino acid, indicating that glycine and alanine could cross the red-cell membrane. Later, Christensen et al. (1952) produced evidence that seemed to show that human erythrocytes could concentrate alanine and glycine. However, glutamate also appeared to accumulate against a concentration gradient, which of course was in seeming contradiction to the results of Ussing (1943). One of the first experiments to provide strong evidence for the involvement of a transporter in red-cell amino-acid translocation was by Rieser (1962), who reported that the transport of valine through the red-cell plasma membrane seemed to be dependent on a carrier.

Winter and Christensen (1964) provided the first detailed account of erythrocyte amino-acid transport with their study of the movement of neutral amino acids across the membrane. The size of the apolar side chain apparently determined the rate of uptake, but the presence of polar groups adversely affected the transport. The membranes were impermeable to glutamate and aspartate, a finding that did not support previous results from that laboratory, yet was consistent with Ussing's observations. Winter and Christensen described two saturable uptake systems: one for all neutral amino acids, but particularly those with large side chains, and a second one of lower capacity preferring alanine and glycine. Amino acids with large hydrocarbon side chains could also enter the cell via a third, nonsaturable route. These early studies laid the groundwork for what we now know of the permeability of human red-cell plasma membranes to amino acids. Christensen (1984) has described seven different amino-acid transporters present in various animal cells, but only four of these, together with an additional pathway specific for

L-tryptophan, have been shown to occur in human erythrocytes. These five different types of amino-acid transport systems so far recognized are now described.

2. Na^+-Independent Flux

2.1. L System

The work of Winter and Christensen (1964) led to the description of the L system in erythrocytes. Later studies confirmed the essential details (Hoare, 1972a,b; Young and Ellory, 1977; Rosenberg and Rafaelson, 1979; Young et al., 1980; Rosenberg, 1981a). This stereospecific transport system is optimal for L-isomers of neutral amino acids with large hydrophobic side chains, particularly leucine and phenylalanine. Smaller amino acids, however, are also transported at a significant rate—alanine and cysteine, for example. Indeed, about half of the total uptake of these latter two amino acids is considered to be through the L system (*see* review by Young and Ellory, 1979). Hoare (1972a,b) found that at 25°C the binding of the amino acid to the transporter increases the carrier reorientation rate, and that this is the rate-limiting step in leucine transport. At equilibrium, moreover, the carrier is equally distributed between the surfaces of the membrane.

A more detailed kinetic analysis of L-leucine transport by human erythrocytes has been carried out by Rosenberg (1981a). Results from net flux, equilibrium-exchange, and infinite experiments were analyzed by the simple pore and carrier theory of Lieb and Stein (1974a,b). These confirmed that leucine flux across the plasma membrane is mediated by a transporter, and it is now apparent that the process displays symmetric properties. In addition, analysis of the data led to the conclusion that the carrier-substrate complex was equally distributed across the membrane at equilibrium, as first reported by Hoare (1972a). Rosenberg also postulated that both the dissociation and the translocation of the carrier-substrate complex is faster than the translocation of the unoccupied carrier, and that no translocation step was rate determining.

The L system has a K_m of about 5 m*M* for leucine. For smaller neutral amino acids, though, the affinity for the carrier is more than 10 times lower. In an early study (Rieser, 1961), evidence seemed to show that sugar transport by red cells shared some of the mechanism of valine transport. If cells were allowed to take up glucose, galactose, or xylose, and were than incubated in valine under conditions in which uptake occurred, a marked inhibition of valine transport was observed. Similar effects were noted if proline or hydroxyproline were used in place of valine. Supposedly, Rieser measured amino-acid uptake by the L system. However, he monitored osmotically induced red cell volume changes and used this as an indirect estimation of uptake. Further, he did not carry out a kinetic analysis, nor did he employ any amino acids other than the three that have been mentioned. Before we can confidently say that the L system and sugar transport share a common mechanism, however limited, more data are needed.

The major route of histidine uptake by human red blood cells is via the L system. Yet this transport is much slower than that of other substrates for this uptake mechanism (Winter and Christensen, 1964; Christensen, 1968). It was shown that, even after 60 min, a steady state had not been achieved. A less significant portion of the histidine uptake occurred through the Ly system, described below. Later work from the same laboratory, in contrast, appears not to confirm that histidine recognizes the L system, since even at 10 m*M* this amino acid was unable to alter the rate of 50 μ*M* [^{14}C]leucine uptake by human red cells (Vadgama and Christensen, 1985). [^{14}C]Histidine was able to be transported by the same cells—presumably by the Ly system.

2.2. Ly System

The first indications that human red cells possess a translocation system specific for dibasic amino acids came from the report of Gardner and Levy (1972). They measured the uptake of labeled lysine, arginine, and ornithine, and concluded that the transport mechanism was Na^+-independent and specific for basic L-amino acids. Furthermore, the uptake was found to be temperature-sensitive and mediated by a carrier. The transport of lysine was shown to involve two components—one with a high affinity and the other with a low affinity for substrate. Unfortunately, with the exception of cystine, other amino acids were not studied. Thus, at that time, it was not obvious if lysine and arginine were transported by a newly discovered system or instead by the L system for neutral amino acids, as had been demonstrated to be the case in sheep erythrocytes (Young et al., 1980).

Studies utilizing rabbit red blood cells led to the proposal that a specific transport process for dibasic amino acids existed (Christensen and Antonioli, 1969; Antonioli and Christensen, 1969). This was later called the Ly system and has since been demonstrated to occur in human red cells (Young and Ellory, 1979; Young et al., 1980). In humans, L-lysine uptake consists of two separate elements; one is a high affinity process whose K_m value has been calculated as 68 μ*M*, whereas a second uptake component exists at high substrate concentrations but does not undergo saturation. When radiolabeled L-arginine was employed as substrate, the existence of a two-component transport system was confirmed, with kinetic properties very similar to those obtained with L-lysine. Competition experiments were performed, and the high affinity component was markedly inhibited by both L-arginine and L-ornithine. D-Lysine, on the other hand, had no effect on uptake, supporting the notion that this system is stereoselective, as first proposed by Gardner and Levy (1972).

2.3. T System

It is now established that tryptophan is taken up by erythrocytes by a carrier that is distinct from that associated with either the L or the Ly system. Christensen (1968) was the first to suggest that tryptophan behaved oddly as a substrate for the L system, but it was not until Rosenberg (1979) and, independently, Young and Ellory (1979) published their work that the existence of a transport system specific for tryptophan in human red blood cells was seriously considered. These researchers then combined their efforts and investigated tryptophan uptake in greater detail (Rosenberg et al., 1980). It was discovered that this so-called T system was optimal for L-tryptophan, but that D-tryptophan could also act as substrate. This latter conclusion was based on the observation that the D-isomer inhibited L-[^{14}C]tryptophan uptake. Subsequent work by Lopez-Burillo et al. (1985), however, has cast doubt on that interpretation since they, too, found that D-tryptophan inhibited L-tryptophan transport, but was *not* itself transported. It thus appears that this isomer binds to the T carrier, but that the resulting complex does not migrate across the plasma membrane. In addition, L-tyrosine and L-phenylalanine can be transported by this newly discovered system, though not as efficiently as by the L system.

The initial rate of L-tryptophan translocation by human erythrocytes has been investigated by Rosenberg (1981b) in zero-trans efflux and equilbrium-exchange efflux experiments. The data confirmed that tryptophan transport at physiological concentrations is predominantly via the T system. In addition, analysis of the results by the theory of Lieb and Stein (1974a,b) supported previous conclusions that the T system employs a carrier-mediated flux.

The T system has also been studied in erythrocytes infected with malaria parasites (Ginsburg and Krugliak, 1983). Evidently, infected cells transported tryptophan more rapidly than did noninfected cells, the effect being on maximum velocity rather than on the affinity of the carrier for substrate. At the same time in these infected cells, the uptake of tryptophan lost its usual sensitivity to inhibition by phloretin and phenylalanine. Certain unidentified factors released by the infected cells were able to markedly increase the V_{max} and K_m of the T system in noninfected cells. No mechanism to explain these membrane alterations has yet been forthcoming. The influence of malaria parasites on other amino-acid transport systems has not yet been studied.

3. Na$^+$-Dependent Flux

3.1. ASC System

In their review of red-cell amino-cell transport, Young and Ellory (1977) concluded that there was very little evidence that human erythrocytes could accumulate amino acids, with the possible exception of L-alanine and glycine. The sodium dependency of even these amino acids was questioned, since at that time, no kinetic analysis of uptake had been performed (Yunis and Arimura, 1965). From the perspective of a decade later, it is now accepted that at least two Na$^+$-dependent amino-acid transport processes are present in the plasma membrane of human red cells, and both are able to accumulate amino acids against a concentration gradient. One of these is known as the ASC system, so named because alanine, serine, and cysteine are the major substrates. The experiments of Young et al. (1979) demonstrated that L-[^{14}C]cysteine could enter human red blood cells via a Na$^+$-dependent, high affinity uptake mechanism (Fig. 1). There appeared to be an accumulation of the amino acid. However, the authors were not prepared to categorically state this was so, since they did not know the degree of cysteine metabolism. Nevertheless, these were the first conclusive data showing that human erythrocytes possess a Na$^+$-dependent transport system for an amino acid.

Subsequent experiments revealed that L-alanine, L-serine, and α-aminobutyrate were also substrates for this relatively high affinity (Km = 27 μM for cysteine), low-capacity transport process (Young et al., 1980; Al-Saleh and Wheeler, 1982; Rosenberg, 1982). Young et al. (1979) calculated that, under physiological conditions, about half the total L-cysteine uptake into a red cell is a direct result of this ASC system. Entry via the L system would account for the major portion of the remainder. Recently, evidence has been presented that glutamine also is a substrate for the ASC transport system (Ellory et al., 1983).

Dibasic amino acids, especially L-arginine, are inhibitors of the ASC in pigeon red cells and rabbit reticulocytes, but are not themselves

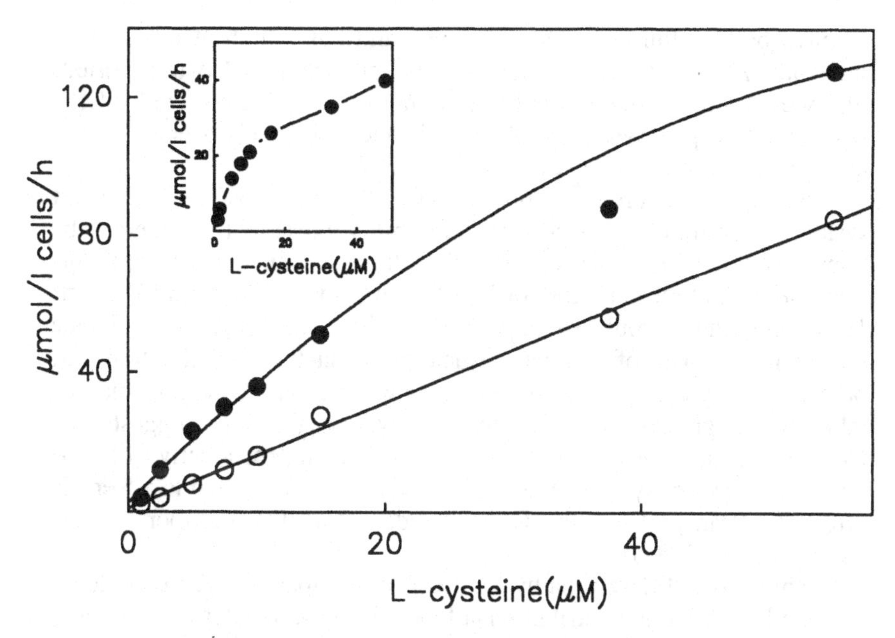

Fig. 1. Effect of concentration on sodium-dependent L-cysteine uptake by human red cells. Closed circles represent uptake in the presence of 150 m*M* NaCl; open circles represent uptake in the presence of choline chloride. The inset is derived from the difference between the two sets of data. Redrawn from Young et al. (1979).

transported (Thomas and Christensen, 1970; Thomas and Christensen, 1971). The explanation put forward was that the positively charged side chain amino or guanidino groups could bind to the Na^+ recognition site on the transporter and, thus, adversely affect translocation of the normal ASC substrates. Young et al. (1988) have recently reported that harmaline inhibited L-alanine uptake by human erythrocytes. This effect was competitive with respect to Na^+, supporting the idea that harmaline binds to the Na^+ site of the ASC system.

3.2. Gly System

Each of the previously described amino-acid transport systems is capable of moving glycine across the human red-cell plasma membrane (Young et al., 1980). The anion transporter also contributes to glycine flux (Young et al., 1981). It is now known, however, that yet another pathway process for glycine exists in these cells (Ellory et al., 1981; Al-Saleh and Wheeler, 1982; Rosenberg, 1982) and is, in fact, the major vehicle of glycine movement across the plasma membrane. This process requires both sodium and chloride ions, and has a high affinity for its substrate (K_m = 25 μ*M*); sarcosine and proline are also transported. Ellory et al. (1981) found that Br^-, but not I^-, could substitute for Cl^-. A glycine transport system similar to this one had already been described in pigeon red cells (Imler and Vidaver, 1972).

Because human red cells possess a Cl^- dependent Na^+/K^+ transport system (Dunham et al., 1980), the possibility exists that glycine can be transported by the same system. The diuretics bumetamide and furosemide are potent inhibitors of Na^+/K^+ transport (Ellory and Stewart, 1982) and, thus, would be expected to inhibit Cl^--dependent glycine uptake if transport of this amino acid was via the Na^+/K^+ transport process. Ellory and colleagues (1981), however, did not observe marked inhibition of glycine uptake by human erythrocytes. This suggests that Cl^--dependent glycine transport occurred independently of the Na^+/K^+ transport system. Moreover, glycine was not found to adversely affect K^+ transport via the Cl^--dependent Na^+/K^+ transport system (Ellory et al., 1981).

Thus, four different pathways for glycine uptake have been identified, and the relative contributions of each have been determined (Ellory et al., 1981). At a concentration of 0.2 m*M* glycine, the uptake by red cells was 42, 16, 15, and 11% by the Gly, band 3, L, and ASC system, respectively. The residual flux accounted for 16% of the total. It is not known if a mediated component contributes to this flux or if, instead, the cell has a basal permeability to glycine. Figure 2 illustrates the relative contribution of each route.

4. Nature of the Amino-Acid Transport Binding Sites

Since none of the amino-acid transporters of red cells has yet been isolated, our knowledge of the molecular nature of these carriers is sparse. From a few indirect experiments that have been reported, it is possible to gain some insight into the features of the various substrate recognition sites on the relevant carriers. For example, a study carried out by Al-Saleh and Wheeler (1982) indicated that Na^+-dependent glycine uptake in human red cells was markedly inhibited by *N*-ethylmaleimide, indicating the involvement of functional thiol groups in the binding of glycine to its carrier, and, thus, in the transport of this amino acid. Because glycine was unable to protect against *N*-ethylmaleimide inhibition, these thiol groups are presumably not at the actual glycine binding site, but could instead be in the vicinity of the site.

L-Tryptophan uptake by the T system, described above, is also

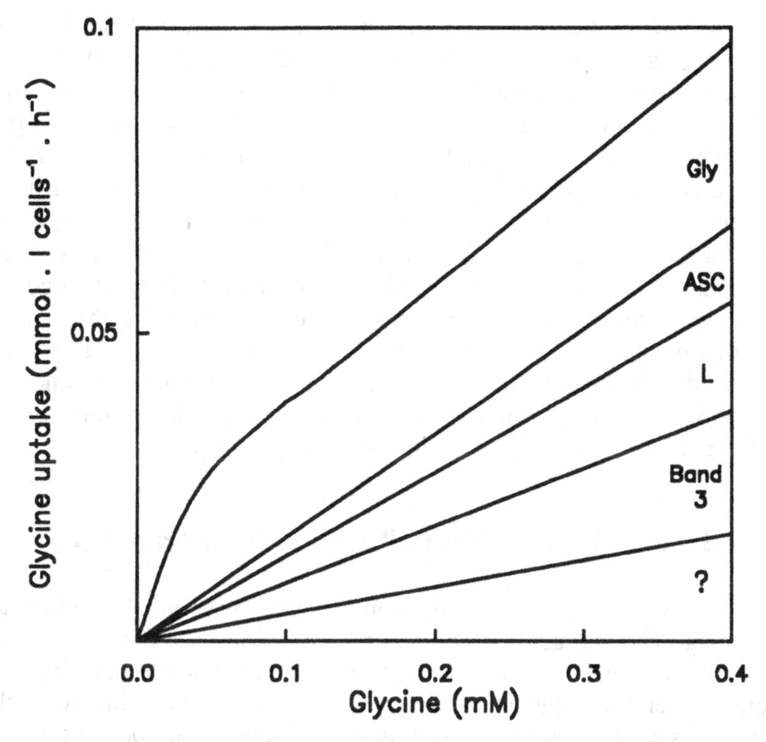

Fig. 2. Summary of the different routes of glycine uptake by human red cells. Physiological concentrations of glycine vary from about 0.2–0.3 mmol/L of plasma. Redrawn from Ellory et al. (1981).

sensitive to inhibition by SH-group reagents. *p*-Chloromethylphenylsulphonate, *N*-ethylmaleimide, and dithiobis-2-nitrobenzoic acid all markedly inhibit the transport of this amino acid by red cells (Deuticke, 1977; Vadgama and Christensen, 1985). In contrast, the uptake of leucine via the L system is only weakly inhibited. Little is known of the ability of substrates to protect against these inhibitory effects of *N*-ethylmaleimide and related inhibitors. In the meantime, it can be tentatively concluded that thiol groups play a role in amino-acid uptake by the T system.

Sheep red cells possess a C system that is selective for small to intermediate-size neutral amino acids. This transport system, however, is Na^+-independent (Rosenberg et al., 1980; Young et al., 1975; Young and Ellory, 1977). This C system is inhibited by thiol-reactive reagents such as $HgCl_2$, *N*-ethylmaleimide, and *p*-chloromethylphenylsulphonate (Young, 1980). As in the human glycine transport system, the C system in sheep erythrocytes appears to depend on the presence of essential thiol groups.

The involvement of metal cations in the transport of amino acids across biological membranes has been considered (Pal and Christensen, 1959). Indeed, Hider and McCormack (1980) have demonstrated that bathophenanthroline sulphonate, a potent copper(II) chelator, has a pronounced inhibitory effect on the L system of human red cells. It had previously been reported that the uptake of Cu^{2+} was stimulated in the presence of various amino acids in such tissues as kidney, liver, brain, and Ehrlich ascites cells (Neumann and Silverberg, 1966; Harris and Sass-Kortsak, 1967; Chutkow, 1978). The idea has been formulated that Cu^{2+} and amino acids form a complex producing a more lipophilic amino acid that would allow it more easily to penetrate membranes. The observations that bathophenanthroline sulphonate can influence amino-acid uptake via the L system promote the notion that Cu^{2+} ions play a functional role in this amino-acid transport pathway.

5. Function of Red-Cell Amino-Acid Transport

It is difficult to assign a function to much of the uptake and release processes of amino acids by human red cells. Because these cells carry out so few metabolic processes, why do they require the ability to sequester certain amino acids? For cysteine, glycine, and glutamine, this question is easier to answer. Erythrocytes require glutathione to maintain their integrity; otherwise oxidative damage can occur. This is nicely illustrated in certain sheep that have a deficit of cysteine red-cell transport. These cells have reduced glutathione levels and a shortened lifespan (Young et al., 1975). The substrates needed for glutathione biosynthesis are glycine, cysteine, and glutamate. Glycine and cysteine uptake from the serum should satisfy this requirement, but of course glutamate cannot significantly enter the human red cell. However, glutamine is readily taken up and can be converted to glutamate within the cell by the catalytic action of glutaminase (Rapoport, 1961; Ellory et al., 1983).

An explanation for the uptake of other amino acids is not so readily at hand. It has been proposed that most of the amino-acid uptake systems are a remnant of those needed during the early stages of erythrocyte development when more abundant metabolic processes existed (e.g., Al-Saleh and Wheeler, 1982). Because their rates of degradation are much lower than the life of the cell, they may simply have outgrown their function. On the other hand, some evidence is forthcoming to sustain the idea that certain amino acids can be carried by erythrocytes to different locations in the body. There may be an intercellular proline cycle, for instance, between red cells and hepatocytes (Phang et al., 1980; Hage-

Table 1
Na^+-Independent Transport Systems in Human Red Cells

System	Specificity	K_m	$*V_{max}$	Inhibitors (type)
L	L-Leucine and other large neutral amino acids	$5 \times 10^{-3} M$ (leu)	119	Bathophenanthroline SO_3 (Cu^{2+} chelator)
Ly	Dibasic L-amino acids	$6.8 \times 10^{-5} M$ (lys)	0.49	
T	L-Tryptophan and other aromatic L-amino acids	$1.5 \times 10^{-3} M$ (trp)	12	D-Tryptophan (competitive) 4-Azidophenylalanine (inactivator), *N*-Ethylmaleimide (inactivator)

*mmol/l cells/hr.

Table 2
Na^+-Dependent Transport Systems in Human Red Cells

System	Specificity	K_m	$*V_{max}$	Inhibitors (type)
ASC	L-Alanine, serine, cysteine, and glutamine	$2.7 \times 10^{-5} M$ (cys)	0.13	*N*-Ethylmaleimide (inactivator) Harmaline (competes with Na^+)
Gly	Glycine and sarcosine	2.5×10^{-5}M (gly)	19	*N*-Ethylmaleimide (inactivator)

*mmol/l cells/hr

dorn et al., 1982). The effect of the cycle is to allow the red cell to gain oxidizing equivalents in the form of NADP, since proline could travel from red cells to hepatocytes for oxidation to pyrroline carboxylate (the enzyme proline oxidase not being present in red cells). Pyrroline carboxylate would then return to the red cell for reduction back to proline.

Experiments with volunteers have demonstrated that infusion of insulin into the brachial artery leads to a significant reduction of glutamate concentration in forearm venous blood (Aoki et al., 1972). The authors of the subsequent report believe that human erythrocytes are able to transport glutamate to skeletal muscle, and that insulin can stimulate the release and subsequent uptake of this amino acid by muscle. Unfortunately, no satisfactory mechanism has been put forward to explain these observations.

6. Conclusions and Summary

There is little doubt that human red cells possess the ability to take up several amino acids from plasma and, in some cases, to accumulate these substrates against a gradient. The transport systems involved can be classified into Na^+-independent mechanisms characterized by carrier-mediated translocation of the facilitated diffusion type and Na^+-dependent mechanisms that are able to concentrate the relevant amino acid. The three distinct systems of the first type are the L, the Ly, and the T systems. The second type are the ASC system and the glycine system. The properties of these five transport pathways are summarized in Tables 1 and 2.

Although it is not clear what function amino-acid transport by human erythrocytes serves, the most compelling evidence points to the support of glutathione biosynthesis within the cell. Another, much more speculative function of uptake is to carry amino acids to various parts of the body where they can be utilized appropriately.

References

Abderhalden, E. and Kurten, H. (1921) Untersuchungen über die Aufnahme von Eiweissabkömmlingen (Peptone, Polypeptide und Aminosäuren) durch rote Blutkörperchen unter bestimmten Bedingungen. *Arch. ges. Physiol. Pflüger's* **189,** 311–323.

Al-Saheh, E. and Wheeler, K. P. (1982) Transport of neutral amino acids by human erythrocytes. *Biochim. Biophys. Acta* **684,** 157–171.

Antonioli, J. A. and Christensen, H. N. (1969) Differences in the schedules of regression of transport systems during reticulocyte maturation. *J. Biol. Chem.* **244,** 1505–1509.

Aoki, T. T., Brennan, M. F., Muller, W. A., Moore, F. D., and Cahill, G. F. (1972) Effect of insulin on muscle uptake. Whole blood versus plasma glutamate analysis. *J. Clin. Invest.* **52,** 2889–2994.

Christensen, H. N. (1968) Histidine transport into isolated animal cells. *Biochim. Biophys. Acta* **165,** 251–261.

Christensen, H. N. (1984) Organic ion transport during seven decades. The amino acids. *Biochim. Biophys. Acta* **779,** 255–269.

Christensen, H. N. and Antonioli, J. A. (1969) Cationic amino acid transport in the rabbit reticulocyte. *J. Biol. Chem.* **244,** 1497–1504.

Christensen, H. N., Cooper, P. F., Johnson, R. D., and Lynch, E. L. (1947) Glycine and alanine concentrations of body fluids: Experimental modifications. *J. Biol. Chem.* **168,** 191–196.

Christensen, H. N., Riggs, T. R., and Ray, N. E. (1952) Concentrative uptake of amino acids by erythrocytes in vitro. *J. Biol. Chem.* **194,** 41–51.

Chutkow, J. G. (1978) Evidence for uptake of nonceruloplasminic copper in the brain: Effect of ionic copper and amino acids. *Proc. Soc. Exp. Biol. Med.* **158,** 113–116.

Constantino, A. (1913a) Untersuchungen über die biologische Bedeutung und den Metabolismus der Eiweissstoffe. *Biochem. Z.* **51,** 91–96.

Constantino, A. (1913b) Die Permeabilität der Blutkörperchen für Aminosäuren. *Biochem. Z.* **55,** 411–418.

Danielson, I. S. (1933) Amino acid nitrogen in blood and its determination. *J. Biol. Chem.* **101,** 505–522.

Deuticke, B. (1977) Properties and structural basis of simple diffusion pathways in the erythrocyte membrane. *Rev. Physiol. Biochem. Pharmacol.* **78,** 1–97.

Dunham, P. B., Stewart, G. W., and Ellory, J. C. (1980) Chloride-activated passive potassium transport in human erythrocytes. *Proc. Natl. Acad. Sci. (USA)* **77,** 1711–1715.

Ellory, J. C. and Stewart, G. W. (1982) The human erythrocyte Cl-dependent Na-K cotransport system as a model for the study of loop diuretics *Br. J. Pharmacol.* **75,** 183–188.

Ellory, J. C., Jones, S. E. M., and Young, J. D. (1981) Glycine transport in human erythrocytes. *J. Physiol.* **320,** 403–422.

Ellory, J. C., Preston, R. L., Osotimehin, B., and Young, J. D. (1983) Transport of amino acids for glutathione biosynthesis in human and dog red cells. *Biomed. Biochim. Acta* **42,** S48–S52.

Gardner, J. D. and Levy, A. G. (1972) Transport of dibasic amino acids by human erythrocytes. *Metabolism* **21,** 413–431.

Ginsberg, H. and Krugliak, M. (1983) Uptake of L-tryptophan by erythrocytes infected with malaria parasites (*Plasmodium falciparum*). *Biochim. Biophys. Acta* **729,** 97–103.

Hagedorn, C. H., Yeh, G. C., and Phang, J. M. (1982) Transfer of

1-pyrroline-5-carboxylate as oxidizing potential from hepatocytes to erythrocytes. *Biochem. J.* **202,** 31–39.

Harris, D. I. M. and Sass-Kortsak, A. (1967) The influence of amino acids on copper uptake by rat liver slices. *J. Clin. Invest.* **46,** 659–667.

Hider, R. C. and McCormack, W. (1980) Facilitated transport of amino acids across organic phases and the human erythrocyte membrane *Biochem. J.* **188,** 541–548.

Hoare, D. G. (1972a) The transport of L-leucine in human erythrocytes: A new kinetic analysis. *J. Physiol.* **221,** 311–329.

Hoare, D. G. (1972b) The temperature dependence of the transport of L-leucine in human erythrocytes. *J. Physiol.* **221,** 331–348.

Imler, J. R. and Vidaver, G. A. (1972) Anion effects on glycine entry into pigeon red blood cells. *Biochim. Biophys. Acta* **288,** 153–165.

Lieb, W. R. and Stein, W. D. (1974a) Testing and characterizing the simple pore. *Biochim. Biophys. Acta* **373,** 165–177.

Lieb, W. R. and Stein, W. D. (1974b) Testing and characterizing the simple carrier. *Biochim. Biophys. Acta* **373,** 178–196.

Lopez-Burillo, S., Garcia-Sancho, J., and Herros, B. (1985) Tryptophan transport through transport system T in the human erythrocyte, the Ehrlich cell and the rat intestine. *Biochim. Biophys. Acta* **820,** 85–94.

Neumann, P. Z. and Silverberg, M. (1966) Active copper transport in mammalian tissues—a possible role in Wilson's disease. *Nature* **210,** 414–416.

Pal, P. R. and Christensen, H. N. (1959) Interrelationships in the cellular uptake of amino acids and metals. *J. Biol. Chem.* **234,** 613–617.

Phang, J. M., Yeh, G. C., and Hagedorn, C. H. (1980) The intercellular proline cycle. *Life Sci.* **28,** 53–58.

Rapoport, S. (1961) Reifung und Alterungsvorgänge in Erythrozyten. *Folia Haematol.* **78,** 364–381.

Rieser, P. (1961) Sugar-inhibited amino acid transport in the human erythrocyte. *Expl. Cell Res.* **24,** 165–167.

Rieser, P. (1962) The kinetics of valine transport in the human erythrocyte. *Expl. Cell Res.* **27,** 577–580.

Rosenberg, R. (1979) Zero-trans uptake of L-tryptophan in the human erythrocyte. *J. Neural Transm.* **(Suppl. 15),** 153–160.

Rosenberg, R. (1981a) L-Leucine transport in human red blood cells: A detailed kinetic analysis. *J. Membr. Transp.* **62,** 79–93.

Rosenberg, R. (1981b) A kinetic analysis of L-tryptophan transport in human red blood cells. *Biochim. Biophys. Acta* **649,** 262–268.

Rosenberg, R. (1982) Na-independent and Na-dependent transport of neutral amino acids in the human red blood cell. *Acta Physiol. Scand.* **116,** 321–330.

Rosenberg, R. and Rafaelsen, O. J. (1979) Transport of neutral amino acids across the human red blood cell membrane. *Progr. Neuropsychopharmacol.* **3,** 377–381.

Rosenberg, R., Young, J. D., and Ellory, J. C. (1980) L-Tryptophan transport in human red blood cells. *Biochim. Biophys. Acta* **598,** 375–384.

Thomas, E. L. and Christensen, H. N. (1970) Indication of spacial relations among structures recognizing amino acids and Na^+ at a transport receptor site. *Biochem. Biophys. Res. Commun.* **40,** 277–283.

Thomas, E. L. and Christensen, H. N. (1971) Nature of the cosubstrate action of Na^+ and neutral amino acids in a transport system. *J. Biol. Chem.* **246,** 1682–1688.

Ussing, H. H. (1943) The nature of the amino nitrogen of red corpuscles. *Acta Physiol. Scand.* **5,** 335–351.

Vadgama, J. V. and Christensen, H. N. (1985) Discrimination of Na^+-independent transport systems L. T. and ASC in erythrocytes. *J. Biol. Chem.* **260,** 2912–2921.

Winter, C. G. and Christensen, H. N. (1964) Migration of amino acids across the membrane of the human erythrocyte. *J. Biol. Chem.* **239,** 872–878.

Young, J. D. (1980) Effects of thiol-reactive agents on amino acid transport by sheep erythrocytes. *Biochim. Biophys. Acta* **602,** 661–672.

Young, J. D. and Ellory, J. C. (1977) Red cell amino acid transport, in *Membrane Transport in Red Cells* (Ellory, J. C. and Lew, V. L., eds.), Academic Press, London, pp. 301–325.

Young, J. D. and Ellory, J. C. (1979) Transport of tryptophan and other amino acids by mammalian erythrocytes. *J. Neural Transm.* **(Suppl. 15),** 139–151.

Young, J. D., Jones, S. E. M., and Ellory, J. C. (1980) Amino acid transport in human and sheep erythrocytes. *Proc. R. Soc. Lond. B.* **209,** 355–375.

Young, J. D., Jones, S. E. M., and Ellory, J. C. (1981) Amino acid transport via red cell anion transport system. *Biochim. Biophys. Acta* **645,** 157–160.

Young, J. D., Mason, D. K., and Fincham, D. A. (1988) Topographical similarities between harmaline inhibition sites on Na^+-dependent amino acid transport system ASC in human erythrocytes and Na^+-independent system asc in horse erythrocytes. *J. Biol. Chem.* **263,** 140–143.

Young, J. D., Ellory, J. C., and Tucker, E. M. (1975) Amino acid transport defect in glutathione-deficient sheep erythrocytes. *Nature* **254,** 156–157.

Young, J. D., Wolowyk, M. W., Jones, S. E. M., and Ellory, J. C. (1979) Sodium-dependent cysteine transport in human red blood cells. *Nature* **279,** 800–802.

Yunis, A. A. and Arimura, G. K. (1965) Amino acid transport in blood cells. II. Patterns of transport of some amino acids in mammalian reticulocytes and mature red cells. *J. Lab. Clin. Med.* **66,** 177–186.

Hexose Transport Across Human Erythrocyte Membranes

Anthony Carruthers

1. Introduction

Human erythrocyte hexose transfer displays a catalytic complexity that is not explained by traditional models for carrier-mediated transport (Naftalin et al., 1985). In this chapter, we shall consider mechanisms for sugar transport and examine the possibility that the anomalous behavior of the red-cell hexose transfer system results from carrier interaction with intracellular environmental factors.

2. Perspective

Human erythrocyte sugar transport is a classic example of a facilitated diffusion-type mechanism. Unidirectional sugar uptake and efflux proceed at rates many orders of magnitude greater than would be expected for the simple diffusion of sugars across protein-free lipid bilayers (Jung, 1971). Sugar transport is passive (net transport occurs in the direction high to low sugar concentration only [Widdas, 1952]), displays saturation kinetics (Widdas, 1952) characterized by a strict stereoselectivity (LeFevre, 1961; Barnett et al., 1975), and is inhibited by relatively low concentrations of inhibitors, such as phloretin (LeFevre and Marshall, 1959) and cytochalasin B (Block, 1973). These features establish that sugar transport is facilitated by a limited number of sugar-selective membrane permeases (carriers).

The human red-cell sugar transporter has been identified using two procedures. Reconstituting chromatographically fractionated red-cell membrane proteins into artificial lipid bilayers (Kasahara and Hinkle, (1977) demonstrated that an integral membrane protein of M_r 46–55 kdaltons catalyzed sugar transport with properties qualitatively similar to those of the intact cell. Exploiting the interaction between the transport system and the transport inhibitor, cytochalasin B (an agent that reacts with the intracellular substrate binding site on the transporter [Basketter and Widdas, 1978]), Sogin and Hinkle (1980a) and Baldwin et al. (1982) have demonstrated that the protein identified as the glucose transporter on the basis of reconstituted transport activity also contains the sugar-sensitive cytochalasin B binding site. The molar stoichiometry of binding sites/protein is very close to unity (Baldwin et al., 1982). Polyvalent (Sogin and Hinkle, 1980b) and monoclonal antisera (Allard and Lienhard 1985) raised against the purified glucose transporter recognize only red-cell membrane proteins of M_r 46–55 kdaltons. These studies refute the arguments that the transporter is a higher mol wt species of approximately 92 kdaltons (Shelton and Langdon, 1983; Carruthers and Melchior, 1984). The rather broad mobility of the sugar carrier on polyacrylamide gel electrophoresis has been ascribed to heterogeneous glycosylation of the transporter following translation (Gorga et al., 1979).

The primary structure of the sugar transporter in Hep G2 cells (Mueckler et al., 1985) and rat brain (Birnbaum et al., 1986) has been deduced by cloning and sequencing the gene encoding for the transporter. The transport proteins share extensive (>92%) homologies (Birnbaum et al., 1986). The limited amino-acid sequence data available for the human red-cell sugar transporter suggest extensive homology between red-cell and Hep G2/rat brain transporter (Mueckler et al., 1985). Circular dichroism studies of purified red-cell sugar transporter (Chin et al., 1986) suggest that at least 70% of transporter mass is α-helix normal to the plane of the bilayer. The deduced sequence data predict 12 bilayer spanning α-helical sequences—five of which may be amphiphilic, forming a transmembrane channel (Mueckler et al., 1985). This view is supported by hydrogen exchange studies (Jung et al., 1986) demonstrating that many more amino-acid side chains can participate in hydrogen exchange than would be expected if most of the protein were buried within the hydrocarbon core of the bilayer. Recent studies have shown that the 19-kdalton tryptic fragment of the carrier containing the intracellular cytochalasin B binding domain (Deziel and Rothstein, 1984; Shanahan and D'Artel Ellis, 1984) also contains the extracellular sugar binding site (Holman et al., 1986). Thus, the postulated membrane spanning segments 7–12, containing three of the five suggested amphipathic α-helices (Mueckler et al., 1985), may also contain the catalytic moiety of the transporter.

3. Catalytic Properties of Human Red-Cell Sugar Transport

Although advances in our understanding of the biochemistry of the sugar transporter have been significant, the catalytic properties of the human red-cell sugar transporter have resisted interpretation within the framework of a number of simple carrier models.

Table 1 summarizes the various theoretically unambiguous sugar transport determinations that may be made with cells (*see* Lieb and Stein, 1974) and the experimental $K_{m(app)}$ and V_{max} derived from such studies with human red cells using either D-glucose or galactose as transported substrate. The salient features of transport at 20°C are the following:

1. Transport is asymmetric. K_m and V_{max} for sugar entry are some 3- – 10-fold lower than the equivalent parameters for sugar exit.
2. Initial rate methods of transport analysis provide lower estimates of K_m for zero trans exit than does application of the integrated Michaelis-Menten equation to exit time course data to obtain K_m (*see* Stein, 1986).
3. When the external surface of the cell is saturated with substrate, K_m for exit is reduced and becomes very close to K_m for entry. K_m for entry is unchanged by the presence of saturating sugar levels at the interior of the cell.
4. Accelerated exchange transport is observed. The unidirectional uptake and efflux of sugar are accelerated by the presence of sugar at the opposite (trans) side of the membrane.

All reactions have an equilibrium constant (K_{eq}), which is the ratio of product (P) formed/substrate (S) remaining at equilibrium (i.e., [P]/[S] when d[P]/dt = d[S]/dt = 0). For a reversible reaction, K_{eq} is given by Segel (1975) as:

$$K_{eq} = \frac{V_{max(f)}\, K_{m(P)}}{V_{max(r)}\, K_{m(S)}} \tag{1}$$

where the subscripts (f) and (r) indicate forward and reverse reactions, respectively. As sugar transport is passive, $K_{eq} = 1$. This means that:

$$\frac{V_{max(f)}}{K_{m(S)}} = \frac{V_{max(r)}}{K_{m(P)}} \tag{2}$$

Table 1
Michaelis and Velocity Parameters for Sugar Transport in Human Red Cells at 20°C

Experiment	Method	Measures	Results[a]	
			Glucose	Galactose
Zero-trans entry	$[S_i] = 0$	K_m entry (K_{oi}^{zt})	1.6 ± 0.1	39,32,12[b]
	$[S_o]$ varied	V_m entry (V_{oi}^{zt})	0.56 ± 0.05	0.3
Zero-trans exit	$[S_i]$ varied	K_m exit (K_{io}^{zt})	(26.4 ± 2.5) 10.7 ± 3.8[c]	(240) 56,58[d]
	$[S_o] = 0$	V_m exit (V_{io}^{zt})	2.8 ± 0.6	4.4 ± 0.2
Infinite-cis entry	$[S_i]$ varied	K_m exit (K_{oi}^{ic})	(3.9 ± 1.1)	(21) 25
	$[S_o]$ saturating	V_m entry (V_{oi}^{ic})		
Infinite-cis exit	$[S_i]$ saturating	K_m entry (K_{io}^{ic})	1.75 ± 0.05	12
	$[S_o]$ varied	V_m exit (V_{io}^{ic})		
Infinite-trans entry	$[S_i]$ saturating	K_m entry (K_{oi}^{it})	1.7	21
	$[S_o]$ varied	V_m exchange (V^{ee})		
Equilibrium exchange	$[S_i] = [S_o]$ varied	K_m exchange (K^{ee})	26.5 ± 3.6	156 ± 18
		V_m exchange (V^{ee})	5.3 ± 0.3	8.7 ± 1.5

[a]Data are taken from (Stein, 1986). All K_m values are in m*M* and V_m values in mmol sugar/L cell water/s. K_m parameters shown in parenthesis were obtained by analysis of time course data using the appropriate integrated rate equation (Stein, 1986). All other values were obtained by measuring initial rates of transport.

[b]Two saturable components of galactose uptake have been detected (Ginsburg and Yeroushalmy, 1978).

[c]Initial rate zero-trans D-glucose exit data are taken from Levine et al. (1965), Miller (1971), Brahm (1983) and Lowe and Walmsley (1986).

[d]Initial rate zero-trans galactose exit data are from Miller (1971) and Ginsburg and Yeroushalmy (1978). $[S_i]$ and $[S_o]$ refer to intra- and extracellular sugar levels, respectively.

$$^{v}21 = \frac{V\frac{S_o}{K} + V^e \frac{S_o(S_i + C_i)}{\alpha\lambda K^2}}{1 + \frac{(S_i + C_i)}{\lambda K} + \frac{(S_o + C_o)}{K} + \frac{(S_i + C_i)(S_o + C_o)}{\alpha\lambda K^2}} =$$

This requirement is satisfied using estimates of $K_{m(P)}$ from initial rate transport measurements, but fails when integrated rate exit data are employed. These considerations suggest that K_m for exit in intact cells is overestimated by the integrated rate analysis (Naftalin and Holman, 1977; Wheeler and Hinkle, 1985; Lowe and Walmsley, 1986).

4. Models for Sugar Transport

Two fundamental models for human red-cell sugar transport have been derived in an attempt to account for the properties of the transfer system. These are the *One-Site model* (variously termed the mobile carrier [Widdas, 1952], the alternating conformer model [Vidaver, 1966], the simple carrier [Lieb and Stein, 1974], the gated pore [Bowman and Levitt, 1977], or the cyclic carrier [Krupka and Devés, 1981]) and the *Two-Site model* (variously termed the two-component carrier [Baker and Widdas, 1973], the simultaneous carrier [Baker and Carruthers, 1981], or linear carrier [Krupka and Devés, 1981]).

Multiple-site models have also been postulated (the tetramer [Lieb and Stein, 1972] and the allosteric pore models [Holman, 1980]). To avoid difficulties, the models presented here are referred to as the One-Site and Two-Site carriers. These models have been extensively analyzed by Britton (1965), Baker and Widdas (1973), Lieb and Stein (1974), and Krupka and Devés (1981). It should be emphasized that physical transformations of the carrier are ignored in the derivation of the kinetic equations describing the transport mechanism. Thus, for the various One-Site carriers noted above, the steady-state kinetic solution for each scheme is identical regardless of the proposed physical mechanism of transport.

The One-Site carrier is illustrated (as a King-Altman diagram) in Fig. 1. The carrier X can exist as either of two stable carrier forms in the absence of sugar (X_i—the sugar binding site facing the interior of the cell or X_o—the binding site facing the cell's exterior) and as either of two central complexes in the presence of substrate ($X{\cdot}S_i$ or $X{\cdot}S_o$). Each of these carrier forms is mutually exclusive. With this model, transport rates are governed by substrate-induced carrier redistributions (*see* legend for details).

The Two-Site carrier is illustrated in Fig. 2. This model differs fundamentally from the One-Site carrier in that the Two-Site carrier can

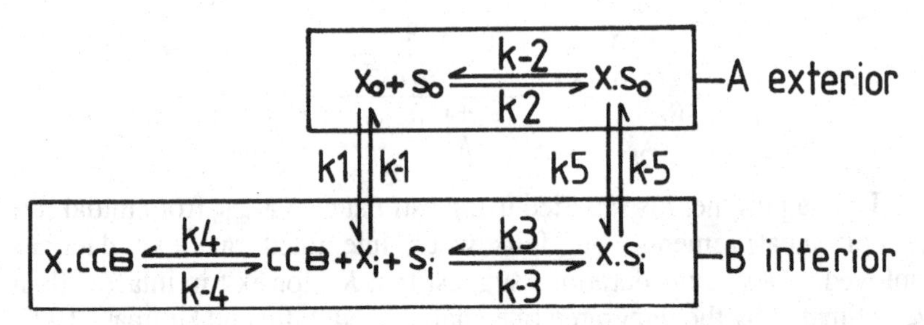

Fig. 1. The One-Site carrier. The carrier, X, can exist in one of either four states. These are: X_o—substrate binding site facing the cell's exterior; X_i—substrate binding site facing the cell's interior; $X{\cdot}S_o$—substrate binding site occupied by extracellular substrate; $X{\cdot}S_i$—substrate binding site occupied by intracellular substrate. Each of these carrier states is mutually exclusive. The rate constants governing carrier form interconversions are indicated in the figure (k_1–k_{-5}). The transport inhibitor cytochalasin B (CCB) binds to X_i. Phloretin binds to X_o (not shown). For a passive transport system, the product of the rate constants for translocation in one direction (e.g., $k_1k_{-2}k_{-3}k_{-5}$) must equal those for translocation in the opposite direction ($k_{-1}k_2k_3k_5$). Transport rates are governed by carrier distributions, which in turn are determined by the various rate constants. For example, in the absence of substrate, that fraction of total carrier in the form X_i is $k_1/(k_1 + k_{-1})$. Using the values for rate constants at 0°C described in Lowe and Walmsley (1986), $X_i/(X_o+X_i)$ is 0.94. If saturated sugar levels are added to the interior of the cell, $X{\cdot}S_i/(X_{total})$ is $k_1/(k_1 + k_{-5}) = 0.12$ (because $[S_i]$ is saturating all of the internal carrier is in the form of $X{\cdot}S_i$) and efflux is proportional to $k_{-5}{\cdot}[X{\cdot}S_i]$. Accelerated exchange transport requires that k_5 and k_{-5} are greater than k_1 and k_{-1}. Thus, if saturating $[S_o]$ is added to the system, $X{\cdot}S_i/(X_{total})$ is given by $k_5/(k_5 + k_{-5}) = 0.92$. Thus, the addition of S_o serves to increase $[X{\cdot}S_i]$ and thereby accelerate sugar efflux. These redistributions are considered in detail by Widdas (1980) and by Helgerson and Carruthers (1987).

bind transport substrate at intra- and extracellular sites to form the ternary complex $S_i{\cdot}X{\cdot}S_o$. With the One-Site carrier, only the binary complexes $X{\cdot}S_i$ and $X{\cdot}S_o$ are permitted, and the availability of binding sites is governed by substrate-induced carrier redistributions. With the Two-Site carrier, both substrate binding sites exist simultaneously, but the affinity of each binding site for substrate could be modified by occupancy of the trans site were coupling to exist between extra- and intracellular sites (*see* Legend for details).

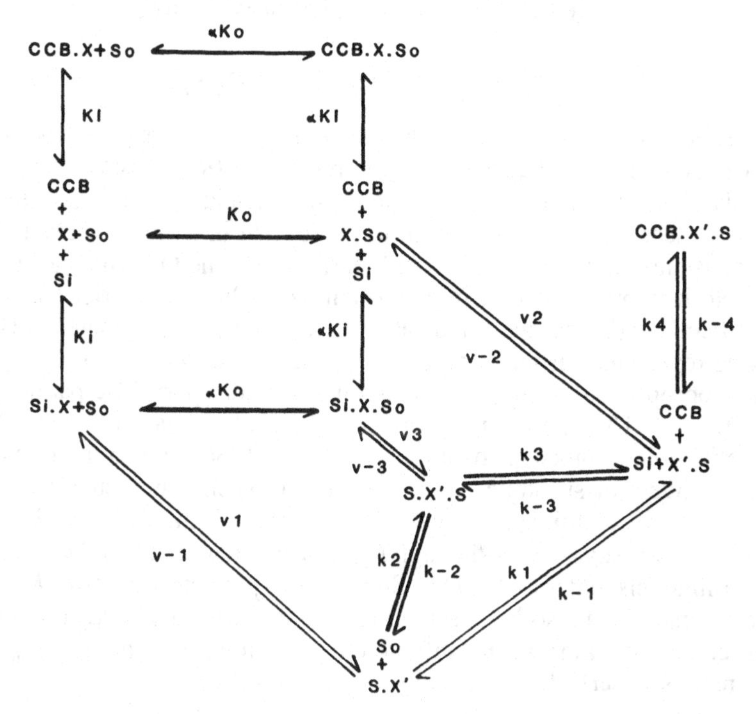

Fig. 2. The Two-Site carrier. This model for hexose transport differs fundamentally from the One-Site carrier—the carrier, X, can bind two molecules of substrate, S, to form $S_i{\cdot}X{\cdot}S_o$. Cytochalasin B (CCB) binds at the internal site. Phloretin binds at the external site (not shown). The various dissociation constants for substrate binding to the carrier are indicated (e.g., K_o, K_i, and so on). This scheme allows for cooperativity in substrate binding. For example, if the parameter α is greater than unity, this means that $K_{s(app)}$ for substrate binding is increased when the opposite, trans site is occupied by substrate. Similarly, if α is less than unity, $K_{s(app)}$ for substrate binding is decreased when the opposite, trans site is occupied by substrate. Translocation steps (e.g., $X{\cdot}S_i$ <—> $S{\cdot}X\dagger$ <—> $X\dagger{\cdot}S$ <—> $X{\cdot}S_o$) are also included in this figure. An efflux cycle is given by $X + S_i$ <—> $X{\cdot}S_i$ <—> $S{\cdot}X\dagger$ <—> $X\dagger{\cdot}S$ <—> $X{\cdot}S_o$ <—> $X + S_o$. The carrier could be a water-filled, substrate-gated channel (*see* Naftalin and Holman, 1977) in which two molecules of substrate are allowed to exchange (pass) during translocation. This could account for accelerated exchange transport. Again, as transport is passive, this means that $v_1v_{-2}k_1 = v_{-1}v_2k_{-1}$. Accelerated exchange transport requires that v_3v_{-3} is greater than $v_1v_{-2}k_1$ and $v_{-1}v_2k_{-k}$, respectively. This system has been extensively analyzed by Helgerson and Carruthers (1987) and assumes that vacant sites are equally affected by substrate occupation of the trans-site.

5. Interpretation of Transport Data

5.1. Testing the One-Site Carrier

Table 2 summarizes the steady-state kinetic solution for the One-Site carrier, and shows the affinity (K) and resistance (R_n) constants derived directly from red-cell transport studies with D-glucose and galactose. Each of these constants may, in turn, be defined in terms of the various rate constants shown in Fig. 1 (*see* Stein, 1986). The One-Site carrier is characterized by a single affinity constant (K). One-Site carrier analysis of red-cell sugar transport indicates at least two K parameters. This finding represents sufficient grounds for rejection of the One-Site carrier as an appropriate description of red-cell sugar transport. The parameter R_{oo} ($1/k_1 + 1/k_{-1}$, i.e., the sum of reciprocals of rate constants for isomerization of unoccupied carrier, *see* Stein, 1986), which in principle is independent of substrate type, displays an experimental substrate dependency. An additional anomaly is the demonstration of two $K_{m(app)}$ parameters for sugar exit—the high K_m for zero-trans exit and the low K_m for infinite cis entry. The One-Site carrier predicts that two $K_{m(app)}$ parameters should also be observed for sugar entry—a low K_m for zero-trans entry and a high K_m for infinite-cis exit. However, only one $K_{m(app)}$ for entry is observed.

5.2. Testing the Two-Site Carrier

The Two-Site carrier also fails to account for the second, low $K_{m(app)}$ site at the interior of the cell. This site could be explained should occupation of the extracellular binding site by substrate exert a positive cooperative effect on substrate binding at the intracellular site. However (assuming vacant sites are affected equally by occupied sites), a similar, positive cooperative effect of intracellular sugar on sugar binding to the external site would be expected, and this is not observed.

5.3. Conclusions

The simple, One-Site and Two-Site carrier models for sugar transport are inconsistent with the experimental data obtained from sugar transport studies with human red cells. Only two possible explanations can account for these findings: (1) The transport system is more complex than allowed for in the transport model or (2) the transport determinations fail to report the "true" kinetic parameters for transfer.

Table 2
Steady-State Solution to the One-Site Carrier[a]

$$V_{io} = \frac{(K + S_o)S_i}{K^2R_{oo} + KR_{io}S_i + KR_{oi}S_o + R_{ee}S_iS_o}$$

where: $R_{io} = \frac{1}{V_{io}^{zt}}$ $\quad K = K_{oi}^{zt}R_{io}/R_{oo}$

$K_{oi}^{zt}R_{oi}/R_{oo}$

$R_{oi} = \frac{1}{V_{oi}^{zt}}$ $\quad K^{ee}R_{ee}/R_{oo}$

$K_{io}^{ic}R_{ee}/R_{io}$

$R_{ee} = \frac{1}{V_{ee}}$ $\quad K_{oi}^{ic}R_{ee}/R_{oi}$

$R_{oo} = R_{io} + R_{oi} - R_{ee}$

Flux in direction o → i is obtained by interchanging subscripts i ⟷ o.

Calculated parameters for sugar transport in red cells[a]

	Glucose	Galactose
R_{io}	0.36 ± 0.07	0.23 ± 0.01
R_{oi}	1.79 ± 0.15	3.33
R_{ee}	0.19 ± 0.01	0.11 ± 0.01
R_{oo}	1.96 ± 0.23	3.45 ± 0.02
K	(4.48 ± 1.27) 1.97 ± 1.63	(16) 3.73
	1.46 ± 0.44	26.7 ± 11.04
	2.57 ± 0.9	4.97
	0.92 ± 0.32	5.74
	0.41 ± 0.2	0.75

[a]Values are in mM and R values in s mM^{-1}. Values shown in parenthesis are these calculated for K_m parameter obtained by the integrated rate approach.

6. A One- or Two-Site Transport Mechanism?

Although the specific One- and Two-Site carrier models for sugar transport are inconsistent with the available hexose transfer data, it is still useful to determine whether the transport system is of the One- or Two-site type. The question we ask here is: Is the carrier capable of binding only one or two sugar molecules at any instant? Three approaches have been employed to answer this question.

6.1. Transport Studies

Krupka and Devés (1981) have developed a kinetic analysis of transport that asks whether competitive inhibitors known to act at opposite faces of the membrane can bind simultaneously to the carrier or whether binding is mutually exclusive. Under conditions where zero-trans sugar exit is monitored in the presence of internal (I_i) and external (I_o) inhibitors, five forms of One-Site carrier would be expected—X_i, $X{\cdot}S_i$, $X{\cdot}I_i$, X_o, and $X{\cdot}I_o$. Of these carrier types, only $X{\cdot}S_i$ can catalyze sugar efflux. With the Two-Site carrier, six carrier forms would be expected—X, $X{\cdot}S_i$, $X{\cdot}I_i$, $X{\cdot}I_o$, $X{\cdot}S_i{\cdot}I_o$ and $X{\cdot}I_i{\cdot}I_o$. Again, only $X{\cdot}S_i$ can catalyze transport. Fractionally less Two-Site carrier can thus participate in efflux than One-Site carrier, and the inhibitions of transport by I_i and I_o would be greater via the Two-Site carrier. Such an inhibitor pair is phloretin (an external competitive inhibitor of sugar transport) and cytochalasin B (an internal competitive inhibitor [Krupka and Devés, 1981]).

Krupka and Devés (1981) observed that experimental inhibition of D-glucose exit most closely resembled the predictions of the One-Site carrier, therefore concluding that the transporter was of the One-Site type. However, should occupation of the Two-Site carrier by phloretin (I_o) displace cytochalasin B and D-glucose from the internal site (i.e., occupied sites display negative cooperativity), the predictions of both models would be indistinguishable (Krupka and Devés, 1981). We will show below that this negative cooperativity between occupied sites has been observed experimentally (Jung and Rampall, 1977; Helgerson and Carruthers, 1987), thereby invalidating the conclusion that transport is necessarily of the One-Site type.

These experiments (Krupka and Devés, 1981) also illustrate an important problem with red-cell sugar transport determinations. $K_{m(app)}$ for D-glucose exit at 20°C (estimated by initial rate measurements) averages 10 mM (Table 1). K_d for cytochalasin B binding to the transporter at 20°C is 140 nM (Jung and Rampal, 1977). The expected $K_{i(app)}$ for cytochalasin B inhibition of glucose (124 mM) exit in these experiments is therefore $K_d{\cdot}(1 + [DG]/K_{m(app)}) = 1.9\ \mu M$. The measured value (0.65 μM, [Krupka and Devés, 1981]) is considerably lower than this, suggesting either that $K_{m(app)}$ is underestimated by initial rate methods and should be in the order of 34 mM—a value close to those found in studies employing integrated rate methods (*see* Table 1)—or that $[\text{D-glucose}_i]_{free}$ is considerably lower (by 71%) than $[\text{D-glucose}_i]_{total}$. We will return to this point in Section 7.1.

An additional procedure for distinguishing One- and Two-Site carrier mechanisms is the hetero-exchange condition. Here, the ability of an extracellular sugar (e.g., galactose) to stimulate the exit of a second sugar (e.g., D-glucose) is monitored. Stein (1986) provides the steady-state One-Site carrier solution for this procedure (*see* Table 3). With the Two-Site carrier, provided that the affinity of the external sugar binding site is independent of the occupancy state of the internal binding site, the $K_{m(app)}$ for stimulation of exit of S_i by P_o should be identical to K_o for zero-trans entry of P_o. Table 3 summarizes hetero-exchange predictions of the One- and Two-Site carrier models and compares them with the available transport data. The conclusion is very clear! $K_{m(app)}$ for galactose-stimulation of glucose exit (at 2°C) is indistinguishable from K_o for galactose zero-trans entry, and the One-Site carrier is unable to accurately predict the experimental data. Transport is more in line with the predictions of the Two-Site carrier provided that the affinity of vacant sites for substrate is unaffected by occupation of trans-sites by transported substrate. This apparent contradiction between cooperativity assignments in order to model the transport data of Krupka and Devés (1981) (assume negative cooperativity) and the heteroexchange data (assume site independence) to the Two-Site carrier is discussed in detail below.

6.2. Ligand Binding Studies

Cytochalasin B binding studies with red cells, ghosts, and purified transporter permit an analysis of how intra- and extracellular sugars and transport inhibitors affect the number of transporters occupied by the ligand. Such studies could indicate whether the various forms of the carrier are mutually exclusive and thus the type of transport mechanism.

Figure 3 summarizes the One- and Two-Site carrier predictions and experimental results of cytochalasin B binding studies performed with a variety of sugar types (Helgerson and Carruthers, 1987). Extracellular D-glucose is without effect on cytochalasin B binding (as predicted by the Two-Site carrier), although it is possible that the increase in binding predicted by the One-Site carrier would be too small to measure. Inhibition of cytochalasin B binding by intracellular D-glucose is less than predicted by the One-Site carrier and, more importantly, extracellular D-glucose does not reduce the inhibition of binding produced by intracellular sugar (*see* Helgerson and Carruthers, 1987). Figure 3 illustrates that these results are predicted by the Two-Site carrier model. The lack of inhibition of cytochalasin B binding by extracellular D-glucose suggests (if binding is described by the Two-Site carrier) that the affinity of the internal site for cytochalasin B is independent of the occupancy state of the external site when the external ligand is the transported sugar, D-glucose.

Table 3
Heteroexchange: Galactose-Stimulated Glucose Exit at 0–4°C[a]

One-Site scheme:

$$V_{io} = \frac{\bar{S}_i + \bar{S}_i P_o}{R_{oo} + R^S_{io}\bar{S}_i + R^P_{oi}\bar{P}_o + R_{ee}\bar{S}_i P_o}$$

Where: $\bar{S}_i = [S_i]/K_s$, $\bar{P}_o = [P_o]/K_p$, $R_{ee}{}^{SP} = 1/V^{ee}$

$[S_i]$ = intracellular [glucose]
$[P_o]$ = extracellular [galactose]
V^{ee} = V_m for galactose stimulated saturated glucose exit

Two-Site scheme:

$$V_{io} = \frac{1}{\alpha}\left[(V^{zt}(1 - \frac{1}{\beta}) + V^{ee}/\beta\right]$$

Where:

$\alpha = 1 + K^S/[S_i]$
$\beta = 1 + K^P/[P_o]$
V^{zt} = V_m for zero-trans glucose exit
K^S = K_m for glucose exit
K^P = K_m for zero-trans galactose entry

Observed	Predicted	
	One-site	Two-Site
K_{sp} 7.5 mM[b]	9.5 (22.4)	6.1 (6.1)
V^{ee} 17.3 ± 1.4 mM·min^{-1}	12.6 (15.5)	17.0 (17.0)

[a]The values in parenthesis employed zero-trans exit data for glucose at 0°C from Lowe and Walmsley (1986); otherwise data from Baker and Naftalin (1979) were employed (0–4°C).

[b]Data taken from (Baker and Naftalin, 1979). The galactose transport data from Ginsburg and Yeroushalmy (1978) at 0°C were employed to calculate R^P_{oi} and K_p for the One-Site carrier. Although two components of galactose uptake are observed in Ginsburg and Yeroushalmy (1978), the component employed here accounts for >90% zero-trans uptake. R_{oo} from Lowe and Walmsley (1986) is 3.25 min mM^{-1} and from Ginsburg and Yeroushalmy (1978) is 5.61 min mM^{-1}. The mean of the two estimates was employed.

The extracellular sugars ethylidene glucose and maltose (non-transported but reactive sugars; Basketter and Widdas, 1978) and the transport inhibitor phloretin (which acts at an external site) inhibit cytochalasin B binding to the carrier. This inhibition is not of the linear competitive type (as predicted by the One-Site carrier) but, rather, is a saturable phenomenon. This can only be explained if the carrier is of the Two-Site type and occupancy of the external site by these nontransported species reduces the affinity of the internal site for cytochalasin B. These results raise the interesting possibility that occupancy of the external site

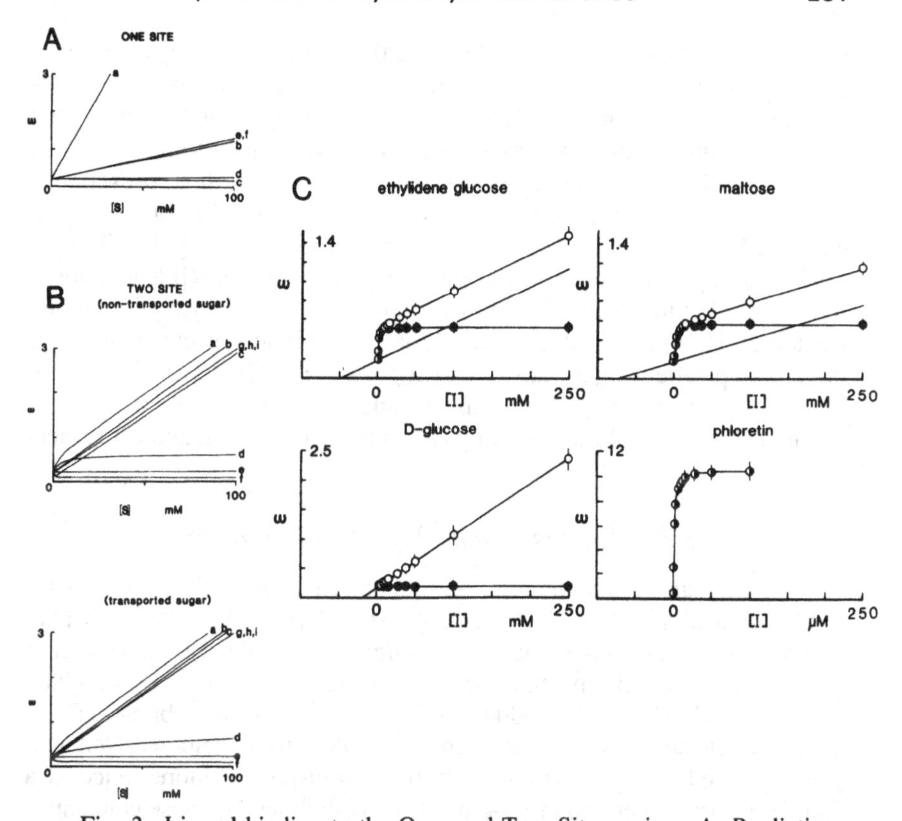

Fig. 3. Ligand binding to the One- and Two-Site carriers. A. Predictions of the One-Site carrier. Ordinate: free/bound [cytochalasin B] (ω, an increase in ω indicates a decrease in bound cytochalasin B). Abscissa: sugar concentration in m*M*. The various affinity and rate constants for this model (*see* Fig. 1) were taken from Lowe and Walmsley (1986) for transport at 0°C. Results for transported sugars are given by curves a, c, and e. Curve a illustrates the effects of S_i alone, curve c the effects of S_o alone, and curve e the effects of S_i and S_o ($[S_i] = [S_o]$) on ω. The analogous experiments with nontransported sugars are shown by curves b, d, and f. B. Predictions of the Two-Site carrier. Ordinate and abscissa as in A. K_o and K_i are taken from Lowe and Walmsley (1986) as 1 and 5 m*M*, respectively. Results are shown for transported (upper graph) and nontransported sugars (lower graph). The substrate conditions are both S_i and S_o present ($[S_i] = [S_o]$, a, b, c), S_o present alone (d, e, f,), and S_i present alone (g, h, i). Results are shown for cooperativity factors (α) of 4 (negative cooperativity, a, d, and g), 1 (no cooperativity, b, e, and h), and 0.25 (positive cooperativity, c, f, and i). C. Experimental results. Abscissa and Ordinate as in A and B. The sugars and inhibitors employed are shown above each graph. Open circles represent data obtained using unsealed ghosts (both S_i and S_o present where $[S_i] = [S_o]$) and closed circles data from intact red cells under conditions where sugars were not permitted to enter the cells. Further details of these theoretical and experimental results are given in Helgerson and Carruthers (1987).

by nontransported (but not by transported) ligand induces negative cooperativity between binding sites. This dependency of cooperativity between binding sites on the ability of an occupying ligand to be transported is precisely that required in order for the Two-Site carrier to be consistent with the findings of Krupka and Devés (1981) and the hetero-exchange experiments (*see* 6.1.). Similar results for phloretin inhibition of cytochalasin B binding in red cells have been obtained by Jung and Rampal (1977). In separate studies, linear inhibition curves for ethylidene glucose inhibition of cytochalasin B binding to alkali-stripped ghosts (Gorga and Lienhard, 1981) and maltose and phloretin inhibition of cytochalasin B binding to purified sugar transporter (Sogin and Hinkle, 1980a) have been observed. The reasons for these discrepancies are unclear at this time. A detailed discussion of these findings is given by Helgerson and Carruthers (1987).

6.3. Fluorescence Quenching Studies

Purified glucose transporter (Gorga and Lienhard, 1982; Carruthers, 1986a) and alkali-stripped red-cell ghosts (Carruthers, 1986b) display emission spectra (when excited at 295 nm) characterized by a scattering peak centered at 295 nm and an emission peak centered at 333 nm. When sugar transport substrate is added to the suspension of membranes, E_{333}–E_{450} is quenched. Gorga and Lienhard (1982) have demonstrated that sugar-induced E_{380} quenching of purified transporter fluorescence is a simple, saturable function in D-glucose and ethylidene glucose concentration. In separate studies, however, it has been shown that D-glucose and ethylidene glucose induced quenching of purified transporter and alkali-stripped ghost membrane E_{333} consists of two saturable components (Carruthers, 1986a,b). The results of Gorga and Lienhard (1982) are consistent with the One-Site carrier model for transport, whereas the results of Carruthers (1986a,b) are consistent with the Two-Site carrier model. The reasons for these divergent findings are not currently understood.

6.4. Conclusions

The available transport data fall into two categories. The hetero-exchange transport data are inconsistent with the One-Site carrier model for transport being more in line with the predictions of the Two-Site carrier. The studies of Krupka and Devés (1981) are consistent with both carrier models provided that intracellular and extracellular inhibitor binding sites in the Two-Site model display negative cooperativity. This latter result does not contradict the conclusions of the hetero-exchange experiments. The kinetic test (Krupka and Devés, 1981) simply cannot distinguish between either model.

Ligand binding studies of D-glucose and phloretin inhibition of cytochalasin B binding to the carrier in native membranes are consistent with the predictions of the Two-Site carrier (Helgerson and Carruthers, 1987), and support the notion that cytochalasin B and phloretin binding sites display strong negative cooperativity. One study has demonstrated that extracellular ethylidene glucose and maltose (nontransported sugars) exert a negative cooperative effect on cytochalasin B binding (Helgerson and Carruthers, 1987). Again these findings are consistent with the Two-Site carrier. Other studies (Gorga and Lienhard, 1981; Sogin and Hinkle, 1980a) did not observe this effect. Fluorescence quenching studies have reported dissonant findings supporting both models. Unlike the ligand binding and fluorescence quenching studies supporting the One-Site carrier, those studies supporting the Two-Site carrier report $K_{d(app)}$ for sugar binding to the carrier very close to $K_{m(app)}$ values for transport.

7. Factors Influencing Transport

In this section, we consider the possibility that red-cell sugar transport complexity results from intracellular factors/environmental conditions extrinsic to the transfer mechanism.

7.1. Compartmentalization of Intracellular Glucose

This hypothesis is described by Naftalin and Holman (1977). D-Glucose and other transported sugars bind nonspecifically to hemoglobin (Hb) and, in doing so, increase the amount of water complexed with Hb. At physiological Hb levels (5 m*M*, as much as 80% (compare with 71%, *see* Section 6.1.) of the sugar within the cell could be complexed with Hb—dissociating only slowly with respect to rates of transport (Naftalin and Holman, 1977). This has serious consequences with regard to transport determinations.

In sugar-loaded cells, intracellular hexose occupies two compartments—free and bound. At the onset of an exit experiment, that amount of intracellular sugar that is initially free to exit is limited to the intracellular free water space. Further exit is rate-limited by sugar dissociation from Hb. Uptake is rate-limited by sugar binding to Hb—the small intracellular water space becoming rapidly filled by sugar penetrating the cell. Using this approach, Naftalin and Holman (1977) and Naftalin et al.

(1985) have approximated the transport system as a symmetric Two-Site carrier with three glucose spaces (extracellular [S_o], free intracellular [Si], and bound intracellular [Hb·S_i]). Simulating a decrease in the rate of equilibration of sugar between free and bound intracellular compartments (but using rate constants for Hb-hexose dissociation close to those measured experimentally) increases both $K_{m(app)}$ and $V_{max(app)}$ for exit when estimated by integrating the time course of sugar efflux, but is without effect on K_m and V_{max} estimated by initial rate methods. This could account for the rather different estimates for K_m for sugar exit obtained by integrated and initial rate methods (*see* Table 1). However, both initial and integrated rate methods of transport analysis would underestimate V_{max} and K_m for infinite-cis D-glucose entry and V_{max} for entry.

Table 4 summarizes the expected results of Hb removal on the apparent kinetics of glucose transport and compares them with the observed transport parameters in substantially Hb-free red-cell ghosts. The findings are broadly in line with the predictions of the model. However, although it has been demonstrated that the effects of cytosol removal on sugar transport are reversible, it appears that the ability of cytosol to restore transport asymmetry to ghosts is lost following removal of low mol wt components (Carruthers and Melchior, 1983; Hebert and Carruthers, 1986). This being the case, intracellular hexose association with Hb cannot be the single explanation of red-cell sugar transport complexity.

The suggested differential transport of α- and β-D-glucose (Gorga and Lienhard, 1981; Wheeler and Hinkle, 1986; Lowe and Walmsley, 1986) by red cells provides a conceptually different, but formally similar explanation of multicomponent D-glucose washout from cells. This cannot be the reason for red-cell sugar transport anomalous behavior because: (1) transport complexity is lost in ghosts, yet equilibrium mixtures of α- and β-D-glucose were employed in the transport determinations (Carruthers and Melchior, 1983; Carruthers, 1986b) and (2) recent studies have demonstrated no significant differences between α- and β-D-glucose transport in intact cells (Carruthers and Melchior, 1985).

7.2. Transport in Fresh, Outdated, and Metabolically Depleted Human Blood

Comparison of sugar transport rates in fresh and outdated (or cold-stored) blood indicates significant differences (Stein, 1986). $K_{m(app)}$ for equilibrium exchange D-glucose transport is roughly doubled upon prolonged cold storage, and V_{max} is decreased slightly (Weiser et al., 1983). K_m for zero-trans exit is increased 1.6-fold and V_{max} is unaffected by cold

Table 4
Effects of Cytosol Removal on Sugar Transport[a]

Parameter	Predicted	Observed[b]
K_{io}^{zt}	Decrease	Decrease
V_{io}^{zt}	No change/decrease	No change
K_{oi}^{zt}	No change/decrease	Increase
V_{oi}^{zt}	Increase	Increase
K_{ci}^{ic}	No change	No change/increase
K_{io}^{ic}	Decrease/no change	Increase/no change

[a]The predicted Michaelis parameters are based on simulations (Naftalin and Holman, 1977; Naftalin et al., 1985) where the K_m parameters for entry and exit are assumed to be identical and low. The predicted results of cytosol removal on these parameters are critically dependent upon the K_m parameter employed in the simulations. The important result is the observed correspondence between predicted and experimental V_{max} changes.
[b]Data taken from Carruthers and Melchior (1983), Carruthers (1986b), Hebert and Carruthers (1986), Jung et al. (1971), and Taverna and Langdon (1973).

storage, whereas deoxygenation reduces $K_{m(app)}$ and V_{max} for exit (Naftalin et al., 1985). A third study has demonstrated that cold-storage of red cells, metabolic depletion by ionophore A 23187 plus Ca^{2+}, or depletion by iodoacetamide plus inosine plus EGTA results in a 1.5-fold decrease in V_{max} for uptake at 5°C (Jacquez, 1983). These results are consistent with the view that compartmentalization of intracellular D-glucose increases in cold-stored and metabolically depleted cells, but that deoxygenation of cells reduces compartmentalization (*see above* for discussion of hexose complexation). Naftalin et al. (1985) suggest that D-glucose binding to deoxyhemoglobin may be less than that to oxyhemoglobin.

7.3. Sugar Transport in Red-Cell Ghosts and Effects of ATP

Sugar transport in human red-cell ghosts at 20°C is symmetric (Table 5). V_{max} and K_m for zero-trans entry K_m for infinite-cis exit and infinite-cis entry are increased upon cell ghosting, whereas K_m for zero-trans exit is decreased. Removal of red-cell cytosol is equivalent to depleting an intracellular source of competitive inhibitor of sugar exit (Carruthers and Melchior, 1983; Carruthers, 1986b; Hebert and Carruthers, 1986). However, One-Site carrier redistributions induced by this putative, intracellular competitive inhibitor cannot account for the effects of cytosol dilution on K_m for zero-trans entry or infinite-cis exit. With Two-site transport, the presence of an inhibitor cannot account for the increase in K_m for infinite-cis entry upon ghosting cells (and loss of inhibitor).

Table 5
D-Glucose Transport Parameters in Intact Cells and Ghosts

	Observed				Predicted			
	Intact cells		Ghosts		2mM ATP		O ATP	
	K_m[a]	V_m[b]	K_m[a]	V_m[b]	K_m[a]	V_m[b]	K_m[a]	V_m[b]
Zero-trans entry	1.6	36	10	185	4	40	10	200
Zero-trans exit	25	190	10	194	20	200	10	200
Infinite-cis entry	2.8	40	11	191	4.4	50	12	200
Infinite-cis exit	1.8	185	11	198	1.6	200	12	200
Equilibrium exchange	31	310			9	300	18	300
Equilibrium exchange	12[c]	360[c]	21[d]	300[d]				

[a]In millimolar.

[b]In millimoles/L of cell water/min.

[c]Data taken from Weiser et al. (1983) for freshly drawn red cells.

[d]Data taken from Weiser et al. (1983) for red cells stored in the cold for 8 wk. The predicted values were calculated from the relationship, influx

$$V_i = \frac{\dfrac{k_1[G_o]}{K_A} + \dfrac{k_2[ATP][G_o]}{\beta K_N K_A} + \dfrac{k_5[G_o][G_i]}{\alpha K_A K_B} + \dfrac{k_6[ATP][G_o][G_i]}{\beta\gamma\delta K_N K_A K_B}}{1 + \dfrac{[G_o]}{K_A} + \dfrac{[G_i]}{K_B} + \dfrac{[ATP]}{K_N} + \dfrac{[G_i][G_o]}{\alpha K_A K_B} + \dfrac{[ATP][G_i]}{\gamma K_B K_N} + \dfrac{[ATP][G_o]}{\beta K_N K_A} + \dfrac{[ATP][G_i][G_o]}{\beta\gamma\delta K_N K_A K_B}}$$

where the various translocation and dissociation constants are illustrated in Fig. 4. Efflux is obtained by interchanging $k_1 \leftrightarrow k_3$, $k_2 \leftrightarrow k_4$ and, in terms lacking $[G_1]$, $[G_o] \leftrightarrow [G_i]$, $K_A \leftrightarrow K_B$, and $\beta \leftrightarrow \gamma$ in the numerator. The values employed as a first approximation were $k_1 = k_3 = k_4 = 200$, $k_5 = k_6 = 300$, and $k_2 = 40$ mmol (L of cell water^{-1}) min^{-1}. $K_A = K_B = 10$ mM, and $\alpha = 1$, $\beta = 0.4$, $\gamma = 2$, and $\delta = 0.25$. For a passive transfer system in the absence of ATP, $K_A/K_B = k_1/k_3$, and in the presence of ATP, $\beta K_A/\gamma K_B = k_2/k_4$. (Taken from Carruthers, 1986b.)

K_m and V_{max} for D-glucose exit from inside-out red-cell vesicles (IOVs) fall dramatically (by up to 10-fold) with increasing concentrations of cytosol in the extravesicular medium (Carruthers and Melchior, 1983). This effect is half-maximal at 45 x dilution of cytosol and is lost if cytosol is extensively dialyzed against buffered saline. These findings support the view that at least one of the factors present in cytosol that modifies sugar transport is a low mol wt species. Hebert and Carruthers (1986) examined this further by characterizing the properties of this factor(s). Their findings indicate that the transport modulating species may be ATP, which acts to increase K_m for exit and decrease K_m and V_{max} for entry. ATP-modulation of transport is half-maximal at about 40–50μM, which is approximately 1/40–1/45 the concentration of cytosolic ATP in the cells employed. The action of ATP on K_m for transport is extremely rapid in onset and reversal, whereas inhibition of V_{max} for entry is more slowly expressed and reversed (Carruthers, 1986b; Hebert and Carruthers, 1986). A rather complete analysis of sugar transport in ghosts indicates that most of the transport properties of intact cells are restored by ATP (Carruthers, 1986b). However, close examination of the deceleration of sugar exit from IOVs by ATP alone (Carruthers, 1986a) indicates that this effect is less marked than that produced by cytosol (Carruthers and Melchior, 1983). This raises the possibility that cytosolic factors other than ATP may also play a role in modulating hexose transfer.

Fluorescence quenching studies (Carruthers, 1986a,b) indicate that ATP binds to the transporter ($K_{d(app)}$ = 40 μM) reducing the low $K_{d(app)}$ for sugar binding to the carrier and increasing the high $K_{d(app)}$ for sugar. These findings are exactly those predicted from the transport studies (Carruthers, 1986b) if the low and high $K_{d(app)}$ correspond to sugar influx and efflux binding sites, respectively. Using these data, a model was proposed ([Carruthers, 1986b], *see* Fig. 4) that could account for most of the catalytic properties of red-cell ghost sugar transport (Table 5). The transporter was proposed to be regulated allosterically by ATP to account for changes in apparent binding constants for sugars and either covalently by γ-phosphoryl transfer or indirectly through modification of regulatory proteins to account for changes in V_{max} parameters for transport.

7.4. Inconsistencies

Although this new model can account for the transport properties of ghosts, a number problems remain. For example, if ATP inhibits sugar entry and increases K_m for exit in ghosts, why do metabolic depletion (Jacquez, 1983) and cold-storage (Weizer et al., 1983; Naftalin et al., 1985) of intact cells have the reverse effects on these parameters? In

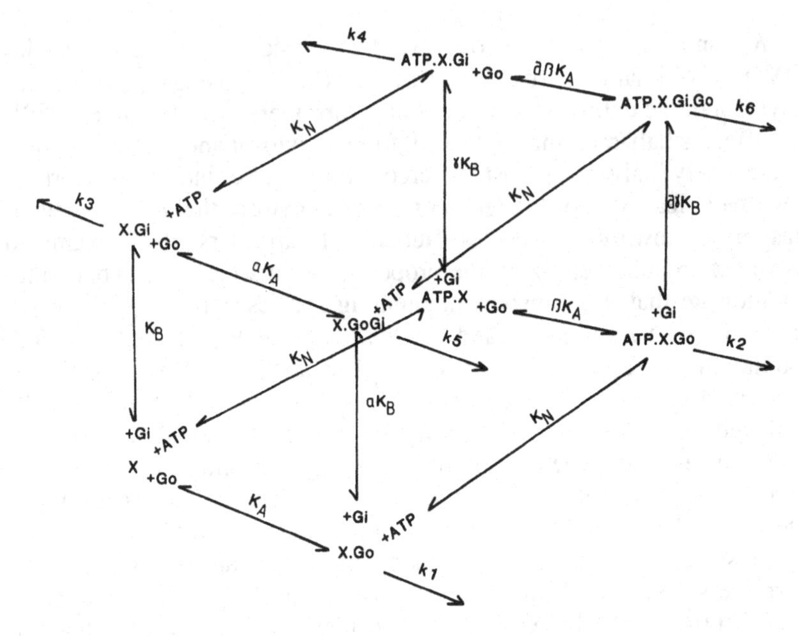

Fig. 4. A model for erythrocyte hexose transfer. The carrier X can combine with G_o, G_i (external and internal D-glucose) and ATP_i to form the various complexes indicated. The rate-limiting steps for efflux, influx, and exchange fluxes are shown by rate constants k_3 and k_4, k_1 and k_2 and by k_5 and k_6, respectively. The predictions of this model are summarized in Table 5. Taken from Carruthers (1986b).

addition, if transport at 20°C in ATP-containing cells is asymmetric, why do the counterflow data of Naftalin et al. (1985) support the view that transport in intact cells is symmetric?

Examination of the counterflow experiments performed by Naftalin et al. (1985) indicates that, in deriving the asymmetry factor λ for transport, the contribution of exchange transport to the counterflow transient was not considered (e.g., *see* Baker and Widdas, 1973). At the peak of the counterflow transient, the rate of radiolabeled sugar entry is identical to the rate of exit. If we adopt the Two-Site carrier model for transport, this means that, at this time:

$$^{v}12 = \frac{\lambda V \dfrac{S_i}{\lambda K} + V^e \dfrac{S_i(S_o + C_o)}{\alpha \lambda K^2}}{1 + \dfrac{(S_i + C_i)}{\lambda K} + \dfrac{(S_o + C_o)}{K} + \dfrac{(S_i + C_i)(S_o + C_o)}{\alpha \lambda K^2}} \tag{3}$$

i.e., influx = efflux where V is V_{max} for entry, V^e is V_{max} for exchange, S_o, S_i (extra-and intracellular labeled sugar), and C_i and C_o (intracellular and extracellular nonradioactive sugar, respectively) are sugar levels at the peak of the counterflow transient, α is the cooperativity between sugar binding sites, and λ is the asymmetry of transport, i.e., λK and λV are K_m and V_{max} for sugar exit, respectively. Since the exchange components of influx and efflux of label are not identical, these terms must be included and the asymmetry factor λ is given by

$$\lambda = \frac{V^e}{\alpha VK} \frac{(S_oC_i - S_iC_o)}{(S_i - S_o)} \tag{4}$$

Assuming α is unity (Helgerson and Carruthers, 1987) and using the counterflow data obtained by Naftalin et al. (1985) at 20°C, and the measured K, V^{ee}, V values of Lowe and Walmsley (1986) and Challis et al. (1980), the asymmetry factor λ is readily shown to be 5.6–10, a value very close to the apparent asymmetry of transport (K_i/K_o, V_{12}/V_{21}, 3–14, Table 1. Thus, the conclusion of apparent symmetry (Naftalin et al., 1985) is unfounded. This analysis breaks down at 0°C, where estimates of λ are in excess of 370 whereas measured values range from 13–100. The demonstrations of multicomponent transport (Naftalin et al., 1985) and the need for activity coefficients of less than unity for intracellular D-glucose in order to simulate counterflow data (Baker and Widdas, 1973) directly support the notion of intracellular complexation of D-glucose. Complexation is presumably lost upon ghost formation.

The differences between transport in ghosts and metabolically depleted cells are less readily explained. Studies with ghosts employ conditions where intracellular solute levels are directly and specifically manipulated. This level of solute control is not possible during metabolic depletion of cells where species in addition to ATP are also modulated. We should also consider whether the agents employed to deplete red cells (Ca^{2+} plus ionophore and iodoacetamide) also directly influence the activity of the transporter. Evidence has been obtained for a direct, inhibitory effect of Ca^{2+} plus A 23187 on avian erythrocyte sugar transport (Simons, 1983), and Ca^{2+} plus ATP have been demonstrated to inhibit transport in human red-cell vesicles (Hebert and Carruthers, 1986). In addition, our studies demonstrate that iodoacetamide treatment of glucose-loaded (100 m*M*), white (nominally ATP-free) ghosts (10 m*M* for 2 h at 37°C) increases the exit time (decreases the initial rate of efflux) for sugar efflux by some 1.7 ± 0.1-fold (n = 7). Thus, inhibitions of

Table 6
Effects of ATP_i and Ca^{2+} on Zero-Trans 3–*O*–Methylglucose Uptake in Red Cells and Red-Cell Ghosts[a]

$[ATP]_i$ m*M*	$[Ca^{2+}]_o$ μ*M*[b]	$[Ca^{2+}]_i$ μ*M*[c]	$K_{m(app)}$ m*M*	V_{max} μmol/L/min	n[d]
Red cells					
1 ± 0.1	0	?	0.4 ± 0.1	114 ± 20	6
0.8 ± 0.1	101	?	0.36 ± 0.03	23 ± 1[e]	4
Red-cell ghosts					
4	0	0	0.42 ± 0.11	110 ± 15	5
4	101	0	0.28 ± 0.04	43 ± 5[e]	4
4	0	15	0.32 ± 0.05	68 ± 6[e]	4
0	0	0	7.1 ± 1.8[e]	841 ± 191[e]	5
0	101	0	2.9 ± 0.4[e]	36 ± 7[e]	4
0	0	11	2.7 ± 0.7[e]	52 ± 5[e]	4

[a]Results are shown mean ± S.E. All transport measurements were made at 0–4°C. All preincubations were at 24°C for 30 min except for ghosts, which additionally included a 40-min resealing step at 37°C. All solutions contained 150 m*M* KCl, 5 m*M* TrisHCl, 2 m*M* $MgCl_2$, 2 m*M* EGTA, pH 7.4.

[b]Ca:EGTA ratio = 2.1:2 at 2 m*M* $MgCl_2$, pH 7.4. For $[Ca^{2+}]_o$ = 101 μ*M*.

[c]Ca:EGTA ratio = 2.2:2 at 2 m*M* $MgCl_2$, 4 m*M* ATP, pH 7.4 for $[Ca^{2+}]_i$ = 15 μ*M* and 2:2 at 2 m*M* $MgCl_2$, 0 ATP, pH 7.4 for $[Ca^{2+}]_i$ = 11 μ*M*.

[d]n is the number of separate experiments employing [30MG] in the range 0.05–20 m*M*.

[e]Indicates that the results are significantly different from control $K_{m(app)}$ and V_{max} values (red blood cell, 0 Ca^{2+}) at the $p<0.05$ level (two-tailed t-test).

transport by Ca^{2+} plus ionophore and by iodoacetamide may not simply reflect the results of ATP-depletion alone, but may also include direct inhibitory actions on transporter function.

Our laboratory observes that Ca^{2+}-treatment of nominally Ca^{2+}-free red cells results in the reversible reduction in V_{max} for 3-*O*-methylglucose (3OMG) uptake (Table 6). $K_{m(app)}$ for 30MG uptake is unaffected by Ca_o^{2+}. Inhibition of V_{max} is maximal within 30 min and reaches 80 ± 4%. Extracellular Ca^{2+} half-maximally inhibits V_{max} for 30MG uptake at 2.4 μ*M* Ca^{2+}. Intracellular Ca^{2+} is equally potent in its ability to reduce V_{max} for 30MG uptake in ATP-containing ghosts. In the absence of ATP_i, Ca_o^{2+} or Ca_i^{2+} produce even greater inhibition of V_{max} for uptake (94–96%, Tables 6,7). This observation suggests that ATP_i blunts the inhibitory actions of intra-and extracellular Ca^{2+} on sugar transport.

Ghosts formed from Ca^{2+}-treated red cells retain lower 30MG transport capacity than do ghosts formed from Ca^{2+}-naive cells (Table 8). However, ATP incorporation into these ghosts restores uptake to levels observed in intact red cells incubated in Ca^{2+}-free medium. The

Table 7
Effects of Ca^{2+} on 30MG Uptake in Red-Cell Ghosts[a]

Condition	30MG uptake μmol/L/min [30MG] of uptake determination 0.2 mM	5 mM	n[b]
Cells	24.0 ± 1.2	112 ± 5	4,4
Cells + 100 μM Ca^{2+}_o[c]	8.1 ± 1.3[d]	26 ± 3[d]	4,4
Ghosts 0 ATP_i	8.1 ± 0.5[d]	236 ± 13[e]	4,4
+ 11 μM Ca^{2+}_i[h]	3.2 ± 0.1[d]	57.5 ± 6.0[d]	4,4
+ 100 μM Ca^{2+}_o[c]	3.3 ± 0.9[d]	43 ± 2.8[d]	4,4
+ Ca^{2+}_i + Ca^{2+}_o	3.4 ± 0.2[d]	26.5 ± 1.8[g]	4,4
Ghosts + 4 mM ATP_i	24.5 ± 0.4	146 ± 6[e]	4,4
+ 15 μM Ca^{2+}_i	15.6 ± 0.4[d]	87 ± 1.4[d]	4,4
+ 100 μM Ca^{2+}_o	14.8 ± 0.2[d]	88.5 ± 1.8[d]	4,4
+ Ca^{2+}_i + Ca^{2+}_o	13.2 ± 0.6[d]	68.1 ± 0.7[g]	4,4
Ghosts + 2mM AMPPNP	23.1 ± 0.7		
+ 100 μM Ca^{2+}_o	9.6 ± 0.4[d]		

[a]Experimental conditions as in Table 6.
[b]n is the number of separate duplicate determinations shown first for uptake at 0.2 mM 30MG and then for uptake at 5 mM 30MG.
[c]Ca:EGTA ratio = 2.1:2 at 2 mM $MgCl_2$, pH 7.4.
[d]Significantly less than control values at the $p<0.01$ level (1-tailed t-test).
[e]Significantly greater than control values at the $p<0.01$ level (1-tailed t-test).
[f]Ca:EGTA ratio = 2.2:2 at 2 mM $MgCl_2$, 4 mM ATP, pH 7.4.
[g]The inhibition by Ca^{2+}_i plus Ca^{2+}_o is significantly greater than that by Ca^{2+}_i or Ca^{2+}_o alone.
[h]Ca:EGTA ratio = 2.2:2 at 2 mM $MgCl_2$, 0ATP, pH 7.4

nonmetabolizable ATP analog adenylyl-imidodiphosphate (AMP-PNP) is unable to mimic the ability of ATP to restore normal transport rates to ghosts formed from Ca^{2+}-treated red cells (Table 8), although it is able to mimic the action of ATP on transport in ghosts formed from Ca^{2+}-naive red cells (Tables 7,8). These findings indicate:

1. That Ca^{2+}_i and Ca^{2+}_o reversibly inhibit sugar transport in human red cells
2. That Ca^{2+} inhibition is not reversed by forming Ca^{2+}-free, nominally ATP-free ghosts
3. That ATP_i can reverse Ca^{2+}-inhibition of transport
4. That the nonmetabolizable analog of ATP (AMP-PNP) cannot substitute for ATP in this action, but can substitute for ATP_i in its ability to modulate transport in Ca^{2+}-naive cells and ghosts.

Table 8
Reversibility of Ca^{2+} Action on 30MG Uptake[a]

	Origin[b]	$[Ca^{2+}]_o$ μM	$[ATP]_i$[c] mM	$[AMPPNP]_i$ mM	30MG uptake at 1 mM μmol/L/min	n[d]
Cells						
	A	0	1.0 ± 0.1	0	53.0 ± 6.2	4
	B	100[e]	0.8 ± 0.1	0	24.6 ± 4.3[f]	4
	B	100 -> 0	?	0	47.9 ± 4.0	4
Ghosts						
	A	0	0	0	29.8 ± 1.0	4
	B	0	0	0	13.7 ± 0.1[f,g]	4
	A	0	4	0	61.5 ± 4.6	4
	B	0	4	0	51.7 ± 4.1	4
	A	0	0	4	55.5 ± 3.2	4
	B	0	0	4	18.1 ± 1.0[f,g]	4

[a]Experimental conditions as in Table 6.
[b]Origin refers to original treatment of cells. A cells were originally Ca-naive. B cells were initially exposed to 100 μM Ca^{2+} for 30 min at 24°C. $[Ca^{2+}]_o$ refers to the concentration of Ca^{2+} present during uptake measurements.
[c]For intact cells $[ATP]_i$ refers to the measured ATP content (mmol/kg cell water) of cells and for ghosts refers to the initial [ATP] of the resealing medium.
[d]n refers to the number of separate measurements made. Results are shown as mean ± S.E.
[e]Ca:EGTA ratio = 2.1:2 at 2 mM $MgCl_2$, pH 7.4.
[f]Indicates that uptake is significantly lower than uptake measured in Ca^{2+}-native intact red cells ($p<0.01$, 1-tailed t-test).
[g]Indicates that uptake by B cell ghosts is significantly less ($p<0.01$, 1-tailed t-test) than uptake by A cell ghosts resealed under identical conditions.

Our studies eliminate the following potential mechanisms for Ca^{2+}-inhibition of transport (Tables 7, 8).

1. Ca^{2+}-activation of cellular phospholipases A_2 and C
2. Ca^{2+}-activation of diacylglycerol-dependent protein kinase C
3. Ca^{2+}-activation of proteases
4. Ca^{2+}-dependent changes in the lipid composition of the red-cell membrane (not shown).

These studies illustrate the ability of ATP to modify sugar transport by a kinase-independent mechanism but also demonstrate a second mechanism of ATP-action on sugar transport—an hydrolysis dependent reversal of Ca^{2+}-inhibition of sugar transport. Our results are consistent with previous studies from this laboratory and additionally with those of Jacquez (1983). Our interpretation of Jacquez's study is that raised Ca_i^{2+} inhibits sugar transport in RBCs and causes ATP_i depletion. The observed reversal of inhibition by ATP_i in ghosts results from ATP-hydrolysis dependent reversal of Ca_i^{2+} inhibition of transport—an effect quite different from the allosteric (hydrolysis independent) effects of ATP_i on transport in the absence of Ca^{2+}.

The stimulation of transport by ghost formation is not limited to the hexose transfer mechanism. The Ca^{2+}-dependent K channel also displays higher catalytic conductivity in ghosts than in metabolically depleted red cells (Simons, 1982); thus, ghosting the red cell may induce the loss of a number of transport-modulating species.

7.5. Conclusions

Early estimates of K_m for sugar exit from intact cells by the integrated rate equation approach have resulted in the overestimation of this parameter. This arises from a multicomponent washout of sugar from the cell, which in turn may result from complexation of intracellular sugar with hemoglobin. Red-cell sugar transport asymmetry appears to be dependent upon intracellular ATP levels. Asymmetry is lost in ghosts, but can be restored upon addition of ATP to the interior of the cell. ATP may act both via allosteric interaction with the transporter, and by covalent modification of either the transporter or transporter-modulating membrane species. Other transport-modulating species may also be lost from red cells upon lysis.

The influence of membrane factors (bilayer lipids) on erythrocyte hexose transporter function has been reviewed elsewhere (Carruthers and Melchior, 1986, 1987).

8. Summary

8.1. The Mechanism of Sugar Transport

This review has described sugar transport and ligand binding studies that support the view that human erythrocyte hexose transfer is mediated via a specific membrane protein capable of binding two transportable sugar molecules simultaneously. Occupancy of either external or internal substrate binding sites by nontransported (but not by transported) substrate reduces the affinity of the vacant, trans site for substrate.

Sugar transport in human erythrocytes and ATP-containing ghosts is asymmetric (K_m and V_{max} for exit > K_m and V_{max} for entry). Asymmetry appears to result from ATP-interaction with the transfer mechanism. The extent of asymmetry has been overestimated in intact cells, possibly because of the complexation of intracellular sugar by hemoglobin. When

Michaelis and velocity parameters for sugar exit from intact cells are obtained by integration of the time course of sugar efflux, the calculated K_m for exit is artificially high because of the delayed washout of residual intracellular sugar. Initial rates of sugar exit provide more accurate estimates of efflux from the intact cell. Transport in nominally ATP-free ghosts is symmetric.

8.2. Regulation of Sugar Transport

Sugar transport in human erythrocyte ghosts is modulated by intracellular ATP, which acts to induce transport asymmetry in ghosts by reducing K_m and V_{max} for entry and increasing K_m for exit of sugar. ATP appears to modulate transport both directly through allosteric interaction and perhaps indirectly through modification of transport modulating species. Significant differences in transport exist between nominally ATP-free ghosts and metabolically depleted intact cells, indicating that transport-modulating factors other than ATP (e.g., Ca^{2+}) may be eluted from cells upon hypotonic lysis. This form of sugar transport regulation is fundamentally different from the proposed mechanism of insulin regulation of sugar transport in adipose and muscle. Insulin appears to induce an energy-dependent recruitment of intracellular sugar transport proteins to the plasma membrane (Simpson and Cushman, 1986), thereby increasing carrier density and membrane transport capacity. Human red-cell sugar transport capacity is modulated by control of the intrinsic activity of carriers at the plasma membrane (Hebert and Carruthers, 1986). This recruitment-independent form of sugar transport control may be of great significance to sugar transport regulation in exercising and metabolically depleted muscle where ATP may act to depress sugar intake (Elbrink and Bihler, 1975).

8.3. Sugar Transport in Other Cell Types

The mechanism of sugar transport in other cell types has not been as extensively studied as that of the human erythrocyte. In many cells (Stein, 1986), sugar transport is symmetric. This finding is not sufficient evidence to reject either the One or Two-Site carrier models for sugar transfer. More sensitive tests, such as the hetero-exchange and infinite-cis procedures, are required to determine the mechanism of transport. Such tests have been performed using the rat erythrocyte hexose transfer system (Helgerson and Carruthers, 1989) and human lymphocytes (Rees and Gliemann, 1985) and are, again, consistent with the Two-Site carrier model for transport. The results of similar studies of sugar transport in molluscan nerve (Baker and Carruthers, 1981, 1984) are consistent with both One and Two-Site carrier models for transport. Thus, where examined, the Two-Site carrier is consistent with the transfer properties of the system under study. Whether this mechanism is a fundamental property of all passive hexose transfer systems (or just those of erythrocytes, lymphocytes, and nerve) remains to be established.

Acknowledgment

This work was supported by N.I.H. grant R01 AM36081.

References

Allard, W. J. and Lienhard, G. E. (1985) Monoclonal antibodies to the glucose transporter from human erythrocytes. *J. Biol. Chem.* **260,** 8668–8675.

Baker, G. F. and Widdas, W. F. (1973) The asymmetry of the facilitated transfer system for hexoses in human red cells and the simple kinetics of a two component model. *J. Physiol (London)* **231,** 143–165.

Baker, G. F. and Naftalin, R. J. (1979) Evidence for multiple operational affinities for D-glucose inside the human erythrocyte membrane. *Biochim. Biophys. Acta* **550,** 474–484.

Baker, P. F. and Carruthers, A. (1981) 3-*O*-Methylglucose transport in internally dialyzed giant axons of *Loligo. J. Physiol. (London)* **316,** 503–525.

Baker, P. F. and Carruthers, A. (1984) Transport of sugars and amino acids. *Curr. Top. Membr. Transp.* **22,** 91–130.

Baldwin, S. A. Baldwin, J. M., and Lienhard, G. E. (1982) The monosaccharide transporter of the human erythrocyte. Characterization of an improved preparation. *Biochemistry* **21,** 3836–3842.

Barnett, J. E. G., Homan, G. D., Chalkey, R. A., and Munday, K. A. (1975) Evidence for two asymmetric conformational states in the human erythrocyte sugar transport system. *Biochem. J.* **145,** 417–429.

Basketter, D. A. and Widdas, W. F. (1978) Asymmetry of the hexose transfer system in human erythrocytes. Comparison of the effects of [cytochalasin B], phloretin and maltose as competitive inhibitors. *J. Physiol. (London)* **278,** 389–401.

Birnbaum, M. J., Haspel, H. C., and Rosen, O. M. (1986) Cloning and characterization of a cDNA encoding the rat brain glucose transporter protein. *Proc. Natl. Acad. Sci. USA* **83,** 5784–5788.

Bloch, R. (1973) Inhibition of glucose transport in the human erythrocyte by cytochalasin *B. Biochemistry* **12,** 4799–4801.

Bowman, R. J. and Levitt, D. G. (1977) Polyol permeability of the human red cell. Interpretation of glucose transport in terms of a pore. *Biochim. Biophys. Acta* **466,** 68–83.

Brahm, J. (1983) Kinetics of glucose transport in human erythrocytes. *J. Physiol. (London)* **339,** 339–354.

Britton, H. G. (1965) Fluxes in passive monovalent and polyvalent carrier systems. *J. Theoret. Biol.* **10,** 28–52.

Carruthers, A. (1986a) ATP regulation of the human red cell sugar transporter. *J. Biol. Chem.* **261,** 11028–11037.

Carruthers, A. (1986b) Anomalous asymmetric kinetics of human red cell hexose transfer: role of cytosolic adenosine 5′-triphosphate. *Biochemistry* **25,** 3592–3602.

Carruthers, A. and Melchior, D. L. (1983) Asymmetric or symmetric? Cytosolic modulation of human erythrocyte hexose transfer. *Biochim. Biophys. Acta* **728,** 254–266.

Carruthers, A. and Melchior, D. L. (1984) A rapid method of reconstituting human erythrocyte sugar transport proteins. *Biochemistry* **24,** 4816–4821.

Carruthers, A. and Melchior, D. L. (1985) Transport of α- and β-D-glucose by the intact human red cell. *Biochemistry* **24,** 4244–4250.

Carruthers, A. and Melchior, D. L. (1986) How bilayer lipids affect membrane protein activity. *TIBS* **11,** 331–335.

Carruthers, A. and Melchior, D. L. (1988) Effect of lipid environment on membrane transport: The human erythrocyte sugar transport protein/lipid bilayer system. *Ann. Rev. Physiol.* **50,** 257–271.

Challis, R. J. A., Taylor, L. P., and Holman, G. D. (1980) Sugar transport asymmetry in human erythrocytes—the effect of bulk haemoglobin removal and the addition of methyl-xanthines. *Biochim. Biophys. Acta* **602,** 155–166.

Chin, J. J., Jung, E. K. Y., and Jung, C. Y. (1986) Structural basis of human erythrocyte glucose transporter function in reconstituted vesicles. *J. Biol. Chem.* **261,** 7101–7104.

Deziel, M. R. and Rothstein, A. (1984) The membrane topology of the human erythrocyte hexose transporter. *Fed. Proc.* **43,** 1375.

Elbrink, J. and Bihler, I. (1975) Membrane transport. Its relation to cellular metabolic rates. *Science* **188,** 1177–1184.

Ginsburg, H. and Yeroushalmy, S. (1978) Effects of temperature on the transport of galactose in human erythrocytes. *J. Physiol (London)* **282,** 399–417.

Gorga, F. R. and Lienhard, G. E. (1981) Equilibria and kinetics of ligand binding to the human erythrocyte glucose transporter. Evidence for an alternating conformation model for transport. *Biochemistry* **20,** 5108–5113.

Gorga, F. R. and Lienhard, G. E. (1982) Changes in the intrinsic fluorescence of the human erythrocyte monosaccharide transporter upon ligand binding. *Biochemistry* **21,** 1905–1908.

Gorga, F. R., Baldwin, S. A., and Lienhard, G. E. (1979) The monosaccharide transporter from human erythrocytes is heterogeneously glycosylated. *Biochem. Biophys. Res. Comm.* **91,** 966–971.

Hebert, D. N. and Carruthers, A. (1986) Direct evidence for ATP modulation of sugar transport in human erythrocyte ghosts. *J. Biol. Chem.* **261,** 10093–10099.

Helgerson, A. L. and Carruthers, A. (1987) Equilibrium ligand binding to the human erythrocyte sugar transporter. Evidence for two sugar binding sites per carrier. *J. Biol. Chem.* **262,** 5464–5475.

Helgerson, A. L. and Carruthers, A. (1989) Analysis of protein-mediated 3-*O*-methylglucose transport in rat erythrocytes. *Biochemistry,* **28,** 4580–4590.

Holman, G. D. (1980) An allosteric pore model for sugar transport in human erythrocytes. *Biochim. Biophys. Acta* **599,** 202–213.

Holman, G. D., Parkar, B. A., and Midgley, P. J. W. (1986) Exofacial photoaffinity labelling of the human erythrocyte sugar transporter. *Biochim. Biophys. Acta* **855,** 115–126.

Jacquez, J. A. (1983) Modulation of glucose transport in human red blood cells by ATP. *Biochim. Biophys. Acta* **727,** 367–378.

Jung, C. Y. (1971) Permeability of bimolecular membranes made from lipid extracts of human red cells to sugar. *J. Memb. Biol.* **5,** 2300–2314.

Jung, C. Y., and Rampal, A. L. (1977) Cytochalasin B binding sites and the glucose transporter in human erythrocyte ghosts. *J. Biol. Chem.* **252,** 5456–5463.

Jung, C. Y., Carlson, L. M., and Whaley, D. A. (1971) Glucose transport carrier activities in extensively washed human red cell ghosts. *Biochim. Biophys. Acta* **241,** 613–627.

Jung, E. K. Y., Chin, J. J., and Jung, C. Y. (1986) Structural basis of human erythrocyte glucose transporter function in reconstituted system. *J. Biol. Chem.* **261,** 9155–9160.

Kasahara, M. and Hinkle, P. C. (1977) Reconstitution and purification of the glucose transporter from human erythrocytes. *J. Biol. Chem.* **253,** 7384–7390.

Krupka, R. M. and Devés, R. (1981) An experimental test for cyclic *versus* linear transport models. The mechanism of glucose and choline transport in erythrocytes. *J. Biol. Chem.* **256,** 5410–5416.

LeFevre, P. G. (1961) Sugar transport in the red blood cell: structure activity relationships in substrates and antagonists. *Pharmacol. Rev.* **13,** 39–70.

LeFevre, P. G. and Marshall, J. K. (1959) The attachment of phloretin and analogues to human erythrocytes in connection with inhibition of sugar transport. *J. Biol. Chem.* **234,** 3022–3027.

Levine, M., Oxender, D. L., and Stein, W. D. (1965) The substrate facilitated transport of the glucose carrier across the human erythrocyte membrane. *Biochim. Biophys. Acta* **109,** 151–163.

Lieb, W. R. and Stein, W. D. (1972) Carrier and non-carrier models for sugar transport in the human red blood cell. *Biochim. Biophys. Acta* **265,** 187–207.

Lieb, W. R. and Stein, W. D. (1974) Testing and characterizing the simple carrier. *Biochim. Biophys. Acta* **373,** 178–196.

Lowe, A. G. and Walmsley, A. R. (1986) The kinetics of glucose transport in human red blood cells. *Biochim. Biophys. Acta* **857,** 146–154.

Miller, D. M. (1971) The kinetics of selective biological transport. V. Further data on the erythrocyte-monosaccharide transport system. *Biophys. J.* **11,** 915–923.

Mueckler, M., Caruso, C., Baldwin, S. A., Panico, M., Blench, I., Morris, H. R., Allard, W J., Lienhard, G. E., and Lodish, H. F. (1985) Sequence and structure of the human glucose transporter. *Science* **229,** 941–945.

Naftalin, R. J. and Holman, G. D. (1977) Transport of sugars in human red cells, in *Membrane Transport in Red Cells* (Ellory, J. C. and Lew, V. L., eds.), Academic Press, New York, pp. 257–300.

Naftalin, R. J., Smith, P. M., and Roselaar, S. E. (1985) Evidence for non-uniform distribution of D-glucose within human red cells during net exit and counterflow. *Biochim. Biophys. Acta* **820,** 235–249.

Rees, W. D. and Gliemann, J. (1985) A kinetic analysis of hexose transport in cultured human lymphocytes (IM-9). *Biochim. Biophys. Acta* **812,** 98–106.

Segel, I. H. (1975) Steady-state kinetics of multireactant enzymes, in *Enzyme Kinetics*, John Wiley and Sons, New York, pp. 527.

Shanahan, M. F. and D'Artel-Ellis, J. (1984) Orientation of the glucose transporter in the human erythrocyte membrane. *J. Biol. Chem.* **259,** 13878–13884.

Shelton, R. L. and Langdon, R. G. (1983) Reconstitution of glucose transport using human erythrocyte band 3. *Biochim. Biophys. Acta* **733,** 25–33.

Simons, T. J. B. (1982) A method for estimating free Ca within human red blood cells, with an application to the study of their Ca-dependent K permeability. *J. Membr. Biol.* **66,** 235–247.

Simons, T. J. B. (1983) The role of calcium in the regulation of sugar transport in the pigeon red blood cell. *J. Physiol. (London)* **338,** 501–526.

Simpson, I. A. and Cushman, S. W. (1986). Hormonal regulation of mammalian glucose transport. *Ann. Rev. Biochem.* **55,** 1059–1089.

Sogin, D. C. and Hinkle, P. C. (1980a) Binding of cytochalasin B to human erythrocyte glucose transport. *Biochemistry* **19,** 5417–5420.

Sogin, D. C. and Hinkle, P. C. (1980b) Immunological identification of the human erythrocyte glucose transporter. *Proc. Natl. Acad. Sci. USA* **77,** 5725–5729.

Stein, W. D. (1986) Facilitated diffusion: The single carrier, in *Transport and diffusion across cell membranes*, Academic Press, New York, pp. 231–306.

Taverna, R. D. and Langdon, R. G. (1973) Glucose transport in white erythrocyte ghosts and membrane derived vesicles. *Biochim. Biophys. Acta* **298,** 422–428.

Vidaver, G. A. (1966) Inhibition of parallel flux and augmentation of counter-flux shown by transport models not involving a mobile carrier. *J. Theor. Biol.* **10,** 301–306.

Weiser, M. B., Razin, M., and Stein, W. D. (1983) Kinetic tests of models for sugar transport in human erythrocytes and a comparison of fresh and cold-stored cells. *Biochim. Biophys. Acta* **727,** 379–388.

Wheeler, T. J. and Hinkle, P. C. (1985) The glucose transporter of mammalian cells. *Ann. Rev. Physiol.* **47,** 503–518.

Widdas, W. F. (1952) Inability of diffusion to account for placental glucose transfer in the sheep and consideration of the kinetics of a possible carrier transfer. *J. Physiol. (London)* **118,** 23–29.

Widdas, W. F. (1980) The asymmetry of the hexose transfer system in the human red cell membrane. *Curr. Top. Membr. Transp.* **14,** 165–223.

Chloride-Dependent Cation Transport in Human Erythrocytes

Gordon W. Stewart and J. Clive Ellory

1. Introduction

When the sodium pump was identified 30 years ago in Glynn's classical experiments on human red-cell Na and K fluxes (Glynn, 1956, 1957), the ouabain-insensitive monovalent cation fluxes were regarded as simply dissipative. It seemed that cell volume regulation could be explained by a two-component pump-leak model (Tosteson and Hoffman, 1960), but even then, the saturation kinetics displayed by the nonpump tracer K and Na influxes precluded a simple description as electrodiffusion. Since that time, it has become clear that ouabain-insensitive movements of Na and K occur via specific membrane transport systems, which show kinetic properties consistent with carriers or channels. A number of pathways for K transport are recognized: the Ca-activated (Gardos) channel, the NaKCl cotransport system, and the KCl cotransporter; for Na, NaKCl cotransport, Na-H and Na-Li exchanges, $NaCO_3^-$ anion exchange via capnophorin, and the Na-dependent amino-acid transporters *asc* and *gly* are significant contributors to transport (Ellory and Tucker, 1983). It is obvious that it is necessary to characterize these transport systems in order to understand the maintenance of intracellular ionic composition and, therefore, volume regulation in red cells (as a model system) and hence other cell types.

In this review, attention will be focused on two of these systems: the NaKCl cotransport and the KCl cotransport. In fact, although these systems have been the subject of increasing recent attention in work on a variety of cell types, it is still not established whether they are separate entities rather than partial functions of the same system, a question that emphasizes the lack of both a good specific inhibitor and of detailed biochemical knowledge of these transporters.

Two main techniques are used to identify these systems. Ion replacement is one, but it is cumbersome. Inhibition by the loop diuretics (furosemide, piretanide, and bumetanide) remains the favored method for identifying Cl-dependent transport systems. The effects are not specific and are discussed in detail below, but we will use the operational definition of the "NaKCl cotransport system" to denote those fluxes that are inhibited by loop diuretics with a high affinity ($K_{1/2} < 10\ \mu M$), and the "KCl system" to denote chloride-dependent loop diuretic-insensitive K transport. This distinction is not absolute, but the thousand-fold difference in loop diuretic sensitivity between the two systems does allow a useful experimental differentiation (*see below*).

2. The NaKCl Cotransport System

Early investigators working on red-cell transport were principally concerned with the NaK pump. When it arose that the ouabain-insensitive Na and K influxes showed saturation kinetics, it seemed possible that these fluxes might simply represent the action of the ouabain-poisoned pump working in some kind of exchange mode without the ability to perform net transport (*see* Beauge and Lew, 1977). Using the diuretic drug ethacrynic acid as an inhibitor, Hoffman and Kregenow (1966) suggested that there was a "second pump," an ouabain-insensitive process capable of net Na extrusion against a gradient, which had kinetic properties and inhibitor sensitivity distinct from the NaK pump. This finding did not receive universal support (Dunn, 1970, 1973), but attention was now focused on this mode of transport as something more interesting than just a "leak." Later workers used the higher affinity inhibitor furosemide (Dunn, 1970, 1973; Sachs, 1971), and it became clear that there was a complex interaction between elements of the Na and K fluxes (e.g., Lubowitz and Whittam, 1969, Beauge and Adragne, 1971). These observations were synthesized by Wiley and Cooper (1974) into the idea of a "NaK cotransport system," a functional unit inhibited by loop diuretics, in which the parallel transport of the two ions was linked. In 1980, two groups independently demonstrated that these mutually interdependent Na and K fluxes were themselves dependent on the chloride anion (Dunham et al., 1980; Chipperfield, 1980), although chloride *transport* via this system could not be measured in the presence of the high-capacity Band 3 anion transporter, capnophorin.

These studies established kinetic and inhibitor properties for the NaKCl cotransport that were distinct from the NaK pump. Residual suspicions that it might represent the behavior of the ouabain-poisoned pump were dispelled by reconstitution studies. Liposomes containing functioning NaK pumps showed neither furosemide-sensitive nor chloride-dependent K fluxes (Karlish et al., 1981).

About this time, similar systems that shared the properties of Na and K transport, Cl dependence (Cl transport could be shown in many types), and loop diuretic sensitivity were being described in a number of cells and tissues, notably the Ehrlich ascites cell (Geck et al., 1980) and the avian red cell (Schmidt and McManus, 1977a,b,c), where these fluxes played a role in volume regulation (*see* Hoffman, 1986; Geck and Heinz, 1986). As we will see, the human red cell is relatively unresponsive and inactive in terms of volume sensitivity. In contrast, the NaKCl cotransport system in avian red cells has a much higher capacity, and shows dramatic regulatory behavior that allows its role in increasing cell volume to be clearly identified and permits experimental approaches not feasible in the human red cell (e.g., McManus et al., 1985).

2.1. Transport Modalities and Substrate Kinetics for Na, K, and Cl

In Na medium, external K activates the loop diuretic sensitive K influx with simple Michaelis-Menten kinetics. The K_m is about 5–8 m*M*. Similar kinetics apply to the activation of the Na influx by external Na in K medium: the K_m is higher at about 15–20 m*M*. In the absence of the co-ion (i.e., in choline or *N*-methyl-*d*-glucamine media), the Na influx virtually disappears, but a loop diuretic-sensitive K influx does persist, with a lower affinity for external K (K_m about 40 m*M*) and a reduced V_{max} at about half of the value found in Na medium. The mutually interdependent fluxes are taken to represent "cotransport," whereas the K influx in non-Na media is interpreted as a K–K exchange mode: it is stimulated by increasing internal K (Canessa et al., 1986a). Curiously, this K–K exchange is stimulated by internal Na and indeed vanishes without it. Internal Na and K stimulate both Na and K effluxes (Garay et al., 1981). The activation by K is hyperbolic, but the activation of furosemide-sensitive Na efflux by internal Na is signoidal. In the absence of internal Na, the loop diuretic-sensitive K efflux vanishes (Garay et al., 1981).

A surprising effect is the inhibition of Na efflux by external K (e.g., Rettori and Lenoir 1972; Glynn et al., 1970, Canessa et al., 1986a). The most likely explanation for this *trans* effect is that external K converts the transporter from an NaK efflux mode into K–K exchange. A corollary of this *trans* effect is that the system cannot be regarded simply as the sum of two NaK cotransporters working in opposite directions, and further, since external K stimulates both Na and K influxes, an increase in its concentration has a double effect on net movements: Na influx is stimulated while efflux is inhibited, a result that is confirmed by net transport measurements in human and avian red cells (Duhm and Gobel, 1984a; Haas et al., 1982).

In contrast to the simple hyperbolic behavior seen for the cations, chloride, whose intra-and extracellular concentrations cannot be varied independently, activates K influx at 7.5 m*M* K with sigmoidal kinetics and low affinity such that saturation does not occur within isotonicity. Chloride activates Na fluxes with a low affinity, but sigmoidicity is not always observed (*see* Chipperfield, 1986).

A number of attempts to make simultaneous measurements of Na and K transport to determine "stoichiometry" have been published, but the results are not conclusive. Tracer K influxes always exceed tracer Na influxes, attributable to K–K exchange. Net studies, measuring intracellular K and Na changes on storage for 24 h (Duhm and Gobel, 1984a) suggested the linkage 1Na:1K, but these measurements cannot be precise because of the low transport rates. The comparison of inward and outward tracer fluxes to give an equilibrium constant implied the linkage 2Na:3K (Brugnara et al., 1986a). Adragna and Tosteson (1984) presented evidence to suggest that the "stoichiometry" of tracer effluxes into Na- and K-free media varied with changing cell water. Thus, there is no consensus on the exact linkage between Na and K: the presence of exchange elements makes tracer measurements difficult to analyze quantitatively. Studies in other cell types also give different results. In ferret red cells where fluxes are higher, tracer studies indicated a coupling of 2Na:1K:3Cl (Hall and Ellory 1985), a ratio also found in squid axon (Altamirano and Russell 1987). In avian and Ehrlich cells, precise net transport measurements, which would not report K–K exchange, point to a simple 1Na:1K:2Cl relationship, which seems inherently the most plausible (e.g., Haas et al., 1982; Geck et al., 1980; Hoffman, 1986).

A satisfactory model to account for these kinetic data on the human system has not yet been proposed. However, Lytle and McManus (1986; *see* also McManus, in Lauf et al., 1987) have recently suggested a reaction scheme for the duck red-cell system, which shares many common properties with the human system, but where fluxes are much greater and net measurements more exact. The model covers inward and outward NaK cotransport, K–K exchange facilitated by internal Na, and inhibition of Na efflux by external K. The avian red cell also shows a prominent Na–Na exchange mode that is facilitated by external K (McManus, in Lauf et al., 1987). Loop diuretic-sensitive Na–Na exchange has been difficult to demonstrate in human red cells, probably for technical reasons (*see* Canessa et al., 1986a).

The model, as we understand it, involves strictly ordered reactions at each face of the membrane, and allows translocation of only the fully loaded or completely empty carrier. Ion binding is governed by simple chemical equilibria, and it is translocation that is rate limiting. Having reached some stage in debinding, the carrier can either rebind or further debind in the dictated order until either it is fully loaded or completely unloaded, when it can translocate. For the inward translocation, the ions bind externally and debind internally in the same order—Na, Cl, K, Cl ("glide symmetry": *see* Stein, 1986). However, for the outward translocation, the order is reversed (*see* Fig. 1).

The requirement that the carrier be either empty or fully loaded for translocation immediately dictates "cotransport" (Panels A and B of Fig. 1). The ordered binding condition neatly explains the sidedness of the co-ion requirement for K–K or Na–Na exchanges in avian red cells: in K–K exchange (Na absent externally), Na must first debind at the inner surface to allow K to debind and therefore be transported into the internal solution; since the Na site is vacated, Na must be present to rebind (Panel C of Fig. 1). At the outer surface, the ordering permits that K can debind and rebind while Na remains on the carrier ready for the return trip: no external Na is necessary in the external medium. The Na simply travels as a passenger on the transporter. A similar argument explains why external K is required for Na–Na exchange (Panel D of Fig. 1). The asymmetry of the carrier is embodied in the binding ordering and is not detected in the cotransport mode. The placing of the anion in the binding order was deduced from duck erythrocyte experiments involving asymetrical anion replacement techniques (*see* McManus, in Lauf et al., 1987). The reactions shown in Panels A–D are extremes: the model allows combinations of all four depending on ionic concentrations. Thus, for instance, if Na and K were present on both sides of the membrane, increasing external K would encourage immediate rebinding of K at the outer surface after debinding of the ion, encouraging immediate inward translocation before debinding of Na, i.e., shifting to the K–K exchange mode instead of full NaKCl cotransport. The experimental result of such a change of mode would be to reduce the tracer efflux of Na, which is indeed the well-known effect of increasing external K in human red cells. No doubt such combinations can explain why tracer techniques give conflicting results in

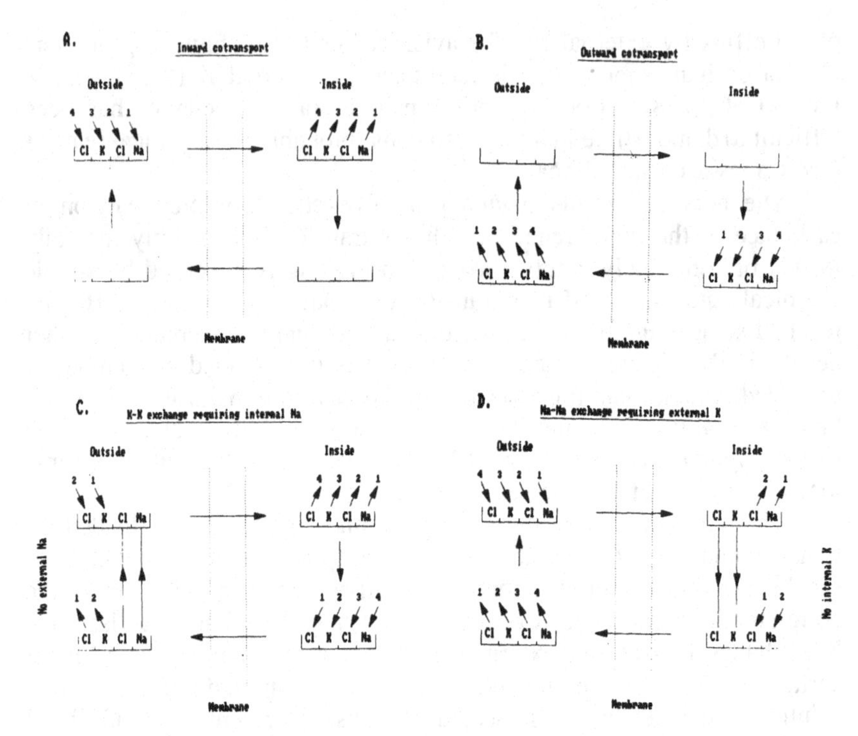

Fig. 1. Ordered binding, glide symmetry model for the NaKCl cotransport system, as proposed by McManus, in Lauf et al., 1987. Ion binding and debinding takes place in the order shown; the carrier can translocate either empty or fully loaded. Panels A and B show how NaKCl cotransport can occur; Panels C and D illustrate the reactions involved in K–K and Na–Na exchange and how the asymmetrical coion requirement results from the binding ordering condition.

stoichiometry experiments (*see above*). Experimental verification in human red cells, particularly of the Na–Na exchange effects, and a full quantitative development of this highly original model, are awaited.

2.2. Inhibitor Sensitivity and Binding

Experimentally, it is particularly useful to have specific high affinity inhibitors for transport systems. The obvious example is the action of cardiac glycosides on the NaK pump. Although the loop diuretics do inhibit the NaKCl system, they also have effects on a number of other systems and enzymes (anion transport via Band 3 [Brazy and Gunn 1976], carbonic anhydrase, adenyl cyclase, glycine transport and the NaK pump [Chipperfield 1986]). Notwithstanding this point, it is clear that the red-cell NaKCl transport system is sensitive to inhibition by loop

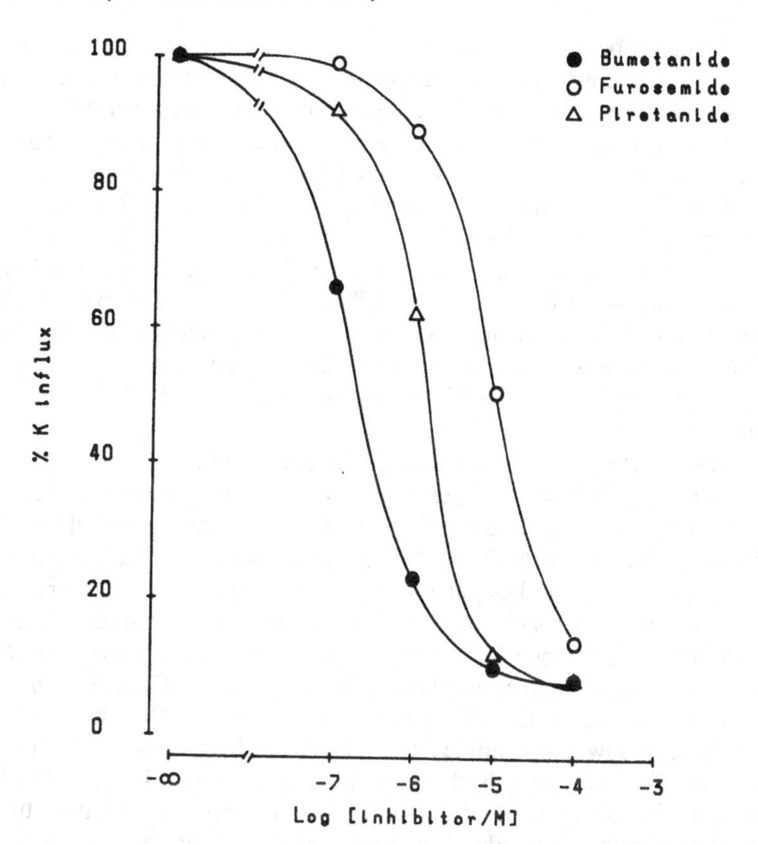

Fig. 2. Dose–response curve showing relative potencies of bumetanide, furosemide, and piretanide as inhibitors of the NaKCl cotransport system in human red cells. K influx was measured in a 142.5-m*M* Na, 7.5-m*M* K, 150-m*M* Cl, 15-m*M* morpholinopropanesulphonic acid (MOPS, pH 7.4), 5-m*M* glucose, 0.1-m*M* ouabain medium.

diuretics, which can be ranked bumetanide > piretanide > furosemide in order of inhibitory potency (*see* Fig. 2; also Ellory and Stewart 1982). This order of potency is characteristic of a number of NaKCl systems in other tissues (Schlatter et al., 1983; O'Grady et al., 1987; Chipperfield 1986), and the demonstration of this ranking is now used as a test to classify ion transport systems (e.g., *see* Aronson, in Warnock et al., 1984). The ranking of these compounds as inhibitors of the red-cell NaKCl system also follows their potency as diuretics in clinical pharmacology (Roberts et al., 1978). Bumetanide is clinically effective at plasma levels of about 0.05 μ*M* (Davies et al., 1974), consistent with its $K_{1/2}$ as shown in Fig. 2 (0.16 μ*M*).

Relations between the univalent ions and loop diuretics have been studied to investigate possible interactions between inhibitor action and substrate binding. No clear conclusion has emerged, but from both human and duck red-cell work, it seems that increasing concentrations of the cations promote binding whereas Cl decreases it, suggesting that bumetanide might inhibit by competing with the anion (Haas and Forbush, in Lauf et al., 1987).

We observed that bumetanide was ineffective in cells treated with the ionophore A23187 and EDTA (Ellory et al., 1983). However, Flatman (in Lauf et al., 1987) has recently presented an alternative hypothesis to interpret these results: that the removal of intracellular Mg^{2+} inhibits NaKCl cotransport but stimulates KCl cotransport, which is relatively bumetanide-insensitive.

The binding of tritiated ouabain has been useful in the study of the sodium pump, for both the enumeration of sites and protein purification. By analogy, labeled loop diuretics (e.g. [^{3}H]bumetanide) would be valuable for investigating the NaKCl cotransport system. In fact, there have been a number of studies on kidney, ascites tumor, and cultured cell lines, but the approach has its limitations. As mentioned earlier, bumetanide is not specific for the cotransport system, but binds to a variety of other proteins. Further, it is reversible and has a high lipid solubility. To overcome this, covalent labeling via UV activation has been used at low bumetanide concentrations (Jorgensen et al., 1984). However, the resultant gels show labels associated with several proteins, with a major component of 34-kdaltons. Independent labeling studies with Ascites tumor cells identified principally a 90-kdalton protein (Hoffman, 1986). In a related study, Saier and Boyden (1984) found that activation of NaKCl transport in MDCK cells was associated with the phosphorylation of a 240-kdalton protein. Cherksey and Zeuthen (1987) have recently published data showing that a 300-kdalton protein with K transport activity could be purified from bovine choroid plexus using a furosemide affinity column.

Mercer and Hoffman (1985) were able to measure 12000 [^{3}H]-bumetanide binding sites/cell on ferret erythrocytes; Haas and Forbush (1986) measured about 1000/cell in avian erythrocytes previously treated with epinephrine and hypertonic shock. Assuming the human system to have a roughly similar turnover rate, in cells with much greater NaKCl transport rates, these results imply that the number of sites/human red cell can be only a few hundred or less.

Finally, there is an intriguing similarity between the anatomical distributions of NaKCl cotransport and the Tamm-Horsfall glycoprotein in the renal tubule, as shown by both loop diuretic binding characteristics (Hartmann et al 1981; Greven et al., 1984) and immunocytochemistry (Alexander et al., 1985). This anatomical mapping suggested a molecular coidentity, but this is unlikely, since the recently published amino-acid sequence of the Tamm-Horsfall protein (Pennica et al., 1987) lacks the hydrophobic segments expected for a membrane-spanning transport protein (cf the sequence for the NaK pump: Shull et al., 1985).

2.3. Effect of Variation In Cell Volume In Vitro

In avian erythrocytes and Ehrlich cells, it is clear that the loop diuretic-inhibited NaKCl cotransport system is involved in the "volume regulatory increase" (VRI) phenomenon, in which the cells restore their cell volume by influx of ions and associated osmotic water after challenge by shrinkage in hypertonic media (McManus et al., 1985; Geck et al 1980). This effect does not occur in human red cells, or if it does, it is muted (e.g., Poznasky and Solomon 1972). Increasing the external NaCl concentration stimulates the human red-cell NaKCl system, but this effect can simply be attributed to the increasing chloride concentration (Dunham et al., 1980), and added sucrose, or added choline nitrate, has little effect (Duhm and Gobel 1982). Adragna and Tosteson (1984) found differing effects depending on whether cell volume was changed by varying the tonicity of the medium or by varyiong the intracellular contents by nystatin treatment. Thus, human red cells show neither consistent volume regulatory behavior nor enhancement of NaKCl flux after shrinkage in vitro.

2.4. Intracellular Energy Sources and Other Ligands

Two groups have examined the effects of energy depletion on this transport system. Both found reductions in furosemide-sensitive transport: Adragna et al. (1985) incubated cells with 2-deoxy-glucose; Dagher et al. (1985) simply incubated in a glucose-free medium. These data are consistent with studies in other tissues (*see* Chipperfield, 1986).

In the nucleated avian red cell, additional activators of the system are catecholamines, which act via cyclic AMP (Schmidt and McManus, 1977b; *see also* Palfrey and Rao, 1983). Human red cells do not respond to this stimulus: indeed, Garay (1982) demonstrated that cyclic-AMP (and intracellular Ca^{2+}) at high doses inhibited NaK cotransport in human red cells.

2.5. Role in the Human Red Cell: Net Flux Measurements and Regulation

It is a curious fact that, although this system mediates significant tracer fluxes in human red cells comparable to those mediated by the NaK pump, a clear role has yet to be demonstrated for it. When loop diuretics are added to cells incubated in vitro over periods up to 24 h in physiological concentrations of Na and K, the effects on intracellular Na and K levels are small in comparison to the dramatic effect of ouabain, for example (Duhm and Gobel 1984a). The inward and outward NaK cotransports seem to balance. If the external K is varied, however, small net fluxes can be measured. As suggested by the tracer flux results described above, reducing external K leads to a net Na efflux, whereas increasing external K gives a net influx. These results are consistent with much more marked effects in avian erythrocytes and in Ehrlich cells (Haas et al., 1982; Hoffman 1986), and are in turn consistent with the notion that the system is "gradient-driven," i.e., the direction of net transport is a function of the electrochemical gradients of Na and K.

In other symmetrical cells where it has been well studied, the NaKCl system acts to increase cell volume. The human red-cell system is atypical in failing to show stimulation after shrinkage of the cell in vitro. However, the capacity of the system does show variation between different normal donors (Duhm and Gobel, 1982). It has been suggested that this interindividual variation is associated with a purported ion transport abnormality in hypertension, but there is no consensus on the results (*see* Parker and Berkowitz, 1983; Chipperfield 1986, for reviews). Aside from blood pressure, higher NaKCl fluxes have been associated with a number of factors: Duhm and Gobel (1984a) noted that it was greater in cells with higher mean cell haemoglobin concentration (MCHC), and in conditions associated with systemic potassium depletion such as mineralocorticoid hypertension (Duhm and Behr, 1986); Wiley (1977) found it to be higher in the dehydrated form of hereditary stomatocytosis, a congenital hemolytic anemia in which the passive leak is abnormally high. One common factor underlying all of these observations is that NaKCl cotransport is high when cell water is low. In the case of the human erythrocyte, it has been difficult to establish a clear cause–effect relationship between these factors, but studies on the rat erythrocyte, in which NaKCl cotransport is directly stimulated in vitro by shrinkage as part of a VRI response (thus, establishing cause and effect in this case; Duhm et al., 1983), show how a similar pattern of red-cell fluxes may result from the same kinds of systemic factors as in the human (Duhm and Gobel, 1984b). By analogy, therefore, it could be that, in human erythrocytes, NaKCl cotransport is likewise upregulated in conditions of relative cellular dehydration, the regulation taking place at an immature stage in cellular development after which the capacity is effectively fixed for the lifetime of the cell.

3. Loop Diuretic-Insensitive Chloride-Dependent K Fluxes

The technique of anion replacement has enabled a further K transporter to be identified. First recognized in sheep red cells (Dunham and Ellory 1981), a chloride-dependent K (but not Na) transporter was described, kinetically distinct from the NaK cotransport system and resistant to loop diuretics. A consistent and identifying feature of this flux is volume sensitivity: thus, swelling the cells by 10% can induce a 3–10-fold increase in K, but not Na, permeability. In the case of sheep red cells, this volume-sensitive K flux was dependent on the presence of Cl or Br in the medium, with other anions (e.g., nitrate, acetate, or methylsulphate) failing to substitute (Dunham and Ellory, 1981). Since red cells have a high Cl permeability, it was likely that this flux was via a KCl cotransport system, rather than representing electrical coupling of fluxes through separate K and Cl channels, which occurs in some cells, e.g., lymphocytes (Grinstein et al., 1983) or ascites tumor cells (Hoffman et al., 1986).

The situation in human red cells regarding this flux was originally controversial: thus, cell swelling has been shown to have no effect on K transport (Brugnara et al., 1985; Ellory et al., 1986; Lauf et al., 1985; Hall and Ellory, 1986a), to decrease (Poznansky and Solomon 1972), or to increase fluxes (Kaji, 1986), or to have complex effects (Adragna and Tosteson, 1984). However, that such a flux was at least elicitable in human red cells was shown by the action of the SH-group alkylating agent, *N*-ethyl maleimide (NEM). This agent had been previously shown to reveal a large chloride-dependent Na-independent bumetanide-insensitive K flux in LK sheep erythrocytes (Lauf and Theg, 1980): it had an identical action in human red cells (Ellory et al., 1982). Thus, it seemed that the transporter was present, but not expressed, in normal human red cells. Recently, a low-capacity KCl flux with similar kinetic properties has been demonstrated in normal human erythrocytes (Blackstock and Stewart, 1987).

The molecular relationship between these loop diuretic-insensitive and loop diuretic-sensitive chloride dependent fluxes is not clear. Under defined substrate conditions, the NaK pump, now defined at the molecular level, can mediate "partial reactions," such as Na–Na or K–K exchange, without modification of its protein structure: on restoration of standard substrate conditions, it will again preform Na–K exchange (Glynn and Karlish, 1975). Could the loop diuretic–insensitive KCl flux represent a partial reaction of the NaKCl system? We know that there is a loop diuretic-sensitive K–K exchange: is this truly distinct from the purported KCl system? Although the two systems do share characteristics and could be closely related at the molecular level, e.g., sharing amino-acid sequences in the K- and Cl-binding regions, we think that there is a strong case for a distinction as separate systems. One of the principal criteria for differentiating the fluxes is sensitivity to loop diuretics. Figure 3 shows the effects of bumetanide on K influx in normal human red cells in Na and non-Na media, and contrasts the high affinity inhibition in these two fluxes with the low affinity effects on the NEM-induced flux in human red cells and a typical volume-sensitive KCl system, in the rabbit erythrocyte.

Other evidence for the separate existence of two systems can be cited. When human cells from donors with different rates of NaKCl cotransport are compared, the loop diuretic-insensitive KCl flux varies *inversely* with the NaKCl fluxes (Stewart, 1988). The loop diuretic-sensitive Na-independent K–K exchange is in direct proportion to both the Na-dependent K influx and the Na influx, as one would expect from a partial reaction (Blackstock and Stewart 1987). Species comparisons of red cells reveals a similar functional separation of the fluxes: LK sheep and rabbit red cells showing marked loop diuretic-insensitive KCl transport but no Na-dependence (Dunham and Ellory, 1981; Stewart and Blackstock, 1989), whereas ferret red cells show huge NaKCl fluxes and minimal loop diuretic-insensitive KCl (Flatman, 1983; Hall and Ellory, 1985); monkey red cells lack the NaKCl system altogether, but do have a minimal volume-sensitive KCl flux, which can be stimulated by NEM (Blackstock et al., 1986; Stewart et al., 1989). The loop diuretic-insensitive KCl is invariably stimulated by cell swelling, whereas the NaKCl system is inhibited, if influenced at all, by this maneuver. Some of these data are summarized in Table 1; but it must be admitted that, in the absence of biochemical evidence to settle the matter, none of the present evidence can be taken as definitive.

There is confusion in the literature over the distinction between these two systems. Thus for instance, Kaji and Kahn (1985) and Kaji (1986) have used anion- and cation-dependence without loop diuretics to categorize K fluxes experimentally and, in our view, confused what we regard as separate systems. Such considerations, along with interindividual differences in transport rates between donors and the low capacity of the systems, have presumably contributed to the apparent contradictions in the literature concerning volume-sensitivity in human red cells.

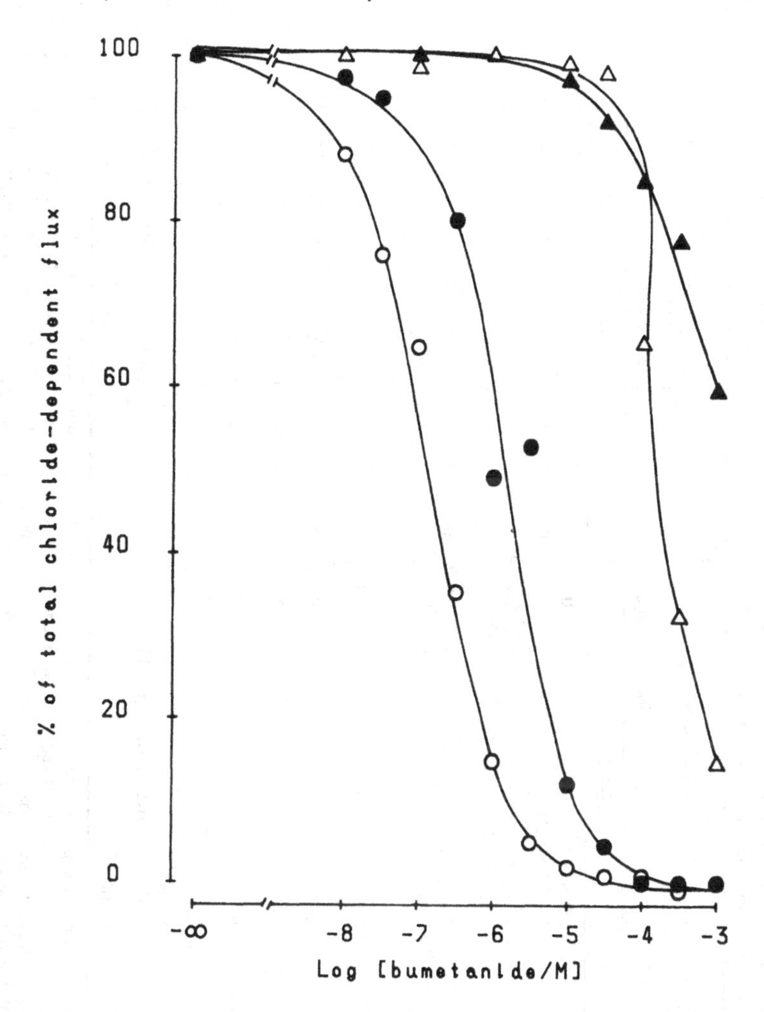

Fig. 3. Effect of bumetanide on different Cl-dependent K influxes. K influx was measured as for Fig. 1. In this case, 100% inhibition was defined as that inhibition evoked by replacement of Cl by NO_3 or methylsulphate (*see* Dunham et al., 1980). The symbols denote: ○, Na-dependent loop-diuretic-sensitive NaKCl cotransport system; ●, Na-independent (*N*-methyl-*d*-glucamine substitution) K–K exchange component of NaKCl cotransport; △, rabbit erythrocyte volume-sensitive Cl-dependent K influx; ▲, NEM-stimulated chloride-dependent K influx in human red cells.

Thus, Kaji (1986) found that cell swelling increased ouabain-insensitive chloride-dependent K influx. Presumably, the donor was one with low loop diuretic-sensitive fluxes, in whom the KCl system was predominant

Table 1
Comparison of Properties of Loop Diuretic-Sensitive Na-Dependent ("NaKCl") and Na-Independent ("K–K") K Influxes and Loop Diuretic-Insensitive ("KCl") Chloride-Dependent Fluxes.

Property	Fluxes*		
	Bumetanide-sensitive		Bumetanide-insensitive
	NaKCl	K–K	KCl
Hypotonic swelling	Inhibits,[1,2] (if any effect)	Inhibits[2] (if any effect)	Stimulates[2,3,4]
Bumetanide-sensitivity $K_{1/2}$ (M)	High[5,6] 2×10^{-7}	High[5,7] 4×10^{-6}	Low[3†, 8‡,9§] Ca 10^{-3}
Furosemide-sensitivity $K_{1/2}$ (M)	Moderate[6] 9×10^{-6}	Moderate[2]	Low[3†, 8‡] Ca 10^{-3}
Kinetic characteristics Approx. K_m for $[K]_o$ (mM)	5[10,11]	40[10,2]	40[3†, 8‡]
V_{max} (mmol l cell^{-1} h^{-1})	0.1–1.0[1]	0.05–0.5[2]	Up to 8[3†, 8‡]
Na dependence	Yes[12]	No[12]	No[3†, 8‡, 13‖]
Effect of hydrostatic pressure	Inhibits[8]	?	Stimulates[8]
Effect of NEM	Inhibits[3,14]	?	Stimulates[3,14]
Species specificity	Ferret[15,16] (large NaKCl, small KCl)		Sheep[13], rabbit[2] (large KCl, absent NaKCl)
Internal Mg^{2+}	Inhibits[17]		Stimulates[17,8‖]

*The data refer to fresh human red cells, except where indicated as follows: †NEM-treated human cells; ‡human cells under high hydrostatic pressure; §abnormal human cells; ‖LK sheep red cells. References: 1Duhm and Gobel, 1982; 2Blackstock and Stewart, unpublished; 3Ellory et al., 1982; 4Hall and Ellory 1986a; 5this article, Fig. 2; 6Ellory and Stewart 1982; 7Ellory et al., 1986; 8Hall et al., 1982; 9Brugnara et al 1986b; 10Wiley and Cooper 1974; 11Dunham et al., 1980; 12 by definition; 13Dunham and Ellory 1981; 14Lauf et al., 1984; 15Flatman 1983; 16Hall and Ellory 1986b; 17Flatman, in Lauf et al., 1987; 18Lauf, 1985.

(*see* Blackstock and Stewart, 1987). Had they made the same measurement on a donor with high loop diuretic-sensitive fluxes, the contribution of K–K exchange via the NaKCl system might have given a different result.

3.1. Other Signals Switching on KCl Cotransport

So far, hypotonic shock has been discussed as the effector for activating this transport pathway. Other routes known to promote this flux are treatment of the cells with NEM (Ellory et al., 1982), compression to high hydrostatic pressures (100–400 atmospheres) (Hall and Ellory, 1986a), low pH, which induces cell swelling (Brugnara et al., 1986b), and ghosting (Dunham and Logue 1986). These maneuvers all induce a transport system with similar kinetic parameters, having K_m values of about 30–60 m*M* and a V_{max} of 5–15 mmol l cells $^{-1}$ h^{-1}. The question immediately arises as to what the common effector could be in these three cases, which raises the more general question of the identity of the signal in volume-change-mediated effects. Experiments with Ca-chelators and Gardos channel inhibitors show that intracellular Ca^{2+} may inhibit the system in the LK sheep erythrocyte (Dunham and Ellory, 1981; Lauf, 1985); intracellular Mg^{2+} may be also be important (*see above*; Flatman, in Lauf et al., 1987; Lauf, 1985)). Metabolism, however, does apparently play a role, since falling intracellular ATP, via metabolic depletion, inhibits the volume-dependent KCl flux in LK sheep red cells (Dunham and Ellory, 1981; Lauf, 1985), and the pressure- and NEM-induced KCl fluxes in human red cells (Lauf et al., 1985; Ellory et al., 1986). There could be a link between the cytoskeleton and the red-cell membrane, and it has been proposed that the transport protein may be associated with the spectrin-ankyrin complex (Ellory et al., 1987).

A further factor could be cell age. Using cell fractionation by density, it was shown that red cells from the youngest, i.e., least dense, fraction showed relatively large amounts of Cl-dependent K transport, although this was absent from the densest cell fraction (Hall and Ellory 1986b). However, in the densest fraction, transport could be induced by NEM treatment, ghosting, or compression, leading to the hypothesis that this system was turned off during red-cell aging in the circulation, but remains in the membrane in a state capable of activation by other signals, in the sequence ghosting = NEM > pressure > hypotonic shock. This might explain the presence of such fluxes in pathological states, such as unstable hemoglobins, since these states are often characterized by high degrees of reticulocytosis. In fact, in a study of 17 cases of high reticulocytosis from a range of defects, from phlebotomy for hemochromatosis to severe stomatocytosis, there was a good linear correlation between measured KCl flux and degree of reticulocytosis, irrespective of its origin (Ellory et al. 1987). Nevertheless, inappropriate activation of this transport pathway may well play a role in cellular dehydration seen in some hemoglobinopathies, notably sickle and hemoglobin CC anemias (Brugnara et al., 1985, 1986b; Ballas et al., 1987; Canessa et al., 1986b). The (admittedly young) cell population in sickle cell donors showed large volume-sensitive KCl cotransport, which played a significant part in determining the intracellular water and characteristic high MCHC in these cells.

4. Summary and Conclusions

Red cells must maintain intracellular K concentrations for enzyme function (Kiernan, 1965), and need transport systems for regulation of volume and ion composition (Sachs et al., 1974). The NaKCl cotransport system plays a major role in the VRI in nucleated red cells, but has been relegated to a minor or obscure role in human red blood cells. The KCl system is likewise of low capacity in human red cells compared to a number of other examples, but is capable of very large fluxes following activation. Inappropriate activation of the KCl system, in defiance of its presumed original role in volume defense against swelling, may be the cause of pathological dehydration in HbSS or HbCC cells, enhancing hemoglobin polymerization and the sickling process.

Acknowledgments

We thank Dr. A. C. Hall for useful discussions, and the Wellcome Trust for financial support.

References

Adragna, N. C., Perkins, C. M., and Lauf, P. K. (1985) Furosemide-sensitive Na^+–K^+ cotransport and cellular metabolism in human erythrocytes. *Biochim. Biophys. Acta* **812,** 293–296.

Adragna, N. C. and Tosteson, D. C. (1984) Effect of volume changes on ouabain-insensitive net outward cation movement in human red cells. *J. Membr. Biol.* **78,** 43–52.

Alexander, D. P., Foster, C. L., and Tan, C. B. (1985) Tamm-Horsfall glycoprotein—its localisation and possible role in the mammalian kidney. *IRCS J. Med. Sci.* **13,** 574–576.

Altamirano, A. A. and Russell J. M. (1987) Coupled Na/K/Cl efflux. "Reverse" unidirectional fluxes in squid giant axons. *J. Gen. Physiol.* **89,** 669–686.

Ballas, S. K., Larner, J., Smith, E. D., Surrey, S., Schwartz, E., and Rappaport, E. F. (1987) The xerocytosis of Hb SC disease. *Blood* **69,** 124–130.

Beauge, L. A. and Adragna, N. C. (1971) The kinetics of ouabain inhibition and the partition of rubidium influx in human red blood cells. *J. Gen. Physiol.* **57,** 576–592.

Beauge, L. A. and Lew, V. L. (1977) Passive fluxes of sodium and potassium across red cell membranes, in *Membrane Transport in Red Cells* (Ellory, J. C. and Lew, V. L., eds.), Academic, London, pp 39–61.

Blackstock, E. J. and Stewart, G. W. (1987) Inverse relationship between bumetanide sensitive and bumetanide-insensitive K influxes in human red cells from different normal donors. *J. Physiol.* **382,** 141P.

Blackstock, E. J., Ellory, J. C., and Stewart, G. W. (1986) K influxes in red cells from two private species. *J. Physiol.* **374,** 34P.

Brazy, P. C. and Gunn, R. B. (1976) Furosemide inhibition of chloride transport in human red blood cells. *J. Gen. Physiol.* **68,** 583–599.

Brugnara, C., Kopin, A. S., Bunn, H. F., and Tosteson, D. C. (1985) Regulation of cation content and cell volume in erythrocytes from patients with homozygous hemoglobin C disease. *J. Clin. Invest.* **75,** 1608–1617.

Brugnara, C., Canessa, M., Cusi, D., and Tosteson, D. C. (1986a) Furosemide-sensitive Na and K fluxes in human red cells. Net uphill transport and equilibrium properties. *J Gen. Physiol.* **87,** 91–112.

Brugnara, C., Bunn, H. F., and Tosteson, D. C. (1986b) Regulation of erythrocyte cation and water content in sickle cell anemia. *Science* **232,** 388–390.

Canessa, M., Brugnara, C., Cusi, D., and Tosteson, D. C. (1986a) Modes of operation and variable stoichiometry of the furosemide sensitive Na and K fluxes in human red cells. *J. Gen. Physiol.* **87,** 113–142.

Canessa, M., Spalvins, A., and Nagel, R. L. (1986b) Volume-dependent and NEM-stimulated K^+,Cl^- transport is elevated in oxygenated SS, SC and CC human red cells. *FEBS Letts* **200,** 197–202.

Cherksey, B. D. and Zeuthen, T. (1987) A membrane protein with a K^+ and a Cl^- channel. *J. Physiol.* **387,** 33P.

Chipperfield, A. R. (1980) An effect of chloride on (Na^+K) cotransport in human red cells. *Nature* **286,** 281–282.

Chipperfield, A. R. (1986) The (Na-K-Cl) cotransport system. *Clin. Sci.* **71,** 465–476.

Dagher, G., Brugnara, C., and Canessa, M. (1985) Effect of metabolic depletion on the furosemide-sensitive Na and K fluxes in human red cells. *J. Membr. Biol.* **86,** 145–155.

Davies, D. L., Lant, A. F., Millard, N. R., Smith, A. J., Ward, J. W., and Wilson, G. M. (1974) Renal action, therapeutic use, and pharmacokinetics of bumetanide. *Clin. Pharm. Ther.* **15,** 141–155.

Duhm, J. and Behr, J. (1986) Role of exogenous factors in alterations of red cell Na^+–Li^+ exchange and Na^+–K^+ cotransport in essential hypertension, primary hyperaldosterionism, and hypokalaemia. *Scand. J. Clin. lab. Invest.* **46 supp 180,** 82–95.

Duhm, J. and Gobel, B. O. (1982) Sodium-lithium exchange and sodium-potassium cotransport in human erythrocytes. *Hypertension* **4,** 468–476.

Duhm, J. and Gobel, B. O. (1984a) Role of the furosemide-sensitive Na^+ K^+ cotransport system in determining the steady state Na^+ and K^+ content and volume of human red erythrocytes in vitro and in vivo. *J. Membr. Biol.* **77,** 243–254.

Duhm, J and Gobel, B. O. (1984b) Na^+-K^+ transport and volume of rat erythrocytes under K^+ deficiency. *Am. J. Physiol.* **246,** C20–C29.

Duhm, J., Gobel, B. O., and Beck, F. X. (1983) Sodium and potassium ion transport accelerations in erythrocytes of DOC, DOC-salt, two-kidney, one clip, and spontaneously hypertensive rats. *Hypertension* **5,** 642–652.

Dunham, P. B. and Ellory, J. C. (1981) Passive potassium transport in LK sheep red cells: dependence upon cell volume and chloride. *J. Physiol.* **318,** 511–530.

Dunham, P. B. and Logue, P. L. (1986) Potassium chloride cotransport in resealed human red cell ghosts. *Am. J. Physiol.* **250,** C578–C583.

Dunham, P. B., Stewart, G. W., and Ellory, J. C. (1980) Chloride-activated passive potassium transport in human erythrocytes. *Proc. Nat. Acad. Sci. USA* **77,** 1711–1715.

Dunn, M. J. (1970) The effects of transport inhibitors on sodium outflux and influx in human red blood cells: evidence for exchange diffusion. *J. Clin. Invest.* **49,** 1804–1814.

Dunn, M. J. (1973) Ouabain-uninhibited sodium transport in erythrocytes. Evidence against a second pump. *J. Clin. Invest.* **52,** 658–670.

Ellory, J. C. and Stewart G. W. (1982) The human erythrocyte C1-dependent Na-K cotransport system as a possible model for the study of loop diuretics. *Br. J. Pharm.* **75,** 183–188.

Ellory, J. C. and Tucker E. M. (1983) Cation transport in red cells. in : *Red Blood Cells of Domestic Animals* (Agar N. S., and Board P. G., eds.), Amsterdam, Elsevier, pp. 291–314.

Ellory, J. C., Flatman, P. W., and Stewart, G. W.(1983) Inhibition of human red cell sodium and potassium influxes by external divalent cations. *J. Physiol.* **340,** 1–17.

Ellory, J. C., Hall, A. C., and Amess, J. A. L. (1987) Passive potassium transport in human red cells. *Biomed. Biochim. Acta* **46,** S31–S35.

Ellory, J. C., Hall, A. C., and Stewart, G. W. (1986) Volume-sensitive cation fluxes in mammalian red cells. *Molecular Physiol.* **8,** 235–246.

Ellory, J. C., Dunham, P. B., Logue, P. J., and Stewart, G. W. (1982) Anion dependent cation transport in erythrocytes. *Phil. Trans. Roy. Soc. Lond. B* **299,** 483–495.

Flatman, P. W. (1983) Sodium and potassium transport in ferret red cells. *J. Physiol.* **341,** 545–567.

Garay, R. P. (1982) Inhibition of the Na^+/K^+ cotransport system by cyclic AMP and intracellular Ca+ in human red cells. *Biochim. Biophys. Acta* **688,** 786–792.

Garay, R. P., Adragna, N., Canessa, M., and Tosteson, D. C. (1981) Outward sodium and potassium transport in human red cells. *J. Membr. Biol.* **62,** 169–174.

Geck, P. and Heinz, E. (1986) The Na-K-2Cl cotransport system. *J. Membr. Biol.* **91,** 97–105.

Geck, P., Pietrzyck, B. C., Pfeiffer, B., and Heinz, E. (1980) Electrically silent cotransport of Na^+, K^+ and Cl^- in Ehrlich cells. *Biochim. Biophys. Acta* **600,** 432–447.

Glynn, I. M. (1956) Sodium and potassium Movements in human red cells. *J. Physiol.* **134,** 278–310.

Glynn, I. M. (1957) The action of cardiac glycosides on sodium and potassium movements in the human red cells. *J. Physiol.* **136,** 148–173.

Glynn, I. M. and Karlish, S. J. D. (1975) The sodium pump. *Ann. Rev. Physiol.* **37,** 13–55.

Glynn, I. M., Lew, V. L., and Luthi, U. (1970) Reversal of the potassium entry mechanism in red cells with and without reversal of the entire pump cycle. *J. Physiol.* **207,** 371–391.

Greven, J., Kolling, B., Bronewski-Schwarzer, B. V., Junker, M., Neffgen, B., and Nilius, R. M. (1984) Evidence for a role of the Tamm-Horsfall protein in the tubular action of furosemide-like loop diuretics, in *Diuretics* (Puschett, J. B., ed.), Elsevier, Amsterdam, pp 203–214.

Grinstein, S., Clarke, C. A., and Rothstein, A. (1983) Activation of Na^+/H^+ exchange in lymphocytes by osmotically-induced volume changes and by cytoplasmic acidification. *J. Gen. Physiol.* **82,** 619–657.

Haas, M. and Forbush, B. (1986) [^{3}H] Bumetanide binding to duck red cells. *J. Biol. Chem.* **261,** 8434–8441.

Haas, M., Schmidt, W. F., and McManus, T. J. (1982) Catecholamine-stimulated ion transport in duck red cells. Gradient effects in electrically neutral [Na+K+2Cl] co-transport. *J. Gen. Physiol.* **80,** 125–147.

Hall, A. C. and Ellory J. C. (1985) Measurement and stoichiometry of bumetanide-sensitive (2Na:1K:3Cl) cotransport in ferret red cells. *J. Membr. Biol.* **85,** 205–213.

Hall, A. C. and Ellory, J. C. (1986a) Effects of high hydrostatic pressure on "passive" monovalent cation transport in human red cells. *J. Membr. Biol.* **94,** 1–17.

Hall, A. C. and Ellory, J. C. (1986b) Evidence for the presence of volume-sensitive KCl transport in 'young' human red cells. *Biochim. Biophys. Acta* **858,** 317–320.

Hall, A. C., Klein, R. J., and Ellory, J. C. (1982) Pressure and temperature effects on human red cell cation transport. *J. Membr. Biol.* **68,** 47–56.

Hartmann, L., Ambert, J. P., Ollier-Hartmann, M. P., Richet, G., and Raynaud, C. (1981) Fixation du chlorure mercurique et de l'acide etacrinique sur l'uromucoide ou proteine de Tamm et Horsfall. *Biomedicine* **35,** 30–34.

Hoffman, E. K. (1986) Anion transport systems in the plasma membrane of Ehrlich ascites cells. *Biochim. Biophys. Acta* **964,** 1–31.

Hoffman, E. K., Lambert, I. H., and Simonsen, L. O. (1986) Separate, Ca^{2+}-activated K^+ and Cl^- transport pathways in Ehrlich ascites tumour cells. *J. Membr. Biol.* **91,** 227–244.

Hoffman, J. F. and Kregenow, F. M. (1966) The characterisation of new energy dependent transport processes in red blood cells. *Ann. NY Acad. Sci.* **137,** 566–576.

Jorgenson, P. L., Peterson, J., and Rees, W. D. (1984) Identification of a Na^+, K^+,Cl^--cotransport protein of Mr 34000 from kidney by photolabelling with [^{3}H] bumetanide. *Biochim. Biophys. Acta* **775,** 105–110.

Kaji, D. (1986) Volume-sensitive K transport in human erythrocytes. *J. Gen. Physiol.* **88,** 719–738.

Kaji, D. and Kahn, T. (1985) Kinetics of Cl-dependent K influx in human erythrocytes with and without external Na; effect of NEM. *Am. J. Physiol.* **249,** C490–C496.

Karlish, S. J. D., Ellory, J. C., and Lew, V. L. (1981) Evidence against sodium pump mediation of calcium-activated potassium transport and diuretic-sensitive sodium/potassium-cotransport. *Biochim. Biophys. Acta* **646,** 353–355.

Kiernan, R. P. (1965) *Cell K* (Butterworths, London).

Lauf, P. K. (1985) K^+:Cl^- cotransport: sulphydryls, divalent cations, and the mechanism of volume activation in a cell. *J. Membr. Biol.* **88,** 1–13.

Lauf, P. K. and Theg, B. E. (1980) A chloride-dependent K^+ flux induced by *N*-ethyl maleimide in genetically low K^+ sheep and goat erythrocytes. *Biochem. Biophys. Res. Commun.* **92,** 1422–1428.

Lauf, P. K., Adragna, N. C., and Garay, R. P. (1984) Activation by *N*-ethylmaleimide of a latent K^+/Cl^- flux in human red cells. *Am. J. Physiol.* **246,** C385–C390.

Lauf, P. K., Perkins, C. M., and Adragna, N. C. (1985) Cell volume and metabolic dependence of NEM-activated K^+Cl^- flux in the human red blood cell. *Am. J. Physiol.* **249,** C124–C128.

Lauf, P. K., McManus, T. J., Haas, M., Forbush, B., Duhm, J., Flatman, P. F., Saier, M. H., and Russell, J. M. (1987) Physiology and biophysics of chloride and cation cotransport across cell membranes. *Fed. Proc.* **46,** 2377–2394.

Lubowitz, H. and Whittam, R. (1969) Ion movements in human red cells independent of the sodium pump. *J. Physiol.* **202,** 111–131.

Lytle, C. and McManus, T. J. (1986) A minimal kinetic model of [Na + K + Cl] cotransport with ordered binding and glide symmetry. *J. Gen. Physiol.* **88,** 96a.

McManus, T. J., Haas, M., Starke, L. C., and Lytle, C. Y. (1985) The duck red cell model of volume-sensitive chloride-dependent cation transport. *Ann. NY Acad. Sci.* **456,** 183–186.

Mercer, R. W. and Hoffman, J. F. (1985) Bumetanide-sensitive Na/K cotransport in ferret red blood cells. *Biophys. J.* **47,** 157a.

O'Grady, S. M., Palfrey, H. C., and Field, M. (1987) Na-K-2Cl cotransport in winter flounder intestine and bovine kidney outer medulla: [^{3}H] bumetanide binding and effects of furosemide analogues. *J. Membr. Biol.* **96,** 11–18.

Palfrey, H. C. and Rao, M. C. (1983) Na/K/Cl cotransport and its regulation. *J. Exp. Biol.* **106,** 43–54.

Parker, J. C. and Berkowitz, L. (1983) Physiologically instructive genetic variants of the human red cell membrane. *Physiol. Rev.* **63,** 261–313.

Pennica, D., Kohr, W. J., Kuang, W. -. J., Glaister, D., Aggarwhal, B. B., Chen, E. Y., and Goeddel, D. V. (1987) Identification of human uromodulin as the Tamm-Horsfall urinary glycoprotein. *Science* **236,** 83–88.

Poznansky, M. and Solomon, A. K. (1972) Regulation of human red cell volume by linked cation fluxes. *J. Membr. Biol.* **10,** 259–266.

Rettori, O. and Lenoir, J. P. (1972) Ouabain-insensitive active sodium transport in erythrocytes: effect of external cation. *Am. J. Physiol.* **222,** 880–884.

Roberts, C. J. C., Homeida, M., Roberts, F., and Bogie, W. (1978) Effects of piretanide, bumetanide and furosemide on electrolyte and urate excretion in normal subjects. *Br. J. Clin. Pharm.* **6,** 129–133.

Sachs, J. R. (1971) Ouabain-insensitive sodium movements in the human red blood cell. *J. Gen. Physiol.* **57,** 259–282.

Sachs, J. R., Knauf, P. L., and Dunham, P. B. (1974) Transport through red cell membranes, in *The Red Blood Cell* (Surgenor, D. M., ed.), Academic, New York, pp 613–703.

Saier, M. H. and Boyden, D. A. (1984) Mechanism, regulation and physiological significance of the loop-diuretic-sensitive NaCl/KCl symport system in animal cells. *Mol. Cell Biochem.* **59,** 11–32.

Schlatter, E., Greger, R., and Weidtke, C. (1983) Effect of 'high ceiling' diuretics on active salt transport in the cortical thick ascending limb of Henle's loop of rabbit kidney. *Pflug. Arch* **396,** 210–217.

Schmidt, W. F. and McManus, T. J. (1977a) Ouabain-insensitive salt and water movements in duck red cells: I. Kinetics of cation transport under hypertonic conditions. *J. Gen. Physiol.* **70,** 59–79.

Schmidt, W. F. and McManus, T. J. (1977b) Ouabain-insensitive salt and water movements in duck red cells: II. Norepinephrine stimulation of sodium plus potassium cotransport. *J. Gen. Physiol.* **70,** 81–97.

Schmidt, W. F and McManus, T. J. (1977c) Ouabain-insensitive salt and water movements in duck red cells: III. The role of chloride in the volume response. *J. Gen. Physiol.* **70,** 99–121.

Shull, G. E., Schwartz, A., and Lingrel, J. B. (1985) Amino-acid sequence of the catalytic subunit of the ($Na^+ + K^+$)ATPase deduced from a complementary DNA. *Nature* **316,** 691–695.

Solomon, A. K., Toon, M. R., and Dix, J. A. (1986) Osmotic properties of human red cells. *J. Membr.* **91,** 259–273.

Stein, W. D. (1986) *Transport and Diffusion Acorss Cell Membranes* (Academic Press, Orlando).

Stewart, G. W. and Blackstock, E. J. (1989) Potassium transport in rabbit erythrocytes. *Exp. Biol.* **48,** 161–165.

Stewart, G. W., Blackstock, E. J., Hall, A. C., and Ellory, J. C. (1989) Potassium transport in monkey erythrocytes. *Exp. Biol.* **48,** 167–172.

Stewart, G. W. (1988) Co-ordinated variations in chloride-dependent potassium transport and cell water in normal human erythrocytes. *J. Physiol.* **401,** 1–16.

Tosteson, D. C. and Hoffman, J. F. (1960) Regulation of cell volume by active cation transport in high and low potassium sheep red cells. *J. Gen. Physiol.* **44,** 169–194.

Warnock, D. G., Greger, R., Dunham, P. B., Benjamin, M., Frizell, R. A., Field, M., Spring, K. R., Ives, H. E., Aronson, P. S., and Seifter, J. (1984) Ion transport in apical membranes of epithelia. *Fed. Proc.* **43,** 2473–2487.

Wiley, J. S. (1977) Genetic abnormalities of cation transport across the human erythrocyte membrane, in *Membrane Transport in Red Cells* (Ellory, J. C. and Lew, V. L., eds.) Academic Press, London, pp 337–361.

Wiley, J. S. and Cooper, R. A. (1974) A furosemide-sensitive cotransport of sodium plus potassium in the human red cell. *J. Clin. Invest.* **53,** 745–755.

3
PHARMACOLOGY

Pharmacological Modification of the Red Cell Ca^{2+}-Pump

B. U. Raess

1. Introduction

When studying drugs that affect Ca^{2+} homeostasis, either directly or indirectly, it is important to consider all factors in the equation determining a precisely controlled level of intracellular Ca^{2+} activity. One such factor in this equation is the highly regulated, intricately controlled plasma membrane (Ca^{2+} + Mg^{2+})-ATPase and Ca^{2+} extrusion mechanism henceforth referred to as the Ca^{2+}-pump. Because of its prominent position in the plasma membrane of all mammalian and perhaps all eukaryotic cells, it is legitimate to assume that this transport mechanism is a potential target for all kinds of drugs and certainly for those that are known to alter the calcium message on the inside of the cell.

This article will briefly describe four of the more common methodologies that can been used in studying the action of drugs on various ion transport aspects in the red cell. Understanding the many experimental advantages but also drawbacks and limitations of each method is deemed important, because this will determine the correct interpretation of the observed results. The focus of the chapter, however, will be a review of our understanding of the modulation of the Ca^{2+}-pump by drugs. This is an extremely rapidly growing field of interest because of the increasing awareness of the role of intracellular Ca^{2+} in numerous pathophysiologic conditions and the realization of the therapeutic potential of pharmacological intervention with these processes.

2. Red-Cell Preparations to Study Molecular Drug Action

Any experimental model, including those to study drug actions, should preferably comprise specificity, sensitivity, and versatility, and should lead to practical, testable, and interpretable propositions. For several decades, the red cell has served an extremely useful role as a model system in various types of investigations, because it comes close to fulfilling most of the mentioned criteria. Other advantages are pointed out in Chapter 1. From a pharmacological point of view, it is important to have this type of structural simplicity and consequent specificity. As is, detailing molecular mechanisms of drug action remains a formidable task and can only be advanced by first understanding relatively simple systems such as the red cell.

2.1. Intact Red Cells

Mature red cells can be used in a variety of ways to answer questions regarding the transport of solute across its plasma membrane. The advantage of using this particular type of preparation, as compared to others discussed here, is that it has the most physiological fidelity. It can be used to study cellular responses to pharmacological and other experimental manipulations of shape and structure, deformability, osmotic resistance, and volume changes. A prerequisite for drug action is the attainment of an effective concentration of the agent at its target site. The whole-red-cell preparation, regardless of the precise locus of the drug effect, is the physiologically most representative experimental tool to study a drug's propensity to cross the plasma membrane and distribute itself intracellularly. Since the turn of the century and the pioneering work by Ernest Overton and Hans H. Meyer, we have come to appreciate the important role of various physicochemical factors governing the passive diffusion of molecules across biological barriers (Meyer, 1899; Overton, 1901; Meyer and Hemmi, 1935), as well as the specialized transport processes responsible for drug distribution in all types of cells (Wilbrandt and Rosenberg, 1961; Csaky, 1965; Stein, 1967, 1986). Thus, the whole-red-cell preparation, especially when suspended in an environment closely resembling blood plasma, becomes an excellent model to examine solute transport in as close to native circumstances as possible. Although static in vitro experiments are invaluable for determining drug action asymmetry *per se*, caution should be exercised, however, in interpreting therapeutic inferences to drug action/penetration and drug-membrane intercalation. Only in cases where the red cell's circulatory and microcirculatory environment is matched, can one speak of experimental systems that approach truly physiological conditions (*see also* Chapter 7).

2.2. Reversibly Hemolyzed Ghosts

Although the whole-cell preparation offers the comforting notion of modeling the most natural state of the cell, investigations can be limited because of a relatively poor experimental control over the intracellular contents of the cell. For this reason, the reversibly hemolyzed and resealed ghost offers several advantages, i.e., it is possible to manipulate the internal milieu of the cell by placing otherwise inaccessible entities at known initial concentrations on the inside of the cell. This is particularly useful for introducing a given concentration of Ca^{2+} to the cell interior without starving or poisoning the cell. Besides inserting large amounts of ions into the cell, which would otherwise be difficult to attain because of extrusion processes, this preparation proves invaluable when there is a need to expose a polar or otherwise poorly diffusible and/or transported drug to the cell interior. By equilibrating a relatively small intracellular space with a large stringent hemolyzing solution, it is possible to expose endo-membrane components to a relatively well-controlled environment containing drugs and even small soluble proteins at known concentrations, while retaining the properties of a tightly sealed preparations (Schwoch and Passow, 1973; Richards and Eisner, 1982).

2.3. White-Ghost Membrane Fragments

The plasma membrane preparation from red cells prepared with the least difficulty is the hypotonic derived white ghost from which most of the hemoglobin has been removed. The amount of hemoglobin and other intracellular proteinaceous components removed is dependent on the osmolality and the presence or the absence of ions, i.e., K^+ (Bjerrum, 1979) or Ca^{2+} in the case of calmodulin (CaM) depletion (Farrance and Vincenzi, 1977b). The literature is deluged with variations of a hypo-osmotic shocked red-cell preparation, usually referred to as Dodge membranes (Dodge et al., 1963), which often produce contradictory or differing results, depending on the ionic and chemical composition of the hemolyzing and washing buffers used (with regard to ($Ca^{2+}+Mg^{2+}$)-ATPase activities; *see* Roufogalis, 1979). On the one hand, this preparation is probably the least sophisticated because it is not suitable for transport measurements or solving questions regarding the asymmetry of the red-cell membrane. On the other hand, the ease of producing large quantities of a particular preparation with a relatively high degree of homogeneity makes it ideal for isolation and purification of membrane-bound proteins and the measurement of membranous enzymic activities. For these reasons, this is the preparation of choice for many types of screening activities, including the screening of drug effects on transport ATPase activities (Raess and Vincenzi, 1980a).

2.4. Inside-Out Vesicularized Membrane Fragments

The attribute of "most ingenious" membrane preparation goes to the invaginated and reanealed membrane portions that have been freed

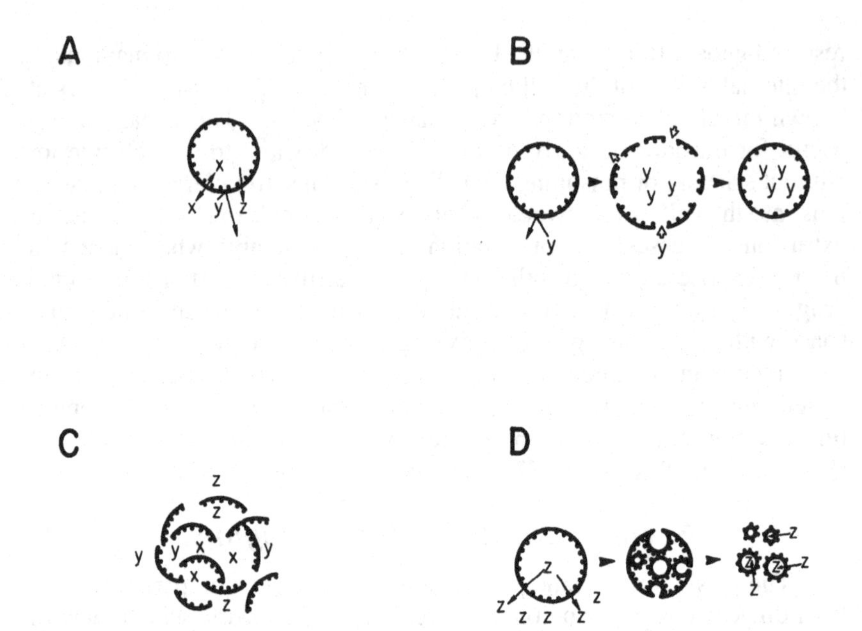

Fig. 1. Schematic representation of experimental red-cell preparations. (A) Whole cell; (B) Reversibly hemolyzed and resealed ghost; (C) White ghost membranes; (D) Inside-out vesicularized membrane fragments (x) Freely diffusible, facilitated, or actively transported solute species or drug; (y) Nonpermeable solute or drug; (z) actively extruded species. The four different experimental red-cell preparations permit distinct manipulation and drug access to either or both sides of the plasma membrane environment. The reversibly hemolyzed ghost (B) can also trap x and z; the same applies for the IOVs (D).

from the original cell-membrane envelope (Fig. 1D). As a result of the reversed sidedness, this preparation can offer several advantages. First, the endo-surface of the membrane is now totally accessible, and its environment is much more easily and more accurately manipulated. Secondly, transport processes that are normally in the outward direction (e.g., Ca^{2+} pumping) are of course also reversed. Therefore, the transported species is accumulated inside the vesicle and measured rather conveniently, because the vesicles can be trapped by filtration before being washed and measured for accumulated solute. Drugs, even those that are not expected to penetrate membranes to any extent, have free access to the endo-surface of the membrane where their effects, especially those on transport, can be studied with ease.

Another important point to keep in mind with this type of preparation is the varied proportion of inside-out and right-side-out vesicles.

Although the early methods (Steck, 1974; Steck and Kant, 1974; Hinds et al., 1981) yielded highly inside-out-oriented vesicle populations (70–>90%), the preparation of these vesicles required various high-speed centrifugation steps and was rather time-consuming. Lew and Seymour (1982) then devised a very rapid and much simpler procedure that resulted in an ion-tight vesicle preparation. Our laboratory has just recently adopted a modified version of this method, which is suitable for using outdated Red Cross CPDA-l packed cells (Raess and Record, unpublished). These vesicles yield a 30–55% IOV population as assessed by acetylcholinesterase accessibility (Steck and Kant, 1974).

In many respects, IOV preparations least resemble a physiological state, since they not only have undergone the membrane inversion (inside-out) change, but have also lost certain membrane components such as the spectrin framework. Nevertheless, this preparation has proved invaluable in demonstrating that a protein such as calmodulin can indeed act on the endo-surface of the red-cell membrane to activate the Ca^{2+} transport process (Macintyre and Green, 1978; Hinds et al., 1978; Larsen et al., 1978; Larsen and Vincenzi, 1979, Macintyre, 1982).

3. Drugs that Modulate the Ca^{2+}-Pump

Drugs are broadly defined as any chemical capable of altering all or parts of a biological system. This includes both stimulatory and inhibitory actions, but in the case of the calcium pump as the target, inhibitory substances are clearly the predominant species. For obvious reasons, it is useful to identify drugs that affect the Ca^{2+}-pump in any way, but it would be a significant advancement to find and recognize chemicals that specifically inhibit or stimulate this particular Ca^{2+} transport process. Not only would a large therapeutic potential be realized, but specific inhibitors in particular would provide a powerful means to differentiate this calcium transport process from other Ca^{2+} regulatory contrivances.

A crucial aspect of delineating drug effects is the description of modes and molecular sites of action. As with many other membrane-bound enzymes, target sites for drug modulation of the $(Ca^{2+}+Mg^{2+})$-ATPase activity can be found on extra- and intracellular as well as on intramembranous sites of the enzyme. The membrane environment, especially in close proximity, but also distal to the Ca^{2+}-pump protein(s), can be a very effective drug target site. Another way to modify $(Ca^{2+}+Mg^{2+})$-ATPase activity is by drug interference with its intracellular regulatory subunit, calmodulin. Again, several different mechanisms have been proposed. One mechanism involves the drug in competition with the (Ca^{2+} + Mg^{2+})-ATPase for a hydrophobic region on calmodulin and thus preventing calmodulin from activating the enzyme. Other modes of inhibition specify that drugs can interact with the enzyme itself and, consequently, hinder calmodulin binding or perhaps impede the postbinding consequences of this intracellular receptor–ligand interaction. Yet another proposal includes the situation where the inhibitor prevents Ca^{2+} from bringing about the conformational change in calmodulin thought to be necessary to activate all calmodulin-regulated processes. Nine distinct potential sites for drug interactions with the Ca^{2+}-pump are listed in Table 1.

At present, we are lacking a truly specific drug of choice for modulating the Ca^{2+}-pump. However, several lines of investigation with potential interest are being pursued. In this section, I shall attempt to organize and summarize the abundant, although at times contradictory and incomplete, information on inhibitors as well as activators of (Ca^{2+} + Mg^{2+})-ATPase activity and Ca^{2+} transport. Though there are well over a hundred compounds cited in the literature as modulators of the plasma membrane Ca^{2+}-pump, I will confine the discussion to those drugs that are either extensively documented, or represent a prototype with respect to chemical structure and/or mechanism of action. Below they are classified according to their primary, but not necessarily only, mechanism of action.

3.1. Inhibitors

3.1.1. Ionic Inhibitors

Lanthanum (La^{3+}) (Weiner and Lee, 1972; Gárdos et al., 1977), holmium, praseodymium (Schatzmann and Tschabold, 1971), samarium and gadolinium (Sarkadi et al., 1977; Schatzmann, 1983) have been shown to inhibit (Ca^{2+} + Mg^{2+})-ATPase activity and Ca^{2+} transport alike. K_i values of these lanthanides are in the low micromolar range, and higher concentrations of La^{3+} have been extremely useful in establishing a number of important details in the reaction mechanisms of the (Ca^{2+} + Mg^{2+})-ATPase. This includes an estimation of total enzyme amounts (Garrahan, 1986). Luterbacher (Luterbacher, 1982; Luterbacher and Schatzmann, 1983), in a series of elegant experiments, has quantitated and identified the site of action of La^{3+}. It was concluded that the lanthanide inhibits dephosphorylation at the step of the Mg^{2+}-facilitated $Ca{\cdot}E_1{\sim}P{\leftrightarrow}E_2\text{-}P$ conformational transformation (for more details of partial reaction mechanisms, *see also* Rega and Garrahan, Chapter 6). Obviously, it is difficult to discern experimentally whether these lanthanum effects are actual ionic competitions for a given site or whether

Table 1
Sites for Drug Modification of the Ca^{2+} Pump

Intercalation of drugs in membrane sites
In the phospholipid phase of the inner membrane leaflet
In the outer membrane leaflet
At the phospholipid—Ca^{2+}-pump interface
Drug binding and consequences on pump sites
At extracellularly accessible sites
At the catalytic site
At regulatory sites
Drug binding to sites on calmodulin
The Ca^{2+}-dependent hydrophobic pump binding domain
Ca^{2+} binding sites ($Ca_{n=1-4}$)
Nonspecific sites (preventing Ca^{2+} binding or conformational changes)

different sites of action are involved noncompetitively. Although experimental data are sparse, one may surmise that the other lanthanides mentioned above inhibit the enzyme in a similar fashion.

Ca^{2+} (Vincenzi and Schatzmann, 1967; Davis and Vincenzi, 1971; Vincenzi et al., 1980; Kratje et al., 1985; Allen et al., 1987) and Mg^{2+} (Schatzmann, 1977; Graf and Penniston, 1981; Villalobo et al., 1986a) at higher concentrations ($\geq 100 \mu M$ for Ca^{2+} and $>2mM$ for Mg^{2+}) exhibit inhibitory effects on the ($Ca^{2+} + Mg^{2+}$)-ATPase activity. Competitive displacement of the two positive effector ions at transport, regulatory, and active sites have been proposed; recently a Mg^{2+}- induced low-affinity state of the ($Ca^{2+} + Mg^{2+}$)-ATPase (Caride et al. 1986) has been also considered as possible mechanisms for this complex Mg^{2+} phenomenon.

In the case of Ca^{2+}-induced inhibition of ($Ca^{2+} + Mg^{2+}$)-ATPase, several proposals, including some stemming from active Ca^{2+} extrusion experiments in resealed ghosts, provide arguments for Ca^{2+} at high concentrations simply becoming a Mg^{2+} competitor (vice versa, the same applies for Mg^{2+}) of free-ion and ion-substrate complex (Kratje et al. 1985; Caride et al., 1986; *see also* Chapter 6). In a series of recent experiments, Allen et al. (1987) provided evidence that Ca^{2+} negatively affects the partial $Ca{\cdot}E_2 - P$ to $E_2 + P$ reaction, i.e., the Ca^{2+} unloading step. This is different from earlier suggestions (Schatzmann, 1982) that high concentrations of Ca^{2+} compete at the preceding Mg^{2+}-specific step in the Ca-loaded phosphorylated intermediate conformation conversion.

Pentavalent vanadate ion, serendipitously discovered as an impurity in commerically available ATP, is now recognized as a fairly universal ATPase inhibitor. Its effects on red-cell (Ca^{2+} + Mg^{2+})-ATPase activity has been studied extensively (Barrabin et al., 1980; Bond and Hudgins, 1980; Rossi et al., 1981; Schatzmann et al., 1986). The mechanism of (Ca^{2+} + Mg^{2+})-ATPase inhibition brought about by this anion is complex, because it is highly dependent on intracellular components such as Na^+, K^+, Mg^{2+}, and ATP. Therefore, the K_i for vanadate can change considerably depending on the experimental conditions. General agreement seems to favor a proposal where vanadate is recognized by the E_2 intermediate and, thus, attenuates $E_2 \rightarrow E_1$ reversion.

3.1.2. Inhibitors of Calmodulin Activation

Long before the initial observation (Bond and Clough, 1973; Farrance and Vincenzi, 1977a,b) and subsequent identification (Gopinath and Vincenzi, 1977; Jarrett and Penniston, 1977) of calmodulin as the positive effector protein of the (Ca^{2+} + Mg^{2+})-ATPase, Schatzmann (1970) described the effects of $10^{-4}M$ chlorpromazine on Ca^{2+} extrusion from Ca^{2+}-loaded and resealed ghosts. This, together with Levin and Weiss' observations (1977, 1978, 1979) that cyclic nucleotide phosphodiesterase activator could bind phenothiazines and other neuroleptics in a Ca^{2+}-dependent manner, led to a series of reports delineating functional drug interactions of calmodulin and neuroleptics with red-cell (Ca^{2+} + Mg^{2+})-ATPase activity (Vincenzi et al., 1978; Levin and Weiss, 1978; Raess and Vincenzi, 1980b; Gietzen et al., 1980; Roufogalis, 1981) and Ca^{2+}-transport by Hinds et al. (1981). Trifluoperazine equivalently antagonized Ca^{2+}-pump stimulation by calmodulin as demonstrated by simultaneous inhibition of (Ca^{2+} + Mg^{2+})-ATPase and Ca^{2+}-transport activities from inside-out vesicularized red-cell preparations. The relatively specific type of antagonism towards calmodulin was in contrast to the highly selective, competitive inhibition by the CaM-binding protein calcineurin, (Larsen et al., 1978), and a nonselective type of inhibition of both basal and CaM-stimulated pump activity by ruthenium red (Hinds et al. 1981). The close correspondence of enzymic and transport parameters in the presence of three apparently different-acting inhibitors can be taken as direct pharmacological evidence in support of a functional singularity of ATPase and transport activities. Other relatively well-established criteria in favor of this idea are the shared ATP requirement, Ca^{2+} affinity, Mg^{2+} dependence, cardiac glycoside insensitivity, transport substrate specificity, and membrane asymmetry (Schatzmann, 1975b).

Rather than listing all the compounds that purportedly inhibit calmodulin-dependent functions, it may be advantageous to ask how some of these so-called calmodulin antagonists may be acting. To answer this question, one has to consider:

1. the conformational state and properties of calmodulin ($CaM{\cdot}Ca_{n_0 \rightarrow 4}$)
2. the physicochemical features of the inhibitor and
3. the calmodulin target itself in the CaM-loaded or depleted state.

Trifluoperazine and many other drugs that bind to calmodulin share the properties of being cationic amphiphiles with varying degrees of hydrophobicity. This is also a salient feature of cellular and other naturally occurring oligopeptide venoms and antibiotics such as gramicidin S, making these compounds ideal candidates for calmodulin interaction (Cox, 1986). Depending on the size and degree of the hydrophobicity, calmodulin-regulated enzymes, peptides and drugs bind to calmodulin with varying affinity. A key requirement for high affinity 1:1 binding (nanomolar) of calmodulin to synthetic model peptide ligands appears to be a high basic α-helical potential in the structure. The optimal length for these helixes is in the 10–32Å range and suggests that basic amphiphilic peptides bind to an acidic, hydrophobic region of the $CaM{\cdot}Ca_{n \geqq 3}$ form (Cox et al., 1985; Cox, 1986). Gramicidin S, a basic amphiphile lacking the helical component but fulfilling the size requirements (15Å), binds to calmodulin with a 2:1 ratio and a diminished K_d of 1–2 μM (Cox, 1986). The antibiotic produces an inhibition resembling that produced by phenothiazines and naphthalenesulfonamide derivatives (Tanaka et al., 1982a,b, 1983; Hidaka and Tanaka, 1983) affecting calmodulin-dependent as well as calmodulin-independent ATPase activities. This suggests additional action loci for such drugs aside from their interaction with calmodulin (Raess and Vincenzi, 1980b; Vincenzi et al., 1982; Iglesias and Rega, 1987).

Compound 48/80, a heterogenous condensation product of *N*-methyl-*p*-methoxy-phenethylamine and formaldehyde, exists as a mixture of homologous polycations. Gietzen et al. (1983a,b) found compound 48/80 to be not only potent, but highly selective for inhibition of the calmodulin-stimulated portion of the ($Ca^{2+}Mg^{2+}$)-ATPase activity. Its selectivity ratio (>824) for calmodulin-dependent functions far exceeds that of any other compound investigated (Gietzen et al., 1983a; Rossi et al., 1985), although calmidazolium (Van Belle, 1981; Gietzen et al., 1983a) and ethoxybutyl berbamine (Xu and Zhang, 1986) also exhibit comparatively good selectivity. Their ratios for inhibition of basal over CaM-dependent ($Ca^{2+} + Mg^{2+}$)-ATPase activities are 29 and 86, respectively. Ca^{2+}-transport stimulated by calmodulin is also inhibited by 48/80. In this case, however, about 10 times more drug is required for half maximal inhibition, a somewhat surprising discrepancy. The mechanism of action of 48/80 appears, as expected from its structure, to involve binding to calmodulin in the presence of Ca^{2+} and, thus, prevents the formation of the calmodulin-target enzyme complex. In keeping with this, kinetic analysis of calmodulin-dependent (Ca^{2+} + Mg^{2+})-ATPase looks competitive up to 1 μg/mL, a value close to the IC_{50} of the drug. CaM-sepharose affinity separation of the various polymers from compound 48/80 revealed the larger components, i.e., octamers to decamers, which make up only a small fraction of the mixture to be the active inhibitory species (Adamcyk-Englemann and Gietzen, 1984). Differences in batch composition and source of the compound may be explanations for relatively large differences in K_i reported by different laboratories (Gietzen et al., 1983a,b; Rossi et al., 1985, 1987; DiJulio and Vincenzi, 1986).

Several attempts to distinguish the mechanisms of (Ca^{2+} + Mg^{2+})-ATPase inhibition by various antagonsits of calmodulin activation have yielded a number of interesting proposals, some including kinetic analysis (Raess and Vincenzi, 1980b; Gietzen et al., 1980, 1982, 1983b; Scharff and Foder, 1984). Scharff and Foder (1984), for instance, have considered one mechanism involving the combining of the drug with the CaM binding site on the enzyme. In addition, the drug reduced CaM affinity for its "receptor" by occupying a distal site. Another mechanism includes the latter possibility, i.e., where the drug reduces CaM affinity allosterically together with a reduction of the free CaM concentration. Unfortunately, a multitude of mechanistic combinations exist. This includes an often overlooked possibility where the drug prevents the binding of Ca^{2+} to CaM and, thus, inpedes its conformational change and exposure of the hydrophobic domain (LaPorte et al., 1980), a transition that appears to be crucial for calmodulin interaction with the target enzyme. It is extremely difficult to dissect the various mechanistic details of drug interactions with the Ca^{2+} pump, such as the determination of dissociation constants (Scharff and Foder, 1984), since in most cases one is dealing with membranous preparations and, at best, only estimates of free drug concentrations.

Many compounds and classes of drugs, possessing diverse structural features, have an inherent propensity to antagonize CaM-mediated functions to various degrees, with different potencies and no apparent specificity with regard to CaM targets. Among them is the relatively new therapeutically useful group of structurally heterogeneous slow Ca^{2+} channel entry inhibitors, commonly referred to as Ca^{2+} antagonists. Reports by Scharff and Foder (1984) comparing verapamil and trifluoperazine effects on CaM activation of $(Ca^{2+} + Mg^{2+})$-ATPase activity, a report by Johnson (1984) proposing a CaM-like Ca^{2+} binding protein as a regulator of the voltage-dependent or slow Ca^{2+} channel, and another by Minocherhomjee and Roufogalis (1984) comparing the actions of various dihydropyridine Ca^{2+} entry blockers on cyclic AMP phosphodiesterase prompted us to investigate a number of Ca^{2+} antagonists on Ca^{2+}, and CaM regulation of $(Ca^{2+} + Mg^{2+})$-ATPase activity and Ca^{2+} transport into IOVs. In accordance with several recent proposals (Spedding, 1985), we found distinct functional differences between the various types of Ca^{2+} antagonists with respect to their effects (or lack thereof) and the red-cell Ca^{2+}-pump. Diltiazem and verapamil at 1 m*M* concentrations inhibited CaM stimulation of $(Ca^{2+} + Mg^{2+})$-ATPase activity with a predominant effect on the maximal velocity, suggesting noncompetitiveness with regard to CaM. Basal, CaM-independent $(Ca^{2+} + Mg^{2+})$-ATPase activity was not affected by these slow channel-selective antagonists, whereas cinnarizine, a type IV, nonselective antagonist (VanHoutte and Paoletti, 1987) appeared to inhibit (Mg^{2+})-ATPase, $(Na^{+} + K^{+})$-ATPase, and basal and CaM-stimulated $(Ca^{2+} + Mg^{2+})$-ATPase activities indiscriminately. Verapamil and diltiazem had no effect on the Ca^{2+} activation of basal activity, nor did they change significantly the dissociation equilibrium constant of Ca^{2+} in CaM activation (Raess and Gersten, 1987; Kim and Raess, 1988). These results are consistent with, but do not prove, the functional removal of the activator form of CaM. Whether this is by prevention of Ca^{2+}-induced conformational changes of CaM or by interaction with the anionic hydrophobic region where it diminishes binding to its target enzyme remains to be shown. The dihydropyridine nifedipine, on the other hand, had no effect on the CaM concentration effect relationship or calmodulin-independent basal activity up to $10^{-5}M$. A combination of 1 m*M* verapamil and 1 m*M* diltiazem was shown to be additive, producing about half of the maximal inhibition of the stimulated activity and a threefold decrease in Ca^{2+} affinity.

Data from IOV preparations suitable to measure both active Ca^{2+} transport and ATPase activities concomitantly produced essentially the same findings. However, these preparations appear to be somewhat more sensitive to drug effects in general. Thus, CaM-stimulated Ca^{2+}-transport into plasma membrane vesicles, too, is inhibited by verapamil and diltiazem (IC_{50} ~1 m*M*), but not by nifedipine. It may well be that nifedipine is simply less potent, but nonspecific interference of solvents needed to keep the drug in solution precludes the use of nifedipine above 10 μ*M* in these preparations. Our findings with a number of different Ca^{2+} antagonists are in agreement with those of Scharff and Foder (1984) and Lamers et al. (1985), who obtained very similar results with cardiac sarcolemma membranes. Moreover, our results are consistent with those of Itoh et al. (1986), who showed a strong correlation between competition of bepridil, naphthalenesulfonamide derivative (W-7), prenylamine, verapamil, and diltiazem, but not dihydropyridine binding and the inhibition of CaM-dependent myosin light-chain kinase phosphorylation. Based on these and other reports (Krebs et al., 1984; Jarrett, 1984, 1986), it appears that there must be at least two drug-binding domains on calmodulin. It will be interesting to see if this knowledge can be exploited to differentiate between binding and functional calmodulin-target enzyme interactions.

3.1.3. Chemical Modification of Enzymic and Transport Sites

Waisman et al. (1981) and Minocherhomjee and Roufogalis (1982) have shown that capnophorin inhibitors 4-acetamido-4′-isothiocyanostilbene 2′-disulfonate and *N*-(4-azido-2-nitrophenyl)-2-aminoethylsulfonate (NAP-taurine) can block the Ca^{2+}-pump rather specifically and at relatively low concentrations. The latter investigators concluded on the basis of asymmetry experiments, IOV transport, lack of probenecid inhibition, and NAP-taurine effects on the purified enzyme that this was a direct inhibition of the enzyme by NAP-taurine rather than a capnophorin-coupled inhibition. They also found an intriguing dissociation of $(Ca^{2+} + Mg^{2+})$-ATPase activity and Ca^{2+} transport brought about by DIDS (4,4,′-diisothiocyano-2-2′-stilbene disulfonate; *see also* Niggli et al., 1982), an observation worthy of further investigation.

Muallem and Karlish (1983) investigated regulatory and catalytic substrate binding sites with the irreversible inactivator fluorescein isothiocyanate (FITC). Both ATP and nonhydrolyzable analogues of the substrate were able to protect against a kinetically complex (biphasic) type of inhibition. FITC appears to be quite specific for various types of ion-transport ATPases, since it inhibits $(Na^{+} + K^{+})$-ATPase and sarcoplasmic reticulum (SR) Ca^{2+}-pump (Karlish, 1980; Pick and Karlish, 1980). Also of considerable interest is a noncovalently acting iodinated fluorescein derivative, erythrosin B, which appears to inhibit noncompetitively the catalytic site by acting as a competitor of ATP binding at the enzyme's regulatory site (Mugica et al., 1984).

Phenylglyoxal, a tool for the qualitative and quantitative description of arginine containing amino-acid sequences in proteins, has been shown to result in a time- and concentration-dependent inactivation of human erythrocyte $(Ca^{2+} + Mg^{2+})$-ATPase (Raess et al., 1985) and Ca^{2+} transport (Raess, 1985). Phenylglyoxal concentrations ranging from 0.3 to 3.0 m*M* over a 60 min preincubation period resulted in rates of inactivation that obeyed apparent pseudo-first-order kinetics until 97% of the initial enzyme activity was lost. From Ca^{2+} concentration effect relationship analysis in nonmodified and phenylglyoxalated membranes, it was clear that the inactivation process did not involve either the Ca^{2+} binding site or the calmodulin binding to the enzyme. The irreversible inactivation of $(Ca^{2+} + Mg^{2+})$-ATPase by phenylglyoxal could be minimized by ATP under a variety of preincubation conditions. Notably, ATP could protect against inactivation in the total absence of either Ca^{2+} or Mg^{2+} ions, providing supportive evidence that free ATP is the ligand at least at the low affinity regulatory site and that an arginine residue plays a key role in this ATP recognition site. This finding has recently been corroborated in a study by Filoteo et al. (1987) identifying the amino-acid sequence with an arginine residue in a FITC-reactive ATP binding region of the Ca^{2+}-pump. Thus, phenylglyoxal and other arginine-specific dicarbonyl compounds can be exploited to investigate various enzymatic and transport details, including the irreversible tagging of ATP binding sites on the Ca^{2+}-pump. Similarly, Ca^{2+} and proton facilitated *N,N'*-dicyclohexylcarbodiimide inhibition of CaM-dependent and CaM-independent $(Ca^{2+} + Mg^{2+})$-ATPase activity (Villalobo et al. 1986b) was proposed to help identify individual reaction steps, as had been previously carried out in the case of SR Ca^{2+}-translocating mechanisms (Scofano et al., 1985; Argaman and Shoshan-Barmatz, 1988).

N-Ethylmaleimide (NEM) inhibition of $(Ca^{2+} + Mg^{2+})$-ATPase is partially protected by ATP, Na^+, and K^+, but facilitated by Ca^{2+} (Richards et al. 1977). Although the inhibitory effects of NEM are not specific for $(Ca^{2+} + Mg^{2+})$-ATPase, they appear to occur at 10 times lower concentrations than those needed to inhibit $(Na^+ + K^+)$-ATPase. Classical oxidative thiol reactants as well as activated oxygen-induced malondialdehyde generation can lead to peroxidation of lipid and membrane-bound protein thiol groups, both of which appear to be detrimental to $(Ca^{2+} + Mg^{2+})$-ATPase activity in red cells (Hebbel et al. 1986). These observations take on significance when considering phenylhydrazine-induced hemolytic anemias (Shalev et al., 1981), sickle cell anemias, and glucose-6-phosphate deficiencies.

Overall, irreversible covalent modification by drugs may be of less importance in terms of pharmacological intervention in the regulation of the Ca^{2+}-pump. However, it may be anticipated that these probes will play a significant role in the identification of other reversible drug binding sites on the protein.

3.2. Activators of the Ca^{2+}-Pump

So far, only a few compounds are known to activate (Ca^{2+} + Mg^{2+})-ATPase activity and Ca^{2+} transport. It is beyond the scope of this article to include details of phospholipid activation, endogenous and exogenous proteolytic cleavage enhancement, organic solvent and detergent activation, ionophore A23187-induced shunt mechanisms, or the intricacies of calmodulin activation of the Ca^{2+}-pump (Al-Jobore and Roufogalis, 1981; Enyedi et al., 1987; Niggli et al., 1981a,b; Rossi and Schatzmann, 1982; Sarkadi et al., 1986, 1987; Au, 1987; Gardos et al., 1977; Carafoli et al., 1982). Instead, I will mention the limited number of drugs and hormones shown to enhance (Ca^{2+} + Mg^{2+})-ATPase activity or Ca^{2+} transport by presumed independent means.

Procaine, belonging to a class of drugs that typically antagonize calmodulin-sensitive processes (Volpi et al., 1981), including the (Ca^{2+} + Mg^{2+})-ATPase activity from red cells, has been proposed by Popescu and coworkers (1987) to activate red-cell (Ca^{2+} + Mg^{2+})-ATPase activity. The stimulatory effect of procaine is apparently CaM independent and thought to, on the basis of phospholipase evidence, take place at the lipid/hydrophobic protein domain of the (Ca^{2+} + Mg^{2+})-ATPase. Yamomoto and Harris (1983), demonstrated that ethanol and pentobarbital increased red-cell $^{45}Ca^{2+}$ efflux and (Ca^{2+} + Mg^{2+})-ATPase activity at concentrations ranging from 25 to 100 m*M*. In addition, four isomeric, nonspecific barbiturates had similar, but much more potent effects. Similarly to the case of procaine, these actions were thought to be the result of membrane disordering effects of the drugs.

On several occasions, we have noticed activation of both (Ca^{2+} + Mg^{2+})-ATPase and Ca^{2+}-transport activities by concentrations of verapamil and diltiazem as low as 0.1–1.0 μM. It is tempting to speculate that this may be part of a therapeutically relevant action of these compounds; however, these effects are very small and difficult to reproduce consistently. At this point, it is not clear if the stimulation is an artifact, or the result of an uncontrolled experimental parameter or variable.

(Ca^{2+} + Mg^{2+})-ATPase activity from human and animal red cells has been shown to be activated in vitro by physiological concentrations of thyroid hormone (Davis and Blas, 1981; Davis et al., 1982, 1983; Constante et al., 1986). This relatively modest response is calmodulin dependent, and produced by both T_3 and T_4. Both bepridil and cetiedil can be shown to reversibly inhibit the stimulation without competing for hormone binding sites (Agre et al., 1984; Dube et al., 1987).

4. Conclusions and Outlook

Pharmacological dogma predicts that there are no drugs or chemicals without more than one effect or action. This short review, using the human red cell as the model of a simple plasma membrane, illustrates the point strikingly. The discovery of a potent, specific, and selective compound for the Ca^{2+} pump, be it inhibitory or activating by nature, has not yet been realized. As far as we can predict, the benefits of such a drug would be countless, not only in determining the forces that govern Ca^{2+} homeostasis of cells, but because potentially useful therapeutic applications could be developed. Alternatively, it may be more realistic to expect that the drugs discussed here, as well as newly discovered ones, will be merely useful devices in delineating both the (Ca^{2+} + Mg^{2+})-ATPase activity and the concomitant transport activities. Whatever the future, with an improved arsenal of biochemical and molecular techniques in membrane protein characterization, we can look forward to an exciting era of discovery of drug actions and mechanisms involving the Ca^{2+} pump and cellular Ca^{2+} homeostasis. On the other hand, we may, as in the past, have to rely on deductive reasoning and on concentration–effect relationship analysis and potency ratios. This possibility has been summarized succinctly some time ago by Theophrastus Bombastus von Hohenheim *alias* Philippus Aureolus Paracelsus (1493–1541): "All things are poisons, for there is nothing without poisonous qualities; it is only the dose which makes a thing a poison."

References

Adamczyk-Engelmann, P. and Gietzen, K. (1984) Histamine release and calmodulin antagonism are two distinct properties of compound 48/80. *Cell Calcium* **5,** 311.

Agre, P., Virshup, D., and Bennett, V. (1984) Bepridil and cetiedil. Vasodilators which inhibit Ca^{2+}-dependent calmodulin interactions with erythrocyte membranes. *J. Clin. Invest.* **74,** 812–820.

Al-Jobore, A. and Roufogalis, B. D. (1981) Phospholipid and calmodulin activation of solubilized calcium-transport ATPase from human erythrocytes: regulation by magnesium. *Can. J. Biochem.* **59,** 880–888.

Allen, B. G., Katz, S., and Roufogalis, B. D. (1987) Effects of Ca^{2+}, Mg^{2+} and calmodulin on the formation and decomposition of the phosphorylated intermediate of the erythrocyte Ca^{2+}-stimulated ATPase. *Biochem. J.* **244,** 617–623.

Argaman, A. and Shoshan-Barmatz, V. (1988) Dicyclohexylcarbodiimide interaction with sarcoplasmic reticulum. Inhibition of Ca^{2+} efflux. *J. Biol. Chem.* **263,** 6315–6321.

Au, K. S. (1987) Activation of erythrocyte membrane Ca^{2+}-ATPase by calpain. *Biochim. Biophys. Acta* **905,** 273–278.

Barrabin, H., Garrahan, P. J., and Rega, A. (1980) Vanadate inhibition of the Ca^{2+}-ATPase from human red cell membranes. *Biochim. Biophys. Acta* **600,** 796–804.

Bjerrum, P. J. (1979) Hemoglobin depleted human erythrocyte ghosts. *J. Membr. Biol.* **48,** 43–67.

Bond, G. H. and Clough, D. L. (1973) A soluble protein activator of $(Ca^{2+} + Mg^{2+})$-dependent ATPase in human red cell membranes. *Biochim. Biophys. Acta* **323,** 592–599.

Bond, G. H. and Hudgins, P. (1980) Inhibition of the red cell Ca^{2+} ATPase by vanadate. *Biochim. Biophys. Acta* **600,** 781–790.

Carafoli, E., Zurini, M., Niggli, V., and Krebs, J. (1982) The calcium-transporting ATPase of erythrocytes. *Ann. N.Y. Acad. Sci.* **402,** 304–328.

Caride, A. J., Rega, A. F., and Garrahan, P. J. (1986) The reaction of Mg^{2+} with the Ca^{2+}-ATPase from human red cell membranes and its modification by Ca^{2+}. *Biochim. Biophys. Acta* **863,** 165–177.

Constante, G., Sand, G., Connart, D., and Glinoer, D. (1986) *In vitro* effects of thyroid hormones on red blood cell Ca^{++}-dependent ATPase activity. *J. Endocrinol. Invest.* **9,** 15–20.

Cox, J. A. (1986) Calcium-calmodulin interaction and cellular function. *J. Cardiovasc. Pharmacol.* **8,** S48–S51.

Cox, J. A., Comte, M., Fitton, J. F., and DeGrado, W. F. (1985) The interaction of calmodulin with amphiphilic peptides. *J. Biol. Chem.* **260,** 2527–2537.

Csaky, T. Z. (1965) Transport through biological membranes. *Ann. Rev. Physiol.* **27,** 415–450.

Davis, P. J. and Blas, S. D. (1981) In vitro stimulation of human red blood cell Ca^{2+}-ATPase by thyroid hormone. *Biochem. Biophys. Res. Commun.* **99,** 1073–1080.

Davis, P. W. and Vincenzi, F. F. (1971) Ca-ATPase activation and NaK-ATPase inhibition as a function of calcium concentration in human red cell membranes. *Life Sci.* **10,** 401–406.

Davis, F. B., Davis, P. J., and Blas, S. D. (1983) Role of calmodulin in thyroid hormone stimulation *in vitro* of human erythrocyte Ca^{2+}-ATPase activity. *J. Clin. Invest.* **71,** 579–586.

Davis, F. B., Kite, J. H., Davis, P. J., and Blas, S. D. (1982) Thyroid hormone stimulation *in vitro* of red blood cell Ca^{2+}-ATPase activity: interspecies variation. *Endocrinology* **110,** 297–302.

DiJulio, D. and Vincenzi, F. F. (1986) Evaluation of trifluoperazine and compound 48/80 as selective antagonists of calmodulin activation of the Ca^{2+}-pump ATPase. *Proc. West. Pharmacol. Soc.* **29,** 445–446.

Dodge, J. T., Mitchell, C., and Hanahan, D. J. (1963) The preparation and chemical characteristics of hemolysate-free ghosts of human erythrocytes. *Arch. Biochem. Biophys.* **100,** 119–130.

Dube, M. P., Davis, F. B., Davis, P. J., and Blas, S. D. (1987) Bepridil and cetiedil reversibly inhibit thyroid hormone stimulation *in vitro* of human red cell Ca^{2+}-ATPase activity. *Mol. Endocrinol.* **1,** 168–171.

Enyedi, A., Flura, M., Sarkadi, B., Gardos, G., and Carafoli, E. (1987) The maximal velocity and the calcium affinity of the red cell calcium pump may be regulated independently. *J. Biol. Chem.* **262,** 6425–6430.

Farrance, M. L. and Vincenzi, F. F. (1977a) Enhancement of (Ca^{2+} + Mg^{2+})-ATPase activity of human erythrocyte membranes by hemolysis in isosmotic imidazole buffer. I. General properties of variously perpared membranes and the mechanism of the isosmotic imidazole effect. *Biochim. Biophys. Acta* **471,** 49–58.

Farrance, M. L. and Vincenzi, F. F. (1977b) Enhancement of (Ca^{2+} + Mg^{2+})-ATPase activity of human erythrocyte membranes by hemolysis in isosmotic imidazole buffer. II. Dependence on calcium and a cytoplasmic activator. *Biochim. Biophys. Acta* **471,** 59–66.

Filoteo, A. G., Gorski, J. P., and Penniston, J. T. (1987) The ATP-binding site of the erythrocyte membrane Ca^{2+} pump. Amino acid sequence of the fluorescein isothiocyanate-reactive region. *J. Biol. Chem.* **262,** 6526–6530.

Gárdos. G., Szász, I., and Sarkadi, B. (1977) Effect of intracellular calcium on the cation transport processes in human red cells. *Acta Biol. Med. Germ.* **36,** 823–829.

Garrahan, P. J. (1986) Inhibitors of the Ca^{2+} pump, in *The Ca^{2+} Pump of Plasma Membranes* (Rega, A. F. and Garrahan, P. J., eds.), CRC Press, Boca Raton, pp. 153–164.

Gietzen, K., Mansard, A., and Bader, H. (1980) Inhibition of human erythrocyte Ca^{++}-transport ATPase by phenothiazines and butyrophenones. *Biochem. Biophys. Res. Commun.* **94,** 674–681.

Gietzen, K., Sadorf, I., and Bader, H. (1982) A model for the regulation of calmodulin-dependent enzymes erythrocyte Ca^{2+}-transport ATPase and brain phosphodiesterase by activators and inhibitors. *Biochem. J.* **207,** 541–548.

Gietzen, K., Adamczyk-Engelmann, P., Wüthrich, A., Konstantinova, A., and Bader, H. (1983a) Compound 48/80 is a selective and powerful inhibitor of calmodulin regulated functions. *Biochim. Biopys. Acta* **736,** 109–118.

Gietzen, K., Sanchez-Delgado, E., and Bader, H. (1983b) Compound 48/80: a powerful and specific inhibitor of calmodulin-dependent Ca^{2+}-transport ATPase. *IRCS Med. Sci.* **11,** 12–13.

Gopinath, R. M. and Vincenzi, F. F. (1977) Phosphodiesterase protein activator mimics red blood cell cytoplasmic activator of (Ca^{2+} + Mg^{2+})-ATPase activity. *Biochem. Biophys. Res. Commun.* **77,** 1203–1209.

Graf, E. and Penniston, J. T. (1981) CaATP: The substrate, at low ATP concentrations, of Ca^{2+}-ATPase from human erythrocyte membranes. *J. Biol. Chem.* **256,** 1587–1592.

Hebbel, R. P., Shalev, O., Foker, W., and Rank, B. H. (1986) Inhibition of erythrocyte Ca^{2+}-ATPase by activated oxygen through thiol- and lipid-dependent mechanisms. *Biochim. Biophys. Acta* **862,** 8–16.

Hidaka, H. and Tanaka, T. (1983) Naphthalenesulfonamides as calmodulin antagonists. *Methods Enzymol.* **102,** 185–194.

Hinds, T. R., Larsen, F. L., and Vincenzi, F. F. (1978) Plasma membrane Ca^{2+} transport: stimulation by soluble proteins. *Biochem. Biophys. Res. Commun.* **81,** 455–461.

Hinds, T. R., Raess, B. U., and Vincenzi, F. F. (1981) Plasma membrane Ca^{2+} transport: antagonism by several potential inhibitors. *J. Membrane Biol.* **58,** 57–65.

Iglesias, R. O. and Rega, A. F. (1987) Gramicidin S inhibition of the Ca^{2+}-ATPase of human red blood cells. *Biochim. Biophys. Acta* **905,** 383–389.

Itoh, H., Tanaka, T, Mitani, Y., and Hidaka, H. (1986) The binding of the calcium channel blocker, bepridil, to calmodulin. *Biochem. Pharmacol.* **35,** 217–220.

Jarrett, H. W. (1984) The synthesis and reaction of a specific affinity label for the hydrophobic drug-binding domains of calmodulin. *J. Biol. Chem.* **259,** 10136–10144.

Jarrett, H. W. (1986) Response of three enzymes to oleic acid, trypsin and calmodulin chemically modified with a reactive phenothiazine. *J. Biol. Chem.* **261,** 4967–4972.

Jarrett, H. W. and Penniston, J. T. (1977) Partial purification of the (Ca^{2+} + Mg^{2+})-ATPase activator from human erythrocytes: its similarity to the activator of 3′:5′-cyclic nucleotide phosphodiesterase. *Biochem. Biophys. Res. Commun.* **77,** 1210–1216.

Johnson, J. D. (1984) A calmodulin-like Ca^{2+} receptor in the Ca^{2+} channel. *Biophys. J.* **45,** 134–136.

Karlish, S. J. D. (1980) Characterization of conformational changes in (Na, K) ATPase labeled with fluorescein at the active site. *J. Bioenerg. Biomembr.* **12,** 111–136.

Kim, H. C. and Raess, B. U. (1988) Verapamil, diltiazem and nifedipine interactions with calmodulin stimulated (Ca^{2+} + Mg^{2+})-ATPase. *Biochem. Pharmacol.* **37,** 917–920.

Kratje, R. B., Garrahan, P. J., and Rega, A. F. (1985) Two modes of inhibition of the Ca^{2+} pump in red cells by Ca^{2+}. *Biochim. Biophys. Acta* **816,** 365–378.

Krebs, J., Buerkler, J., Guerini, D., Brunner, J., and Carafoli, E. (1984) 3-(Trifluoromethyl)-3-(*m*-[125I]iodophenyl)diazirine, a hydrophobic, photoreactive probe, labels calmodulin and calmodulin fragments in a Ca^{2+}–dependent way. *Biochemistry* **23,** 400–403.

Lamers, J. M. J., Cysouw, K. J., and Verdouw, P. D. (1985) Slow calcium channel blockers and calmodulin. Effect of felodipine, nifedipine, prenylamine and bepridil on cardiac sarcolemmal calcium pumping ATPase. *Biochem. Pharmacol.* **34,** 3837–3843.

LaPorte, D. C., Wierman, B. M., and Storm, D. R. (1980) Calcium-induced exposure of a hydrophobic surface on calmodulin. *Biochemistry* **19,** 3814–3819.

Larsen, F. L. and Vincenzi, F. F. (1979) Calcium transport across the plasma membrane: stimulation by calmodulin. *Science* **204,** 306–309.

Larsen, F. L., Raess, B. U., Hinds, T. R., and Vincenzi, F. F. (1978) Modulator binding protein antagonizes activation of (Ca^{2+} + Mg^{2+})-ATPase and Ca^{2+} transport of red blood cell membranes. *J. Supramolec. Struct.* **9,** 269–274.

Levin, R. M. and Weiss, B. (1977) Binding of trifluoperazine to the calcium-dependent activator of cyclic nucleotide phosphodiesterase. *Mol. Pharmacol.* **13,** 690–697.

Levin, R. M. and Weiss, B. (1978) Selective inhibition of an activatable ATPase by trifluoperazine. *The Pharmacologist* **20,** 195.

Levin, R. M. and Weiss, B. (1979) Selective binding of antipsychotics and other psychoactive agents to the calcium-dependent activator of cyclic nucleotide phosphodiesterase. *J. Pharmacol. Exp. Ther.* **208,** 454–459.

Lew, V. and Seymour, C. A. (1982) Preparation of sealed vesicles and transport measurements. Cation transport in one-step inside-out- vesicles from red cell membranes, in *Techniques in Lipid and Membrane Biochemistry* B415, Elsevier, Amsterdam, pp. 1–13.

Luterbacher, S. (1982) Die Teilreaktion der ATP-Spaltung durch das isolierte Protein der Ca^{2+}-Pumpe aus der Erythrozytenmembran. Dissertation, University of Bern, Switzerland.

Luterbacher, S. and Schatzmann, H. J. (1983) The site of action of La^{3+} in the reaction cycle of the human red cell membrane Ca^{2+}-pump ATPase. *Experientia* **39,** 311–312.

Macintyre, J. D. (1982) Properties and uses of human erythrocyte membrane vesicles, in *Red Cell Membranes: A Methodological Approach* (Ellory, J. C. and Young, J. D., eds.), Academic Press, London, pp. 199–217.

Macintyre, J. D. and Green, J. W. (1978) Stimulation of calcium transport in inside-out vesicles of human erythrocyte membranes by a soluble cytoplasmic activator. *Biochim. Biophys. Acta* **510,** 373–377.

Meyer, H. H. (1899) Zur Theorie der Alkoholnarkose. Erste Mitteilung. *Arch. Exp. Path. Pharmak.* **42,** 109–118.

Meyer, H. H. and Hemmi, H. (1935) Beiträge zur Theorie der Narkose. III. *Biochem. Z.* **277,** 39–54.

Minocherhomjee, A. and Roufogalis, B. D. (1982) Selective antagonism of the Ca^{2+} transport ATPase of the red cell membrane by *N*-(4-azido-2-nitrophenyl)-2-amino-ethylsulfonate (NAP-taurine). *J. Biol. Chem.* **257,** 5426–5430.

Minocherhomjee, A. M. and Roufogalis, B. D. (1984) Antagonism of calmodulin and phosphodiesterase by nifedipine and related calcium entry blockers. *Cell Calcium* **5,** 57–63.

Muallem, S. and Karlish, S. J. D. (1983) Catalytic and regulatory ATP-binding sites of the red cell Ca^{2+} pump studied by irreversible modification with fluorescein isothiocyanate. *J. Biol. Chem.* **258,** 169–175.

Mugica, H., Rega, A. F., and Garrahan, P. J. (1984) The inhibition of the calcium dependent ATPase from human red cells by erythrosin B. *Acta Physiol. Pharmacol. Lat.* **34,** 163.

Niggli, V., Adunyah, E. S., and Carafoli, E. (1981a) Acidic phospholipids, unsaturated fatty acids, and limited proteolysis mimic the effect of calmodulin on the purified erythrocyte Ca^{2+}-ATPase. *J. Biol. Chem.* **256,** 8588–8592.

Niggli, V., Adunyah, E. S., Penniston, J. T., and Carafoli, E. (1981b) Purified (Ca^{2+}-Mg^{2+})-ATPase of the erythrocyte membrane. Reconstitution and effect of calmodulin and phospholipids. *J. Biol. Chem.* **256,** 395–401.

Niggli, V., Sigel, E., and Carafoli, E. (1982) Inhibition of the purified and reconstituted calcium pump of erythrocytes by μM levels of DIDS and NAP-taurine. *FEBS Lett.* **138,** 164–166.

Overton, E. (1901) *Studien über die Narkose, zugleich ein Beitrag zur allgemeinin Pharmakologie.* Fischer, Jena.

Pick, U. and Karlish, S. J. D. (1980) Indications for an oligomeric structure and for conformational changes in sarcoplasmic reticulum Ca^{2+}-ATPase labelled selectively with fluorescin. *Biochim. Biophys. Acta* **626,** 255–261.

Popescu, L. M., Hinescu, M. E., Spiru, L., Popescu, M., and Diculescu, I. (1987) The effect of procaine on Ca^{2+}-Mg^{2+}-ATPase of erythrocyte membrane. *Rev. Roum. Biochim.* **24,** 119–124.

Raess, B. U. (1985) Inhibition of human erythrocyte ($Ca^{2+}+Mg^{2+}$)-ATPase and Ca^{2+}-transport by phenylglyoxal. *Proc. Fed. Amer. Soc. Exp. Biol.* **44,** 1595.

Raess, B. U. and Gersten, M. H. (1987) Calmodulin-stimulated plasma membrane (Ca^{2+} + Mg^{2+})-ATPase: inhibition by calcium channel entry blockers. *Biochem. Pharmacol.* **36,** 2455–2459.

Raess, B. U. and Vincenzi, F. F. (1980a) A semi-automated method for determination of multiple membrane ATPase activities. *J. Pharmacol. Methods.* **4,** 273–283.

Raess, B. U. and Vincenzi, F. F. (1980b) Calmodulin activation of red blood cell (Ca^{2+} + Mg^{2+})-ATPase and its antagonism by phenothiazines. *Mol. Pharmacol.* **18,** 253–258.

Raess, B. U., Record, D. M., and Tunnicliff, G. (1985) Interaction of phenylglyoxal with the human erythrocyte (Ca^{2+} + Mg^{2+})-ATPase. Evidence for the presence of an essential arginyl residue. *Mol. Pharmacol.* **27,** 444–450.

Richards, D. E. and Eisner, D. A. (1982) Preparation and use of resealed red cell ghosts, in *Red Cell Membranes - A Methodological Approach* (Ellory, J. D. and Young, J. D., eds.), Academic Press, London, pp. 165–177.

Richards, D. E., Rega, A. F. and Garrahan, P. J. (1977) ATPase and phosphatase activities from human red cell membranes. I. The effects of *N*-ethylmaleimide *J. Membr. Biol.* **35,** 113–124.

Rossi, J. P. F. C. and Schatzmann, H. J. (1982) Trypsin activation of the red cell Ca^{2+}-pump ATPase is calcium-sensitive. *Cell Calcium* **3,** 583–590.

Rossi, J. P. F. C., Garrahan, P. J., and Rega, A. F. (1981) Vanadate inhibition of active Ca^{2+} transport across human red cell membranes. *Biochim. Biophys Acta* **648,** 145–150.

Rossi, J. P. F. C., Garrahan, P. J., and Rega, A. F. (1987) Differential effects of compounds 48/80 on the ATPase and phosphatase activities of the Ca^{2+} pump of red cells. *Biochim. Biophys. Acta* **902,** 101–108.

Rossi, J. P. F. C., Rega, A. F., and Garragan, P. J. (1985) Compound 48/80 and calmodulin modify the interaction of ATP with the (Ca^{2+} + Mg^{2+})-ATPase of red cell membranes. *Biochim. Biophys. Acta* **816,** 379–386.

Roufogalis, B. D. (1979) Regulation of calcium translocation across the red blood cell membrane. *Can. J. Physiol. Pharmacol.* **57,** 1331–1349.

Roufogalis, B. D. (1981) Phenothiazine antagonism of calmodulin: a structurally-nonspecific interaction. *Biochem. Biophys. Res. Commun.* **98,** 607–613.

Sarkadi, B., Szász, I., Gerloczy, A., and Gárdos, G. (1977) Transport parameters and stoichiometry of active calcium ion extrusion in intact human red cells. *Biochim. Biophys. Acta* **464,** 93–107.

Sarkadi, B., Enyedi, A., Foldes-Papp, Z., and Gárdos, G. (1986) Molecular characterization of the *in situ* red cell membrane calcium pump by limited proteolysis. *J. Biol. Chem.* **261,** 9552–9557.

Sarkadi, B., Enyedi, A., and Gárdos, G. (1987) Conformational changes of the in situ red cell membrane calcium pump affect its proteolysis. *Biochim. Biophys. Acta* **899,** 129–133.

Scharff, O. and Foder, B. (1984) Effect of trifluoperazine, compound 48/80, TMB-8 and verapamil on the rate of calmodulin binding to erythrocyte Ca^{2+}-ATPase. *Biochim. Biophys. Acta* **772,** 29–36.

Schatzmann, H. J. (1970) Transmembrane calcium movements in resealed human red cells, in *Calcium and Cellular Function*. (Cuthbert, A. W., ed.), Martin's Press, New York, pp. 85–95.

Schatzmann, H. J. (1975a) Active calcium transport across the plasma membrane of erythrocytes, in *Calcium Transport in Contraction and Secretion*. (Carafoli, E., Clementi, F., Drabikowski, W., and Margreth, A., eds.) North-Holland Publishing Company, Amsterdam, pp. 45–49.

Schatzamnn, H. J. (1975b) Active calcium transport and Ca^{2+}-activated ATPase in human red cells, in *Current Topics in Membranes and Transport* (Bronner, F. and Kleinzeller, A., eds.), Academic Press, New York and London, pp. 125–161.

Schatzmann, H. J. (1977) Role of magnesium in the (Ca^{2+} + Mg^{2+})-stimulated membrane ATPase of human red blood cells. *J. Membrane Biol.* **35,** 149–158.

Schatzmann, H. J. (1982) The plasma membrane calcium pump of erythrocytes and other animal cells, in *Membrane Transport of Calcium* (Carafoli, E., ed.), Academic Press, New York, pp. 41–108.

Schatzmann, H. J. (1983) The red cell calcium pump. *Ann. Rev. Physiol.* **45,** 303–312.

Schatzmann, H. J. and Rossi, G. L. (1971) (Ca^{2+} + Mg^{2+})-activated membrane ATPase in human red cells and their possible relations to certain transport. *Biochim. Biophys. Acta* **241,** 379–392.
Schatzmann, H. J. and Tschabold, M. (1971) The lanthanides Ho^{3+} and Pr^{3+} as inhibitors of calcium transport in human red cells. *Experientia* **27,** 59–61.
Schatzmann, H. J. and Vincenzi, F. F. (1969) Calcium movements across the membrane of human red cells. *J. Physiol.* **201,** 369–395.
Schatzmann, H. J., Luterbacher, S., Stieger, J., and Wüthrich, A. (1986) Red blood cell calcium pump and its inhibition by vanadate and lanthanum. *J. Cardiovasc. Pharmacol.* **8,** S33–S37.
Schwoch, G. and and Passow, H. (1973) Preparation and properties of human erythrocyte ghosts. *Mol. Cell Biochem.* **2,** 197–218.
Scofano, H. M., Barrabin, H., Lewis, D., and Inesi, G. (1985) Specific dicyclohexylcarbodiimide inhibition of the E-P + H_2O = E+Pi reaction and ATP = Pi exchange in sarcoplasmic reticulum adenosinetriphosphatase. *Biochemistry* **24,** 1025–1029.
Shalev, O., Leida, M. N., Hebbel, R. P., Jacob, H. S., and Eaton, J. W. (1981) Abnormal erythrocyte calcium homeostasis in oxidant-induced hemolytic disease. *Blood* **58,** 1232–1235.
Spedding, M. (1985) Calcium antagonist subgroups. *Trends Pharmacol. Sci.* **6,** 109–114.
Steck, T. L. (1974) Preparation of impermeable inside-out and right-side-out vesicles from human erythrocyte membranes, in *Methods in Membrane Biology*, vol. 2 (Korn, E. D., ed.), Plenum, New York, pp. 245–281.
Steck, T. L. and Kant, J. A. (1974) Preparation of impermeable ghosts and inside-out vesicles from human erythrocyte membranes. *Methods Enzymol.* **31,** 172–180.
Stein, W. D. (1967) *The Movement of Molecules Across Cell Membranes.* Academic, New York.
Stein, W. D. (1986) *Transport and Diffusion Across Cell Membranes.* Academic, London.
Tanaka, T., Ohmura, T., and Hidaka, H. (1982a) Hydrophobic interaction of the Ca^{2+}-calmodulin complex with calmodulin antagonists. Naphthalenesulfonamide derivatives. *Mol. Pharmacol.* **22,** 403–407.
Tanaka, T., Ohmura, T., Yamakado, T., and Hidaka, H. (1982b) Two types of calcium-dependent protein phosphorylations modulated by calmodulin antagonists. Naphthalenesulfonamide derivatives. *Mol. Pharmacol.* **22,** 408–412.
Tanaka, T., Ohmura, T., and Hidaka, H. (1983) Calmodulin antagonists' binding sites on calmodulin. *Pharmacology* **26,** 249–257.
Van Belle, H. (1981) R-24-571: A potent inhibitor of calmodulin-activated enzymes. *Cell Calcium* **2,** 483–494.
VanHoutte, P. M. and Paoletti, R. (1987) The WHO classification of calcium antagonists. *Trends Pharmacol. Sci.* **8,** 4–6.
Villalobo A., Brown, L., and Roufogalis, B. D. (1986a) Kinetic properties of

the purified Ca^{2+}-translocating ATPase from human erythrocyte plasma membrane. *Biochim. Biophys. Acta* **854,** 9–20.

Villalobo, A., Harris, J. W., and Roufogalis, B. D. (1986b) Calcium-dependent inhibition of the erythrocyte Ca^{2+} translocating ATPase by carbodiimides. *Biochim. Biophys. Acta* **858,** 188–194.

Vincenzi, F. F. and Schatzmann, H. J. (1967) Some properties of Ca-activated ATPase in human red cell membranes. *Helv. Physiol. Acta* **25,** CR233–CR234.

Vincenzi, F. F., Raess, B. U., Larsen, F. L., Jung, N. S. G. T., and Hinds, T. R. (1978) The Plasma membrane calcium-pump: A potential target for drug action. *The Pharmacologist* **20,** 195.

Vincenzi, F. F., Hinds, T. R., and Raess, B. U. (1980) Calmodulin and the plasma membrane calcium pump. *Ann. NY Acad. Sci.* **356,** 232–244.

Vincenzi, F. F., Adunyah, E. S., Niggli, V., and Carafoli, E. (1982) Purified red blood cell Ca^{2+}-pump ATPase: evidence for direct inhibition by presumed anticalmodulin drugs in the absnece of calmodulin. *Cell Calcium* **3,** 545–559.

Volpi, M., Sha'afi, R. I., Epstein P. M., Andrenyak, D. M., and Feinstein, M. B. (1981) Local anesthetics, mepacrine and propranolol are antagonists of calmodulin. Proc. *Natl. Acad. Sci. USA* **78,** 795–799.

Waisman, D. M., Gimble, J. M., Goodman, D. B. P., and Rasmussen, H. (1981) Studies of the Ca^{2+} transport mechanism of human inside-out vesicles. II. Stimulation of the Ca^{2+} pump by phosphate. *J. Biol. Chem.* **256,** 415–419.

Weiner, M. L. and Lee, K. S. (1972). Active calcium ion uptake by inside-out and right-side out vesicles of red blood cell membranes. *J. Gen. Physiol.* **59,** 462–475.

Wilbrandt, W. and Rosenberg, T. (1961) The concept of carrier transport and its corollaries in pharmacology. *Pharmacol. Rev.* **13,** 109–129.

Xu, Y. and Zhang, S. (1986) A derivative of bisbenzylisoquinoline alkaloid is a new and potential calmodulin antagonist. *Biochem. Biophys. Res. Commun.* **140,** 461–467.

Yamamoto, H. -A. and Harris, R. A. (1983) Effects of ethanol and barbiturates on Ca^{2+}-ATPase activity of erythrocyte and brain membranes. *Biochem. Pharmacol.* **32,** 2787–2791.

Irreversible Modification of the Anion Transporter

P. J. Bjerrum

1. Introduction

1.1. Physiological Aspects

The major fraction of CO_2 formed by metabolism is transformed to bicarbonate ions in the blood. The catalyzed hydration of CO_2 takes place inside red cells, and the major fraction of the produced bicarbonate is then transported across the membrane in exchange for extracellular chloride. This transfer ensures that the "CO_2" transporting capacity of the blood is fully utilized.

The facilitated exchange diffusion is mediated by an abundant integral membrane protein, identified by Cabantchik and Rothstein (1974) and Passow et al. (1975). The protein has a relative molecular mass of 10^5 dalton and constitutes about 30% of the total membrane polypeptide (Fairbanks et al., 1971; Steck, 1974). This corresponds to $0.8–1.2 \times 10^6$ molecules/cell, as determined by covalent binding of "specific" stilbene disulfonate inhibitors of the anion exchange process (Passow et al., 1975; Zaki et al., 1975; Lepki et al., 1976; Halestrap, 1976; Ship et al., 1977; Funder et al., 1978, Wieth et al., 1982c). The electro-silent one-for-one-exchange of chloride for bicarbonate across the membrane (Hunter, 1971; Hunter, 1977) is very rapid (10^5 anions/transport molecule/s at 38°C; Brahm, 1977), but despite this high transport rate, which is among the highest for any facilitated transport process, the translocation is an important rate-limiting step in the transfer of CO_2 from tissues to lungs (Wieth et al., 1982a).

1.2. Structure and Function

The amino-acid sequence of the transport molecule, also known as band 3 protein (Fairbanks et al., 1971) or capnophorin (Wieth and Bjerrum, 1983), has recently been determined from the DNA sequence of the gene coding for murine band 3 protein (Kopito and Lodish, 1985a,b). Despite structural exploration of the molecule (as reviewed by Passow [1986] and Jay and Cantley [1986], and by Jennings in Chapter 8 of this book) and an increased understanding of the transport kinetics (as reviewed by Knauf in Chapter 9), the molecular transport mechanism is still not very well characterized. Further structural information is clearly needed before the molecular mechanism can be understood in detail.

This chapter will focus on irreversible chemical modification of capnophorin using group specific amino-acid reagents for identification of functionally important amino-acid residues. Modification of arginyl residues and carboxyl groups is emphasized, because they appear to be essential groups involved directly in the transport process, as demonstrated by titration of the transport function (Wieth and Bjerrum, 1982; Milanick and Gunn, 1982). Chemical modification of other important residues, such as lysine residues (Jennings, 1982a; Passow et al., 1980; Nanri et al., 1983; Matsuyama et al. 1983; Kempf et al., 1981; Sigrist and Zahler, 1982; Rudloff et al., 1983; Jennings and Nicknish, 1985; Jennings et al., 1985), histidine residues (Chiba et al., 1986), and tyrosine residues (Craik et al., 1986), and modification with dansyl chloride of yet unknown amino-acid groups (Lepke and Passow, 1982; Berghout et al., 1985; Raida and Passow, 1985) are dealt with in more detail by Jennings in Chapter 8, particularly the importance of the various functional lysine groups.

2. Functional Groups Identified by Acid-Base Titration

The anion transport process can only be examined in the pH range from 5.5–10.5 when intracellular pH equals extracellular pH (Funder and Wieth, 1976). The functional deterioration seen at intracellular pH values above 11 is apparently related to release (alkali stripping) of peripheral membrane proteins (Steck and Yu, 1973), whereas titration at low pH leads to changes in the distribution of the cytoskeleton proteins followed by an increase in the permeability to ions and small hydrophilic molecules (Bjerrum et al., 1980). If titration is performed solely on the extracellular side of the membrane for a short period of time (less than 1–2 min), the transport function can surprisingly be measured in the entire pH range from 3.5–13 (Wieth and Bjerrum, 1982).

2.1. Extracellular Titration

Anion exchange depends on several classes of exofacial titratable groups. Deprotonation of one class of groups with apparent pK around 5 activates the transport of monovalent anions (Wieth et al., 1982a) and decreases the transport of divalent anions (Milanick and Gunn, 1982, 1984). Transport of monovalent anions (chloride) is nearly constant in the extracellular pH range from 6–10. Translocation is then increased by titration of the "extracellular modifier group" having an apparent p*K* of 11. After stimulation in the narrow pH interval (10–11.5), the transport system is inactivated by titration of group(s) with an apparent p*K* around 12 (Wieth and Bjerrum, 1982).

2.2. Effects of Substrate Ions

The nearly sigmoidal exofacial titration curves does not necessarily imply that the pH titration unveils distinct types of groups directly or allosterically involved in anion binding and/or translocation. The observed relation between pH and anion transport could be the result of titration of clusters of groups acting in concert. Nevertheless, this possibility appears unlikely, mainly because the titration curve is simply displaced on changing the concentration of substrate ions. Thus protonation of the exofacial groups with apparent pK around 5 increases the sulfate binding, and conversely, sulfate binding increases the p*K* of the groups (Milanick and Gunn, 1982). The p*K* of the alkaline groups (with apparent pK of 12) depends on the chloride concentration. The p*K* of the groups decreases about 1 U when the extracellular chloride concentration is reduced 10-fold, with only minor changes occurring in the shape of the titration curve (Wieth and Bjerrum, 1982; Bjerrum, 1986).

2.3. Effects of Transport Coupled Conformational Changes

Anion transport can be described by a simple ping-pong model involving a reciprocating anion binding site (Gunn and Fröhlich, 1979; Jennings, 1980; Jennings and Adams, 1981; Fröhlich, 1982; Jennings, 1982b; Knauf et al., 1984; Knauf and Mann, 1984; Furuya et al., 1984; Hautmann and Schnell, 1985; Falke and Chan, 1985). According to the simple ping-pong kinetics, only one transported anion is bound to the transport protein at a time, and the translocation is coupled to conformational changes of the binding region (Knauf, 1979; Knauf, 1986; Passow, 1986). The distribution between outward- and inward-facing "substrate binding sites" involved in the transport process should, according to the model, be changed and more sites be recruited inward, solely by a reduction in the intracellular chloride concentration. This redistribution should alter the apparent kinetic parameters on the exofacial side of the membrane.

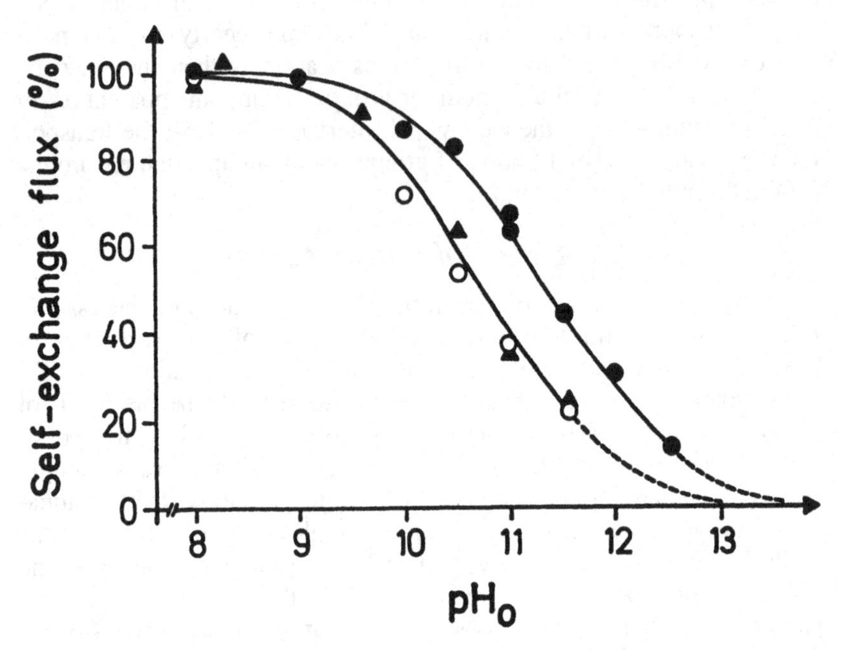

Fig. 1. Extracellular titration of the chloride self-exchange in the presence of transmembrane chloride gradients at 0°C. Experiments (▲, ○) were performed on resealed membranes (pH 7.8) containing 165 KCl in accordance with Wieth and Bjerrum (1982). Efflux medium: 16 m*M* KCl, 22 m*M* potassium citrate, 180 m*M* sucrose, 0°C.

Experiment (●) was performed on resealed membranes (pH 7.8) containing 4.5 m*M* KCl. Preparation: 1 vol packed erythrocytes were hemolyzed in 40 vol, 20 m*M* potassium citrate, 2 m*M* $MgSO_4$, pH 6.2, 0°C. After 5 min, 4 vol, 25 m*M* Tris was added and the suspension was titrated to pH 7.3 with 0.1*M* KOH. Resealing and ^{36}Cl-efflux was performed as described by Wieth and Bjerrum (1982). The intracellular chloride concentration was calculated from the ^{36}Cl distribution ratio $R = Cl_i/Cl_o$ in accordance with Wieth and Bjerrum (1982) determined after washing and packing the resealed membranes in 3 m*M* KCl, 25 m*M* potassium citrate, pH 7.8 (44.000*g* for 5 min). Efflux medium: 16 m*M* KCl, 20 m*M* potassium citrate. The magnitude of the chloride fluxes was calculated relative to the values found at pH 8.0. These references fluxes were 331 (○), 308 (▲), and 59 (●) pmol $cm^{-2}s^{-1}$. The apparent p*K* in the presence of an outward and inward chloride gradient was 11.4 and 10.8, respectively.

Wieth and Bjerrum (1982) found no significant effect on the exofacial alkaline p*K* when the intracellular chloride concentration was decreased from 165 to 16 m*M*. In more recent experiments using a larger change in the chloride gradient across the membrane and a constant intra- and extracellular ionic strength, a change in the extracellular p*K* was observed. The experiment (Fig. 1) shows that the apparent p*K* of the alkaline titratable groups is increased approximately 0.6 pH U when the intracellular KCl concentration is decreased from 165 to 4.5 m*M*. This is the expected direction for the change in the exofacial p*K* when more sites are recruited inward.

Kinetic studies by Milanick and Gunn (1984) demonstrate that the acidic groups, besides affecting chloride and sulfate binding, also appear to be directly involved in the cotransport of protons (Jennings, 1976) when extracellular sulfate exchanges with intracellular chloride. Moreover, it was found that this translocation of protons apparently can be separated from chloride translocation, so that the chloride- and the proton "binding sites" can be on opposite sides of the membrane at the same time (Milanick and Gunn, 1986).

2.4. *Summary*

The various kinetic results demonstrate that both the alkaline titratable groups and the acidic groups are intimately involved in the transport process. On the basis of the p*K* values of the two classes of groups and their ionization enthalpies, it was proposed by Wieth et al. (1982a,b) that these groups correspond to arginyl residues and carboxyl groups, respectively.

3. Modification of Arginine Residues

Arginyl residues in proteins can be modified selectively with α-dicarbonyl reagents (Meanes and Feeney, 1971; Lundblad and Noyes, 1984). These arginine-modifying compounds have played a major role in the identification of arginyl residues as essential binding sites for negatively charged substrates in numerous enzymes (Takahasi, 1968; Riordan et al., 1977; Riordan, 1979) and have also been used for identification of essential arginyl residues in membrane proteins (Grisham, 1979; Raess et al., 1985)

Most of the work carried out hitherto on modification of capnophorin has employed the dicarbonyl reagent PG* (Wieth et al., 1982b; Zaki, 1982; Bjerrum et al., 1983; Zaki, 1984; Zaki and Julien, 1985; Bjerrum 1986, 1989), but modification can also be performed with 1,2-cyclohexanedione and 2,3-butanedione and derivatives of PG, (Zaki, 1981, 1982; Zaki and Julien, 1986). PG reacts preferentially with the nonionized form of the arginine residue, and only slowly with α- and ϵ-amino groups (Takahasi, 1968; Cherung and Fonda, 1979a,b). The rate of reaction with arginine groups is found to increase with increasing pH. The essential arginine groups in the active center of many enzymes appear to have a higher reactivity than the other arginine residues in the enzymes (Riordan, 1979). Patthy and Thész (1980) explained this modification selectivity at anion recognition sites by suggesting that the p*K* of functional arginyl residues is generally lower than that of other arginyl residues in the enzymes.

The PG–arginine interaction is not completely understood. It has been estimated from modification studies that the reaction occurs with a 2:1 stoichiometry (Takahasi, 1968; Riordan, 1979). In some investigations, however, a 1:1 stoichoimetry has been reported (Borders and Riordan, 1975; Weber et al., 1975; Kazarinof and Snell, 1976; Vandenbunder et al., 1981). Most of these results were obtained in the presence of borate buffer, which can apparently complex the cis-diol of the reversible 1:1 complex. A primary, stable 1:1 2,3-butanedione-arginine product, stabilized by a pinacol-type rearrangement, has also been reported (Riordan, 1973). The analogous PG-arginine adduct 3 (Fig. 2) could also be a possibility, as suggested by Bjerrum (1989). The stoichiometry of the reaction with important arginyl residues in the red-cell anion transport protein is therefore not established with certainty.

3.1. Phenylglyoxal Modification of the Transport System

Time-dependent inactivation of chloride transport (Wieth and Bjerrum, 1983; Bjerrum, 1989) and sulfate transport (Zaki, 1982, 1983; Zaki and Julian 1985) with PG is obtained if capnophorin is exposed to the reagent at neutral or alkaline pH from both sides of the membrane.

PG rapidly permeates the erythrocyte membrane with a half-time of 40 ms at 25°C (Wieth et al., 1982c). The possibility of obtaining selective

*The abbreviations used are: CAPS (3-(*N*-cyclohexylamino)-1-propanesulfonic acid), CHES (2-(*N*-cyclohexylamino)ethanesulfonic acid), CMC (1-cyclohexyl-3-(2-morpholino-ethyl)carbodiimide), DIDS (4.4′-diisothiocyano-2.2′-stilbenedisulfonic acid), DNDS (4.4′-dinitrostilbene-2.2′-disulfonic acid), EAC (1-ethyl-3-(4-azonia-4,4-dimethylpentyl)carbodiimide), EDC (1-ethyl-3-(3-dimethyl-aminopropyl)carbodiimide), GME (glycin methyl ester), MES (2-*N*-morpholinoethanesulfonic acid), PG (phenylglyoxal), and TEE (tyrosine ethyl ester).

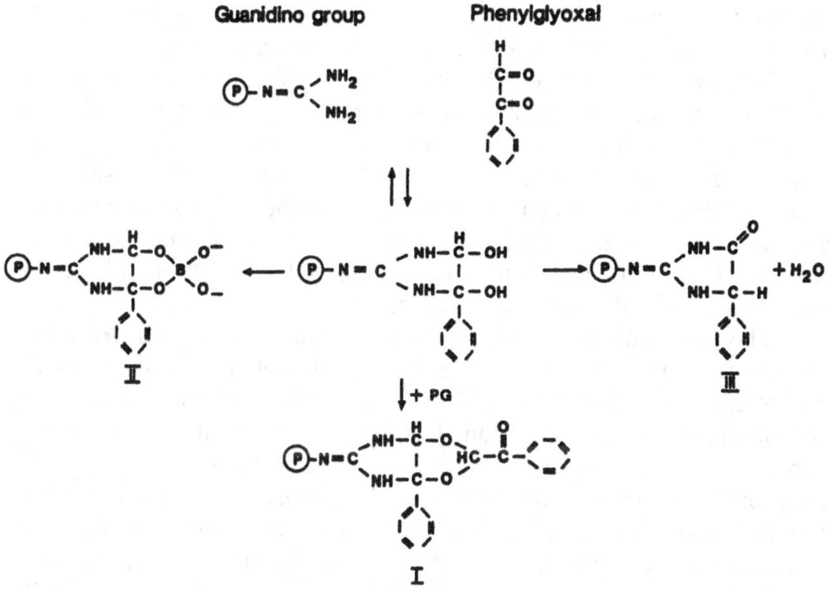

Fig. 2. Scheme for reaction of guanidine free-base of an arginine residue with PG (from Bjerrum, 1989). PG is proposed (Takahasi, 1968) to form a 2:1 adduct with arginine (I). A 1:1 adduct stabilized by borate, shown as product II, has been been suggested (Borders and Riordan, 1975). Finally, product III shows another stable 1:1 adduct proposed by Bjerrum (1989) in accordance with Riordan (1973).

inactivation from one side of the membrane therefore depends on how long a stable proton gradient can be maintained across the membrane. The erythrocyte membrane is not very permeable to hydroxyl ions and protons, and even less during PG inactivation because of the concomitant reversible inhibition of the transport process (Wieth et al., 1982b). Selective modification from the extracellular side of the membrane was obtained under conditions of neutral intracellular pH and alkaline extracellular pH (9–11) as reported by Wieth et al. (1982b) and Bjerrum et al. (1983). Selective modification of intracellular arginines was obtained at neutral extracellular and alkaline intracellular pH (9.6), as reported by Bjerrum (1989).

3.2. Characterization of the Exofacial Modification

PG inactivation is a pseudo first-order process, when inactivation is performed from both sides of the membrane (Zaki and Julien, 1985; Bjerrum, 1989) and when performed solely from the exofacial side of the membrane at more alkaline pH values (Wieth et al., 1982b; Bjerrum, 1989), indicating reaction with a functional essential group.

In characterization of the exofacial modification, Wieth et al. (1982b) found that the pseudo first-order rate constant for inactivation was a linear function of the PG concentration used, no saturation occurring at the highest concentrations examined (30 m*M*). This result indicates that PG does not behave as an affinity label (with high reversible affinity for the substrate binding groups), but acts as a directly modifying reagent. The rate of inactivation can, in accordance with this, be determined as a second-order rate constant.

The inactivation of capnophorin at various extracellular pH values demonstrates that PG only, or at least preferentially, reacts with the deprotonated arginine groups (Wieth et al., 1982b; Bjerrum, 1989). The second-order rate constants plotted as function of pH, at a given chloride concentration, give a nearly sigmoidal titration curve. The pK of this function decreases with increase in the extracellular chloride concentration in a manner that parallels the chloride-dependent change in p*K* of the titratable extracellular groups (Wieth and Bjerrum, 1982; Bjerrum, 1986). This result indicates that the functional titratable groups and the PG-reacting groups are likely to be identical. The remarkable shift in the apparent pK of the functional titratable arginine residues when chloride is removed from the "transport site" may be the result of electrostatic repulsion from some other positively charged amino-acid group(s), e.g., lysine or arginine residues located in (or in the vicinity of) the substrate binding region. This interpretation provides a simple explanation for the increased PG reactivity observed at reduced chloride concentration (Wieth et al., 1982b; Bjerrum et al., 1983; Bjerrum, 1989).

The rate of inactivation decreases approximately 10-fold when the temperature is lowered from 38°C to 25°C, corresponding to an apparent Arrhenius activation energy of 33 kcal/mol. This marked temperature dependence can partly be accounted for by the high ionization enthalpy of the apparently modified guanidino group(s) (Cohn and Edsall, 1943). The somewhat higher PG inactivation pK of 12.6 at 165 m*M* KCl, obtained by extrapolation to 0°C (using the high ionization enthalpy for arginine groups), compared to the p*K* of 12.0 obtained from the titration studies at 0°C, could be because of changes in the hydrophobic binding region with temperature (Wieth et al., 1982b), but may also reflect changes induced by reversible binding of PG, as discussed below.

3.2.1. The Nature of the Residual Flux

A residual chloride exchange flux of about 10% is normally obtained after extracellular PG inactivation. This residual transport activity is not the result of 90% reduction in the transport capacity of the individual transport molecules (Wieth et al., 1982b), but represents release of PG from a fraction of the modified transport molecules. The PG release, which is time and temperature dependent, takes place shortly after modification (Bjerrum, 1989).

3.2.2. Effects of Chloride Gradient

The rate of irreversible PG-inactivation at a given pH was found to depend entirely on the extracellular chloride concentration (Wieth et al., 1982b). This observation could indicate that the modified groups are insensitive to recruitment of the "substrate binding sites," and thus not directly involved in anion binding and translocation. Another explanation (discussed below) could be that the transport system is "locked" in a particular protein conformation by instantaneous reversible phenyglyoxal binding, before the irreversible inactivation takes place.

3.3. Instantaneous Reversible Phenylglyoxal Inhibition

PG instantaneously and reversibly inhibits anion transport. The degree of inhibition depends on concentration and pH. The irreversible PG-inactivation reaction and the reversible PG-inhibition of the anion transport system are "unrelated processes" as pointed out by Wieth et al. (1982b). It was found that the rate of irreversible inactivation increases linearly with PG at concentrations where the anion transport is nearly "fully" arrested by reversibly bound PG and that the reversible inhibition shows positive cooperativity with a Hill coefficient of 1.7 (whereas the irreversible inactivation is first order with respect to PG). Moreover, reversible inhibition appears rather independent of the extracellular chloride concentration (Bjerrum, 1989) in comparison to the significant chloride dependence of irreversible inactivation.

The reaction kinetics and chloride dependence of the two types of reactions make it unlikely that reversible PG binding is the first step in the irreversible inactivation, as suggested by Zaki and Julien (1985); instead it is more likely that the instantaneously reversible inhibition involves binding to some other groups in the protein. This reversibly bound PG appears to arrest all transport molecules in an outward-facing (chloride-binding) conformation (Bjerrum, 1989). Irreversible exofacial PG-inactivation of the functional essential arginine residue(s) (at concentrations of PG where the anion transport is "abolished" and a stable proton gradient obtained) is therefore entirely dependent on the extracellular chloride concentration. In contrast to PG, butanedione and cyclohexanedione show little, if any, reversible inhibition of anion transport (Bjerrum

P. J., unpublished), but these reagents are difficult to use in irreversible exofacial inactivation studies because of problems with establishing a stable proton gradient.

3.4. PG Inactivation and Inhibitor Binding

The affinity as well as the irreversible reaction of the tritiated anion transport inhibitor DIDS were found to be decreased in PG-treated resealed membranes (Bjerrum et al., 1983). Irreversible binding of $^3H_2$-DIDS was also found to be reduced in cyclohexanedione modified membranes (Zaki, 1981). Moreover, a reduction of reversible binding of the bulky stilbene disulfonate fluorescent probe DBDS to PG-modified membranes has been demonstrated (Passow, 1986). The modified arginine residues, therefore, appear to be either directly or at least indirectly (allosterically) involved in binding of stilbene disulfonates.

The effect of anion transport inhibitors on the irreversible exofacial PG modification was examined by Wieth et al. (1982b). Only DNDS, salicylate, and trinitrocresolate appear to protect against irreversible PG inactivation.

The competitive anion transport inhibitor DNDS was found to be only a weak inhibitor of irreversible PG inactivation (Wieth et al., 1982b), making the PG binding less selective (Bjerrum et al., 1983). The DNDS concentration corresponding to half-maximal protection against PG inactivation can be calculated to be about 50-fold higher than that for half-maximal transport inhibition, under the same alkaline conditions, using a value for K_I estimated from Wieth and Bjerrum (1983). This difference in effect may suggest that DNDS only interferes with PG inactivation allosterically. However, another interpretation can be arrived at from the Dixon plots in Fig. 3, which demonstrates the mutual influence of reversible DNDS- and PG-binding on chloride transport. The position (to the left and below the abscissa) of the intersection points between the line obtained without PG in the medium and the two lines with PG present, appear to demonstrate: first, that DNDS and PG can be bound reversibly to the protein at the same time, and secondly, that DNDS binding is reduced when the transport protein is saturated with reversibly bound PG. Determination of the reduction in DNDS affinity is complicated by the positive cooperativity in reversible PG binding. The apparent reduction at 15 m*M* PG can nevertheless be roughly estimated from the intersection with the abscissa to be 15–30-fold. This result provides a simple explanation for the weak DNDS inhibition of irreversible PG inactivation.

It was reported by Wieth et al. (1982b) that DNDS (2 m*M*) has no

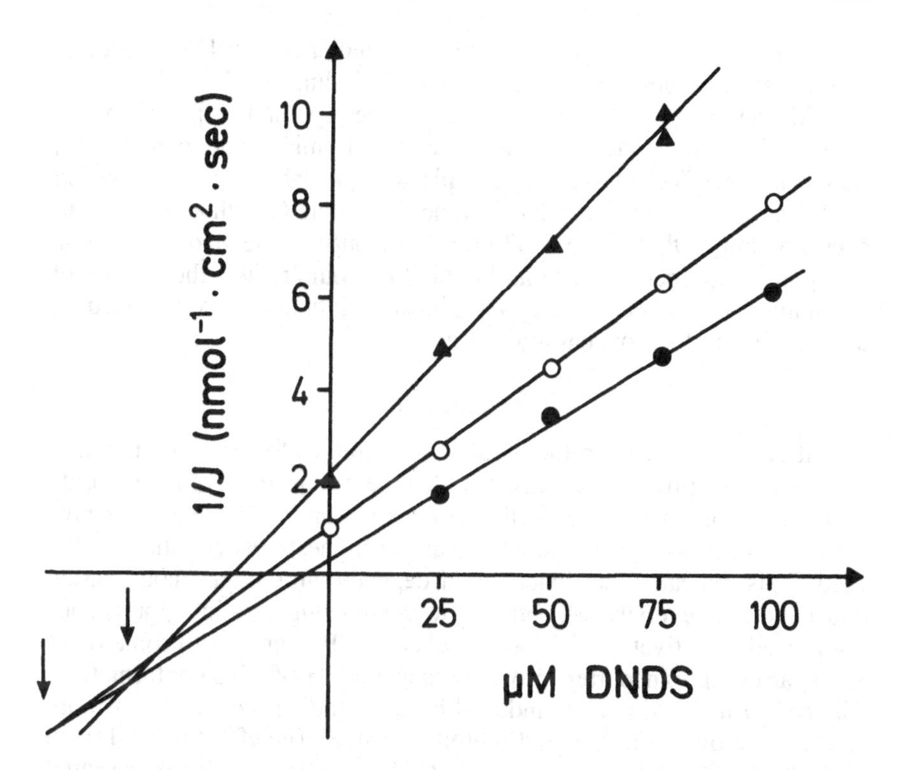

Fig. 3. Dixon plots of the chloride self-exchange flux at different concentrations of DNDS and PG. The ^{36}Cl-efflux from resealed membranes (prepared as described by Bjerrum et al., 1983) and containing 165 m*M* KCl was measured for 20 s in an efflux medium: 16 m*M* NaCl, 22 m*M* sodium citrate, 180 m*M* sucrose, 1 m*M* CHES*, pH 9.0, 25°C at various concentrations of DNDS, using the Millipore-Swinnex technique (Dalmark and Wieth, 1972). PG concentrations: (●) O m*M*, (○) 15 m*M*, and (▲) 25 m*M*. The regression lines intersect to the left below the abscissa. The concommitant irreversible PG inactivation ($t_{1/2}$ approximately 180 s, as estimated from Wieth et al., 1982b) can be neglected because of the short efflux time.

protecting effect against exofacial PG inactivation at alkaline pH and high extracellular chloride concentration (165 m*M*). This loss of effect is apparently the result of chloride competition with DNDS binding. In order to observe DNDS protection in 165 m*M* KCl, higher DNDS concentrations are needed. The value for half-maximal DNDS protection under these conditions was found to be 11 m*M* (Bjerrum, unpublished). The apparent decrease in DNDS affinity at high chloride concentration parallels the reduction in the apparent DNDS affinity of untreated cells with increasing chloride concentration, and indicates that DNDS protection is owing to binding to the "substrate binding site."

All the other inhibitors examined, namely phloretin, phlorizin, dipyridamol, tetrathionate, niflumic acid, furosemide, and bumetanide, were without effect at the alkaline pH and the inhibitor concentration used. The lack of protection by the other inhibitors (Wieth et al., 1982b) does not imply that these inhibitors bind outside the exofacial anion transport site proper. Reversible PG binding could reduce the affinity of the inhibitors so much that they only have a slight effect on the inactivation process at the concentration used.

3.5. Summary

It is concluded from the modification studies that the PG-modified groups and the titratable groups (the alkaline groups involved in chloride transport) appear to be identical. PG inactivation is a second-order process, showing no sign of saturation at the highest concentration of PG used. This indicates that inactivation depends on a single modification event and that reversible PG binding to the reacting arginine groups does not precede inactivation. However, reversible PG binding to some other site(s) arrest all transporters probably in an outward-facing conformation. The conformational change induced by this binding reduces the affinity for the specific anion transport inhibitor DNDS. The effect of DNDS on irreversible PG inactivation demonstrates that the modified essential groups appear to be located close to the exofacial inhibitor domain.

4. Selectivity of Arginine Modification

DNDS at high concentrations (2 m*M*) effectively protects against irreversible PG inactivation when modifications are performed from both sides of the membrane at pH 8.3 in a low-chloride medium. Under these conditions, approximately one-third of all the arginine residues in the protein can be modified without serious effect on anion translocation (Wieth and Bjerrum, 1983; Bjerrum, 1989). A very small difference in covalent binding, i.e., only a few PG molecules/band 3 protein, was observed with and without DNDS present. These results imply that only very few arginine groups are essential for anion translocation and that PG inactivation does not take place from the intracellular side of the membrane, when DNDS is bound.

Direct measurements of ^{14}C-phenylglyoxal incorporation in capnophorin show that approximately six PG molecules are incorporated/

transport molecule when the reaction proceeds at both sides of the membrane under optimum conditions (Zaki, 1984). Inactivation from the extracellular side is even more selective. Extrapolation to full inhibition indicates that only 2–3 arginyl residues/band 3 protein are modified at high extracellular chloride concentrations (assuming a stoichiometry of 2 PG/arginine residue). This number is reduced to only one single arginine, when inactivation is performed at low extracellular chloride concentration, as demonstrated by Bjerrum et al. (1983)

4.1. Localization of the Exofacial Essential Arginine Groups

Membrane-bound capnophorin can by cleaved by various enzymes into a number of well-defined peptide fragments (for a detailed survey of the peptide mapping studies, *see* Passow [1986]. Capnophorin is thus cleaved by extracellular chymotrypsin to produce a 35-kdalton C-terminal and a 60-kdalton *N*-terminal fragment. When extracellularly selectively modified resealed membranes are exposed to extracellular chymotrypsin treatment, most of the ^{14}C-phenylglyoxal is located in the C-terminal 35-kdalton chymotryptic fragment (Bjerrum et al., 1983). This fragment appears as a diffused band in SDS-polyacrylamide gel electrophoresis as a result of its heterogeneous carbohydrate content (Yu and Steck, 1975; Mueller et al., 1979; Golovtchenko-Matsumoto and Osawa, 1980). The carbohydrate of the band 3 molecule can be removed before electrophoresis with a suitable glycosidase (Fukuda et al., 1979; Jennings et al., 1984). In the experiment shown in Fig. 4, the glycans was removed with endo- β -*N*-acetylglucosaminidase F (Endoglycosidase F) isolated by Elder and Alexander (1982). This endoglycosidase cleaves both high mannose and complex types of glycans linked through asparagine to the protein backbone. After endoglycosidase treatment, about two-thirds of the ^{14}C-activity in the 35-kdalton fragment was distinctly located at the 32-kdalton position of the carbohydrate-deprived chrymotryptic fragment (Fig. 4). It is not clear whether this lower content in the carbohydrate-free C-terminal chymotryptic fragment indicates that PG binds with a 1:1 rather than a 2:1 stoichiometry or whether the low content is because of incomplete removal of carbohydrate from the band 3 protein. The results imply that the 35-kdalton chymotryptic fragment contains at least one essential arginine residue. However, more arginines may be of importance, since at least two arginines in the 35-kdalton chymotrypsin fragment were modified under the less selective conditions at high extracellular chloride (Bjerrum et al., 1983). Papain cleavage of the chymotrypsin-treated membranes demonstrates that the activity follows the 28-kdalton

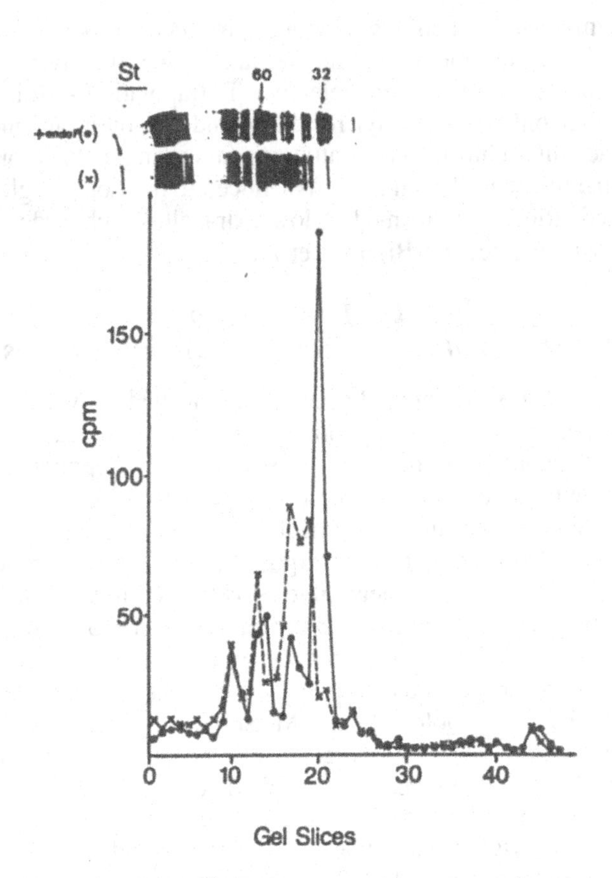

Fig. 4. Distribution of ^{14}C-PG in red-cell membrane proteins after extracellular chymotrypsin cleavage of capnophorin. The selective exofacial modification was performed for 8.8 s in an isotonic sucrose citrate medium containing 20 m*M* PG, 8 m*M* KCl, 5 m*M* CAPS*, pH 10.3, 25°C (as described by Bjerrum et al., 1983). ^{14}C-labeled membrane material (300 μg) was solubilized in 1% Triton X-100, pH 6.0, and incubated for 20 h at room temperature with two units of endoglycosidase F. The membrane material was after incubation solubilized by SDS and used for gel electrophoresis as described by Bjerrum (1989). The stained gels with (•) and without (X) endoglycosidase treatment are shown in the upper part of the figure. These gels and the counted gels demonstrate that the diffuse 35-kdalton fragment becomes a distinct fragment with a position of 32-kdalton after endoglycosidase F treatment. Based on the inhibition (69%) and binding of ^{14}C-PG (1.8×10^6 mol/cell) and the fraction of PG in the 32-kdalton fragment (37%), the binding at maximal inhibition (90%) was determined to be about 1 mol of PG/32-kdalton fragment mol.

papain fragment, and that PG inhibition and the inhibition following papain treatment are additive (Bjerrum 1989).

Degradation of the transport protein from both sides of the membrane with chymotrypsin demonstrates that about 50% of the ^{14}C-PG ends up in a diffuse 9-kdalton membrane-bound peptide fragment. The localization of this peptide fragment in the 28-kdalton papain fragment has been proposed to involve the stretch of amino acids from about 770 to 830, including the long hydrophobic sequence from 779 to 823 (Bjerrum, 1989).

4.2. Modification from the Intracellular Side

A maximum of only 40% inactivation can apparently be obtained when modification of the anion transporter is performed selectively from the intracellular side of the membrane at low intracellular chloride (5 m*M*), high intracellular pH (9.6), and neutral extracellular pH (Bjerrum, 1989). The accompanying binding of PG under these conditions is 15–20 $\times 10^6$ PG molecules/cell and, thus, not very selective. About one-quarter of these molecules were located in band 3, equally distributed between the 35-kdalton and the 60-kdalton extracellular chymotryptic fragments. The major fraction of ^{14}C-phenylglyoxal in the 35-kdalton segment appears to be located in the 10-kdalton peptic fragment (Bjerrum, 1983). This fragment is nearly identical to the 7-kdalton papain fragment (Jennings et al., 1984).

Since only partial inactivation can be obtained from the intracellular side (Bjerrum, 1989), it is not possible to establish whether this inactivation at alkaline intracellular pH is the result of a specific reaction with one single group in the protein, apparently reducing the transport capacity by 40%, or whether this inactivation includes several different arginyl residues that may be hidden and unreactive at neutral pH. In favor of the last hypothesis is the observation that the kinetics of inactivation are second-order when resealed membranes are subjected to reaction from both sides of the membrane under symmetrical conditions and at neutral pH (Zaki and Julian, 1985; Bjerrum, 1989). Such simple reaction kinetics should not be expected if inactivation proceeds from both sides of the membrane, the intracellular modification giving only about half-maximal inhibition. These arguments therefore support the suggestion (Bjerrum, 1989) that modification of functionally essential arginyl residues takes place only from the exofacial side of the membrane.

4.3. Summary

Anion translocation is inactivated by modification of only one single exofacial arginine residue per band 3 protein. This group, which is in the

28-kdalton extracellular papain fragment, is likely to be located in the stretch of amino-acid residues from about 770 to 830.

Under conditions of DNDS protection of the substrate "site," as many as one-third of all arginines can be modified in the protein without allosteric effects on substrate binding and translocation. This result indicates that the molecular translocation mechanism containing the functional arginine group(s) apparently is located deep inside the transport protein molecule and that the conformational changes of this part of the protein during the translocation process is rather insensitive to modification of arginines on the "surface" of the protein.

5. Modification of Carboxyl Groups

Carboxyl groups in proteins can be modified with various chemical reagents such as *p*-bromophenylbromide (Takahasi et al., 1967), diazo compounds (Wilcox, 1967, 1972), the carbodiimides (Hoare and Koshland, 1967; Carraway and Koshland, 1972), and the trimethyloxionium fluoroborate salts (Paterson and Knowles, 1972) and the related reagent for activation of carboxyl groups for synthetic purposes known as Woodward's reagent K, *N*-ethyl-5-phenylisoxazolium-3-sulfonate (Woodward et al., 1961; Woodward and Olofson, 1966). The water-soluble carbodiimide reagents, introduced by Hoare and Koshland (1966) are the most widely used reagents for studying functional carboxyl groups.

None of the various reagents is completely selective. Thus, Woodward's reagent K can also react with lysine groups (Bodlaender et al., 1969), and carbodiimides react readily with the hydroxyl group of tyrosine residues (Carraway and Koshland, 1968). Carbodiimides have also been reported to react with an active site serine in α-chymotrypsin (Banks et al., 1969) as well as with an active site cysteinyl group in papain (Perfetti et al., 1976). The reaction with essential serine and tyrosine residues proceeds faster with increasing pH. Such pH dependencies are inconsistent with carboxyl group modification, which requires a protonated carboxyl group for the reaction to proceed (Lundblad and Noyes, 1984). Besides pH-dependence of the reaction, stimulation of inactivation in the presence of a suitable nucleophile is another indication of reaction with a carboxyl group (Hoare and Koshland, 1967).

Carbodiimide also reacts with groups other than amino-acid residues, such as various acids, including the weak acid buffers containing phosphate and carboxyl groups. The reactivity towards these groups depends on the particular carbodiimide used, and on the accessibility and reactivity (nucleophilicity) of the involved groups (Khorana, 1953, 1961;

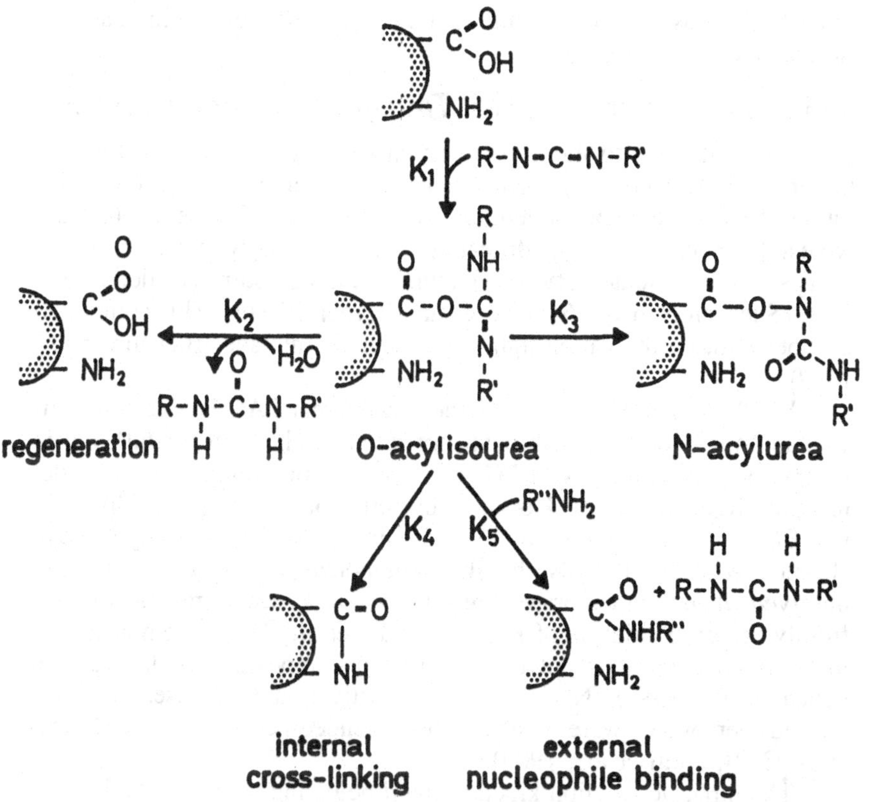

Fig. 5. The chemical reaction between carbodiimide reagents and carboxyl groups in accordance with Hoare and Koshland (1967). The reaction scheme adopted from Pedemonte and Kaplan (1986) is explained in more detail in the text.

Smith et al., 1958; Kurzer and Douraghi-Zadeh, 1967; George and Borders, 1979).

The proposed chemical reaction between carbodiimide reagents (Hoare and Koshland, 1967) and carboxyl groups is demonstrated in Fig. 5. The figure shows that the carbodiimide reaction is initiated by formation of an unstable *O*-acylisourea by reaction with a protonated carboxyl group from the protein (K_1). The reaction then follows several routes. The carboxyl group can be regenerated by hydrolysis of the carbodiimide adduct (K_2). The *O*-acylisourea can rearrange to a stable *N*-acylurea (K_3) or react with a nearby endogenous nucleophile (most likely a lysine group) forming an internal crosslink (K_4). This reaction can also take place with an external nucleophile having access to the reaction region

(K_5). In that case, the carbodiimide molecule itself will be eliminated as an inactive urea compound.

5.1. Inactivation of Anion Transport by Carbodiimides

Early attempts to modify the erythrocyte membrane with the carbodiimide EDC were only partially successful, demonstrating some inhibition of sulfate transport (Deuticke, 1977). Later, Wieth et al. (1982a) were able to demonstrate modification of extracellularly located carboxyl groups using a quaternary alkyl ammonium carbodiimide derivative, EAC (Sheehan et al., 1961; George and Borders, 1979). This compound cannot permeate the cation-tight erythrocyte membrane (Bjerrum et al., 1989).

Wieth et al. (1982a) have demonstrated that EAC is able to inactivate anion transport in erythrocytes or resealed ghosts at pH 6.0, but in contrast to inactivation with PG, this inactivation is not a second-order reaction. Reduction to 50% of the transport capacity occurs rapidly at 30 m*M* EAC, whereas inhibition of the remaining transport capacity is much slower (*see* Fig. 6). Only the slow component of the modification is observed when the membranes are subjected to a second treatment with a freshly prepared solution of EAC (Bjerrum et al., 1989), demonstrating that this component is not the result of decomposition of the reagent (Chan et al., 1988). No reversal of inactivation was observed when reacted membranes were incubated for extended periods of time (hours) at 38°C (Bjerrum et al., 1989).

The biphasic reaction kinetics could be owing to steric or allosteric interactions between adjacent protomers of the dimeric transport molecules when binding the rather bulky carbodiimide molecule, as suggested by Wieth et al. (1982a) and Andersen et al. (1983). This suggestion appears unlikely for several reasons (Bjerrum et al., 1989). It was found that the fraction of rapidly reacting transporters was strongly dependent on the concentration of EAC used. This fraction is about 30% at 12 m*M* EAC, increasing to about 75% at 80 m*M* EAC. The level of inactivation obtained by the rapid phase seems to be directly dependent on the free EAC concentration in the medium, and increases when extra EAC is added during inactivation. Partial transport inactivation with covalent bound DIDS (which appear to be randomly incorporated between protomers of the transport protein) performed before EAC inactivation has no influence on the pattern of EAC inactivation of the remaining transport molecules. These results make cooperative interaction between the adjacent protomers unlikely, but do not exclude the possibility of random modification of several carboxyl groups within the single anion transport molecule as an explanation for the biphasic reaction kinetics. Some of the

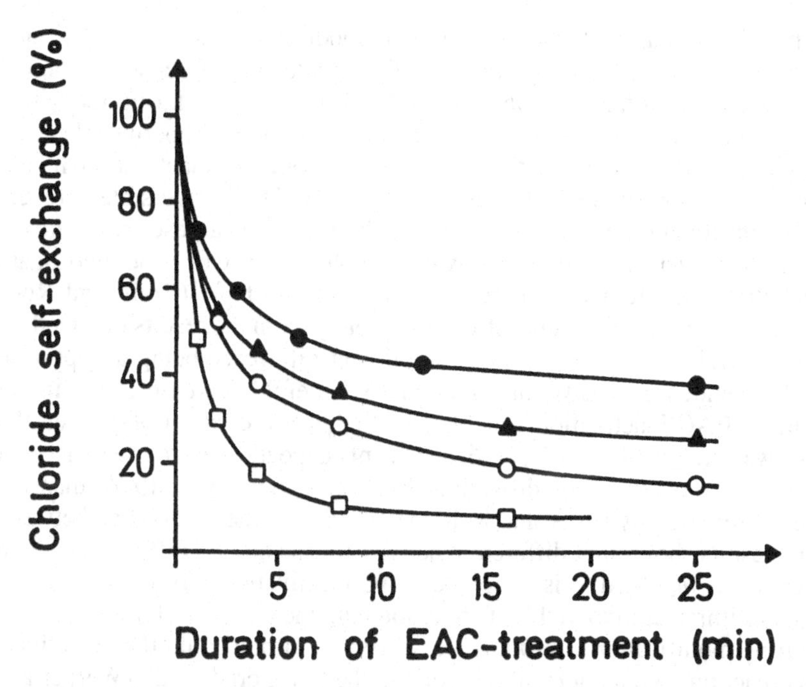

Fig. 6. Time course of anion transport inactivation by EAC in the presence of the reversible anion transport inhibitor DNDS. Modification and ^{36}Cl-efflux was performed as described by Bjerrum et al., 1989 (method B). EAC reaction medium: 165 m*M* NaCl, 30 m*M* EAC, 10 m*M* MES*, pH 5.8, 38°C. Final DNDS concentrations (m*M*) after addition of packed resealed membranes: 0 (●), 0.17 (▲), 1.7 (○), and 17 (□). The inactivation closely obeys pseudo first-order kinetics at the highest DNDS concentration.

modified groups could be essential for anion transport, and some could protect against inactivation when modified (Bjerrum et al., 1989).

Biphasic reaction kinetics were also observed by Perfetti et al. (1976), for inactivation of papain by the carbodiimide EDC. These authors suggest that the initial fast inactivation of papain corresponds to the build-up by the route K_1 (Fig. 5), reaching a maximum when its pseudo first-order rate of formation is equal to that of disappearance via the K_2 and K_3 paths. The subsequent, slower inactivation was suggested to proceed by the slow irreversible formation of the *N*-acylurea derivative by the K_3 step. This explanation, however, cannot be used directly for EAC modification of capnophorin because inactivation of this protein is irreversible, even after sudden removal of EAC (Bjerrum et al., 1989).

Craik and Reithmeier (1984, 1985) examined carbodiimide inactivation of the anion transporter in red cells, but at slightly higher pH (7).

They found that inactivation with the carbodiimides CMC and EDC also proceeds by a biphasic reaction. A slight protecting effect was obtained when the anion transport inhibitor DNDS was present during modification. Inactivation performed with the membrane-impermeable carbodiimide EAC showed the same pattern, but the inactivation in the presence of phosphate and/or chloride was found to be reversible. Irreversible inactivation was found only in a chloride-free sucrose-citrate medium. Moreover, the rate of inactivation in the presence of either phosphate or citrate was found to increase as a function of pH (Craik and Reithmeier, 1985). This result differs completely from the results obtained by Wieth and coworkers, the most significant difference being an opposite pH-dependence of EAC inactivation. A threefold decrease in the initial rate of EAC inactivation was observed (Bjerrum et al., 1989) when the pH was increased from 5.8 to 6.8. This pH dependence is more consistent with carboxyl group modification (Lundblad and Noyes, 1984) than the result reported by Craik and Reithmeier (1985). The divergence between the results shows that different reactive groups and/or a different reaction scheme is involved: it is likely that EAC inactivation performed by Craik and Reithmeier involved EAC-reaction with the citrate buffer ions present during modification. Carbodiimide-activated citrate could then inactivate by reaction with a nucleophile group in the transport domain (Werner and Reithmeier, 1988).

5.2. Stimulation by Nucleophile Reaction

The nucleophile TEE stimulates the initial EAC modification by a factor of about 2 when the anion transport protein is modified at 25–30 m*M* EAC and 40 m*M* TEE, pH 5.8, 38°C (Bjerrum et al., 1989). This stimulation indicates that the EAC reaction involves carboxyl group modification because only carbodiimide-activated carboxyl groups are likely to react with a nucleophile (Hoare and Koshland, 1967; Lundblad and Noyes, 1984). The polar nucleophile GME has, in contrast to the aromatic TEE, no stimulating effect on EAC inactivation (Bjerrum et al., 1989). Similar effects of the two nucleophilic reagents were reported by Pho et al. (1977) in modification of yeast hexokinase and probably indicate that the reacting carboxyl group is located in a nonpolar environment.

5.3. Inactivation and Stilbene Disulfonate Binding

From experiments where EAC-modified membranes were reacted with DIDS, it could be concluded that the reaction with EAC alone appears to divide the transport molecules into at least two EAC-modified species. Since the different species can react covalently with DIDS, it could also be deduced that the extracellular "anion binding sites" are left essentially intact after EAC modification (Bjerrum et al., 1989).

When the reversible inhibitor of anion transport DNDS was present in the EAC-reaction medium, the rate of inactivation was found surprisingly to increase. DNDS, which is not a nucleophile, increases the rate of initial inactivation as well as the maximum level of inactivation achieved (Fig. 6). At the higher DNDS concentration (>2 m*M*), a maximum of approximately 90% inhibition was obtained. Inactivation with EAC at very high DNDS concentrations (>20 m*M*) displays nearly pseudo first-order kinetics, suggesting that the inactivation, which is likely to involve several carboxyl groups, now depends only on one single homogeneous modification event (Bjerrum et al., 1989).

The stimulation of inactivation obtained when DNDS is present (Fig. 6) indicates that the reacting group cannot be part of the exofacial anion binding site, at least not when DNDS is bound. Also, the stimulation cannot be analogous to the syncatalytic modification of aspartate aminotransferase (Christen and Riordan, 1970) owing to the high concentration of inhibitor needed. The stimulation is thus unlikely to be correlated with conformational changes induced by specific DNDS binding to the transport site. A possible explanation for the activation of the reaction might be that the negatively charged DNDS acta as a catalyst for the EAC-carboxyl interactions. DNDS could also bind to a low affinity DNDS binding site, changing the conformation of the transport protein or the surface charge density of the protein so that EAC has better access for reaction with a functional carboxyl group (Bjerrum et al., 1989).

5.4. Extracellular Titration After Modification

Complete inhibition of the transport system with EAC is difficult to obtain, because inactivation at high concentrations of EAC cannot be performed for extended periods of time without hemolysis of the cells (Wieth et al., 1982a). Maximal inactivation can, on the other hand, be obtained when DNDS is present in the medium. The smallest residual transport obtained under these conditions is about 10%. The residual flux that can be inhibited by DNDS is not the result of 10% unreacted transporters, but appears to correspond to a modified transport system (Bjerrum et al., 1989).

Titration studies of "maximal" EAC-inactivated resealed membranes demonstrate that both the exofacial titratable groups with a pK around 5 (Fig. 7), and the intracellular titratable groups with a pK around

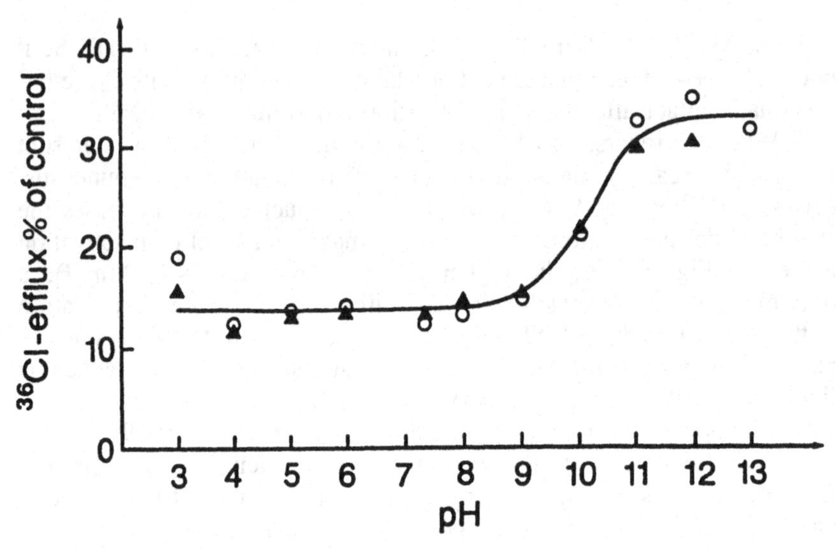

Fig. 7. Chloride transport of "maximally" EAC-treated ghosts (90% inhibition) as a function of the extracellular pH. The EAC modification and ^{36}Cl-efflux were performed as described by Bjerrum et al. (1989). The modification was performed for 30–40 min in: 165 m*M* KCl, 2 m*M* DNDS, 10 m*M* MES, pH 5.8, 38°C. The EAC concentrations after addition of the packed resealed membranes was 30 m*M* (▲) and 48 m*M* (○). The acidic titratable groups with a p*K* around 5 have apparently disappeared. A moderate transport stimulation is seen at alkaline pH. This effect is not discussed in the text.

6 have disappeared (Bjerrum et al., 1989). This alteration of the titration "profile" is not because of major changes in the extracellular "chloride binding site," since the chloride and DNDS affinities of EAC-modified membranes were found to be very similar to the affinities observed for untreated membranes. It is therefore put forward that EAC modifies the extracellular titratable group, and that this modification apparently abolishes extracellular and intracellular pH-titration and reduces anion transport to about 10% of its normal value. Since it is unlikely that the modified group is related to the "anion binding site," it is proposed (Bjerrum et al., 1989) that the modification only influences anion exchange allosterically and that the groups could be directly involved in the cotransport of protons across the membrane. This assumption is in accordance with the suggestion by Milanick and Gunn (1986) that there can exist asyncrony between anion and proton movements across the membrane.

5.5. Summary

Anion transport in human red-cell membranes can be inactivated with carbodiimides. The biphasic inactivation is stimulated by the hydrophobic nucleophile TEE. This indicates that inactivation is a result of carboxyl group modification. Inactivation is enhanced by high concentrations of the reversible inhibitor of anion exchange DNDS (to a maximum of 10% residual flux). The extracellular "substrate binding site" was found to be essentially intact after complete carbodiimide inactivation in the presence of DNDS, indicating that groups outside the inhibitor binding region were modified. Modification of these groups makes the anion exchange insensitive to pH in the range from 4–7.

6. Identification of Functional Important Carboxyl Group(s)

6.1. Carbodiimide-Modified Carboxyl Groups

The aromatic nucleophile TEE stimulates EAC-inactivation of the anion transport protein. The stimulation of inactivation is accompanied by incorporation of ^{3}H-TEE into capnophorin. Only a few percent of the total TEE incorporated, or approximately 10^6 molecules of TEE, were found to be located in capnophorin of trypsin-treated membranes on extrapolation to full inhibition. A larger fraction was located in the PAS-positive glycoproteins and a very large fraction at the position of the glycolipids in the gel (Bjerrum et al., 1989). The small amount of ^{3}H-TEE bound to the transport protein indicates that TEE incorporation appears to be a rather selective reaction.

When ^{3}H-TEE-modified membranes were exposed to extracellular chymotrypsin, the major fraction of the ^{3}H-TEE activity at the band 3 position was moved to the position of the diffuse C-terminal 35-kdalton-chymotryptic fragment (Bjerrum, 1983). (Since this reported experiment was not performed with trypsin-treated membranes, some of the activity at the position of the 35-kdalton fragment may correspond to labeling of PAS I glycoproteins degraded by chymotrypsin). After papain treatment the major part of the activity at the position of the 35-kdalton fragment was removed without reappearing at the position of the 28-kdalton papain fragment (Bjerrum, 1983). A functional carboxyl group could therefore be located in the "released" 7-kdalton papain fragment. This fragment has been characterized by Jennings et al. (1984).

Approximately one EAC molecule is incorporated into capnophorin when inactivation is performed with ^{14}C-EAC in the presence of DNDS at high concentration. DNDS increases the rapid inactivation and also the amount of ^{14}C-EAC taken up by band 3 as compared to the total incorporation in the membrane. Nearly all the ^{14}C-EAC activity in capnophorin is found in the 60-kdalton chymotrypsin fragment (Fig. 8) after extracellular chymotrypsin treatment (Bjerrum et al., 1989).

6.2. Modification with Woodward's Reagent K

In addition to the above-mentioned carboxyl groups, still further exofacial carboxyl groups are of importance for anion transport, as demonstrated by Jennings and Anderson (1987) and Jennings and Al-Rhaiyel (1988) using the negatively charged carboxyl group modifying reagent known as Woodward's reagent K. Inactivation with this reagent followed by reduction with $^{3}H\text{-}BH_4^-$ has shown that two glutamate residues are labeled in capnophorin, one in the 60-kdalton and the other in the 35-kdalton chymotrypsin fragments. Both of these groups in contrast to the EAC-modified groups can be protected by stilbene disulfonates and are thus likely to be located close to the anion binding site. The different modification pattern of EAC and Woodward's reagent K (as denoted by Jennings and Anderson) could be a consequence of the opposite charges of these two reagents.

6.3. Summary

Carboxyl groups both in the 60-kdalton and the 35-kdalton chymotrypsin fragments are modified with EAC. It is not known whether the groups identified by ^{3}H-TEE or ^{14}C-EAC incorporation both represent different functionally important carboxyl groups located in different parts of the protein or whether only one of them is responsible for inactivation. Modification of the two groups could also be responsible for the biphasic inactivation kinetics. EAC may react randomly with the carboxyl groups, one of which could be functional and the other a group that protects the transport molecule against inactivation when modified.

7. Transport Pathway Models

The primary sequence of murine capnophorin is 929 amino acids long, and the homology with known sequences of human capnophorin is very high especially with the membrane-bound sequences, indicating a functional role of this part of the molecule (Brock et al., 1983; Brock and Tanner, 1986). Analysis of the hydrophobicity of the membrane-bound

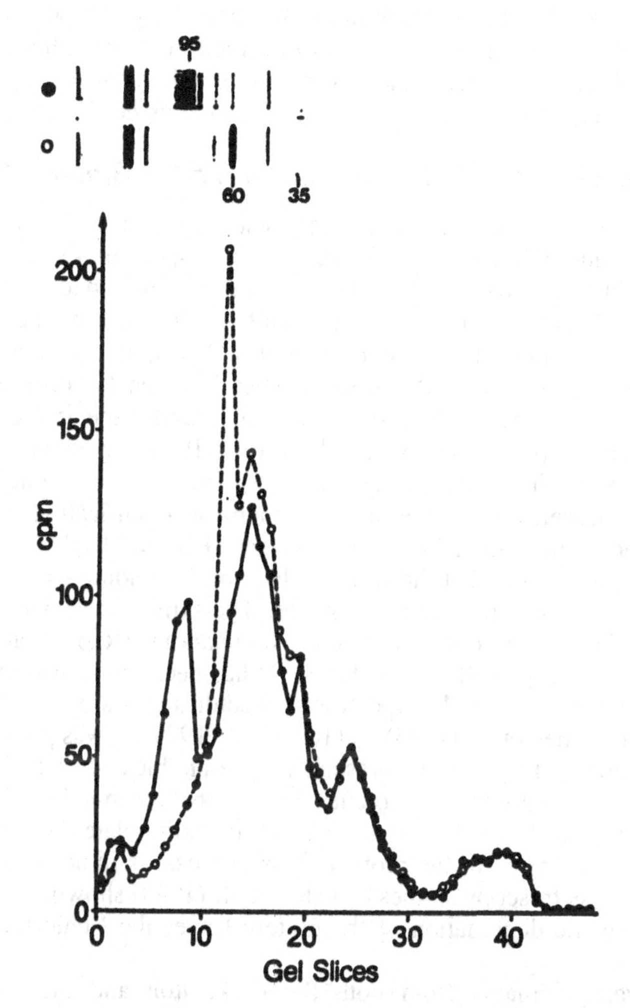

Fig. 8. Distribution of ^{14}C-EAC in red-cell membrane proteins before and after chymotrypsin cleavage. Extracellular trypsin-treated resealed ghosts were reacted for 2.5 min in: 165 m*M* KCl, 10 m*M* MES, 36 m*M* EAC, 2 m*M* DNDS, pH 5.8, 38°C (experimental as described by Bjerrum et al., 1989). The chloride self-exchange was inhibited 55%, with 4.1 × 10^6 EAC mol bound/cell. The distribution of ^{14}C activity in the membrane proteins before (●) and after (○) chymotrypsin treatment demonstrates that nearly all the activity located in band 3 (17.5%, corresponding to 1.2 EAC mol/transport mol when extrapolating to full inhibition) was moved to a position of 60 kdaltons after chymotrypsin treatment.

fractions of the molecule, using the hydrophobicity plot analysis (Eisenberg, 1984) in combination with chemical modification studies and peptide mapping, has revealed that the membrane domain of the molecule spans the membrane at least eight times (Jennings et al., 1986).

7.1. The Exofacial Substrate and Inhibitor Binding "Sites"

The anion transport system is very selective towards different types of anions and inhibitors, particularly with respect to the number of charges, but also with respect to the size and hydrophobicity of the molecules (Barzilay et al. 1979; Knauf 1979; Cousin and Motais, 1982a,b). This indicates the existence of a ''substrate binding site'' or ''anion recognition site'' that discriminates between the different ions.

The ''substrate binding site,'' which is under the influence of a positive surface potential (Wieth and Bjerrum, 1982), apparently contains at least one positive-charged essential arginine as demonstrated by the alkaline extracellular titration and the chemical modification studies. It can moreover be deduced from the chloride, PG, and DNDS interaction studies reported here that the stilbene disulfonate binding site apparently overlaps, with the chloride binding site. The same conclusion has been obtained in several other inhibitor examinations (Rao et al., 1979; Fröhlich, 1982; Schnell et al., 1983) and has been confirmed in nuclear magnetic resonance (NMR) spectroscopy studies (Falke et al., 1984). In studies of Schnell et al. (1983) and Rao et al. (1979), it was proposed that the inhibitor binding region represents a hydrophobic cleft, which extends some distance into the membrane. The ''substrate binding site'' thus appears to be ''buried'' in the membrane and related to the trans-membrane segments of the protein. This proposal is in accordance with the NMR-spectroscopy studies of Falke et al. (1985) showing that extensive enzymatic degradation of the protein leaves the ''chloride binding site'' intact.

Several domains from both the 60-kdalton and the 35-kdalton chymotrypsin fragments form the basis for the stilbene disulfonate ''binding site'' or inhibitor ''binding pocket'' (Passow, 1986). The function of the ''pocket'' is apparently to create a local electrostatic field that attracts anions for the translocation process. The inhibitor binding region therefore contains a number of adjacent positively charged amino-acid residues, including the several lysine groups involved in stilbene disulfonate binding (Passow, 1986). Several of these groups interact directly or allosterically with the groups in the ''anion transport site,'' but they also function as extra attachment points increasing the selectivity and affinity for the different overlapping inhibitors that can bind in the ''pocket''

(Fröhlich and Gunn, 1987; Falke et al., 1984). In contrast to the extracellular anion binding region, almost nothing is known about the internally facing orifice (Passow, 1986).

7.2. Translocation Models

Several mechanisms have been proposed to explain the kinetic behavior of the transport protein (Passow, 1986; Jay and Cantley, 1986). The initiating event in most models is that the negative anion binds to a positive functional essential group in the protein, either by breaking a salt-bridge or by direct interaction with the free positive group destabilizing the local conformation of the protein.

In the zipper mechanism proposed by Wieth et al. (1982a) and by Brock et al. (1983), it is suggested that the penetrating anion opens a series of salt-bridges between negative and positive groups positioned in the transport path. The model appears less probable after the elucidation of the amino-acid sequence, because pairing of positive and negative charges from different hydrophobic segments is not obvious from the distribution of charged groups in the hydrophobic areas of the sequence (Fig. 9). A more plausible way of moving the anion across the membrane is therefore as proposed by the "alternating binding site model" (Knauf, 1979; Knauf and Law, 1980; Bjerrum, 1986, 1989). Assuming this model, it is not known whether the "binding site complex" containing the anion is directly exposed to the other side of the membrane or whether the anion is transferred via other amino-acid groups located in relation to the transport path. It is not necessary, however, in any case to assume that the anion is transported the whole distance across the membrane, it only needs to be moved a few Å to overcome the diffusion barrier (Läuger, 1980).

7.3. Possible Location and Function of the Modified Groups

A schematic view of the primary structure of capnophorin is shown in Fig. 9. According to the scheme, adapted from Kopito and Lodish (1985a) and Passow (1986), the exofacial modified arginine could be arginine number 664, 675, 748, 778, or 800. Arginines 664 and 675 are located in a segment confined to the extracellular side of the membrane (Passow, 1986; Jay and Cantley, 1986). In contrast, arginines 748 and 778 have recently been shown to be located differently (Jennings et al., 1986). The last arginine, number 800, has a special residence. Binding and/or translocation could involve this arginine, which has been proposed to be the essential arginine in the 9-kdalton chymotrypsin fragment (Bjerrum, 1989).

The arginine is placed close to the middle of a hydrophobic segment (779–823), which is just long enough to span the membrane twice in α-helical conformation. The arginine can be related to the exofacial side of the membrane, but this configuration has no preference over one in which the arginine is exposed to the intracellular side (Fig. 9). This is so, even if the two ends of the segment derive from the intracellular side of the membrane (Jennings et al., 1986). The hydrophobic segment that is highly conserved between mouse and humans (Brock and Tanner, 1986) includes five glycine and two proline residues. These residues, regarded as α-helix breaks in hydrophilic proteins, are arranged orderly in the segment. Arginine 800 is placed between the two prolines, namely, residues 796 and 802. Proline residues give much less rotational freedom to peptide chains than, e.g., glycine residues. They could be of importance for the stabilization and relative position of this part of the segment, and the conformational position of arginine 800. All other bending points are glycine groups, apparently rendering the hydrophobic sequence rather flexible and allowing it to undergo conformational changes. The possible loops and/or the helix amino terminus of these binding points moreover contains serine or threonine residues. Such α-helix ends, capable of hydrogen bond formation, have recently been suggested (Quiocho et al., 1987; Warshel, 1987) to provide enough highly localized partial countercharge to neutralize and stabilize charged groups localized in an otherwise hydrophobic interior of a protein.

Calculations and direct measurements of the electrostatic potentials in an enzyme active site (Gilson and Honig, 1987; Sternberg et al., 1987) have demonstrated that several charged amino-acid groups normally contribute to the local potential. It is therefore very likely, as suggested by Wieth and Bjerrum (1982) and Bjerrum (1983), that the p*K* shift or the change in electrostatic potential of the essential arginine residues observed at low extracellular chloride concentrations is the result of electrostatic interaction from one or a few other charged groups. In this connection, it is noticeable that the extracellular alkaline titration (including the modifier stimulation) at various chloride gradients across the membrane can be accounted for quantitatively by a simple ping-pong model in combination with a simple extracellular binding site model that involves only two titratable groups, one or both of which are arginine (Bjerrum, P. J., article in preparation).

The hydrophobic segment containing arginine 800 is proposed to interact with and be surrounded by the other membrane-spanning α-helices, forming the environment for the hydrophobic segment in the mem-

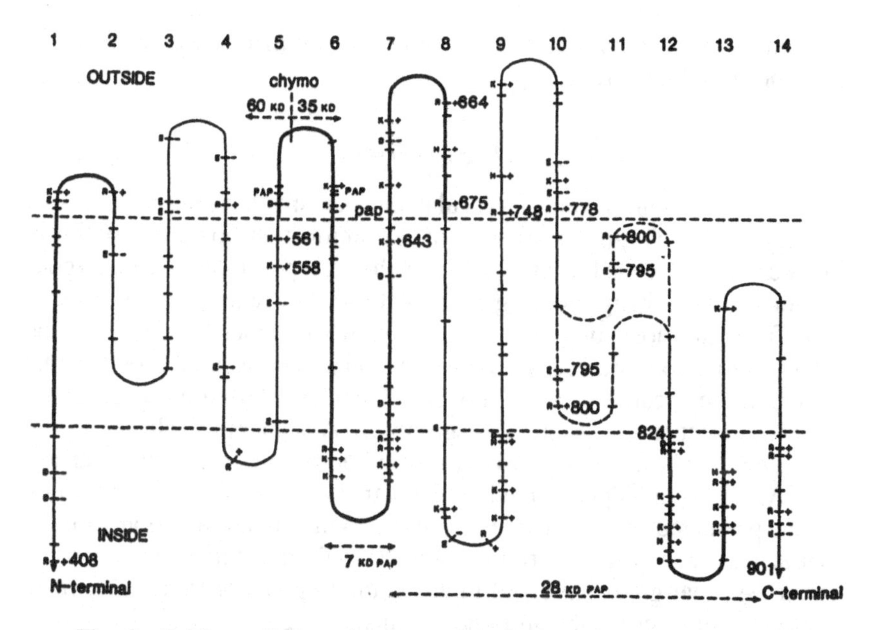

Fig. 9. Folding of the primary structure of murine capnophorin in the membrane bilayer (from Bjerrum, 1989). The dashed line shows the two possible configurations suggested by Bjerrum (1989). The scheme shows the location of the charged amino-acid residues in the C-terminal part of the molecule responsible for transport function. The exofacial cleavage points for chymotrypsin (chymo) and papain (pap) are shown, together with the primary binding site for covalently bound DIDS (lysine 558 or 561). The heavily drawn lines show the part of the sequence where extra- and intracellular localization has been established experimentally. R = arg, K = lys, H = his, D = asp, and E = glu. Prolines and glycines are also indicated by breaks in the sequence chain (–). For a detailed description of the sequence, *see* Passow (1986).

brane. This segment binding the anion could move by a dilatation of the space between the enfolding α-helices. This arrangement would be consistent with the finding of an extraordinarily large activation volume for the anion transporter, as recently reported by Canfield and Macey (1984). The energy needed for this expansion may derive from transient formation of hydrogen bonds or from ion pair formation between the charged group (795) in the hydrophobic segment containing arginine 800 and/or between groups in the enclosing α-helices. Some of these groups could be accessible for chemical modification. Such an arrangement would provide an explanation for the apparent small energy difference between the two different "binding site" conformations (Knauf et al., 1986). A model where the hydrophobic segments are assumed to be flexible enough to allow the arginine to move across the membrane has recently been proposed (Bjerrum, 1986).

7.4. Summary

In conclusion, several molecular mechanisms have been suggested to explain the ping-pong kinetics of the anion transport system. From kinetic studies, it has become clear that the transport molecule undergoes conformational changes during translocation of the anion. It is substantiated from titration studies and chemical modification studies that the transported anion interacts with positively charged essential group(s) that appear to be located in a hydrophobic "pocket." Binding of an anion alters the electric field around the group(s). This alteration allows a local conformational change to move the anion across the membrane barrier.

Although no documented molecular mechanism for translocation can be presented, it is concluded from the chemical modification with PG that at least one arginine group in the 28-kdalton papain fragment could be an essential group involved in anion binding and/or translocation. In contrast to the essential arginine groups, the EAC-modified carboxyl groups (probably located in the 60-kdalton fragment) have only allosteric influence on anion translocation. The groups are, however, likely to be directly involved in the cotransport of protons across the membrane.

Acknowledgments

A. Valsted is thanked for valued technical assistance, and P. A. Knauf for critical reading and constructive criticism of the manuscript. I am greateful to Ingeborg and Leo Dannins Foundation for Medical Research for support by a fellowship and to the NOVO foundation for a research grant.

References

Andersen, O. S., Bjerrum, P. J., Borders, C. L., Jr., Broda, T., and Wieth, J. O. (1983) Essential carboxyl groups in the anion exchange protein of human red blood cell membranes. *Biophysical J.* **41,** 164a.

Banks, T. E., Blossey, B. K., and Shafer, J. A. (1969) Inactivation of α-chymotrypsin by a water-soluble carbodiimide. *J. Biol. Chem.* **244,** 6323–6333.

Barzilay, M., Ship, S., and Cabantchik, Z. I. (1979) Anion transport in red blood cells I. Chemical properties of anion recognition sites as revealed by structure-activity relationships of aromatic sulfonic acids. *Membr. Biochem.* **2,** 227–254.

Berghout, A., Raida, M., Romano, L., and Passow, H. (1985) pH dependence of phosphate transport across the red blod cell membrane after modification by dansyl chloride. *Biochim. Biophys. Acta* **815,** 281–286.

Bjerrum, P. J. (1983) Identification and location of amino acid residues essential for anion transport in red cell membranes, in *Structure and Function of Membrane Proteins* (Quagliariello, E. and Palmieri F., eds.), Elsevier Science Publishers, Amsterdam, pp. 107–115.

Bjerrum, P. J. (1986) Titration and chemical modification of the anion transport system in red cells, in *Eight School on Biophysics of Membrane Transport*, School Proceedings, Vol. 1 Wroclaw, Poland, Publ. Co. Argicultural University, pp. 9–29.

Bjerrum, P. J. (1989) Chemical modification of the anion transport system with phenylglyoxal, (Fleicher, S. B.,) *Meth. Enzymol.* **173**, 466–494.

Bjerrum, P. J., Tranum-Jensen, J., and Møllgård, K. (1980) Morphology of erythrocyte membranes and their transport function following aggregation of membrane proteins, in *Membrane Transport in Erythrocytes*, (Lassen, U. V., Ussing, H. H., and Wieth, J. O., eds.), Munksgaard, Copenhagen (Alfred Benzon symposium 14), pp. 51–72.

Bjerrum, P. J., Wieth, J. O., and Borders, C. L., Jr. (1983) Selective phenylglyoxalation of functionally essential arginyl residues in the erythrocyte anion transport protein. *J. Gen. Physiol.* **81,** 453–484.

Bjerrum, P. J., Andersen, O. S., Borders, C. L., Jr., and Wieth, J. O. (1989) Functional carboxyl groups in the red cell anion transport protein. *J. Gen. Physiol.* **93,** 813–839.

Bodlaender, P., Feinstein, G., and Shaw, E. (1969) The use of *isoxazolium* salts for carboxyl group modification in proteins. *Biochemistry* **8,** 4941–4949.

Borders, C. L., Jr. and Riordan, J. R. (1975) An essential arginyl residue at the nucleotide binding site of creatine kinase. *Biochemistry* **14,** 4699–4704.

Brahm, J. (1977) Temperature-dependent changes of chloride transport kinetics in human red cells. *J. Gen. Physiol.* **70,** 283–306.

Brock, C. J. and Tanner, M. J. A. (1986) The human erythrocyte anion-transport protein. Further amino acid sequence from the integral membrane domain homologous with the murine protein. *Biochem. J.* **235,** 899–901.

Brock, C. J., Tanner, M. J. A., and Kempf, C. (1983) The human erythrocyte anion transport protein. Partial amino acid sequence, confirmation and possible molecular mechanism for anion exchange. *Biochem. J.* **213,** 577–586.

Cabantchik, Z. I. and Rothstein, A. (1974) Membrane protein related to anion permeability of human red blood cells I. Localization of disulfonic stilbene binding sites in proteins involved in permeation. *J. Memb. Biol.* **15,** 207–226.

Canfield, V. A. and Macey, R. I. (1984) Anion exchange in human erythrocytes has a large activation volume. *Biochim. Biophys. Acta* **778,** 379–384.

Carraway, K. L. and Koshland, D. E., Jr. (1968) Reaction of tyrosine residues in proteins with carbodiimide reagents. *Biochim. Biophys. Acta* **160,** 272–274.

Carraway, K. L. and Koshland, D. E., Jr. (1972) Carbodiimide modification of proteins. *Meth. Enzymol.* **25,** 616–632.

Chan, V. W. F., Jorgensen, A. M., and Borders, C. L., Jr. (1988) Inactivation of bovine thrombin by water-soluble carbodiimides: Evidence for an essential carboxyl group with pK_a of 5.51 *Biochem. Biophys. Res. Commun.* **151,** 709–716.

Cherung, S. T. and Fonda, M. L. (1979a) Kinetics of the inactivation of Escherichia coli glutamate apodecarboxylase by phenylglyoxal. *Arch. Biochem. Biophys.* **198,** 541–547.

Cherung, S. T. and Fonda, M. L. (1979b) Reaction of phenylglyoxal with arginine. The effect of buffers and pH. *Biochem. Biophys. Res. Commun.* **90,** 940–947.

Chiba, T., Sato, Y., and Suzuki, Y. (1986) Amino acid residues complexed with eosin 5-isothiocyanate in band 3 protein of the human erythrocyte. *Biochim. Biophys. Acta* **858,** 107–117.

Christen, P. and Riordan, J. F. (1970) Syncatalytic modification of a functional tyrosyl residue in aspartate aminotransferase. *Biochemistry* **9,** 3025–3033.

Cohn, E. J. and Edsall, J. T. (1943) *Proteins, Amino Acids and Peptides as Ions and Dipolar Ions* (Reinhold Publishing Co. New York).

Cousin, J. L. and Motais, R. (1982a) Inhibition of anion transport in the red blood cell by anionic amphiphilic compounds I. Determination of the flufenemate-binding site by proteolytic dissection of the band 3 protein. *Biochim. Biophys. Acta* **687,** 147–155.

Cousin, J. L. and Motais, R. (1982b) Inhibition of anion transport in red blood cell by anionic amphiphilic compounds II. chemical properties of the flufenemate-binding site on the band 3 protein. *Biochem. Biophys. Acta* **687,**156–164.

Craik, J. D. and Reithmeier, R. A. T. (1984) Inhibition of phosphate transport in human erythrocytes by water-soluble carbodiimides. *Biochim. Biophys. Acta* **778,** 429–434.

Craik, J. D. and Reithmeier, R. A. T. (1985) Reversible and irreversible inhibition of phosphate transport in human erythrocytes by a membrane impermeant carbodiimide. *J. Biol. Chem.* **260,** 2404–2408.

Craik, J. D., Grouden, K., and Reithmeier, R. A. T. (1986) Inhibition of phosphate transport in human erythrocytes by 7-chloro-4-nitrobenzo-2-oxa-1,3-diazole (NBD-Cl). *Biochim. Biophys. Acta* **856,** 602–609.

Dalmark, M. and Wieth, J. O. (1972) Temperature dependence of chloride, bromide, iodide, thiocyanate and salicylate transport in human red cells. *J. Physiol. (Lond)* **244,** 583–610.

Deuticke, B. (1977) Properties and structural basis of simple diffusion pathways in the erythrocyte membrane. *Rev. Physiol. Biochem. Pharmacol.* **78,** 1–97.

Eisenberg, D. (1984) Three-dimensional structure of membrane and surface proteins. *Ann. Rev. Biochem.* **53,** 595–623.

Elder, J. H. and Alexander, F. (1982) Endo-beta-*N*-acetylglucosaminidase F: Endoglycosidase from Flavobacterium meningosepticum that cleaves both high mannose and complex glucoproteins. *Proc. Natl. Acad. Sci. USA* **79,** 4540–4544.

Fairbanks, G., Steck, T. L., and Wallach, D. F. H. (1971) Electrophoretic analysis of the major polypeptides of the human erythrocyte membrane. *Biochemistry* **10,** 2606–2617.

Falke, J. J. and Chan, S. I. (1985) Evidence that anion transport by band 3 proceeds via a ping-pong mechanism involving a single transport site. A ^{35}Cl NMR study. *J. Biol. Chem.* **260,** 9537–9544.

Falke, J. J., Kanes, K. J., and Chan, S. I. (1985) The minimal structure containing the band 3 anion transport site. A ^{35}Cl NMR study. *J. Biol. Chem.* **260,** 13294–13303.

Falke, J. J., Pace, R. J., and Chan, S. I. (1984) Chloride binding to the anion transport binding sites of band 3. A ^{35}Cl NMR study. *J. Biol. Chem.* **259,** 6472–6480.

Frölich, O. (1982) The external anion binding site of the human erythrocyte anion transporter: DNDS binding and competition with chloride. *J. Memb. Biol.* **65,** 111–123.

Frölich, O. and Gunn, R. B. (1987) Interactions of inhibitors and anion transporter of human erythrocyte. *Am. J. Physiol.* **252,** C153–C162.

Fukuda, M. N., Fukuda, M., and Hakamori, S. (1979) Cell surface modification by endo-beta-galactosidase. *J. Biol. Chem.* **254,** 5458–5465.

Funder, J. and Wieth, J. O. (1976) Chloride transport in human erythrocytes and ghosts: a quantitative comparison. *J. Physiol* **262,** 679–698.

Funder, J., Tosteson, D. C., and Wieth, J. O. (1978) Effects of bicarbonate on lithium transport in human red cells. *J. Gen. Physiol.* **71,** 721–746.

Furuya, W., Tarshis, T. Few, F. Y., and Knauf, P. A. (1984) Transmembrane effects of intracellular chloride on the inhibitory potency of extracellular H_2DIDS: Evidence for two conformations of the transport site of the human erythrocyte anion exchange protein. *J. Gen. Physiol.* **83,** 657–681.

George, A. L., Jr. and Borders, C. L., Jr. (1979) Essential carboxyl residues in yeast enolase. *Biochem. Biophys. Res. Comm.* **87,** 59–65.

Gilson, M. K. and Honig, B. H. (1987) Calculation of electrostatic potentials in an enzyme active site. *Nature* **330,** 84–86.

Golovtchenko-Matsumoto, A. M. and Osawa, T. (1980) Heterogeneity of band 3, the major intrinsic protein of human erythrocyte membranes. Studies by crossed immunoelectrophoresis and crossed immuno-affinoelectrophoresis. *J. Biochem.* **87,** 847–854.

Grisham, C. M. (1979) Characterization of essential arginyl residues in sheep kidney (Na^+ + K^+)-ATPase. *Biochim. Biophys. Res. Comm.* **88,** 229–236.

Gunn, R. B. and Fröhlich, O. (1979) Assymetry in the mechanism for anion exchange in human red blood cell membranes. Evidence for reciprocating sites that react with one transported anion at a time. *J. Gen. Physiol.* **74,** 351–374.

Halstrap, A. P. (1976) Transport of pyruvate and lactate in human erythrocytes. Evidence for the involvement of the chloride carrier and a chloride-indipendent carrier. *Biochem. J.* **156,** 193–207.

Hautmann, M. and Schnell, K. F. (1985) Concentration dependence of the chloride self-exchange and homo exchange fluxes in human red cell ghosts. *Pflügers Arch.* **405,** 193–201.

Hoare, D. G. and Koshland, D. E., Jr. (1966) A procedure for the selective modification of carboxyl groups in proteins. *J. Am. Chem. Soc.* **88,** 2057–2059.

Hoare, D. G. and Koshland, D. E., Jr. (1967) A method for the quantitative modification and estimation of carboxylic acid groups in proteins. *J. Biol. Chem.* **242,** 2447–2453.

Hunter, M. J. (1971) A quantitative estimate of the non exchange-resticted chloride permeability of the human red cell. *J. Physiol.* **218,** 49–50.

Hunter, M. J. (1977) Human erythrocyte anion permeabilities measured under conditions of net charge transfer. *J. Physiol.* **268,** 35–49.

Jay, D. and Cantley, L. (1986) Structural aspects of the red cell anion exchange protein. *Ann. Rev. Biochem.* **55,** 511–538.

Jennings, M. L. (1976) Proton fluxes associated with erythrocyte membrane anion exchange. *J. Memb. Biol.* **28,** 187–205.

Jennings, M. L. (1980) Apparent ''recruitment'' of SO_4 transport sites by the Cl gradient across the human erythrocyte membrane, in *Membrane Transport in Erythrocytes* (Lassen, U. V., Ussing, H. H., and Wieth, J. O., eds.), Munksgaard, Copenhagen (Alfred Benzon Symposium 14) pp. 450–463.

Jennings, J. M. (1982a) Reductive methylation of the two H_2DIDS-binding lysine residues of band 3, the human erythrocyte anion transport protein. *J. Biol. Chem.* **257,** 7554–7559.

Jennings, M. L. (1982b) Stoichiometry of a half-turnover of band 3, the chloride transport protein of human erythrocytes. *J. Gen. Physiol.* **79,** 169–185.

Jennings, M. L. and Adams, M. F. (1981) Modification by papain of the structure and function of band 3, the erythrocyte anion transport protein. *Biochemistry* **20,** 7118–7123.

Jennings, M. L. and Al-Rhaiyel, S. (1988). Modification of carboxyl groups that appears to cross the permeability barrier in the red blood cell anion transporter. *J. Gen. Physiol.* **92,** 161–178.

Jennings, M. L. and Anderson, M. P. (1987) Chemical modifiaction and labeling of glutamate residues at the stilbenedisulfonate site of human red blood cell band 3 protein. *J. Biol. Chem.* **262,** 1691–1697.

Jennings, M. L. and Nicknish, J. S. (1985) Localization of a site of intermolecular cross-linking in human red blood cell band 3 protein. *J. Biol. Chem.* **260,** 5472–5479.

Jennings, M. L., Adams-Lackey, M., and Denney, G. H. (1984) Peptides of human erythrocyte band 3 protein produced by extracellular papain cleavage. *J. Biol. Chem.* **259,** 4652–4660.

Jennings, M. L. Anderson, M. P., and Mornaghan, R. (1986) Monoclonal antibodies agains human erythrocyte band 3 protein. *J. Biol. Chem.* **261,** 9002–9010.

Jennings, M. L., Mornaghan, R., Douglas, S. M., and Nicknish, J. S. (1985) Functions of extracellular lysine residues in the human erythrocyte anion transport protein. *J. Gen. Physiol.* **86,** 653–669.

Kazarinoff, M. N. and Snell, E. E. (1976) D-serine dehydratase from *Escherichia Coli*. Essential arginine residue at the 5-phosphate binding site. *J. Biol. Chem.* **251,** 6179–6182.

Kempf, C., Brock, C., Sigrist, H., Tanner, M. J. A., and Zahler, P. (1981) Interaction of phenylisothiocyanate with human erythrocyte band 3 protein. *Biochim. Biophys. Acta* **641,** 88–98.

Khorana, H. G. (1953) The chemistry of carbodiimides. *Chem. Revs.* **53,** 145–166.

Khorana, H. G. (1961) *Some Recent Developments in the Chemistry of Phosphate Esters of Biological Interest* (John Wiley & Sons, New York).

Knauf, P. A. (1979) Erythrocyte anion exchange and the band 3 protein: transport kinetics and molecular structure. *Curr. Top. Memb. Transp.* **12,** 249–363.

Knauf, P. A. (1986) Anion transport in eyrthrocytes, in *Membrane Transport Disorders* (Andreoli, T., Hoffmann, J. F., Schultz, S. G., Fanenstil, D. D, eds.) Plenumm, New York, pp. 191–220.

Knauf, P. A. and Law, F.-Y. (1980) Relation of net anion flow to the anion exchange system, in *Membrane Transport in Erythrocytes* (Lassen, U. V., Ussing, H. H., and Wieth, J. O., eds.) Munksgaard, Copenhagen, (Alfred Benson Symposium 14), pp. 488–493.

Knauf, P. A. and Mann, N. A. (1984) Use of niflumic acid to determine the nature of the asymmetry of the human erythrocyte anion exchange system. *J. Gen. Physiol.* **83,** 703–725.

Knauf, P. A., Brahm, J., Bjerrum, P. J., and Mann, N. (1986) Kinetic asymmetry of the human erythrocyte anion exchange system, in *Eighth School on Biophysics of Membrane Transport,* School Proceedings vol. 1, Wroclaw, Poland, Publ. Co. Agricultural University, pp. 157–169.

Knauf, P. A., Law, F.-Y., Tarshis, T., and Furuya, W. (1984) Effects of the transport site conformation on the binding of external Nap-taurine to the human erythrocyte anion exchange system. *J. Gen. Physiol.* **83,** 683–701.

Kopito, R. R. and Lodish, H. F. (1985a) Primary structure and transmembrane orientation of the murine anion exchange protein. *Nature* **316,** 234–238.

Kopito, R. R. and Lodish, H. F. (1985b) Structure of the murine anion exchange protein. *J. Cell Biochem.* **29,** 1–17.

Kurzer, F. and Douraghi-Zadeh, K. (1967) Advances in the chemistry of carbodiimides. *Chemical Reviews* **67,** 107–152.

Läuger, P. (1980) Kinetic properties of ion carriers and channels. *J. Memb. Biol.* **57,** 163–178.

Lepke, S. and Passow, H. (1982) Inverse effects of dansylation of red blood cell membrane on band 3 protein-mediated transport of sulfate and chloride. *J. Physiol.* **328,** 27–48.

Lepke, S., Fasold, H., Pring, M., and Passow, H. (1976) A study of the relationship between inhibition of anion exchange and binding to the red blood cell membrane of 4,4′-diisothiocyano-stilbene-2,2′-disulfonic acid (DIDS) and of its dehydroderivative (H_2DIDS). *J. Memb. Biol.* **29,** 147–177.

Lundblad, R. L. and Noyes, C. M., eds. (1984) *Chemical Reagents for Protein Modification* vol. I and II (CRC Press Inc., Boca Raton).

Matsuyama, H., Kawano, Y., and Hamasaki, N. (1983) Anion transport activity in the human erythrocyte membrane modulation by proteolytic digestion of the 38.000 dalton fragment in band 3. *J. Biol. Chem.* **258,** 15376–15381.

Means, G. E. and Feeney, R. E., eds. (1971) *Chemical Modification of Proteins* (Holden-Day Inc., San Francisco).

Milanick, M. A. and Gunn, R. B. (1982) Proton-sulfate cotransport: mechanism of H^+ and sulfate addition to the chloride transporter of human red blood cells. *J. Gen. Physiol.* **79,** 87–113.

Milanick, M. A. and Gunn, R. B. (1984) Proton-sulfate cotransport: external proton activation of sulfate influx into human red blood cells. *Am. J. Physiol.* **247,** C247–C259.

Milanick, M. A. and Gunn, R. B. (1986) Proton inhibition of chloride exchange: asynchrony of band 3 proton and anion transport sites. *Am. J. Physiol.* **250,** C955–C969.

Mueller, T. J., Li, Y. T., and Morrison, M. (1979) Effects of Endo-β-galactiosidase, on intact human erythrocytes. *J. Biol. Chemistry* **254,** 8103–8106.

Nanri, H., Hamasaki, N., and Minakawi, S. (1983) Affinity labeling of erythrocyte band 3 protein with pyridoxal 5-phosphate, involvement of the 35.000 dalton fragment in anion transport. *J. Biol. Chem.* **258,** 5985–5989.

Passow, H. (1986) Molecular aspects of band 3 protein-mediated anion transport across the red blood cell membrane. *Rev. Physiol. Biochem. Pharmacol.* **103,** 61–203.

Passow, H., Fasold, H., Zaki, L. Schuhmann, B., and Lepke, S. (1975) Membrane proteins and anion exchange in human erythrocytes, in *Bio-*

membranes: Structure and Function vol. 35 (Gardos, G. and Szasz, I., eds.) Budapest, Publishing House of the Hungarian Academy of Sciences, pp. 197–214.

Passow, H., Fasold, H., Gartner, M. E., Legrum, B., Ruffing, W., and Zaki, L. (1980) Anion transport across the red blood cell membrane and the conformation of the protein in band 3. *Ann. NY Acad. Sci.* **341,** 361–383.

Paterson, A. K. and Knowles, J. R. (1972) The number of catalytically essential carboxyl groups in pepsin. Modification of enzyme by trimethyloxonium fluroborate. *Eur. J. Biochem.* **31,** 510–517.

Patthy, L. and Thész, J. (1980) Origin of selectivity of α-dicarbonyl reagents for arginyl residues of anion-binding sites. *Eur. J. Biochem.* **105,** 387–393.

Pedemonte, C. H. and Kaplan, J. G. (1986) Carbodiimide interaction of Na,K-ATPase. A consequence of internal cross-linking and not carboxyl group modification. *J. Biol. Chem.* **261,** 3632–3639.

Perfetti, R. B., Anderson, C. D., and Hall, P. L. (1976) The chemical modification of papain with 1-ethyl-3-(3-dimethylaminopropyl)-carbodiimide. *Biochemistry* **15,** 1735–1743.

Pho, D. B., Roustan, C., Tot, A. N. T., and Pradel, L. -A. (1977) Evidence for an essential glutamyl residue in yeast hexokinase. *Biochemistry* **16,** 4533–4537.

Quiocho, F. A., Sack, J. S., and Vyas, N. K. (1987) Stabilization of charges on isolated ionic groups sequestred in proteins by polarized peptide units. *Nature* **329,** 561–564.

Raess, B. U., Record, D. M., and Tunnicliff, G. (1985) Interaction of phenylglyoxal with the human erythrocyte (Ca^{2+},Mg^{2+})-ATPase. *Molecular Pharmacol.* **27,** 444–450.

Raida, M. and Passow, H. (1985) Enhancement of divalent anion transport across the human red blood cell membrane by the water-soluble dansyl chloride derivative 2-(*N*-piperidine)ethylamine-1-naphthyl-5-sulfonylchloride (PENS-Cl). *Biochim. Biophys. Acta* **812,** 624–632.

Rao, A., Martin, P., Reithmeier, R. A. F., and Cantley, L. C. (1979) Location of the stilbenedisulfonate binding site of the human erythrocyte anion-exchange system by resonance energy transfer *Biochemistry* **18,** 4505–4516.

Riordan, J. F. (1973) Functional arginyl residues in carboxypeptidase A. Modification with butanedione. *Biochemistry* **12,** 3915–3923.

Riordan, J. F. (1979) Arginyl residues and anion binding sites in proteins. *Prot. Mol. Cell. Biochem.* **26,** 71–92.

Riordan, J. F., McElvang, K. D., and Borders, C. L., Jr. (1977) Arginyl residues: Anion recognition sites in enzymes. *Science* **195,** 884–886.

Rudloff, V., Lepke, S., and Passow, H. (1983) Inhibition of anion transport across the red cell membrane by dinitrophenylation of a specific lysine residue at the H_2-DIDS binding site of the band 3 protein. *FEBS Lett.* **163,** 14–21.

Schnell, K. F., Elbe, W., Kasbauer, J., Kaufmann, E. (1983) Electron spin resonance studies on the inorganic-anion-transport system of the human red blood cell. Binding of a disulfonatostilbene spin label (NDS-Tempo) and inhibiton of anion transport. *Biochim. Biophys. Acta* **732,** 266–275.

Sheehan, J. C., Cruickshank, P. A., and Boshhart, G. L. (1961) A conveinient synthesis of water soluble carbodiimides. *J. Org. Chem.* **26,** 2525–2528.

Ships, S., Shami, Y., Breuer, W., and Rothstein, A. (1977) Synthesis of tritiated 4.4′-diisothiocyano-2.2′-stilbene disulfonic acid (^{3}H-DIDS) and its covalent reaction with sites related to anion transport in human red blood cells. *J. Memb. Biol.* **33,** 311–323.

Sigrist, H. and Zahler, P. (1982) Hydrophobic labeling and crosslinking of membrane proteins, in *Membrane and Transport*, vol. I. (Martonosi, A. N., ed.), Plenum, New York, pp. 173–184.

Smith, M., Moffatt, J. G., and Khorana, H. G. (1958) Observations on the reaction of carbodiimides with acids and some new applications in the synthesis of phosphoric acid esters. *J. Am. Chem. Soc.* **80,** 6204–6212.

Steck, T. L. (1974) The organization of proteins in the human red cell membrane. *J. Cell. Biol.* **62,** 1–19.

Steck, T. L. and Yu, J. (1973) Selective solubilization of proteins from red blood cell membranes by protein pertubants. *J. Supra. Mol. Struct.* **1,** 220–232.

Sternberg, M. J. E., Hayes, F. R. F, Russell, A. J., Thomas, P. G., and Fersht, A. R. (1987) Prediction of electrostatic effects of engineering of protein charges. *Nature* **330,** 86–88.

Takahashi, K. (1968) The reaction of phenylglyoxal with arginine residues in proteins. *J. Biol. Chem.* **243,** 6171–6179.

Takahashi, K., Stein, W. H., and Moore, S. (1967) The identification of a glutamic acid residue as part of the active site of ribonuclease T_1. *J. Biol. Chem.* **242,** 4682–4690.

Vandenbunder, B., Dreyfus, M., Bertrand, O., Dognin, M. J., Sibilli, L., and Buc, H. (1981) Mechanism of allosteric activation of glycogen phosphorylase probed by the reactivity of essential arginyl residues. Identification of an arginine residue involved in the binding of glucose 1-phosphate. *Biochemistry* **20,** 2354–2360.

Warshel, A. (1987) What about protein polarity? *Nature* **330,** 15–16.

Weber, M. M., Moldovan, M., and Sokolovsky, M. (1975) Modification of arginyl residues in porcine carboxypeptidase B. *Eur. J. Biochem.* **53,** 207–216.

Werner, P. K. and Reithmeier, R. A. F. (1988) The mechanisms of inhibition of anion exchange in human erythrocytes by 1-ethyl-3-[3-(tri-methylammonio)propyl]carbodiimide. *Biochem. Biophys. Acta* **942,** 19–32.

Wieth, J. O. and Bjerrum, P. J. (1982) Titration of transport and modifier sites in the red cell anion transport system. *J. Gen. Physiol.* **79,** 253–282.

Wieth, J. O. and Bjerrum, P. J. (1983) Transport and modifier sites in Capnophorin, the anion transport protein of the erythrocyte membrane, in

Structure and Function of Membrane Proteins (Quagliariello, E. and Palmieri, F. eds.), Elsevier Sciences Publishers, Amsterdam, pp. 95–106.

Wieth, J. O., Bjerrum, P. J., and Borders, C. L., Jr. (1982b) Irreversible inactivation of red cell chloride exchange with phenylglyoxal, an arginine specific reagent. *J. Gen. Physiol.* **79,** 283–312.

Wieth, J. O., Bjerrum, P. J., Brahm, J., and Andersen, O. S. (1982c) The anion transport protein of the red cell membrane. A zipper mechanism of anion exchange. *Tokai J. Exp. Clin. Med.* **7 suppl.,** pp. 91–101.

Wieth, J. O., Andersen, O. S., Brahm, J., Bjerrum, P. J., and Borders, C. L., Jr. (1982a) Chloride-bicarbonate exchange in red blood cells: Physiology of transport and chemical modification of binding sites. *Phil. Trans. R. Soc. London* **299,** 383–399.

Wilcox, P. E. (1967) Esterification. *Meth. Enzymol.* **11,** 605–626.

Wilcox, P. E. (1972) Esterification. *Meth. Enzymol.* **25,** 596–617.

Woodward, R. B. and Olofson, R. A. (1966) The reaction of isoxazolium salts with nucleophiles. Tetrahedron **suppl. 7,** 415–440.

Woodward, R. B., Olofson, R. A., and Mayer, H. (1961) A new synthesis of peptides. *J. Am. Chem. Soc.* **83,** 1010–1012.

Yu, J. and Steck, T. L. (1975) Associations of band 3, the predominant polypeptide of the human erythrocyte membrane. *J. Biol. Chem.* **250,** 9176–9184.

Zaki, L. (1981) Inhibition of anion transport across red blood cells with 1,2-cyclohexanedione. *Biochem. Biophys. Res. Comm.* **99,** 243–251.

Zaki, L. (1982) The effect of arginine specific reagents on anion transport across the red blood cells, in *Protides Biol. Fluids* **29,** (Peeters, H., ed.,) 29th Collogium 1981, Pergamon, Oxford, pp. 279–282.

Zaki, L. (1983) Anion transport in red blood cell and arginine specific reagents (1) Effects of chloride and sulfate ions on phenylglyoxal sensitive sites in the red blood cell membrane. *Biochim. Biophys. Res. Comm.* **110,** 616–624.

Zaki, L. (1984) Anion transport in red blood cells and arginine specific reagents. The location of ^{14}C-phenylglyoxal binding sites in the anion transport protein in the membrane of human red cells. *FEBS Lett.* **169,** 234–240.

Zaki, L. and Julien, T. (1985) Anion transport in red blood cells and arginine-specific reagents. Interaction between the substrate-binding site and the binding of arginine specific reagents. *Biochim. Biophys. Acta* **818,** 325–332.

Zaki, L. and Julien, T. (1986) Chemical properties of anion binding site in red blood cell membrane, in *Eight School on Biophysics of Membrane Transport. School Proceedings*, vol. II, Wroclaw, Poland, Pub. Co. Agricultural University, pp. 239–259.

Zaki, L., Fasold, H., Schuhmann, B., and Passow, H. (1975) Chemical modification of membrane proteins in relation to inhibition of anion exchange in human red blood cells. *J. Cell Physiol.* **86,** 471–494.

Drug Actions on Potassium Fluxes in Red Cells

Balázs Sarkadi and George Gárdos

1. Introduction

Investigations of the effects of drugs on transport phenomena have two parallel and interrelated achievements. On the one hand, they help in defining, characterizing, and sometimes even isolating a given transport system and thus allow the clarification of the molecular nature of the transporter. On the other hand, such studies will ultimately explain the mechanism of action of a given therapeutic compound and allow one to determine or foresee its clinical applicability, possible side-effects, and toxicity.

A good example of this relationship is the discovery of the effects of digitalis alkaloids (strophanthin or ouabain) on the Na^+–K^+ pump in human red cells (Schatzmann, 1953, *see* Chapters 2,3). As has been established in the past 30 yr or more, this drug is extremely specific in its action on the Na^+–K^+ pump protein, and probably all of its therapeutic and toxic effects can be explained on the basis of this single molecular interaction. In membrane biochemistry, the use of ouabain helped to clarify the molecular basis and the various modes of operation of the ubiquitous, ATP-dependent active Na^+–K^+-transport system. Although in this chapter we do not discuss the digitalis effects in detail, ouabain will continuously appear in the following paragraphs as a restrictive clause: we shall deal with ouabain-resistant potassium fluxes, since in most experiments discussed below, ouabain was the tool to eliminate Na^+–K^+ pump fluxes, which allows one to explore the "nonpump" potassium transport pathways.

The "nonpump" ion fluxes, that is, the ones not directly fueled by the energy of cellular ATP, can be categorized into two major groups: one is the group of "membrane-mediated" processes, and the other is the so-called "passive" ion movements. As discussed earlier by Sarkadi and Tosteson (1979), from an energetic point of view, such a transport for a given ion can be downhill and dissipative or uphill and "active", whereas from a mechanistic point of view, it can be transport-coupled either with a substance(s) moving in the same direction (cotransport, T^o) or in the opposite direction (countertransport, T^r). There are also known examples of ion movements when a chemical reaction in the membrane with an activating agent opens specific transport pathways.

Membrane-active drugs may modify these membrane-mediated pathways through specific interaction with a given transport system or its activating mechanism, but they may also alter "passive" movements of substances by disturbing the structure of the lipid bilayer or by changing lipid–protein interactions in the cell membrane. In the following, we try to concentrate on the actions of various chemicals that modify the "non-pump" fluxes of potassium ions through the red blood cell membrane. Such a restriction of the topic is certainly unavoidable if we wish to deal with these questions in a relatively detailed way.

It has to be stated here in the Introduction that there is no "ouabain" for any of the potassium transport systems presented in the following sections, and this fact contributes to the uncertainty in their terminology and characterization. Therefore, instead of concentrating on the types or effects of the drugs used in these studies, we approach our subject by discussing the activation and inhibition of various potassium transport pathways, distinguished not only by their drug sensitivities, but also by several other criteria.

2. Chemically Induced Potassium Transport in Red Cells

2.1. General Increase in the Cation Permeability of the Red Cell Membrane—SH Group Reagents

A variety of hemolytic agents act by producing a nonspecific increase in the cation permeability of the red cell membrane (for reviews, *see* Orringer and Parker, 1973; Parker and Berkowitz, 1986). A classic example for such an action is the blockade of membrane sulfhydryl groups by organic mercurials, PCMB,* PCMBS, or DTNB, which all

*CCCP = carbonylcyanide m-chlorophenylhydrazone; DIDS = 4,4′-diisothiocyano-2,2′-stilbene disulfonate; diS-C_3-(5) = 3,3′-dipropylthiadicarbocyanine; DTNB = 5,5′-dithyobis (2-nitrobenzoate); HK cells = high potassium red cells; IAA = iodoacetamide; LK cells = low potassium red cells; NEM = *N*-ethylmaleimide; PCMB = *p*-chlormercuribenzoate; PCMBS = *p*-chlormercuribenzenesulfonic acid; RVD = regulatory volume decrease; SITS = 4-acetamido-4′-isothiocyano-2,2′-stilbene sulfonate; TEA = tetraethylammonium; TMA = tetramethylammonium.

inhibit active Na^+–K^+ pumping while greatly facilitate passive cation movements (*see* Jacob and Jandl, 1962; Sutherland et al., 1967; Rothstein, 1981). Under these conditions, the red cell cannot compensate any more for the osmotic activity of cellular hemoglobin and of other impermeable or poorly permeable intracellular substances, such as organic phosphate esters. As a result, relatively hypoosmotic saline enters the cell, leading to swelling, disruption of the membrane integrity, and eventually the leakage of hemoglobin: a "colloid osmotic" hemolysis.

The membrane action of PCMB and PCMBS (the latter being more water soluble at physiological pH values) has been examined in detail. They were shown to block SH groups at the internal membrane surface, and this action is completely reversible upon the addition of compounds containing free SH groups, such as cysteine or dithiothreitol (Sutherland et al., 1967; Garrahan and Rega, 1967). PCMBS travels through the anion exchanger of the red cell membrane to reach the internal surface; thus, specific inhibitors of the anion exchanger, such as SITS or DIDS, inhibit the development of PCMBS-induced cation leak, but have no effect on these cation movements once already evoked (Rothstein, 1981; Haas and Schmidt, 1985). The latter authors also demonstrated that, at relatively high PCMBS concentration (above 0.5 m*M*), the drug-induced cation movement depends on the anion composition of the incubation medium. Recent studies indicated a direct role of the anion exchanger (the Band 3 protein) in the formation of the cation leak pathway (Lukačovič et al., 1984); however, this suggestion awaits further experimental support.

Cell lysis by mercurials including PCMB can be prevented by adding osmotically active nonpenetrating substances (such as sucrose) to the incubation media, which balance the osmotic activity of the internal nonpenetrable solutes. By using such conditions, the reversible membrane permeabilizing action of PCMB has been applied to change the intracellular composition of red cells both for monovalent and divalent cations (Garrahan and Rega, 1967; Schatzmann, 1973). This technique has been widely applied for in vitro red cell studies in various clinical conditions (*see* Garay et al., 1980), but since it may drastically change the function of SH groups in other transport systems (*see* section 3.), such an applicability is questionable.

2.2. Alkali Ion Transport Mediated by Ionophores

Certain compounds facilitate the passage of ions through the lipid barriers of cellular membranes by forming mobile carriers or membrane-spanning permeable structures called "channels." These agents do not act through specific receptors or endogenous protein structures of the membranes and were termed "ionophores" by Pressman et al. (1967).

Valinomycin and some other related antibiotics are extremely specific for functioning as mobile carriers for potassium ions (*see* Pressman, 1976). The ion transport rate (the transference number) by these antibiotics is relatively small, as compared to that of the channel formers, such as gramicidin. Also, the large hydrophobic complex of the mobile carriers cannot be introduced in large quantities into cellular membranes. Moreover, valinomycin was shown to inhibit some naturally occurring ion transport pathways (Sarkadi et al., 1985a). Still, it was the valinomycin-induced K^+ transport that could be first used to demonstrate that the net anion permeability of the human red cell membrane is about four orders of magnitude smaller than the permeability calculated from tracer Cl^- exchange, catalyzed by the obligatory anion exchanger pathway (Hunter, 1971, 1977). In these experiments, valinomycin supported a rapid tracer K^+ exchange, but net K^+ movement was limited by the relatively low net anion permeability of the membrane.

Another mobile carrier ionophore, nigericin, carries an obligatory, electroneutral exchange of small alkali cations and/or protons. The divalent cation ionophores, such as X-537 A, A23187, or ionomycin, are relatively poor transporters for monovalent cations, but by changing intracellular divalent cation concentrations, they may induce secondary K^+ movements (*see* section 2.3.). Gramicidin dimers form membrane-spanning monovalent cation selective channels with almost equal permeability for K^+ or Na^+, but relatively low permeability for H^+ (for detailed reviews of the ionophore actions, *see* Pressman, 1976). A short, reversible exposure of the red cells to ionophore antibiotics has been successfully applied to introduce tracer cations into the cell interior or to produce a rapid change in the cellular divalent cation concentrations (Sarkadi et al., 1976).

Nystatin and Amphotericin B, antibiotics produced by fungi, form relatively large pores across the cell membrane and, through an interaction with membrane cholesterol, allow the passage of small nonelectrolytes, cations, and anions (Cass et al., 1970; Deuticke et al., 1973; Freedman and Hoffman, 1979). Since the action of nystatin can be easily reversed by extracting the ionophore from the membrane (washings at 37°C, in the presence of albumin), this compound is often used for altering the intracellular salt composition of animal cells (*see* Cass and Dalmark, 1973; Ikehara et al., 1986). In fact, at present, this is probably the method of choice for studying the kinetics of cation transport pathways in health and disease (*see* Tosteson, 1981).

A nonselective increase in the cation permeability of the red cell membrane and a colloid-osmotic type lysis is most probably the basis of the killer lymphocyte- or the complement-mediated destruction of the target erythrocytes. It is suggested that a membrane-spanning, ionophore-like polypeptide complex, such as that formed by the C9 fragment of complement, is inserted into the target cell membrane (Podack and Dennert, 1983).

2.3. Selective Increase in the Potassium Permeability—The Role of Intracellular Calcium

Metabolic depletion of red cells, through the elimination of ATP supply for active Na^+–K^+ transport, eventually leads to salt and water entry, although the normal leak permeability of the red cell membrane is small and the development of such a colloid osmotic swelling is relatively slow (*see* Tosteson and Hoffman, 1960; Tosteson, 1964). Moreover, inhibitors of red cell metabolism were shown to produce KCl and water loss by a selective increase in the membrane potassium permeability. Wilbrandt (1940) first observed that the glycolytic inhibitors fluoride or iodo-acetate induced such an effect, and Gárdos (1958a,b, 1959) demonstrated that inhibition of cell metabolism must be accompanied by the presence of calcium in order to obtain rapid K^+ transport. By now, it is generally accepted that access to the internal surface of the red cell membrane by calcium ions activates a K^+-selective "channel," which allows a diffusion-type K^+ transport downhill, driven by the electrochemical potential gradient. The drugs that evoke this phenomenon either just allow calcium ions to reach the cell interior or even further increase the calcium sensitivity of the opening mechanism. Probably the latter is the case with fluoride, propranolol, or the divalent cation ionophore A23187 (*see* Ekman et al., 1969; Lew and Ferreira, 1976, 1978; Gárdos et al., 1977; Szász et al., 1977; Schwarz and Passow, 1983; Sarkadi and Gárdos, 1985). Some divalent and trivalent cations may replace calcium in its role of activating the K^+ channels, and probably this is the explanation for the selective K^+ loss after lead exposure (observed as early as 1935 by Ørskov) or of the rapid K^+ transport found in ghosts reconstituted with low concentrations of lanthanides (Wood and Mueller, 1985). It is interesting to mention that trypsin treatment of the internal membrane surface of human red cells first eliminates the K^+ selectivity and then the calcium-sensitivity of the transport pathway (Wood and Mueller, 1984).

A selective increase in the red cell membrane K^+ permeability, which may or may not be related to the Ca^{2+}-induced pathway, has been shown to be present in various hematological conditions, such as pyruvate kinase deficiency, Heinz-body formation, sickle-cell disease, or favism (*see* Orringer and Parker, 1973; Lew and Bookchin, 1980; Berkowitz and Orringer, 1985; DeFlora et al., 1985; Parker and Berkowitz, 1986). Some animal toxins may also act through the activation of this Ca^{2+}-dependent K^+ transport pathway (Nagase et al., 1984).

2.4. Potassium-Chloride Cotransport Induced by NEM

As discussed previously, most sulfhydryl reagents increase the cation permeability of the red cell membrane nonselectively. A unique feature of the compound NEM is the induction of a selective K^+ transport pathway in most types of red cells. This effect was first noted in duck (Tosteson and Johnson, 1957) and then in human red cells (Jacob and Jandl, 1962). In a NaCl medium, low concentrations of the alkylating compound NEM for a short exposure caused cell shrinkage and K^+ loss, whereas further treatment resulted in cell swelling and Na^+ uptake, just as observed with other SH-group reagents. The experiments carried out in the last few years further extended these observations and revealed several basic features of the NEM-induced K^+ transport.

A unique characteristic of the NEM-induced K^+ transport is its Cl^--dependence. This means that the downhill movement of K^+ requires the presence, and most probably the cotransport of Cl^- or Br^- ions, whereas these anions are only poorly replaced in this role by sulfate, thiocyanate, or nitrate (Lauf and Theg, 1980; Ellory et al., 1982; Smalley et al., 1982; Lauf, 1985). The NEM-induced, Cl^--dependent K^+ transport was found to be present in LK, but not in HK sheep red cells (Lauf and Theg, 1980), whereas the reticulocytes of both LK and HK sheep possessed the pathway (Lauf, 1983a). The pathway is present in human red cells (Ellory et al., 1982; Lauf et al., 1984; Wiater and Dunham, 1983), in LK goat red cells (Lauf and Theg, 1980), and in the red cells of several animal species (Ellory et al., 1982). Both ATP depletion and an increase in cellular calcium inhibit the NEM-induced, Cl^--dependent K^+ transport (Lauf, 1983b, Lauf and Mangor-Jensen, 1984; Ellory et al., 1983). According to our present view, the NEM-induced pathway is most probably a "constitutive" activation of a K^+/Cl^- cotransport system, naturally occurring in the red cell membrane (*see* section 3.3.).

3. Inhibitors of "Nonpump" Potassium Transport Pathways in Red Cells

3.1. Inhibitors of the Ca^{2+}-Induced K^+ Transport

The activation of this selective K^+ transport pathway requires the interaction of calcium ions with the internal surface of the red cell membrane, and since cellular calcium is normally kept at a very low level, this original state has to be disturbed before obtaining a rapid K^+ transport (*see* section 2.3.). Several compounds were shown to interfere with this "prephase" of the opening of the K^+ pathway, that is, with the metabolic depletion or the calcium entry. Ouabain, which inhibits Ca^{2+}-dependent K^+ transport in ATP-depleted cells (Blum and Hoffman, 1971) evokes this effect by sparing ATP when blocking the ATP consumption of the Na^+–K^+ pump (Lew, 1971, 1974). Verapamil was shown to inhibit calcium entry, and consequently, it also prevented an increase in K^+ movement when cellular Ca^{2+} was allowed to increase after the inhibition of the ATP-dependent calcium extrusion by vanadate (Varečka and Carafoli, 1982). If the ionophore A23187 is applied to increase cell calcium and initiate K^+ transport, Mn^{2+} or Co^{2+} ions effectively compete with the ionophore-catalyzed calcium entry and, thus, eliminate rapid K^+ transport (*see* Lew and Ferreira, 1978; Sarkadi et al., 1976).

Another possibility of an indirect drug effect on Ca^{2+}-induced K^+ transport is the fact that such a K^+ permeability may significantly override the net Cl^- permeability of the red cell membrane, and thus, net KCl movement becomes limited by the Cl^- movement (Glynn and Warner, 1972; Hoffman and Knauf, 1973; Gárdos et al., 1975; Schubert and Sarkadi, 1977). Owing to this restriction, all the inhibitors of the Cl^- transport (e.g., SITS, DIDS, furosemide, or dipyridamole) significantly reduce net K^+ efflux into normal (low) K^+ incubation media. In similar media, because of an increase in membrane hyperpolarization, anion transport inhibitors accelerate tracer K^+ influx. These compounds have no effect on tracer K^+ exchange when K^+ ions are in equilibrium between the two sides of the membrane (Gárdos et al., 1975).

When calcium entry or anion movements do not limit Ca^{2+}-dependent K^+ transport, the effects of the true inhibitors of the pathway can be examined. Ca^{2+}-induced K^+ transport in red cells is significantly modified by alkali cations: external K^+ is required to obtain rapid K^+ transport (maximum activation at about 3 m*M* K_o^+ (*see* Blum and Hoffman, 1971; Knauf et al., 1974, 1975; Gárdos et al., 1975) and K^+-free media irreversibly inactivate the pathway (Heinz and Passow, 1980). High extracellular K^+ inhibits Ca^{2+}-induced K^+ transport (Blum and Hoffman, 1971, 1972; Yingst and Hoffman, 1984), and the same is true for increasing concentrations of intracellular Na^+ (Riordan and Passow, 1971; Knauf et al., 1975; Yingst and Hoffman, 1984; Simons, 1976a). From among divalent cations, Sr^{2+} can replace Ca^{2+} in activating the pathway, Pb^{2+} in low concentrations activates, while in higher concentrations inhibits, and Mg^{2+} or Ba^{2+} only inhibit rapid K^+ transport (Lew and Ferreira, 1978; Passow, 1981; Schwarz and Passow, 1983; Shields et al., 1985; Alvarez et al., 1986a). There are indications that high cellular calcium concentrations produce a self-inhibition of K^+ movement.

The trivalent lanthanides at the external membrane surface inhibit Ca^{2+} entry and thus the activation of the pathway at limiting cellular calcium levels (Szász et al., 1978a,b), but may also have a direct inhibitory action on the K^+ movement (Wood and Mueller, 1985). Intracellular lanthanides in low concentrations activate and in higher concentration inhibit rapid K^+ transport (Szász et al., 1978b; Wood and Mueller, 1985).

One class of inhibitors, relatively selective for the Ca^{2+}-induced K^+ transport, includes amphiphylic cations, such as quinine and quinidine or the carbocyanine dyes, e.g., diS-C_3-(5), often applied for measuring membrane potential. These compounds react with the external K^+-dependent site of the transport system, their action is competitively antagonized with external K^+, and is reversible if the drugs are eliminated through binding, e.g., to albumin (Armando-Hardy et al., 1975; Reichstein and Rothstein, 1981; Simons, 1976b, 1979). Cetiedil, an antisickling agent, that has been reported by Berkowitz and Orringer (1981) to block Ca^{2+}-induced K^+ transport, may belong to the same class of inhibitors.

SH group reagents, such as NEM, mersalyl, or PCMB, directly inhibit Ca^{2+}-induced K^+ transport, but the site of interaction is not clarified as yet (*see* Gárdos et al., 1975; Sarkadi and Gárdos, 1985). Inhibition of the pathway by high concentrations (up to 5 m*M*) of iodoacetic acid was shown by Plishker (1985), and the carboxymethylation of a specific protein was indicated to correlate with this inhibition. However, this reaction may not be specific enough to identify convincingly the protein responsible for rapid K^+ transport.

Several lipophylic drugs that antagonize calmodulin action also strongly inhibit Ca^{2+}-induced K^+ transport in red cells (Szász and Gárdos, 1974; Szász et al., 1978a; Lackington and Orrego, 1981;

Nagase et al., 1984; Alvarez et al., 1986b). However, this effect may not be achieved through calmodulin antagonism, since calmodulin is not required to elicit Ca^{2+}-dependent K^+ transport in one-step inside-out red cell membrane vesicles (Lew et al., 1982; Garcia-Sancho et al., 1982) and the efficacy of these drugs in inhibiting K^+ transport is not strictly parallel to their calmodulin-inhibitory potential (Alvarez et al., 1986b).

TEA and TMA, known inhibitors of various K^+ transport pathways, are also effective against the Ca^{2+}-induced K^+ transport in red cells. They seem to interact with the K^+ channel at the external surface of the cell membrane (Gárdos et al., 1975; Hermann and Gorman, 1981; Iwatsuki and Petersen, 1985). A highly effective inhibitor of the Ca^{2+}-induced K^+ transport is oligomycin (Blum and Hoffman, 1971, 1972; Riordan and Passow, 1971; Gárdos et al., 1975), especially the separated compound oligomycin A (Sarkadi et al., 1985a). Since oligomycin also inhibits the Na^+–K^+ pump, this effect was suggested to link the two transport pathways (Blum and Hoffman, 1971, 1972). By now, it seems to be clear that neither the Na^+–K^+ pump, nor the Ca^{2+} pump is involved in catalyzing the Ca^{2+}-induced K^+ transport (Richhardt et al., 1979; Karlish et al., 1981; Parker, 1983; Verma and Penniston, 1985). In several tissues, apamin, a bee venom toxin, was shown to inhibit Ca^{2+}-dependent K^+ transport and was applied to identify the molecular basis of this transport system (Banks et al., 1979; Hugues et al., 1982). However, in red cells and in other blood cells, such as lymphocytes, apamin is ineffective against Ca^{2+}-induced K^+ transport (Burgess et al., 1981; Sarkadi and Gárdos, 1985).

Here we mention that, in a recent work (Sarkadi et al., 1985b), we demonstrated the applicability of a simple experimental system for following drug actions on the Ca^{2+}-induced K^+ transport in human red cells. The method utilizes cell volume measurements in a K^+-propionate medium supplemented with a protonophore (CCCP). Under these conditions, anion or H^+ transport do not limit K^+ movements and a Ca^{2+}-induced K^+ transport is directly reflected in a rapid volume increase. The method may offer some help in the further characterization of the Ca^{2+}-induced K^+ pathway.

In mammalian or avian red cells, volume changes apparently have no significant effect on the Ca^{2+}-induced K^+ pathway. However, in several animal cell types, such as lymphocytes, monocytes, platelets, CHO tumor cells, Ehrlich ascites cells, and so on; regulatory volume decrease (RVD) induced by hypoosmotic swelling was shown to involve a Ca^{2+}-activated rapid K^+ transport that is sensitive to the same inhibitors as the Ca^{2+}-induced K^+ transport in red cells (*see* Grinstein et al., 1982, 1984; Sarkadi et al., 1985a, Hoffmann, 1986). During volume regulation a concomitant induction of an independent, conductive Cl^- transport system has also been demonstrated (Grinstein et al., 1984; Sarkadi et al., 1984a,b; Hoffmann et al., 1984; Hoffmann, 1986). A careful survey of the data in the literature indicates that amphibian or fish red blood cells may have a similar, swelling-activated, Ca^{2+}-dependent K^+ transport system. In *Amphiuma* red cells, Cala's detailed studies suggested that an RVD in hypoosmotic solutions involves an electroneutral K^+–H^+ exchange, coupled to a Cl^-–HCO_3^- exchange system (Cala, 1980, 1983a,b, 1985a,b). This supposed K^+–H^+ exchange is stimulated by intracellular calcium; the addition of the A23187 ionophore + calcium in the medium increases, whereas A23187 + EGTA strongly decreases the K^+ loss during RVD. In addition to this, quinine and phenothiazine calmodulin antagonists inhibit volume-induced K^+ loss, just as observed in lymphocytes or Ehrlich ascites cells (Cala et al., 1986). Cala's hypothesis for an obligatory, electroneutral K^+–H^+ exchange in *Amphiuma* red cells is predominantly based on observing high ratios of H^+ over the Cl^- efflux (up to 5:1), on microelectrode measurements showing no change in the membrane potential during RVD, and on the dissociation of alkali metal ion and net anion movements after exposing the cells to DIDS, an inhibitor of the Cl^-–HCO_3^- exchange system. Moreover, addition of the K^+ ionophore valinomycin was shown to induce a hyperpolarization in *Amphiuma* red cells, regardless of the osmolarity of the medium. The calcium dependence and the drug sensitivity of RVD in *Amphiuma* red cells may suggest a close relationship of this K^+ movement with the Ca^{2+}-induced K^+ transport in mammalian red cells. In fact, such a K^+ transport was shown to be present in *Amphiuma* erythrocytes (Lassen et al., 1973, 1976; Gárdos et al., 1976), and a detailed analysis of Cala's data by Lew and Bookchin (1986) also indicates that it may not be necessary to suppose an obligatory, electroneutral K^+–H^+ exchange during RVD. If the net anion permeability is relatively high in these cells (or hypotonic swelling induces rapid net anion transport, just as in lymphocytes or Ehrlich cells), the anion permeability does not limit K^+ transport and there could be little change, if any, in the membrane potential when measured by microelectrodes. The large amount of H^+ uptake, instead of Cl^- loss, during RVD is easily explained if the *Amphiuma* red cells have a high buffering capacity for H^+ ions and a high rate of Cl^-–HCO_3^- exchange, resulting in a high effective H^+ permeability. In fact, an early hypothesis for the mechanism of the Ca^{2+}-induced K^+ transport in human red cells also considered a K^+–H^+ exchange pathway, based on the observed alkalization in bicarbonate-containing media (Szönyi, 1960). In lymphocytes or Ehrlich cells, where Cl^-–HCO_3^- exchange is negligible, the contribution of H^+

transport and buffering is less significant than in an erythrocyte, which contains the anion exchanger.

The finding that valinomycin has the same hyperpolarizing effect in resting or in hypoosmotically swollen cells may be explained by a heterogeneous distribution of the ionophore, as suggested by Lew and Bookchin (1986) or by the finding that valinomycin in surprisingly low concentrations (0.1–0.5 μM) is an effective inhibitor of the volume-induced Cl^- transport in lymphocytes (Sarkadi et al., 1985a). By promoting K^+ transport and inhibiting volume-induced Cl^- transport, valinomycin can easily produce the same hyperpolarization in isoosmotic or hypoosmotic media, respectively. Thus, according to this analysis, H^+ movement may not be strictly coupled to K^+ transport during RVD in *Amphiuma* red cells. Further studies with the inhibitors of the Ca^{2+}-induced K^+ transport may offer a considerable help for resolving these questions.

3.2. Inhibitors of the $Na^+/K^+/Cl^-$ Cotransport

It was first noted in human red cells that uphill Na^+ efflux was greater than could be accounted for by the ouabain-sensitive Na^+–K^+ pump, and ethacrynic acid inhibited this ouabain-resistant Na^+ transport (Hoffman and Kregenow, 1966). This Na^+ transport was also found to be inhibited by furosemide (Dunn, 1970; Sachs, 1971) and Wiley and Cooper (1974) demonstrated that a ouabain-resistant but furosemide-sensitive Na^+ transport, working both in outward and inward direction, was associated with K^+ movement. They postulated the existence of a symmetrical Na^+/K^+ cotransport system. Several studies have since examined this coupled Na^+/K^+ transport system, which is sensitive to cellular ATP (Adragna et al., 1985; Dagher et al., 1985) and is strongly inhibited by low concentrations of the loop diuretics, e.g., furosemide, piretanide, or bumetanide (for recent reviews, *see* Hoffmann, 1986, Chipperfield, 1986). The pathway seems to be independent of a Na^+/Na^+ (Li^+) countertransport, which is insensitive to K^+, inhibited by phloretin, but almost unaffected by furosemide (Haas et al., 1975; Sarkadi et al., 1978; Tosteson, 1981).

It has also been demonstrated that the diuretic-sensitive, coupled Na^+/K^+ transport is strictly Cl^- dependent: anion substitution by nitrate, sulfate, or thiocyanate abolishes this transport, and only Br^- is a partially effective substituent (Chipperfield, 1980, 1981; Dunham et al., 1980; Dunham and Ellory, 1981). In human red cells, Cl^- exchange is extremely fast, and even net Cl^- movement through this exchanger is more rapid than Na^+/K^+ transport; therefore, in this type of cell, it cannot be decided if anion dependence means a coupled $Na^+/K^+/Cl^-$ cotransport. Moreover, the Cl^- exchange system is also sensitive to low doses of furosemide (Brazy and Gunn, 1976). Still, all the kinetic evidence supports a coupled electroneutral movement of Na^+, K^+, and Cl^- through a diuretic-sensitive cotransport. In the red cells of ferrets, where this transport system is abundant, a coupling of Cl^- movement to the Na^+/K^+ cotransport could be convincingly demonstrated (Ellory and Hall, 1984; Hall and Ellory, 1985).

In human red cells, according to the affinities of the $Na^+/K^+/Cl^-$ cotransport system to the respective ions, under physiological conditions the pathway does not carry net ion movement in either direction. The significance of the system may be that, in the case of abnormal high Na^+ levels, downhill K^+ movement by the cotransport carries an uphill Na^+ extrusion (*see* Tosteson, 1981). There is no evidence that changes in cell volume in human red cells could activate $Na^+/K^+/Cl^-$ cotransport, although the stimulation of this pathway in hyperosmotically shrunken rat red blood cells has been demonstrated (Duhm and Gobel, 1984).

A more suitable model for studying this transport pathway could be the avian red cell, in which hyperosmotic shrinkage or the addition of receptor agonists, such as norepinephrine, induces an electroneutral cotransport of $Na^+/K^+/2Cl^-$, which results in a net salt uptake (*see* Schmidt and McManus, 1974; Palfrey et al., 1980; Kregenow, 1981; Haas et al., 1982; McManus et al., 1985). Extensive studies on the drug sensitivity of this transport system have shown that bumetanide completely blocks at a concentration about 100 times smaller than furosemide. Bumetanide in this low concentration is relatively selective for the $Na^+/K^+/Cl^-$ cotransport, but it may still inhibit another K^+ and/or Cl^- transport (*see below*). The bumetanide sensitivity of the avian $Na^+/K^+/Cl^-$ cotransport is influenced by the ionic composition of the incubation medium: rising external K^+ or Na^+ concentrations greatly increase the inhibitory potential of the drug. Both bumetanide inhibition and binding are saturable functions of external K^+ and Na^+. The apparent affinities by which these cations support bumetanide action and binding closely correlate to their respective affinities for the cotransport pathway (Haas and McManus, 1983; Haas and Forbush, 1986). In contrast to this, external Cl^- concentration has a biphasic effect on bumetanide action: increasing Cl^- from 0 to about 30 m*M* increases bumetanide inhibition and binding, whereas greater Cl^- concentrations compete with bumetanide. In fact, Cl^- and bumetanide seem to compete for a common site, kinetically similar to the second Cl^- binding site of the cotransport system (Haas and McManus, 1983; Haas and Forbush, 1986).

Recent data on the kinetics of the cotransport in duck red cells (Lytle and McManus, 1986) may help to explain both the mode of action and the bumetanide-sensitivity of the pathway. Although the stoichiometry of the net cotransport seems to be always 1 Na^+/1 K^+/2 Cl^-, that is, no independent NaCl or KCl movements are observed, the partial reactions can provide K^+–K^+(Rb) or Na^+–Na^+(Li^+) exchange. In order to obtain a K^+–K^+ (Rb) exchange, Na^+, K^+, and Cl^- all must be present inside, but only K^+ and Cl^- are required outside the cells. Na^+–Na^+ exchange requires only Na^+ ions inside, but all the three ions (Na^+, K^+, and Cl^-) in the outside medium. The so-called "glide-symmetry" model, worked out by the Durham group (*see* Lytle and McManus, 1986) using the proposal by W. D. Stein, gives a good explanation for all these phenomena: a specifically ordered reaction of the respective ions with the transporter and the interaction of bumetanide with a certain "loaded" species satisfactorily describes the probable molecular events (Fig. 1.). According to the model, only the empty and the fully loaded transporters may oscillate between the two sides of the membrane, the latter being the faster species. At the internal surface, the order of the ion binding is the same as the order of release of these ions at the external surface ("first on-first off" symmetry). As seen in Fig. 1, the model easily explains that K^+–Rb^+ exchange does not require external, only internal Na^+, whereas Na^+–Li^+ exchange needs external, but not internal K^+. Bumetanide is supposed to act at the external surface by binding to the transporter loaded with Na^+, K^+, and one Cl^-, thus, the saturation of the transporter with these ions will favor bumetanide binding and inhibition. If the site of bumetanide binding is the same as the binding site for the second Cl^- ion, the competition of high Cl^- concentration with bumetanide is also explained.

The above model seems to be very helpful for studying the mechanism and the drug-interactions of the Na^+/K^+/Cl^- cotransport system in various cell types. However, the restriction applied to this model, that is, only the movement of the fully loaded transporter is allowed across the membrane, may not be generally valid for other cell membranes. According to recent data by Brugnara et al. (1986) and Canessa et al. (1986a), in human red cells, the ouabain-resistant, furosemide-sensitive, Cl^--dependent Na^+/K^+ cotransport may have a variable stoichiometry. Depending on the prevailing ionic conditions, ratios ranging from 2Na^+/1K^+–1 Na^+/2 K^+ were observed. In the model of Fig. 1., this would mean that partial reactions providing net NaCl or KCl transport may occur; thus, the partially loaded transporter can also oscillate. Another explanation of the observed variable stoichiometry may be that, in

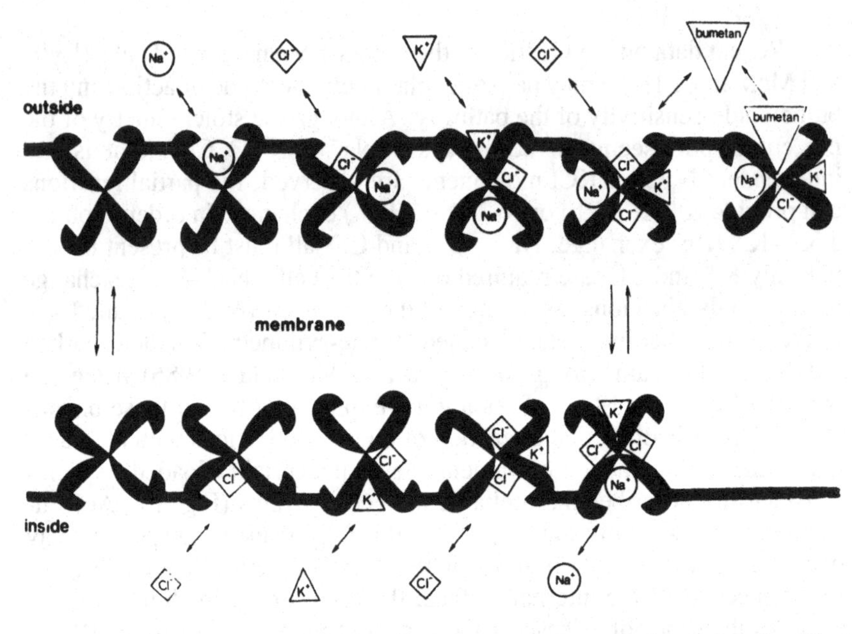

Fig. 1. Model for the $Na^+/K^+/Cl^-$ cotransport in avian red cells. The transport system reacts in an ordered way with the respective ligands as depicted in the figure. Based on Lytle and McManus, 1986—adapted by the permission of the above authors.

human red cells, the $Na^+/K^+/Cl^-$ cotransport is not the only Cl^--dependent Na^+ and/or K^+ pathway (*see below*). Thus, the interference of these latter pathways may result in an apparently variable stoichiometry of the $Na^+/K^+/Cl^-$ system. The dependence of furosemide or bumetanide inhibition of the cotransport system on the cation and anion composition of the incubation media may also contribute to such an apparent variation of the coupling ratio. Clearly, further experimental data are necessary to decide these questions.

The diuretic sensitivity of the $Na^+/K^+/Cl^-$ cotransport in red cells indicates its close relationship to similar pathway(s) in the kidney tubules (*see* Ellory and Stewart, 1982), where this pathway probably functions as a basis for Na^+ and Cl^- reabsorption and multiplies the thermodynamic efficiency of the ATP-dependent Na^+–K^+ pump (*see* Geck and Heinz, 1986). In red cells, where Cl^- transport is relatively fast, the cotransport can only be important for changing cellular cation concentrations, e.g., during the volume regulatory response of avian red cells to hyperosmotic media.

3.3. Inhibitors of K^+/Cl^- Cotransport

Several data in the literature indicate that red cells possess a K^+/Cl^- cotransport system different from the quaternary $Na^+/K^+/2Cl^-$ cotransport discussed in the previous section. In avian red cells, swelling in hypoosmotic media induces a coupled, electroneutral KCl efflux resulting in net salt and water loss and, thus, a regulatory volume decrease. This Cl^--dependent, but Na^+ independent K^+ transport is insensitive to ouabain or the inhibitors of the Cl^-–HCO_3^- exchange, such as SITS or DIDS, but inhibited (although with a lower sensitivity than the $Na^+/K^+/Cl^-$ pathway) by furosemide or bumetanide (*see* Kregenow, 1971, 1977, 1981). In dog red cells, hypoosmotic swelling induces a furosemide-sensitive, Cl^--dependent K^+ movement (Parker, 1983), and the same is true for LK sheep (Dunham and Ellory, 1981) and fish (Lauf, 1982) red blood cells. Such a volume-sensitive K^+/Cl^- cotransport was also observed in human red cells, but only under high hydrostatic pressure (Ellory et al., 1985).

It seems feasible that the same volume-sensitive K^+/Cl^- cotransport pathway is activated constitutively when red cells are exposed to NEM and specific SH groups are alkylated by this reagent (*see* section 2.3.). The NEM-induced Cl^--dependent K^+ transport is also partially inhibited by furosemide and bumetanide. The sensitivity of the system to the diuretics is lower than that of the $Na^+/K^+/Cl^-$ system, but the presence of K^+ or Rb^+ at the cell surface significantly increases this sensitivity (Lauf, 1984; Ellory et al., 1982). The NEM-induced K^+/Cl^- cotransport is insensitive to the inhibitors of the rapid anion exchange, such as SITS or DIDS (Lauf and Theg, 1980), but blocked by iodoacetamide (Bauer and Lauf, 1983), which indicates that the alkylation of the SH groups responsible for blocking the pathway can be prevented by IAA. A close relationship between the volume-, and NEM-induced K^+/Cl^- cotransport is suggested by kinetic data (Kaji and Kahn, 1985), by the inhibition of both kinds of ion movements by anti-L antibody in sheep red cells (Lauf, 1985; *see* Fig. 2.), and by the inhibitory effect of intracellular calcium on both pathways (Ellory et al., 1983; Lauf, 1985).

Recent data indicate that "young" human red cells (that is, the lightest fractions obtained by centrifugation) have a much greater swelling-induced K^+/Cl^- cotransport than the more dense cell fractions (Hall and Ellory, 1986), and NEM also induces a greater K^+/Cl^- cotransport in these younger cell populations (Brugnara and Tosteson, 1986; Berkowitz and Orringer, 1986). According to the findings of Canessa et al. (1986b,c), volume-stimulated K^+/Cl^- cotransport is especially abundant in sickle cells, and these cells also have a more pro-

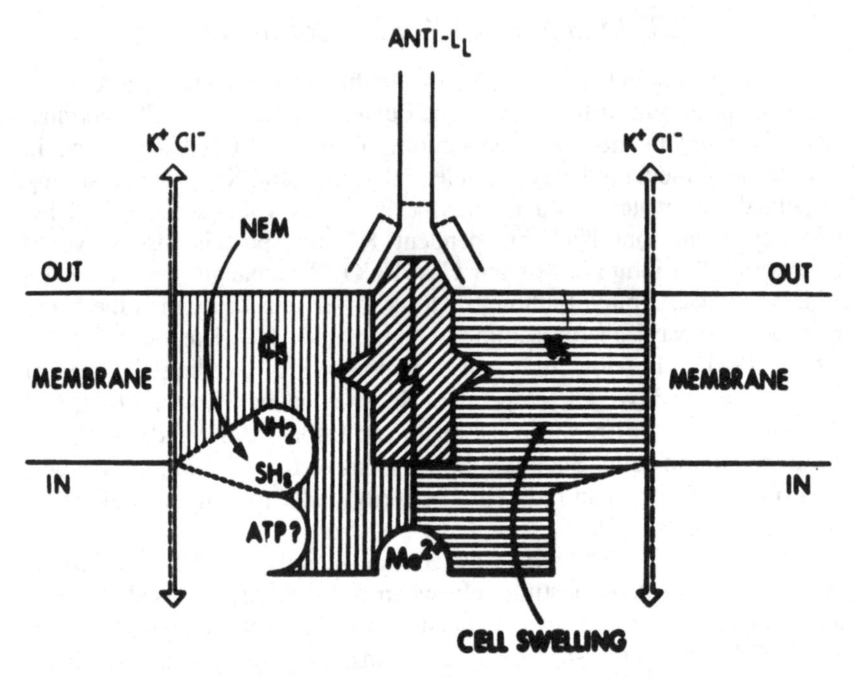

Fig. 2. Model for the NEM-, or swelling-induced K^+/Cl^- cotransport. The two-domain model shows the functional heterogeneity and immunologic homogeneity of thiol- and volume-stimulated K^+/Cl^- fluxes in LK sheep red cells. C_s and V_s are the chemically and volume-stimulated domains that interface with L_L, the antigen binding to the anti-L_L antibody. NEM acts by irreversibly binding to a stimulatory SH group, which owes its low pK_a to the presence of a nearby amino group that may be protonated and thus influence the metal–site interaction at the transporter. This process may be also volume-dependent and require the presence of ATP.

Reproduced from Lauf, 1985, by the permission of the author and the publisher.

nounced NEM-stimulated K^+/Cl^- transport. A K^+/Cl^- cotransport, independent of the $Na^+/K^+/Cl^-$ transport system, has been suggested to play an important role in the salt transport in epithelial cells as well (*see* Kinne et al., 1985; Geck and Heinz, 1986).

4. Summary and Conclusions

In this review, we have tried to summarize the data concerning various chemical modifiers of K^+ transport pathways in the red cell membranes. Since the ouabain-sensitive, ATP-fueled active $Na^+–K^+$

Table 1
Ouabain-Resistant Membrane-Mediated K^+ Transport Pathways in Red Cells*

Effectors	Ca^{2+}-Induced K^+ Transport	$Na^+/K^+/Cl^-$ Cotransport	K^+/Cl^- Cotransport
Cl^- dependence	No	Yes	Yes
Intracellular Ca^{2+}	Activates	Inhibits	Inhibits
Activators	A23187 + Ca^{2+}	β-receptor agonists (increased cAMP)	Volume increase
	Propranolol		NEM
	Pb^{2+}, La^{3+}	Volume decrease (avian cells)	
	Volume increase in Amphiuma (?)		
Inhibitors	Quinine, quinidine	Furosemide	Furosemide (higher conc.)
		Bumetanide	
	Cetiedil	Piretanide	
	Calmodulin antagonists	Ethacrynic acid	
	Oligomycin(A)	cAMP (human RBC?)	
	Ba^{2+}, La^{3+}		
	SH-reagents		
Cellular ATP	Not required	Required	Required (with NEM)
Probable mechanism	Conductive $K^{\pm}$ selective channels	Electrically silent (obligatory?) cotransport of $1Na^+$, $1K^+$ and 2 Cl^- ions	Electrically silent cotransport of $1K^+$ and $1Cl^-$ ion

*For details and references, *see* text.

transport is dealt with in great detail in other chapters of this book, we concentrated on the ouabain-resistant K^+ fluxes. Table 1 provides a backbone summary of the major features and effectors of the three major membrane-mediated, but ouabain-resistant K^+ transport systems described in several types of red cells. Of course, there can be major differences between the nucleated avian or fish and the nonnucleated mammalian red cells, but we tried to concentrate on the general characteristics and common basic features. We believe that examining the drug sensitivity of a given transport pathway may give a key to understand its molecular nature and to an evaluation of the specificity of the system as well. Both the Ca^{2+}-induced and the Cl^--dependent K^+ transport pathways examined in this chapter have a general prevalence in many tissues and cell types. They are involved in cell stimulation, hormonal regulation, and in sustaining the general electrolyte balance of the organisms (for recent reviews, *see* Kinne et al., 1985; Palfrey and Rao, 1983; Saier and Boyden,1984; Petersen, 1986; Lauf, 1985; Hoffmann, 1986; Geck and Heinz, 1986; Chipperfield, 1986). Our task here was to provide a survey of these basic systems in the simplest model of the animal cells, the red blood cell.

Acknowledgments

The authors are grateful for discussing recent experimental findings and interpretations with Drs. T. J. McManus, J. C. Parker, M. Canessa, C. Brugnara, and D. C. Tosteson.

Note: This manuscript was submitted in 1987, thus it reflects the data available at that time.

References

Adragna, N. D., Perkins, C. M., and Lauf, P. K. (1985) Furosemide-sensitive Na-K cotransport and cellular metabolism in human erythrocytes. *Biochim. Biophys. Acta* **812,** 293–296.

Alvarez, J., Garcia-Sancho, J., and Herreros, B. (1986a) Inhibition of Ca-dependent K channels by lead in one-step inside-out vesicles from human red cell membranes. *Biochim. Biophys. Acta* **857,** 291–294.

Alvarez, J., Garcia-Sancho, J., and Herreros, B. (1986b) The role of calmodulin on Ca-dependent K transport regulation in the human red cell. *Biochim. Biophys. Acta* **860,** 25–34.

Armando-Hardy, M., Ellory, J. C., Ferreira, H. G., Fleminger, S., and Lew, V. L. (1975) Inhibition of the calcium-induced increase in the potassium permeability of human red blood cells by quinine. *J. Physiol. (London)* **250,** 32–33P.

Banks, B. E. C., Brown, C., Burgess, G. M., Burnstock, G., Claret, M., Coks, T. M., and Jenkinson, D. A. (1979) Apamin blocks certain neurotransmitter-induced increases in potassium permeability. *Nature (London)* **282,** 415–417.

Bauer, J. and Lauf, P. K. (1983) Thiol-dependent passive K/Cl transport in sheep red cells: III. Differential reactivity of membrane SH groups with *N*-ethylmaleimide and iodoacetamide. *J. Membrane Biol.* **73,** 257–261.

Berkowitz, L. R. and Orringer, E. P. (1981) Effect of cetiedil, an *in vitro* antisickling agent on erythrocyte membrane cation permeability. *J. Clin. Invest.* **68,** 1215–1220.

Berkowitz, L. R. and Orringer, E. P. (1985) Passive sodium and potassium movements in sickle erythrocytes. *Am. J. Physiol.* **249,** C208–C214.

Berkowitz, L. R. and Orringer, E. P. (1986) The effect of *N*-ethylmaleimide (NEM) on K transport in human erythrocytes of different ages. *J. Gen. Physiol.* **88,** 12a.

Blum, R. M. and Hoffman, J. F. (1971) The membrane locus of Ca-stimulated K transport in energy-depleted human red blood cells. *J. Membrane Biol.* **6,** 315–328.

Blum, R. M. and Hoffman, J. F. (1972) Ca-induced K transport in human red cells: Localization of the Ca-sensitive site to the inside of the membrane. *Biochem. Biophys. Res. Commun.* **46,** 1146–1151.

Brazy, P. C. and Gunn, R. B. (1976) Furosemide inhibition of chloride transport in human red blood cells. *J. Gen. Physiol.* **68,** 583–599.

Brugnara, C. and Tosteson, D. C. (1986) Cell volume, K transport and cell density in human erythrocytes. *J. Gen. Physiol.* **88,** 14a.

Brugnara, C., Canessa, M., Cusi, D., and Tosteson, D. C. (1986) Furosemide-sensitive Na and K fluxes in human red cells. Net uphill Na extrusion and equilibrium properties. *J. Gen. Physiol.* **87,** 91–112.

Burgess, G. M., Claret, M., and Jenkinson, D. H. (1981) Effects of quinine and apamin on the calcium-dependent potassium permeability of mammalian hepatocytes and red cells. *J. Physiol. (London)* **317,** 67–90.

Cala, P. M. (1980) Volume regulation by Amphiuma red blood cells: The membrane potential and its implications regarding the nature of the ion-flux pathways. *J. Gen. Physiol.* **76,** 683–708.

Cala, P. M. (1983a) Cell volume regulation by Amphiuma red blood cells. *J. Gen. Physiol.* **82,** 761–784.

Cala, P. M. (1983b) Volume regulation by red blood cells: Mechanisms of ion transport. *Mol. Physiol.* **4,** 33–52.

Cala, P. M. (1985a) Volume regulation by Amphiuma red blood cells: characteristics of volume-sensitive K/H and Na/H exchange. *Mol. Physiol.* **8,** 199–214.

Cala, P. M. (1985b) Volume regulation by Amphiuma red blood cells: strategies for identifying alkali metal/H transport. *Fed. Proc.* **44,** 2500–2507.

Cala, P. M. Mandel, L. J., and Murphy, E. (1986) Volume regulation by Amphiuma red blood cells: cytosolic free Ca and alkali metal-H exchange. *Am. J. Physiol.* **250,** C423–C429.

Canessa, M., Brugnara, C., Cusi, D., and Tosteson, D. C. (1986a) Modes of operation and variable stoichiometry of the furosemide-sensitive Na and K fluxes in human red cells. *J. Gen. Physiol.* **87,** 113–142.

Canessa, M., Fabry, M. E., Spalvins, A., Blumenfeld, N., and Nagel, R. L. (1986b) A volume-dependent KCl transporter is highly expressed in young human red cells with AA and SS hemoglobin: Physiological and pathophysiological implications. *J. Gen. Physiol.* **88,** 14a.

Canessa, M., Spalvins, A., and Nagel, R. L. (1986c) Volume-dependent and NEM-stimulated K, Cl transport is elevated in oxygenated SS, SC and CC human red cells. *FEBS Letters,* **200,** 197–202.

Cass, A. and Dalmark, A. (1973) Equilibrium dialysis of ions in nystatin-treated red cells. *Nature* **244,** 47–49.

Cass, A., Finkelstein, A., and Krespi, V. (1970) The ion permeability induced in thin lipid membranes by the polyene antibiotics nystatin and amphotericin B. *J. Gen. Physiol.* **556,** 100–124.

Chipperfield, A. R. (1980) An effect of chloride on (Na+K) cotransport in human red cells. *Nature* **286,** 281–282.

Chipperfield, A. R. (1981) Chloride dependence of furosemide and phloretin-sensitive passive sodium and potassium fluxes in human red cells. *J. Physiol. (London)* **312,** 435–444.

Chipperfield, A. R. (1986) The (Na-K-Cl) co-transport system. *Clin. Sci.* **71,** 465–476.

Dagher, G., Brugnara, C., and Canessa, M. (1985) Effect of metabolic depletion on the furosemide-sensitive Na and K fluxes in human red cells. *J. Membrane Biol.* **86,** 145–155.

DeFlora, A., Benatti, U., Guida, L., Forteleoni, G., and Meloni, T. (1985) Favism; disordered erythrocyte calcium homeostasis. *Blood* **66,** 294–297.

Deuticke, B., Kim, M., and Zollner, C. (1973) The influence of amphotericin B on permeability of mammalian erythrocytes to nonelectrolytes, anions and cations. *Biochim. Biophys. Acta* **318,** 345–359.

Duhm, J. and Gobel, B. O. (1984) Na-K transport and volume of rat erythrocytes under dietary K deficiency. *Am. J. Physiol.* **246,** C20–C29.

Dunham, P. B. and Ellory, J. C. (1981) Passive potassium transport in low potassium sheep red cells: dependence upon cell volume and chloride. *J. Physiol. (London)* **318,** 511–530.

Dunham, P. B., Stewart, G. W., and Ellory, J. C. (1980) Chloride-activated passive potassium transport in human erythrocytes. *Proc. Natl. Acad. of Sci. USA* **77,** 1711–1715.

Dunn, M. J. (1970) The effects of transport inhibitors on sodium outflux and influx in red blood cells: Evidence for exchange diffusion. *J. Clin. Invest.* **52,** 1804–1814.

Ekman, A., Manninen, V., and Salminen, S. (1969) Ion movements in red cells treated with propranolol. *Acta Physiol. Scand.* **75,** 333–344.

Ellory, J. C. and Hall, A. C. (1984) Na, K, Cl-cotransport stoichiometry: measurement of bumetanide-sensitive fluxes in ferret red cells. *J. Physiol. (London)* **357,** 63.

Ellory, J. C. and Stewart, G. W. (1982) The human erythrocyte Cl-dependent Na-K cotransport system as a possible model for studying the action of loop diuretics. *Br. J. Pharmacol.* **75,** 183–188.

Ellory, J. C., Flatman, P. W., and Stewart, G. W. (1983) Inhibition of human red cell sodium and potassium transport by divalent cations. *J. Physiol. (London)* **340,** 1–17.

Ellory, J. C., Hall, A. C., and Stewart, G. W. (1985) Volume-sensitive cation fluxes in mammalian red cells. *Mol. Physiol.* **8,** 235–246.

Ellory, J. C., Dunham, P. B., Logue, P. J., and Stewart, G. W. (1982) Anion-dependent cation transport in erythrocytes. *Trans. Roy. Soc. B* **299,** 483–495.

Freedman, J. C. and Hoffman, J. F. (1979) Ionic and osmotic equilibria of human red blood cells treated with nystatin. *J. Gen. Physiol.* **74,** 157–185.

Garay, R. P., Dagher, G., Pernollet, M. G., Devynck, M. A., and Meyer, P. (1980) Inherited defect in a Na, K-cotransport system in erythrocytes from essential hypertensive patients. *Nature* **284,** 281–283.

Garcia-Sancho, J., Sanchez, A., and Herreros, B. (1982) All-or-none response of the Ca-dependent K channel in inside-out vesicles. *Nature* **296,** 744–746.

Gárdos, G. (1958a) Effect of ethylenediamine-tetraacetate on the permeability of human erythrocytes. *Acta Physiol. Hung.* **14,** 1–5.

Gárdos, G. (1958b) The function of calcium in the potassium permeability of human erythrocytes. *Biochim. Biophys. Acta* **30,** 653–654.

Gárdos, G. (1959) The role of calcium in the potassium permeability of human erythrocytes. *Acta Physiol. Hung.* **15,** 121–125.

Gárdos, G., Lassen, U. V., and Pape, L. (1976) Effect of antihistamines and chlorpromazine on the calcium-induced hyperpolarization of the Amphiuma red cell membrane. *Biochim. Biophys. Acta* **448,** 599–606.

Gárdos, G., Szász, I., and Sarkadi, B. (1975) Mechanism of Ca-dependent K transport in human red cells. *FEBS Proc.* **35,** 167–180.

Gárdos, G., Szász, I., and Sarkadi, B. (1977) Effect of intracellular calcium on the cation transport processes in human red cells. *Acta Biol. Med. Germ.* **36,** 823–829.

Garrahan, P. J. and Rega, A. F. (1967) Cation loading of red blood cells. *J. Physiol. (London)* **193,** 459–466.

Geck, P. and Heinz, E. (1986) The Na-K-2Cl cotransport system. *J. Membrane Biol.* **91,** 97–105.

Glynn, I. M. and Warner, A. E. (1972) Nature of the calcium dependent potassium leak induced by (+)-propranolol, and its possible relevance to the drug's antiarrhythmic effect. *Br. J. Pharmacol.* **44,** 271–278.

Grinstein, S., DuPre, A., and Rothstein, A. (1982) Volume regulation by human lymphocytes: role of calcium. *J. Gen. Physiol.* **79,** 849–868.

Grinstein, S., Rothstein, A., Sarkadi, B., and Gelfand, E. W. (1984) Responses of lymphocytes to anisotonic media: volume-regulating behavior. *Am. J. Physiol.* **246,** C204–C215.

Haas, M. and Forbush, B., III. (1986) [3 H]Bumetanide binding to duck red cells. *J. Biol. Chem.* **261,** 8434–8441.

Haas, M. and McManus, T. J. (1983) Bumetanide inhibits (Na + K + 2Cl) cotransport at a chloride site. *Am. J. Physiol.* **245,** C353–C355.

Haas, M. and Schmidt, W. F., III. (1985) *p*-Chloromercuribenzenesulfonic acid stimulation of chloride-dependent sodium and potassium transport in human red blood cells. *Biochim. Biophys. Acta* **814,** 43–49.

Haas, M., Schmidt, W. F., III, and McManus, T. J. (1982) Catecholamine-stimulated ion transport in duck red cells. Gradient effects in electrically neutral (Na + K + 2Cl) co-transport. *J. Gen. Physiol.* **80,** 125–147.

Haas, M. J., Schooler, J., and Tosteson, D. C. (1975) Coupling of lithium to sodium transport in human red cells. *Nature* **258,** 425–427.

Hall, A. C. and Ellory, J. C. (1985) Measurements and stoichiometry of bumetanide-sensitive (2Na,1K:3Cl) cotransport in ferret red cells. *J. Membrane Biol.* **85,** 205–213.

Hall, A. C. and Ellory, J. C. (1986) Evidence for the presence of volume-sensitive KCl transport in 'young' human red cells. *Biochim. Biophys. Acta* **858,** 317–320.

Heinz, A. and Passow, H. (1980) Role of external potassium in the calcium-induced potassium efflux from human red blood cell ghosts. *J. Membrane Biol.* **57,** 119–131.

Hermann, A. and Gorman, A. L. (1981) Effects of tetraethylammonium on potassium currents in a molluscan neuron. *J. Gen. Physiol.* **78,** 87–110.

Hoffman, J. F. and Knauf, P. A. (1973) The mechanism of the increased K transport induced by Ca in human red blood cells, in *Erythrocytes, Thrombocytes, Leukocytes* (Gerlach, E., Moser, K., Deutsch, E., and Wilmanns, W., eds.), Georg Thieme, Stuttgart, pp. 66–70.

Hoffman, J. F. and Kregenow, F. M (1966) The characterization of new energy dependent cation transport processes in red blood cells. *Ann. New York Acad. Sci.* **137,** 566–576.

Hoffmann, E. K. (1986) Anion transport systems in the plasma membrane of vertebrate cells. *Biochim. Biophys. Acta* **864,** 1–31.

Hoffmann, E. K., Simonsen, L. O., and Lambert, I. H. (1984) Volume-induced increase of K and Cl permeabilities in Ehrlich ascites tumor cells. Role of internal Ca. *J. Membrane Biol.* **78,** 211–222.

Hugues, M., Schmid, H., and Lazdunski, M (1982) Identification of a protein component of the Ca-dependent K channel by affinity labeling with apamin. *Biochem. Biophys. Res. Commun.* **107,** 1557–1582.

Hunter, M. J. (1971) A quantitative estimate of the non-exchange restricted chloride permeability of the human red cell. *J. Physiol. (London)* **218,** 49P.

Hunter, M. J. (1977) Human erythrocyte anion permeabilities measured under conditions of net charge transfer. *J. Physiol. (London)* **268,** 35–49.

Ikehara, T., Yamaguchi, H., Hosokawa, K., Yonezu, T., and Miyamoto, H. (1986) Effects of nystatin on intracellular contents and membrane transport of alkali cations and cell volume in HeLa cells. *J. Membrane Biol.* **90,** 231–240.

Iwatsuki, N. and Petersen, O. H. (1985) Action of tetraethylammonium on calcium-activated potassium channels in pig pancreatic acinar cells studied by patch-clamp single-channel and whole-cell current recording. *J. Membrane Biol.* **86,** 139–144.

Jacob, H. S. and Jandl, J. H. (1962) Effects of sulfhydryl inhibition on red blood cells. I. Mechanism of hemolysis. *J. Clin. Invest.* **41,** 779–792.

Kaji, D. and Kahn, T. (1985) Kinetics of Cl-dependent K influx in human erythrocytes with and without external Na: effect of NEM. *Am. J. Physiol.* **249,** C490–C496.

Karlish, S. J. D., Ellory, J. C., and Lew, V. L. (1981) Evidence against Na-pump mediation of Ca-activated K transport and diuretic-sensitive Na/K cotransport. *Biochim. Biophys. Acta* **646,** 353–355.

Kinne, R., Hannafin, J. A., and Konig, B. (1985) Role of the NaCl-KCl cotransport system in active chloride absorption and secretion. *Ann. New York Acad. Sci.* **156,** 198–206.

Knauf, P. A., Riordan, J. R., Schuhmann, B., and Passow, H. (1974) Effects of external potassium on calcium-induced potassium leakage from human red blood cell ghosts, in *Membranes: Comparative Biochemistry and Physiology of Transport* (Bolis, L., Bloch, K., Luria, S. E., and Lynen, F., eds.), North Holland, Amsterdam, pp. 305–309.

Knauf, P. A., Riordan, J. R., Schuhmann, B., Wood-Guth, I., and Passow, H. (1975) Calcium-potassium stimulated net potassium efflux from human erythrocyte ghosts. *J. Membrane Biol.* **25,** 1–22.

Kregenow, F. M. (1971) The response of duck erythrocytes to nonhemolytic hypotonic media. Evidence for a volume-controlling mechanism. *J. Gen. Physiol.* **58,** 372–395.

Kregenow, F. M. (1977) Transport in red cells, in *Membrane Transport in Red Cells* (Ellory, J. C. and Lew, V. L., eds.), Academic Press, New York, pp. 384–426.

Kregenow, F. M. (1981) Osmoregulatory salt transporting mechanisms: Control of cell volume in anisotonic media. *Ann. Rev. Physiol.* **45,** 493–505.

Lackington, I. and Orrego, F. (1981) Inhibition of calcium-activated potassium conductance of human erythrocytes by calmodulin inhibitory drugs. *FEBS Lett.* **133,** 103–106.

Lassen, U. V., Pape, L., and Vestergaard-Bogind, B. (1973) Membrane potential of Amphiuma red cells: Effect of calcium, in *Erythrocytes, Thrombocytes, Leukocytes* (Gerlach, E., Moser, K., Deutsch, E., and Wilmanns, W., eds.), Georg Thieme, Stuttgart, pp. 33–36.

Lassen, U. V., Pape, L., and Vestergaard-Bogind, B. (1976) Effect of calcium on the membrane potential of Amphiuma red cells. *J. Membrane Biol.* **26,** 51–70.

Lauf, P. K. (1982) Evidence for chloride-dependent potassium and water transport induced by hyposmotic stress in erythrocytes of the marine teleost, Opsanus tau. *J. Comp. Physiol.* **146,** 9–16.

Lauf, P. K. (1983a) Thiol-dependent passive K/Cl transport in sheep red cells. I. Dependence on chloride and external K/Rb/ions. *J. Membrane Biol.* **73,** 237–246.

Lauf, P. K. (1983b) Thiol-dependent passive K/Cl transport in sheep red cells: V. Dependence on metabolism. *Am. J. Physiol.* **245,** C445–C448.

Lauf, P. K. (1984) Thiol-dependent passive K/Cl transport in sheep red cells: IV. Furosemide inhibition as a function of external Rb, Na, and Cl. *J. Membrane Biol.* **77,** 57–62.

Lauf, P. K. (1985) K:Cl cotransport: Sulfhydryls, divalent cations, and the mechanism of volume activation in a red cell. *J. Membrane Biol.* **88,** 1–13.

Lauf, P. K. and Mangor-Jensen, A. (1984) Effects of A23187 and Ca on volume- and thiol-stimulated ouabain-resistant KCl fluxes in low K sheep erythrocytes. *Biochem. Biophys. Res. Commun.* **125,** 790–796.

Lauf, P. K. and Theg, B. E. (1980) A chloride dependent K flux induced by *N*-ethylmaleimide in genetically low K sheep and goat erythrocytes. *Biochem. Biophys. Res. Commun.* **92,** 1422–1428.

Lauf, P. K., Adragna, N., and Garay, R. P. (1984) Activation by *N*-ethylmaleimide of a latent K Cl flux in human red blood cells. *Am. J. Physiol.* **246,** C385–C390.

Lew, V. L. (1971) Effect of ouabain on the Ca-dependent increase in K permeability in ATP depleted guinea-pig red cells. *Biochim. Biophys. Acta* **249,** 236–239.

Lew, V. L. (1974) On the mechanism of the Ca-induced increase in K permeability observed in human red cell membranes, in *Comparative Biochemistry and Physiology of Transport* (Bolis, L., Bloch, K., Luria, S. E., and Lynen, F., eds.), North Holland, Amsterdam, pp. 310–316.

Lew, V. L. and Bookchin, R. M. (1980) A Ca-refractory state of the Ca-sensitive K permeability mechanism in sickle cell anaemia red cells. *Biochim. Biophys. Acta* **602,** 196–200.

Lew, V. L. and Bookchin, R. M. (1986) Volume, pH, and ion-content regulation in human red cells: analysis of transient behaviour with an integrated model. *J. Membrane Biol.* **92,** 57–74.

Lew, V. L. and Ferreira, H. G. (1976) Variable Ca sensitivity of a K selective channel in intact red cell membranes. *Nature* **263,** 336–338.

Lew, V. L. and Ferreira, H. G. (1978) Calcium transport and the properties of a Ca-activated potassium channel in red cell membranes, in *Current Topics in Membranes and Transport*, Vol 10 (Bronner, F. and Kleinzeller, A., eds.), Academic Press, New York, pp. 217–277.

Lew, V. L., Muallem, S., and Seymour, C. A. (1982) Properties of the Ca-activated K channel in one-step inside-out vesicles from human red cell membranes. *Nature* **296,** 742–744.

Lukačovič, M. F., Toon, M. R., and Solomon, A. R. (1984) Site of red cell cation leak induced by mercurial sulfhydryl reagents. *Biochim. Biophys. Acta* **772,** 313–320.

Lytle, C. and McManus, T. J. (1986) A minimal kinetic model of Na + K + 2Cl cotransport with ordered binding and glide symmetry. *J. Gen. Physiol.* **88,** 36a.

McManus, T. J., Haas, M., Starke, L. C., and Lytle, C. Y. (1985) The duck red cell model of volume-sensitive chloride-dependent cation transport. *Ann. New York Acad. Sci.* **456,** 183–186.

Nagase, H., Ozaki, H., and Urakawa, N. (1984) Inhibitory effect of calmodulin inhibitors on polytosein-induced K release from rabbit erythrocytes. *FEBS Letters* **178,** 44–46.

Orringer, E. P. and Parker, J. C. (1973) Ion and water movements in red blood cells, in: *Progress in Hematology*, Vol. 8. (Brown, E. B., ed.), Grune and Stratton, New York, pp. 1–23.

Ørskov, S. L., (1935) Untersuchungen über den Einfluss von Kohlensäure und Blei auf die Permeabilität der Blutkörperchen für Kalium und Rubidium. *Biochem. Z.* **279,** 250–261.

Palfrey, H. C. and Rao, M. C. (1983) Na/K/Cl co-transport and its regulation. *J. Exp. Biol.* **106,** 43–54.

Palfrey, H. C., Feit, P. W., and Greengard, P. (1980) cAMP-stimulated cation transport in avian erythrocytes: inhibition by 'loop' diuretics. *Am. J. Physiol.* **238,** C130–C148.

Parker, J. C. (1983) Hemolytic action of potassium salts on dog red blood cells. *Am. J. Physiol.* **244,** C313–C317.

Parker, J. C. and Berkowitz, L. R. (1986) Genetic variants affecting the structure and function of the human red cell membrane, in *Physiology of Membrane Disorders* (Andreoli, T. E., Hoffman, J. F., Fanestil, D. D., and Schultz, S. G., eds.) Plenum Press, New York, pp. 785–814.

Passow, H. (1981) Selective enhancement of potassium efflux from red blood cells by lead, in *The Functions of Red Blood Cells: Erythrocyte Pathobiology* (Wallach, D. F., ed.) Alan R. Liss, New York, pp. 80–104.

Petersen, O. H. (1986) Calcium-activated potassium channels and fluid secretion by exocrine glands. *Am. J. Physiol.* **251,** G1–G13.

Plishker, G. A. (1985) Iodoacetic acid inhibition of calcium-dependent potassium efflux in red blood cells. *Am. J. Physiol.* **248,** C419–C424.

Podack, E. R. and Dennert, G. (1983) Assembly of two types of tubules with putative cytolytic function by cloned natural killer cells. *Nature* **302,** 442–445.

Pressman, B. C. (1976) Biological application of ionophores. *Ann. Rev. Biochem.* **45,** 323–352.

Pressman, B. C., Harris, E. J., Jagger, W. S., and Johnson, J. H. (1967) Antibiotic-mediated transport of alkali ions across lipid barriers. *Proc. Natl. Acad. Sci. USA* **58,** 1949–1956.

Reichstein, E. and Rothstein, A. (1981) Effects of quinine on Ca-induced K efflux from human red blood cells. *J. Membrane Biol.* **59,** 57–63.

Richhardt, H. W., Fuhrmann, G. F., and Knauf, P. A. (1979) Dog red blood cells exhibit a Ca-stimulated increase in K permeability in the absence of (Na, K) ATPase activity. *Nature* **279,** 248–250.

Riordan, J. R. and Passow, H. (1971) Effects of calcium and lead on potassium permeability of human erythrocyte ghosts. *Biochim. Biophys. Acta* **249,** 601–605.

Rothstein, A. (1981) Mercurials and red cell membranes, in *The Functions of Red Blood Cells: Erythrocyte Pathobiology* (Wallach, D. F., ed.), Alan R. Liss, New York, pp. 105–131.

Sachs, J. R. (1971) Ouabain-insensitive sodium movements in the human red blood cell. *J. Gen. Physiol.* **57,** 259–282.

Saier, M. H., Jr. and Boyden, D. A. (1984) Mechanism, regulation and physiological significance of the loop diuretics-sensitive NaCl/KCl symport system in animal cells. *Mol. Cell. Biochem.* **59,** 11–32.

Sarkadi, B. and Gárdos, G. (1985) Calcium-induced potassium transport in cell membranes, in *The Enzymes of Biological Membranes*, 2nd edition, vol. 3 (Martonosi A. N., ed.), Plenum Press, New York, pp. 193–234.

Sarkadi, B. and Tosteson, D. C. (1979) Active cation transport in human red cells, in *Membrane Transport in Biology*, Vol. 2 (Giebisch, G., Tosteson, D. C., and Ussing, H. H., eds.), Springer Verlag, Berlin, pp. 117–160.

Sarkadi, B., Szász, I., and Gárdos, G. (1976) The use of ionophores for rapid loading of human red cells with radioactive cations for cation pump studies. *J. Membrane Biol.* **26,** 357–370.

Sarkadi, B., Mack, E., and Rothstein, A. (1984a) Ionic events during the volume response of human peripheral blood lymphocytes to hypotonic media. I. Distinction between volume-activated Cl and K conductance pathways. *J. Gen. Physiol.* **83,** 497–512.

Sarkadi, B., Mack, E., and Rothstein, A. (1984b) Ionic events during the volume response of human peripheral blood lymphocytes to hypotonic media. II. Volume- and time-dependent activation and inactivation of ion transport pathways. *J. Gen. Physiol.* **83,** 513–527.

Sarkadi, B., Alifimoff, J. K., Gunn, R. B., and Tosteson, D. C. (1978) Kinetics and stoichiometry of Na-dependent Li-transport in human red blood cells. *J. Gen. Physiol.* **72,** 249–265.

Sarkadi, B., Cheung, R., Mack, E., Grinstein, S., Gelfand, E. W., Rothstein, A. (1985a) Characterization of cation and anion transport pathways in the volume regulatory response of human lymphocytes to hypoosmotic media. *Am. J. Physiol.* **248,** C480–C487.

Sarkadi, B., Grinstein, S., Rothstein, A., and Gárdos, G. (1985b) Analysis of Ca-induced K transport by human erythrocytes in propionate media. *Acta Biochim. Biophys. Acad. Sci. Hung.* **20,** 193–202.

Schatzmann, H. J. (1953) Herzglykoside als Hemmstoffe für den aktiven Kalium and Natrium-transport durch die Erythrocytenmembran. *Helv. Physiol. Acta* **11,** 346–354.

Schatzmann, H. J. (1973) Dependence on calcium concentration and stoichiometry of the Ca-pump in human red cells. *J. Physiol. (London)* **235,** 551–569.

Schmidt, W. F. and McManus, T. J. (1974) A furosemide-sensitive cotransport of Na plus K into duck red cells activated by hypertonicity or catecholamines. *Fed. Proc.* **33,** 1457.

Schubert, A. and Sarkadi, B. (1977) Kinetic studies on the calcium-dependent potassium transport in human red cells. *Acta Biochim. Biophys. Acad. Sci. Hung.* **12,** 207–216.

Schwarz, W. and Passow, H. (1983) Ca-activated K channels in erythrocytes and excitable cells. *Ann. Rev. Physiol.* **45,** 359–374.

Shields, M., Grygorczyk, R., Fuhrmann, G. F., Schwarz, W., and Passow, H. (1985) Lead-induced activation and inhibition of potassium-selective channels in the human red blood cell. *Biochim. Biophys. Acta* **815,** 223–232.

Simons, T. J. B. (1976a) Calcium-dependent potassium exchange in human red cell ghosts. *J. Physiol. (London)* **256,** 227–244.

Simons, T. J. B. (1976b) Carbocyanide dyes inhibit Ca-dependent K efflux from human red cell ghosts. *Nature* **267,** 467–469.

Simons, T. J. B. (1979) Actions of a carbocyanine dye on calcium-dependent potassium transport in human red cell ghosts. *J. Physiol. (London)* **288,** 481–507.

Smalley, C. E., Tucker, E. M., Dunham, P. B., and Ellory, D. C. (1982) Interaction of L antibody with low potassium-type sheep red cells: Resolution of two separate functional antibodies. *J. Membrane Biol.* **64,** 167–174.

Sutherland, R. M., Rothstein, A., and Weed, R. I. (1967) Erythrocyte membrane sulfhydryl groups and cation permeability. *J. Cell. Physiol.* **69,** 185–198.

Szász, I. and Gárdos, G. (1974) Mechanism of various drug effects on the Ca-dependent K-efflux from human red blood cells. *FEBS Lett.* **44,** 213–216.

Szász, I., Sarkadi, B., and Gárdos, G. (1977) Mechanism of Ca-dependent selective rapid K-transport induced by propranolol in red cells. *J. Membrane Biol.* **35,** 75–93.

Szász, I., Sarkadi, B., and Gárdos, G. (1978a) Effects of drugs on calcium-dependent rapid potassium transport in calcium-loaded intact red cells. *Acta Biochim. Biophys. Acad. Sci. Hung.* **13,** 133–141.

Szász, I., Sarkadi, B., Schubert, A., and Gárdos, G. (1978b) Effects of lanthanum on calcium-dependent phenomena in human red cells. *Biochim. Biophys. Acta* **512,** 331–340.

Szönyi, S. (1960) Wirkung von Fluorid auf die Verteilung von Kalium und Natrium sowie auf die CO_2-Bindung in menschlichen Blut. *Acta Physiol. Hung.* **17,** 9–13.

Tosteson, D. C. (1964) Regulation of cell volume by sodium and potassium transport, in *The Cellular Functions of Membrane Transport* (Hoffman, J. F., ed.), Prentice Hall, Englewood Cliffs, pp. 3–22.

Tosteson, D. C. (1981) Cation countertransport and cotransport in human red cells. *Fed. Proc.* **40,** 1429–1433.

Tosteson, D. C. and Hoffman, J. F. (1960) Regulation of cell volume by active cation transport in high and low potassium sheep red cells. *J. Gen. Physiol.* **44,** 169–194.

Tosteson, D. C. and Johnson, J. (1957) The coupling of potassium transport with metabolism in duck red cells, I. The effect of sodium fluoride and other metabolic inhibitors. *J. Cell. Comp. Physiol.* **50,** 169–184.

Varečka, L. and Carafoli, E. (1982) Vanadate-induced movements of Ca and K in human red cells. *J. Biol. Chem.* **257,** 7414–7421.

Verma, A. K. and Penniston, J. T. (1985) Evidence against involvement of the human erythrocyte plasma membrane Ca-ATPase in Ca-dependent K transport. *Biochim. Biophys. Acta* **815,** 135–138.

Wiater, L. A. and Dunham, P. B. (1983) Passive transport of K and Na in human red blood cells: sulfhydryl binding agents and furosemide. *Am. J. Physiol.* **245,** C348–C356.

Wilbrandt, W. (1940) Die Abhängigkeit der Ionenpermeabilität der Erythrocyten vom glykolytischen Stoffwechsel. *Pflügers Arch.* **243,** 519–536.

Wiley, J. S. and Cooper, R. A. (1974) A furosemide-sensitive cotransport of sodium plus potassium in the human red cells. *J. Clin. Invest.* **53,** 745–755.

Wood, P. G. and Mueller, H. (1984) Modification of the cation selectivity filter and the calcium receptor of the Ca-stimulated K channels in resealed ghosts of human red blood cells by low level of incorporated trypsin. *Eur. J. Biochem.* **141,** 91–95.

Wood, P. G. and Mueller, H. (1985) The effects of terbium (III) on the Ca-activated K channel found in the resealed human erythrocyte membrane. *Eur. J. Biochem.* **146,** 65–69.

Yingst, D. R. and Hoffman, J. F. (1984) Ca-induced K transport in human red blood cell ghosts containing arsenazo III: Transmembrane interactions of Na, K and Ca and the relationship to the functioning Na-, K-pump. *J. Gen. Physiol.* **83,** 19–45.

4
PATHOPHYSIOLOGY

Membrane Lipid Changes in the Abnormal Red Cell

Paolo Luly

1. Introduction

The relevance of a lipid component influencing the physiological properties of cell membranes was established at the end of the 19th century by the pioneering work of Overton (Overton, 1902). A structural role for lipids was suggested by the work of Gortel and Grendel (Gortel and Grendel, 1925), who demonstrated that the lipid extracted from erythrocyte membrane was sufficient to form a continuous bilayer.

In the subsequent decades, plasma membrane models have evolved (Bretscher and Raff, 1975) until the current "fluid mosaic model" was proposed by Singer and Nicolson (Singer and Nicolson, 1972). The lipid bilayer constitutes a dynamic structure in which, under physiological conditions, the lipid components show a certain freedom to diffuse in the two-dimensional framework. This diffusion can be translational—a phenomenon that can be resolved into lateral or vertical motions—or rotational, which in turn can be resolved into uniaxial and flip-flop rotation (Shinitzky, 1984). On such a basis, lipid fluidity as a general term indicates the relative motional freedom of lipid molecules in the two-dimensional lipid framework, including both rotational and lateral motions as well as rate and extent of movement. The estimation of lipid fluidity can be obtained by several spectroscopic methods, such as Nuclear Magnetic Resonance (NMR), Electron Spin Resonance (ESR), and Fluorescence Polarization (FP) (Chapman and Hayward, 1985; Shinitzky and Barenholz, 1978).

The lipid fluidity of biological membranes can be modulated by physical and chemical effectors (Shinitzky, 1984). Physical effectors, such as temperature, pH, membrane potential, and calcium, are definitely beyond the objectives of the present chapter; chemical effectors, which will now be briefly summarized, receive greater attention in the section 2.

as to their role in the abnormal erythrocyte plasma membrane. The main chemical effectors that modulate plasma membrane lipid fluidity are: the cholesterol level, which is often reported as cholesterol:phospholipid ratio; the degree of unsaturation of phospholipid acyl chains; the sphingomyelin:lecithin (or phosphatidylcholine) ratio; and the protein concentration, which can be referred to as protein:lipid ratio (Shinitzky and Henkart, 1979).

Cholesterol may be regarded as the main modulator of plasma membrane lipid fluidity. In a phospholipid system that is above the phase transition temperature—the normal condition for the nonhibernating mammal—cholesterol increases the order of the lipid bilayer, thereby reducing the flexibility of lipid hydrocarbon chains, or in other words, hampering their mobility and thus reducing the overall lipid fluidity (Chapman, 1973). If, on one side, cholesterol increases the packing efficiency of phospholipid molecules (Demel and De Kruyff, 1976), the introduction of a *cis*-double bond in a phospholipid acyl chain increases its specific volume (Shinitzky, 1984), thus decreasing the phospholipid packing. For these reasons, the relevance of cholesterol:phospholipid (C:PL) ratio as an indicator of lipid fluidity conditions of plasma membranes is now clear; in the normal human erythrocyte, the C:PL molar ratio is between 0.9 and 1.0.

Sphingomyelin (SM), because of its high phase transition temperature, acts as a rigidifier of membrane fluidity, whereas phosphatidylcholine (PC) has an opposite effect. For this reason, the ratio of these two phospholipids can also be a good indicator of membrane lipid fluidity (Shinitzky, 1984), and in some experimental conditions, it has been altered in order to modify the physiological properties of mammalian erythrocytes: in particular, an increase of PC:SM molar ratio increases plasma membrane fluidity as well as osmotic fragility in sheep red cells (Borochov et al., 1977).

The presence of a high content of proteins in a plasma membrane has a marked rigidifying effect (Singer and Nicolson, 1972). The volume occupied by a single protein molecule can surpass by up to two orders of magnitude the volume of a phospholipid, thus increasing the overall rigidity of the bilayer. It has been hypothesized that the protein effect on membrane lipid dynamics is rather similar to that of cholesterol (Shinitzky and Inbar, 1976).

Alterations of plasma membrane lipid fluidity reportedly affect membrane-located physiological functions, such as enzymatic activities, transport, and permeability. Changes in the fluidity of the phospholipid hydrocarbon chains may influence membrane-bound enzymatic activities in different ways: the lipid fluidity could modulate the diffusion or rotation of the enzyme molecules in the bilayer, but could also affect the substrate accessibility through the lipid domain. Alternatively, the membrane lipid physical state could, in some way, alter the cooperative activity of enzyme subunits and the catalytic properties. The investigation on membrane enzymatic activities as a function of lipid fluidity can be carried out at isothermic conditions, both by changing phospholipid composition or saturation as well as cholesterol levels, and by inducing phase transition in the bilayer in a temperature range in which the enzyme is fully operating and not inactivated (Kimelberg, 1977). The results of these modifications can be investigated by Arrehnius plots in search of possible "discontinuities" or changes in slope (breaks) of the curve relating the logarithm of reaction velocity to 1/T; any break in the Arrehnius plot usually indicates a phase transition of membrane lipids depending on the fatty acid composition.

The phospholipid immediately surrounding a membrane-bound enzyme (or annulus) can be an efficient modulator of catalytic activity depending on its qualitative composition. Cholesterol seems to be excluded from the boundary region circumventing the catalytic molecule (Cooper and Strauss, 1984), but its insertion into the boundary lipid with the ensuing block of lipid phase transitions may dramatically affect the enzymatic activity (Incerpi et al., 1983). For instance, the dietary modulation of lipid fluidity of rat red cells, both increase and decrease, has been shown to alter the cooperative behavior of membrane-bound enzymes (Farias, 1980).

The influence of membrane lipid fluidity on permeability and transport phenomena has been investigated in mammalian red cells mainly by the modulation of cholesterol levels. It seems that cholesterol enrichment of the red-cell membrane reduces passive permeability and facilitated diffusion of several electrolytes and nonelectrolytes, whereas sterol depletion has the opposite effect (Cooper and Strauss, 1984).

2. Lipid Composition of Erythrocyte Plasma Membrane in Normal and Abnormal Conditions

2.1. Lipid Composition of the Normal Erythrocyte Plasma Membrane

2.1.1. Total Lipids

The main types of lipids detectable in mammalian red cells are: neutral lipids, phospholipids, and glycosphingolipids. Total lipids of the human erythrocyte, according to Nelson (Nelson, 1972), are 5.20 mg/mL packed cells or $5.09 \cdot 10^{-13}$ g/cell. Neutral lipids, made up by cholesterol for more than 98%, account for about 25% of total lipids; phospholipids represent about 60% of total lipids, whereas the remaining 15% is represented by glycosphingolipids. In Table 1 the overall lipid composition of erythrocytes from different mammalian species is reported. The data of Table 1 indicate a considerable homogeneity of mammalian erythrocyte lipid composition: the most striking feature seems to be the wide variability of glycosphingolipids.

2.1.2. Neutral Lipids

As noted before, red-cell neutral lipids are represented almost completely by cholesterol (≧98%). The cholesterol of human erythrocyte plasma membrane can be considered as a single pool in equilibrium with plasma sterol (Cooper and Strauss, 1984); it is present in both leaflets of the bilayer, all being exchangeable with plasma (Lange and D'Alessandro, 1978): 10% of human erythrocyte cholesterol is more readily exchangeable than the remainder. The transbilayer movement of cholesterol (flip-flop movement) occurs very rapidly, with a half-time shorter than 10 s (Lange et al., 1981).

2.1.3. Phospholipids

In the mammalian erythrocyte, the main phospholipid species are phosphatidylcholine, phosphatidylserine, phosphatidylethanolamine, and sphingomyelin, which cover more than 90% of total phospholipids; phosphatidylinositol, phosphatidic acid, and unidentified phospholipids are also present (*see* Table 2). Red-cell phospholipids also contain ether-bonded components (plasmalogens) that are virtually present only in the phosphatidylethanolamine fraction: human red-cell ethanolamine plasmalogen covers about 7% of phosphatidylethanolamine content (Rao et al., 1978; White, 1973).

It is evident from Table 2 that the phospholipid composition of red-cell membrane shows wide variations, with special reference to phosphatidylcholine and sphingomyelin. This feature can be ascribed not only to metabolic reasons, but also to dietary habits that are well known to affect the overall lipids composition of erythrocyte plasma membrane and, in particular, the fatty acid distribution (Farias, 1980).

Studies on the transversal distribution of phospholipids in the mammalian erythrocyte plasma membrane carried out with chemical and immunological methods, with phospholipase treatment, and with physicochemical techniques, such as NMR, have given the most convincing evidence of an asymmetric distribution (Op den Kamp, 1979) (Table 3). In particular, about 80% of the phosphatidylethanolamine and nearly all

Table 1
Total Lipids, Cholesterol, Phospholipid and Glycosphingolipid Content of Erythrocytes from Different Mammalian Species*

	Total lipids	Cholesterol	Phospholipids	Glycosphingolipids
Human	5.20	1.34	3.26	0.60
Cow	4.21	1.23	2.64	0.34
Rat	5.06	1.43	3.21	0.42
Sheep	4.42	1.24	2.68	0.50
Dog	5.20	1.43	2.66	1.11
Rabbit	4.48	1.32	3.01	0.25
Guinea pig	5.57	1.54	3.03	1.00
Cat	6.04	1.62	3.70	0.72
Pig	4.31	1.15	2.58	0.58

*Data derived from Nelson (1972) and White (1973), as well as from the author's laboratory (Spinedi, A. and Luly, P., 1986 unpublished), are reported as mg/mL packed cells.

of the phosphatidylserine are in the inner leaflet, whereas about 76% of phosphatidylcholine and 82% of sphingomyelin are in the outer leaflet.

The physiological relevance of phospholipid asymmetry in erythrocyte plasma membrane is largely unknown, even if there are studies relating membrane-bound enzymatic activities to phospholipid composition (Roelofsen and van Deenen, 1973; Roelofsen and Schatzmann, 1977). In addition, there is evidence that, when the negatively charged phosphatidylserine is located in the outer leaflet (in inside-out vesicles from human erythrocytes), the coagulation velocity is markedly increased (Zwaal et al., 1977). This observation gives, at least partially, reasonable and convincing support to the physiological relevance of asymmetry in phospholipid distribution, considering the different fluidity of the plasma membrane leaflets and that the inner leaflet is more fluid than the outer one (Emmelot and van Hoeven, 1975; Rimon et al., 1984).

The fatty acid composition of red-cell plasma membrane (Table 4) largely reflects that of phospholipids, with a minor contribution from glycosphingolipids. In view of the main topic of this chapter, which is basically devoted to the human red cell, compositional data concerning overall fatty acid content as well as fatty acid content of different phospholipid classes will be limited to this cellular type. It appears, from data reported in Table 5, that phosphatidylcholine is more saturated than phosphatidylethanolamine and phosphatidylserine. This fact, together with the significant difference of fatty acid composition of phospholipid

Table 2
Phospholipid Composition of Mammalian Erythrocytes*

	PC	PI	PE	PS	SM	LPC	PA	Others
Human	30.5	1.0	27.5	13.5	25.5	1.0	1.0	1.0
Cow	9.5	3.5	28.6	15.3	46.2	–	0.3	2.0
Rat	47.8	3.5	20.6	13.2	16.2	3.8	0.2	–
Sheep	Traces	3.4	30.0	13.2	51.0	0.1	0.3	3.0
Dog	46.9	2.2	22.4	15.4	10.8	1.8	0.5	1.0
Rabbit	33.8	1.5	32.0	12.6	20.8	0.2	0.2	–
Guinea pig	41.1	2.4	21.6	16.8	11.1	0.3	4.2	2.0
Cat	30.5	7.4	22.2	13.2	26.1	0.3	0.8	–
Pig	27.4	2.3	29.5	17.0	25.0	0.9	0.3	–

*Abbreviations: PC, phosphatidylcholine; PI, phosphatidylinositol; PE, Phosphatidylethanolamine; PS, phosphatidylserine; SM, sphingomyelin; LPC, lysophosphatidylcholine; PA, phosphatidic acid. Data derived from Nelson (1972) and White (1973), as well as from the author's laboratory (Spinedi, A. and Luly P., 1986 unpublished), are reported as percentages of total phospholipids.

classes, clearly indicates the importance of a well-defined lipid environment for the physiological functions of red-cell plasma membrane.

2.1.4. Glycosphingolipids

Glycosphingolipids constitute between 10 and 20% of total lipids of the mammalian erythrocyte plasma membrane. These compounds are usually divided into three groups:

1. Neutral glycosphingolipids, which contain only sphingosine, fatty acids, and unsubstituted hexoses (usually glucose or galactose), and which are also called ceramide mono or polyhexosides
2. Hexosamine glycosphingolipids, also called globosides, which contain two or more unsubstituted residues (always hexosamine and, in addition, galactosamine or glycosamine)
3. Gangliosides, which contain several hexose residues and one or two sialic acid residues (Nelson, 1972; Mohandas et al., 1980).

2.2. Lipid Composition of the Abnormal Erythrocyte Plasma Membrane

2.2.1. Diabetes

It has recently been suggested that the plasma membrane of the diabetic erythrocyte is somewhat different with respect to its normal counterpart being a possible target in diabetes (McMillan, 1983). Actu-

Table 3
Lipid Topology of Mammalian Erythrocyte Plasma Membrane*

Outer leaflet	Inner leaflet
Phosphatidylcholine	Phosphatidylethanolamine
Sphingomyelin	Phosphatidylserine
Glycolipids	Phosphatidylinositol
Cholesterol	Cholesterol

*This table gives the preferential distribution of lipids between the two leaflets of the plasma membrane bilayer.

Table 4
Total Fatty Acid Composition of the Normal Human Erythrocyte*

Fatty acid	% of total fatty acids
C12:0	0.4
C14:0	0.7
C15:0	Traces
C16:0	24.7
C16:1	1.7
C17:0	Traces
C18:0	14.7
C18:1	17.9
C18:2	10.7
C18:3	0.8
C20:0	0.8
C20:1	Traces
C20:2	0.4
C20:3	2.0
C20:4	13.2
C22:0	2.5
C22:1	Traces
C22:2	Traces
C22:4	2.0
C22:5	2.4
C22:6	3.7
C24:0	3.8
C24:1	4.0

*Data are derived from Nelson (1972) and from the author's laboratory (Spinedi, A. and Luly, P., 1986 unpublished).

Table 5
Fatty Acid Composition of Different Phopholipid Classes of the Normal Human Erythrocyte Membrane*

	% of total fatty acids					
Fatty acid	PC	PE	PS	PI	SM	LPC
C12:0	0.4	Traces	Traces	0.7	Traces	2.7
C14:0	0.7	0.9	1.3	0.8	1.9	3.8
C15:0	0.3	Traces	Traces	2.7	Traces	3.1
C16:0	33.4	19.3	4.8	10.5	33.9	47.0
C16:1	1.5	1.5	1.0	0.8	2.3	3.2
C17:0	0.5	0.6	1.3	1.0	0.6	1.6
C18:0	11.0	11.0	33.4	24.3	10.0	16.8
C18:1	19.8	20.4	13.1	13.5	7.1	13.7
C18:2	18.6	6.1	3.3	4.6	2.4	7.5
C18:3	Traces	–	–	–	–	–
C20:0	0.7	1.0	1.3	8.0	1.5	4.5
C20:1	0.5	0.8	Traces	–	0.9	–
C20:2	0.5	0.8	0.8	0.9	–	–
C20:3	1.9	1.5	2.0	2.1	–	1.1
C20:4	6.3	21.6	21.0	21.1	–	Traces
C22:0	Traces	Traces	Traces	0.4	7.7	Traces
C22:1	–	Traces	Traces	Traces	Traces	–
C22:2	Traces	Traces	Traces	–	–	–
C22:4	0.6	6.0	4.3	3.4	–	–
C22:5	0.7	4.0	4.6	1.7	–	–
C22:6	1.6	6.1	7.1	2.3	–	–
C24:0	–	2.3	2.4	1.1	15.0	Traces
C24:1	Traces	–	–	–	18.1	1.3

*Data are derived from Nelson (1972) and from the author's laboratory (Spinedi, A. and Luly, P., 1986 unpublished). For abbreviations, *see* the footnote to Table 2.

ally, the first evidence of an involvement of the lipid component of the red-cell membrane was offered about 10 yr ago in streptozotocin-treated rats, in which the induced diabetes mellitus produced in significant drop in membrane cholesterol from 0.62–0.53 μmol/mg membrane protein; the total membrane phospholipid remained unchanged, and the cholesterol:phospholipid molar ratio dropped from 0.91–0.79 (Chandramouli and Carter, 1975).

More recently, Otsuji and coworkers in Japan (Baba et al., 1979), working on diabetic patients, observed a decreased membrane fluidity using the steady-state fluorescence polarization technique. This group later confirmed this observation by spin labeling and ESR measurements

(Kamada and Otsuji, 1983). They also observed a small but significant decrease in phosphatidylethanolamine content paralleled by an increase of sphingomyelin (from 35.9–32.6% and from 23.9–26.4% respectively, as percentages of total phospholipid molar composition), all other lipid compositional parameters being unchanged, including cholesterol:phospholipid molar ratio (Kamada and Otsuji, 1983). The data reported by Otsuji and coworkers are to be interpreted, in our opinion, with some caution for the simple reason that the diabetic population studied contained insulin-dependent and noninsulin-dependent diabetic patients, a circumstance that to a certain extent affects the relevance of the observation. In fact, very recently Bryszewska and coworkers in Poland (Bryszewska et al., 1986), working on red cells from insulin-dependent patients, observed a dramatic increase of cholesterol:phospholipid molar ratio (from 0.88–1.41).

In conclusion, data available so far on human diabetic red cells are all in favor of an increased rigidity of the plasma membrane lipid fluidity, and are well in line with published observations pointing to a decreased deformability (i.e., increased rigidity) of the human red-cell membrane in diabetes (McMillan et al., 1978; Juhan et al., 1981).

2.2.2. Muscular Dystrophy

The involvement of possible membrane defects in muscular dystrophies has been suggested years ago, and morphological alterations have been reported in erythrocytes from patients with myotonic and Duchenne dystrophy (Lucy, 1980). Studies on membrane lipid fluidity of erythrocytes from carriers of myotonic muscular dystrophy and of Duchenne muscular dystrophy, a dystrophy without myotonia, and patients with congenital myotonia, a myotonia without dystrophy, have shown the following: using ESR spectroscopy, red-cell membranes from patients with myotonic dystrophy have an increased lipid fluidity, whereas this is not the case in nonmyotonic dystrophic patients (Butterfield et al., 1976). An abnormal distribution, with respect to normal red cells, of spin labels in erythrocyte membranes from Duchenne dystrophy patients was interpreted as a result of an increased membrane ridigity (Wilkerson et al., 1978). Increased rigidity of erythrocyte plasma membrane was also reported in hereditary muscle dystrophy in chicken, a model similar to the human Duchenne dystrophy (Butterfield and Leung, 1978). Increased lipid fluidity of erythrocyte membrane in myotonic muscular dystrophic patients was also more recently observed (Butterfield, 1981). The altered fluidity characteristics observed in erythrocyte membrane from dystrophic patients may be responsible for the altered Ca^{2+} permeability and

Na^+-pumping activities shown in these cells. These reports, which shall not be dealt with in this chapter, are rather contradictory (Lucy, 1980).

Alterations of the lipid pattern of dystrophic erythrocytes have sometimes been reported (Lucy, 1980). These basically indicate a possible increase in sphingomyelin and a decrease in phosphatidylserine and phosphatidylethanolamine, with no changes of cholesterol levels and fatty acid pattern. These early observations have been confirmed recently with special reference to the fatty acid profile—which appears to be not modified—of red cells from myotonic dystrophic patients (Antoku et al., 1985). Recent reports on dystrophic chicken erythrocytes have shown an increased phosphatidylserine and a decreased phosphatidylethanolamine level; in addition, these membranes have an increased presence of ethanolamine plasmalogen with respect to matched controls (Kester and Privitera, 1984a,b). The reported increased fluidity (Butterfield et al., 1976; Butterfield, 1981) of erythrocyte membranes from myotonic dystrophic patients is well supported by the increased phosphatidylethanolamine methylation rate observed in these cells, which reduces phosphatidylethanolamine by increasing phosphatidylcholine content (Eda and Takeshita, 1984).

2.2.3. LCAT Deficiency

Plasma Lecithin-Cholesterol Acyl Transferase (LCAT) deficiency is a rare hereditary disorder in which cholesterol level of erythrocyte membrane is increased. Patients with the hereditary disease, but also patients in which the disease appears following hepatic abnormalities, show an increased (10–20%) cholesterol:phospholipid ratio and a decreased sphingomyelin:phospholipid ratio (from 0.75–0.26 or 0.54 in the hereditary or secondary deficiency, respectively); in addition, the arachidonic acid (C20:4) content of membrane phospholipids decreases by an average 15% in both hereditary and secondary deficiency, linoleic acid (C18:2) being increased by 50% only in the first case (Beynen et al., 1984). It was suggested that, as a compensatory mechanism, the increased membrane rigidity resulting from the increased cholesterol:phospholipid ratio could be counteracted by a fluidizing decrease in the sphingomyelin:phosphatidylcholine ratio and by the increased C18:2 percentage, which is three times larger than the decrease of C20:4. These compensatory mechanisms seem to be impaired in the LCAT deficiency secondary to liver disease, in which linoleic acid levels remain unchanged.

On the same topic, a thorough study was carried out in Japan by Yawata and coworkers on a single middle-aged male patient (Yawata et al., 1984). Erythrocyte membrane lipids showed an increased cholesterol level counteracted by an increased phosphatidylcholine, and decreased sphingomyelin and phosphatidylethanolamine content; moreover, in phosphatidylcholine, C16:0 was increased and C18:0 as well as C24:0 decreased. Thus, the decreased plasma membrane fluidity induced by the metabolic defect seems to be counterbalanced by changes in phospholipid content and increasing levels of shorter chain fatty acids together with an increase of unsaturation level (C18:2 was increased). All these facts clearly indicate the natural trend to restore normal fluidity conditions in the red-cell membrane in spite of the inherited metabolic defect.

2.2.4. Huntington's Disease

It has been suggested that Huntington's Disease, an inherited disease of the nervous system characterized by progressive involuntary movements and dementia, could involve a generalized membrane defect (Butterfield and Markesbery, 1981). Early studies pointed to some lipid involvement in red cells (Nelson, 1972), but more recent studies seem to rule out this possibility (Butterfield and Markesbery, 1981; Abood and Butler, 1979; Sumbilla and Lakowicz, 1982). ESR studies indicated an altered conformation of membrane proteins in erythrocytes, but no alteration of membrane fluidity (Butterfield and Markesbery, 1981); the same indication was obtained using fluorescent probes (Sumbilla and Lakowicz, 1982). Furthermore, studies carried out using quenching of membrane intrinsic tryptophan fluorescence ruled out any alteration of erythrocyte membrane lipid fluidity, and excluded any modification of the fatty acid pattern of membrane lipids (Abood and Butler, 1979).

2.2.5. Acanthocytosis

The term acanthocyte indicates an irregularly spiculated red cell that is not able to revert to a normal condition both in fresh plasma and after washing (Mohandas et al., 1980). Acanthocytosis is generally the result of abetalipoproteinemia (but is also present after splenectomy or liver disease), in which red cells that were released basically normal from marrow change their lipid composition during in vivo aging. The very reason for the morphological change is still unclear, but is definitely related to an increased sphingomyelin:lecithin ratio that seems to take place without changes in the total lipid concentrations of the red-cell membrane (Nelson, 1972; Mohandas et al., 1980; Owen et al., 1982), thus inducing a decrease of membrane lipid fluidity as also indicated by early work on phospholipid fatty acid composition, which apparently showed a decreased saturation level (Nelson, 1972). It is worth mentioning, in this connection, that Flamm and Schachter (1982) have demonstrated a decrease of lipid fluidity that is localized in the outer leaflet of the membrane bilayer where sphingomyelin and lecithin (phosphatidylcholine) are usually located. Recent work on acanthocytosis-inducing disorders indicates a normal phospholipid composition and distribution in the red-cell membrane, but a markedly enhanced exchangeability of phosphatidylcholine, which reaches a maximum (about 100%) within 8 h of incubation with respect to the usual 75% exchange shown by normal red cells (Kuypers et al., 1985).

2.2.6. Anemias

Early reports appearing in the 1960s and 1970s relating different types of anemias to red-cell lipid abnormalities have furnished contradictory indications, which have been reviewed previously (Nelson, 1972; Mohandas et al., 1980).

2.2.6.1. Spur Cell Anemia. Cooper described spur cell anemia (Cooper, 1969), in which patients with severe liver disease (generally following alcoholic cirrhosis) show a primary disorder of plasma lipoproteins for which serum LDL are enriched in cholesterol. The sterol is then transferred to red-cell membranes, increasing their cholesterol:phospholipid ratio by 25–65% up to a value of 1.6; this syndrome, which can be induced in rodents and dogs fed with a cholesterol-rich atherogenic diet, produces cholesterol accumulation not only in red cells, but also in platelets, macrophages, and liver cells (Cooper and Strauss, 1984). The osmotic fragility of these red cells is increased following an alteration of cationic equilibria (decrease of Na^+, increase of K^+, both in the cellular compartments) because of both decreasing ouabain-sensitive Na^+ efflux and K^+ influx, as well as reduced electrolyte passive permeability (Cooper et al., 1975). These observations, concerned with an increased cholesterol:phospholipid ratio, have been recently confirmed in the poorly studied transient form of spur cell anemia associated with infantile cholestatic liver disease (Cynamon et al., 1985).

2.2.6.2. Hereditary Spherocytosis. Hereditary spherocytosis is a red-cell defect caused by alterations of cytoskeletal structure, in which the morphology of the cell as well as cationic fluxes and ion-pumping activity are affected (Mohandas et al., 1980). Alterations of membrane lipid composition were proposed by some authors and refuted by others (Nelson, 1972). Recent work suggests that the primary membrane defect resides in an altered cytoskeleton-membrane protein interaction, which, in turn, affects lateral mobility of the phospholipids in the bilayer: the lateral diffusion of phospholipids of the inner leaflet is decreased in spherocytic red cells, whereas the outer leaflet is unaltered (Rimon et al., 1984).

2.2.6.3. Rh Null Red Cells. Rh Null Red Cells are a rare blood type characterized by the absence of the Rh blood group system. These cells have an abnormal morphology, an increased fragility, and are associated to a mild hemolytic anemia; a decreased membrane-bound ATPase activity was also observed (Mohandas et al., 1980). Recent studies on the phospholipid asymmetry of these peculiar red cells have shown, as previously described for acanthocytes, a total exchangeability of phosphatidylcholine with respect to normal controls (100 vs 75%). This feature is accompanied by a doubled accessibility of phosphatidylethanolamine to phospholipase A_2 treatment of intact cells (Kuypers et al., 1984). It was thus concluded that the absence of the two antigenic polypeptides markedly affects both phospholipid asymmetric distribution and mobility, offering a possible molecular reason for the impairment of membrane-located ion-pumping activities in light of a correct interplay between plasma membrane and the cytoskeleton (Kuypers, et al., 1984).

2.2.6.4. Sickle Cell Anemia. Red-cell membrane phospholipid organization in sickle cell anemia has been recently reviewed in detail (Wagner et al., 1985). Contradictory evidence has been reported as to a possible loss of phospholipid asymmetry in sickle erythrocytes. Raval and Allan (1984) showed an unchanged asymmetric distribution (also in the region that corresponds to extended spicules on the cells and that seems to be free of cytoskeletal proteins), whereas Schwartz and coworkers (Schwartz et al., 1985) observed an abnormal exposure of phosphatidylserine and phosphatidylethanolamine in the outer leaflet of these cells. Schwartz and coworkers (Schwartz et al., 1985) suggest that the abnormal presence of phosphatidylserine in the membrane outer leaflet could serve to trigger the recognition of sickle cells by circulating monocytes, thus shortening their survival. Furthermore, Franck and coworkers (Franck et al., 1985), in experiments on free spicules released from sickle cells as well as on the remnant despiculated cells, demonstrated that an enhanced transbilayer movement of phosphatidylcholine and the exposure of phosphatidylserine in the outer leaflet occur predominantly in the spiculated parts of sickle cell plasma membrane.

2.2.7. Cystic Fibrosis

Patients suffering from cystic fibrosis with pancreatic exocrine insufficiency, fed with a regular diet supplemented with pancreatic enzymes, have a reduced linoleic acid content in erythrocyte membrane phosphatidylcholine (Parsons et al., 1986), whereas all other lipid compositional parameters are basically unchanged. This situation, which can be worsened in the absence of a pancreatic enzyme supplementation allowing a proper uptake of the essential fatty acid, is related—although not yet clearly—to a massive change of erythrocyte morphology (Parsons et al., 1986) consisting of a significant presence (30 vs 15%, i.e., twice the normal percentage) of echinocytes, i.e., of crenated red cells (Mohandas et al., 1980).

2.2.8. Ethanol Treatment

Ethanol insertion in the membrane lipid bilayer enhances lipid fluidity (Freund, 1979). As an apparent response to this alteration, compensatory changes take place in membrane lipid composition (Chin and Goldstein, 1977). Restricting our attention to the mammalian red cell, we note that, in mice, chronic ethanol treatment-induced increased membrane fluidity is counteracted by an increase of membrane cholesterol (Chin et al., 1978). This observation was not confirmed in rat erythrocytes where, on the other hand, a decrease of the overall membrane lipid fluidity was achieved by a reduction of the saturation level of total phospholipids (La Droitte et al., 1984). In particular, in rat erythrocytes, C18:1 increases, whereas C18:2 and C22:4 decrease, all other lipid compositional parameters being unchanged (La Droitte et al., 1984). These observations have received support by more recent research on human erythrocytes: in chronic alcoholics an unchanged cholesterol:phospholipid ratio together with a very high saturated:unsaturated fatty acid ratio in phospholipids were reported (Clemens et al., 1986a,b). Plasmatic linoleic acid decreased from 32.3–21.8% of total fatty acids, and this change was paralleled by a marked decreases of this fatty acid in red-cell membrane phosphatidylcholine (Clemens et al., 1986a,b). These results were interpreted as a compensatory mechanism (decreased membrane fluidity following increased saturation as a consequence of the fluidizing effect of ethanol) and as a direct result of plasma lipid alterations.

The reported altered erythrocyte morphology with an increased cell volume (macrocytosis) in chronic alcoholism was experimentally related in mice to an uneven distribution of fatty acid changes between the two membrane leaflets (Wing et al., 1984). Changes of saturation levels were reported to be mostly (about 90%) restricted to the inner leaflet phospholipids, such as phosphatidylethanolamine (80%) and phosphatidylserine-phosphatidylinositol (10%). Other phospholipids were less affected. This observation seems somewhat at variance with respect to data reported on human red cells (Clemens et al., 1986a).

In conclusion, the reported in vitro ethanol tolerance observed after chronic ethanol treatment in rodent erythrocytes (Chin and Goldstein,

1977; Chin et al., 1978) was interpreted, on an adaptive basis, as depending on changes of membrane lipid composition. Alteration of cholesterol:phospholipid ratio appears to be of minor relevance, whereas the increased membrane phospholipid saturation, also obtained through an appropriate feeding, seems to play a pivotal role in the development of ethanol tolerance (Rottenberg, 1986).

2.2.9. Malaria-Infected Red Cells

The infection of red cells by the malarial parasite *Plasmodium* markedly changes their lipid composition (Sherman, 1979). The total lipid content of malaria-infected red cells is always higher with respect to the lipid content of normal red cells. An average 60% increase was estimated, being mostly attributed to membrane phospholipids of the parasite, whereas the cholesterol:phospholipid ratio is decreased (Sherman, 1979). As far as phospholipid composition is concerned, phosphatidylethanolamine and phosphatidylinositol increase, whereas sphingomyelin decreases upon infection. Concerning red-cell total fatty acids, there is a marked increase in saturated fatty acids, such as palmitic and oleic acid, whereas unsaturated linoleic and arachidonic acid decline (Sherman, 1979).

Information reported so far, taken from an extensive review that appeared in 1979 (Sherman, 1979), is related to studies carried out in the 1970s on mammalian and avian red cells, when the then current hypothesis (not modified since then) was that changes in host-cell plasma membrane lipids were the result of an intercalation of newly parasite-derived membrane material. Focusing our attention on more recent studies carried out on mammalian erythrocytes, we will report observations concerning infection by *Plasmodium berghei* (in rodents), by *Plasmodium knowlesi* (in monkeys) and by *Plasmodium falciparum* in humans.

Rat or mouse red cells infected by *Plasmodium berghei* have a decreased cholesterol content, a decreased cholesterol:phospholipid ratio, increased saturated fatty acid levels (C16:0 in particular), and decreased arachidonic acid (Howard and Sawyer, 1980; Butler et al., 1984). An extremely thorough study has been published very recently concerning the modification of the lipid pattern of simian (*Macaca fascicularis*) red-cell membranes infected by *Plasmodium knowlesi* (Beaumelle and Vial, 1986). In this study, a significant increase in phosphatidylcholine and phosphatidylethanolamine content was observed, whereas sphingomyelin and phosphatidylserine decreased. The total fatty acid content of these red cells showed a sixfold increase. Basically, the level of saturated species was higher in some cases (C16:0, +23%) and lower in other cases (C18:0, −27%; C22:0 and C24:0, −50%). The level of the major polyunsaturated fatty acid (C20:4) was decreased by 40%, but the unsaturation index was unchanged owing to the nearly twofold increase of linoleic acid (C18:2). Also in this very interesting publication by Beaumelle and Vial (Beaumelle and Vial, 1986), specific alterations of the fatty acid pattern of individual phospholipids and of neutral lipids were observed in infected cells together with a peculiar decrease of total plasmalogen (−60%), which showed a new redistribution between phosphatidylethanolamine (decrease) and phosphatidylinositol (increase).

The infection of human red cells by *Plasmodium falciparum* has proved to have an increasingly disordering action in the phospholipid bilayer at different stages of the infection, as studied by ESR spectroscopy (Taraschi et al., 1986), as well as by the use of phospholipase A_2 and Merocyanin 540 as external membrane probes (Joshi and Gupta, 1988). This decrease in the overall membrane lipid fluidity in human red cells is basically in agreement with the reported decrease of membrane cholesterol level (Dluzewski et al., 1985).

2.2.10. Miscellaneous Diseases

Alterations of red-cell membrane lipid composition in other diseases have been reported in the past with reference to studies carried out in the 1960s and 1970s (Nelson, 1972; Mohandas et al., 1980). In the following, a brief account concerning red-cell lipid composition in some abnormal situations, not quoted in previous subsections, will be given. The lipid fluidity of red-cell membrane from uremic, hemodialyzed patients was investigated with a spin label probe. Lower levels of membrane fluidity were detected and related to a decreased presence of phosphatidylcholine as well as to a decreased phosphatidylcholine:sphingomyelin ratio (Komidori et al., 1985). Experimental hyperthyroidism in rats decreased plasma cholesterol levels, at the same time increasing its presence in the red-cell membrane, whereas total phospholipids increased both in plasma and in the erythrocyte membrane, thus leaving the cholesterol:phospholipid ratio unaltered (Ruggiero et al., 1984). Studies on erythrocyte membrane fluidity in human obesity, carried out with the fluorescence polarization technique, found a strong correlation between decreased fluidity and increasing body weight. Lipid estimation in red-cell membranes demonstrated an increase of cholesterol and a decrease of phospholipids (Beguinot et al., 1985).

An interesting observation was reported in nephropathic patients by Panasenko and coworkers (Panasenko et al., 1985) working with spin labels on red-cell membranes. They found an increase of order parameters (i.e., a decrease of lipid fluidity) that correlated with the gravity of the disease and of edema in patients. In addition, they observed that the alterations reported in red cell membranes were comparable to those induced by lipid peroxidation, so they suggested a lipid peroxidation involvement in the pathogenesis of nephropathy at the membrane level.

3. Conclusions

In the past 15 yr, after the proposal of the "fluid mosaic model" of plasma membranes (Singer and Nicolson, 1972), increasing attention has been paid to the membrane lipid microenvironment, as well as to lipid-protein functional interplay in a normal status or in situations that could be in some way linked to a membrane pathology. The mammalian erythrocyte, for its limited and recognized metabolic activities, represents a unique model system in which most of plasma membrane-located physiological functions can be studied, reducing the artifactual drawback of liposomes. A very good opportunity is offered, in this context, by the methodology established years ago, which allows the experimental modulation of red-cell membrane lipids by dietary (Farias et al., 1975) or in vitro treatment (Incerpi et al., 1983; Shinitzky, 1978). It is now possible to reproduce, in the laboratory, some abnormal conditions in which the red-cell membrane has an altered lipid composition, a situation that—as is now rather clear—is likely to alter membrane physiology profoundly (Shinitzky, 1984). The modulation of erythrocyte membrane lipids has represented a good model for studies concerned with membrane alterations in atherosclerosis (Kummerow, 1983). On the other hand, studies of a normal phenomenon like aging—carried out on human red-cell membranes (Pfeffer and Swislocki, 1976, 1982; Terranova et al., 1985) but also in other systems (Schroder, 1984)—did not give unequivocal results as far as the plasma membrane lipid composition is concerned.

In conclusion, the "simple" mammalian erythrocyte still represents the living model of choice to investigate an array of diseases that induce a membrane pathology. Further elucidations of the role of lipids in normal and abnormal plasma membranes will certainly open new horizons as to what has been recently defined as "membrane lipid engineering" (Shinitzky, 1984).

References

Abood, M. E. and Butler, M. (1979) Membrane fluidity and fatty acid composition of phospholipids in erythrocyte membranes of patients with Huntington's Disease. *J. Neurosci.* **4,** 183–187.

Antoku, Y., Sakai, T., Goto, I., Iwashita, H., and Kuroiwa, Y. (1985) Fatty acid composition of erythrocytes, mononuclear cells and blood plasma of patients with myotonic dystrophy. *J. Neurol. Sci.* **71,** 387–393.

Baba, Y., Kai, M., Kamada, T., Setoyama, S., and Otsuji, S. (1979) Higher levels of erythrocyte membrane microviscosity in diabetes. *Diabetes* **28,** 1138–1140.

Beaumelle, B. D. and Vial, H. J. (1986) Modification of the fatty acid composition of individual phospholipids and neutral lipids after infection of the simian erythrocyte by *Plasmodium knowlesi. Biochim. Biophys. Acta* **877,** 262–270.

Beguinot, F., Tramontano, D., Duilio, C., Formisano, S., Beguinot, L., Mattioli, P., Mancini, M., and Aloj, S. M. (1985) Alteration of erythrocyte membrane lipid fluidity in human obesity. *J. Clin. Endocrinol. Metab.* **60,** 1226–1230.

Beynen, A. C., Schouten, J. A., and Popp-Snijders, C. (1984) Compensatory changes in the lipid composition of the erythrocyte membrane. *Trend. Biochem. Sci.* **9,** 474.

Borochov, H., Zahler, P., Wilbrandt, W., and Shinitzky, M. (1977) The effect of phosphatidylcholine to sphingomyelin mole ratio on the dynamic properties of sheep erythrocyte membranes. *Biochim. Biophys. Acta* **470,** 382–388.

Bretscher, M. S. and Raff, M. (1975) Mammalian plasma membranes. *Nature* **258,** 43–49.

Bryszewska, M., Watala, C., and Torzecka, W. (1986) Changes in fluidity and composition of erythrocyte membranes and in composition of plasma lipids in Type 1 diabetes. *Br. J. Haematol.* **62,** 111–116.

Butler, K. W., Deslauriers, R., and Smith, I. C. (1984) *Plasmodium berghei*: electron spin resonance and lipid analysis of infected mouse erythrocyte membranes. *Exp. Parasitol.* **57,** 178–184.

Butterfield, D. A. (1981) Myotonic muscular dystrophy. Time-dependent alterations in erythrocyte membrane fluidity. *J. Neurol. Sci.* **52,** 61–67.

Butterfield, D. A. and Leung, P. K. (1978) Erythrocyte membrane fluidity in chicken muscular dystrophy. *Life Sci.* **22,** 1783–1788.

Butterfield, D. A. and Markesbery, W. R. (1981) Huntington's Disease: a generalized membrane defect. *Life Sci.* **28,** 1117–1131.

Butterfield, D. A., Chesnut, D. B., Appel, S. H., and Roses, A. D. (1976) Spin label study of erythrocyte membrane fluidity in myotonic and Duchenne muscular dystrophy and congenital myotonia. *Nature* **263,** 159–161.

Chandramouli, V. and Carter, J. R. (1975) Cell membrane changes in chronically diabetic rats. *Diabetes* **24,** 257–263.

Chapman, D. (1973) Recent studies of lipids, lipid-cholesterol, and membrane systems, in *Biological Membranes*, vol. 2 (Chapman, D. and Wallach, D. F. H., eds.), Academic Press, New York, pp. 91–144.

Chapman, D. and Hayward, J. A. (1985) New biophysical techniques and their application to the study of membranes. *Biochem. J.* **228,** 281–295.

Chin, J. H. and Goldstein, D. B. (1977) Drug tolerance in biomembranes: a spin label study of the effects of ethanol. *Science* **196,** 684–685.

Chin, J. H., Parsons, L. M., and Goldstein, D. B. (1978) Increased cholesterol content of erythrocytes and brain membranes in ethanol-tolerant mice. *Biochim. Biophys. Acta* **513,** 358–363.
Clemens, M. R., Kessler, W., Schied, H. W., Schupmann, A., and Waller, H. D. (1986a) Plasma and red cell lipids in alcoholics with macrocytosis. *Clin. Chim. Acta* **156,** 321–328.
Clemens, M. R., Schied, H. W., Daiss, W., and Waller, H. D. (1986b) Lipid abnormalities in plasma and red cell membranes of chronic alcoholics. *Klin. Wochenschr.* **64,** 181–185.
Cooper, R. A. (1969) Anemia with spur cells: a red cell defect acquired in serum and modified in the circulation. *J. Clin. Invest.* **48,** 1820–1831.
Cooper, R. A. and Strauss, III, J. F. (1984) Regulation of cell membrane cholesterol, in *Physiology of Membrane Fluidity*, vol. 1, (Shinitzky, M., ed.), C.R.C. Press, Boca Raton, pp. 73–97.
Cooper, R. A., Arner, E. C., Wiley, J. S., and Shattil, S. J. (1975) Modification of red cell membrane structure by cholesterol-rich lipid dispersion. A model for the primary spur cell defect. *J. Clin. Invest.* **55,** 115–126.
Cynamon, H. A., Isenberg, J. N., Gustavson, L. P., and Gourley, W. K. (1985) Erythrocyte lipid alteration in pediatric cholestatic liver disease: spur cell anemia of infancy. *J. Pediatr. Gastroenterol. Nutr.* **4,** 542–549.
Demel, R. A. and De Kruyff, B. (1976) The functions of sterols in membranes. *Biochim. Biophys. Acta* **457,** 109–132.
Dluzewski, A. R., Rangachari, K., Wilson, R. J., and Gratzer, W. B. (1985) Relation of red cell membrane properties to invasion by *Plasmodium falciparum*. *Parasitology* **91,** 273–280.
Eda, I. and Takeshita, K. (1984) Methylation of erythrocyte membrane phospholipid in patients with early-onset myotonic dystrophy. *Brain Dev.* **6,** 284–288.
Emmelot, P. and van Hoeven, R. P. (1975) Phospholipid unsaturation and plasma membrane organization. *Chem. Phys. Lipids* **14,** 236–246.
Farias, R. N. (1980) Membrane cooperative enzymes as a tool for the investigation of membrane structure and related phenomena. *Adv. Lipid Res.* **17,** 251–282.
Farias, R. N., Bloj, B., Morero, R. D., Sineriz, F., and Trucco, R. E. (1975) Regulation of allosteric membrane-bound enzymes through changes in membrane lipid composition. *Biochim. Biophys. Acta* **415,** 231–251.
Flamm, M. and Schachter, D. (1982) Acanthocytosis and cholesterol enrichment decrease lipid fluidity of only the outer human erythrocyte membrane leaflet. *Nature* **298,** 290–292.
Franck, P. F., Bevers, E. M., Lubin, B. H., Comfurius, P., Chiu, D T., Op den Kamp, J. A., Zwaal, R. F., van Deenen, L. L. M., and Roelofsen, B. (1985) Uncoupling of the membrane skeleton from the lipid bilayer. The cause of the accelerated phospholipid flip-flop leading to an enhanced procoagulant activity of sickled cells. *J. Clin. Invest.* **75,** 183–190.
Freund, G. (1979) Possible relationship of alcohol in membranes to cancer. *Cancer Res.* **39,** 2899–2901.

Gortel, E. and Grendel, F. (1925) On bimolecular layers of lipoids on the chromocytes of the blood. *J. Exp. Med.* **41,** 439–443.

Howard, R. J. and Sawyer, W. H. (1980) Changes in the membrane microviscosity of mouse red blood cells infected with *Plasmodium berghei* detected using N-(anthroyloxy)-fatty acid fluorescent probes. *Parasitology* **80,** 331–342.

Incerpi, S., Baldini, P., and Luly, P. (1983) Cholesterol modulation *in vitro* of plasma membrane-bound enzymes in human erythrocytes. *Cell Mol. Biol.* **29,** 285–289.

Joshi, P. and Gupta, C. M. (1988) Abnormal membrane phospholipid organization in *Plasmodium falciparum*-infected human erythrocytes. *Br. J. Haematol.* **68,** 255–259.

Juhan, P., Vague, P., Buonocore, M., Moulin, J. P., Calas, M. F., Vialettes, B., and Verdot, J. J. (1981) Effects of insulin on erythrocyte deformability in diabetes—relationship between erythrocyte deformability and platelet aggregation. *Scand. J. Clin. Lab. Invest.* **41 (Suppl. 156),** 159–164.

Kamada, T. and Otsuji, S. (1983) Lower levels of erythrocyte membrane fluidity in diabetic patients. *Diabetes* **32,** 585–591.

Kester, M. and Privitera, C. A. (1984a) Phospholipid composition of dystrophic chicken erythrocyte plasmalemma. I. Isolation of a unique lipid in dystrophic erythrocyte membranes. *Biochim. Biophys. Acta* **778,** 112–120.

Kester, M. and Privitera, C. A. (1984b) Phospholipid composition of dystrophic chicken erythrocyte plasmalemma. II. Characterization of a unique lipid from dystrophic erythrocyte membranes as ethanolamine plasmalogen. *Biochim. Biophys. Acta* **778,** 121–128.

Kimelberg, H. K. (1977) The influence of membrane fluidity on the activity of membrane-bound enzymes, in *Dynamic Aspects of Cell Surface Organization* (Poste, G. and Nicolson, G. L., eds.), North Holland Publishing, Co., Amsterdam, pp. 205–293.

Komidori, K., Kamada, T., Yamashita, T., Harada, R., Otsuji, Y., Hashimoto, S., Chuman, Y., and Otsuji, S. (1985) Erythrocyte membrane fluidity decreased in uremic hemodialyzed patients. *Nephron* **40,** 185–188.

Kummerow, F. A. (1983) Modification of cell membrane composition by dietary lipids and its implications for atherosclerosis. *Ann. N Y. Acad. Sci.* **414,** 29–43.

Kuypers, F. A., van Linde Sibenius Trip, M., Roelofsen, B., Op den Kamp, J. A., Tanner, M. J., and Anstee, D. J. (1985) The phospholipid organization in the membrane of the McLeod and Leach phenotype erythrocytes. *F.E.B.S. Lett.* **184,** 20–24.

Kuypers, F. A., van Linde-Sibenius Trip, M., Roelofsen, B., Tanner, M. J., Anstee, D. J., and Op den Kamp, J. A. (1984) Rh null human erythrocytes have an abnormal membrane phospholipid organization. *Biochem. J.* **221,** 931–934.

La Droitte, P., Lamboeuf, Y., and De Saint-Blanquat, G. (1984) Lipid composition of the synaptosome and erythrocyte membrane during chronic ethanol treatment and withdrawal. *Biochem. Pharmacol.* **33,** 615–624.

Lange, Y. and D'Alessandro, J. S. (1978) The exchangeability of human erythrocyte membrane cholesterol. *J. Supramol. Struct.* **8,** 391–397.

Lange, Y., Dolde, J., and Steck, T. L. (1981) The rate of transmembrane movement of cholesterol in the human erythrocyte. *J. Biol. Chem.* **256,** 5321–5323.

Lucy, J. A. (1980) Is there a membrane defect in muscle and other cells? *Br. Med. Bull.* **36,** 187–192.

McMillan, D. E. (1983) Insulin, diabetes, and the cell membrane: An hypothesis. *Diabetologia* **24,** 308–310.

McMillan, D. E., Utterback, N. G., and La Puma, J. (1978) Reduced erythrocyte deformability in diabetes. *Diabetes* **27,** 895–901.

Mohandas, N., Weed, R. I., and Bessis, M. (1980) Red cell membrane pathology, in *Pathobiology of Cell Membranes*, vol. 2, (Trump, B. F. and Arstila, A. V., eds.), Academic Press, New York, pp. 41–88.

Nelson, G. J. (1972) Lipid composition and metabolism of erythrocytes, in *Blood Lipids and Lipoprotein: Quantitation, Composition and Metabolism* (Nelson, G. J., ed.), Wiley Interscience, New York, pp. 317–386.

Op den Kamp, J. A. F. (1979) Lipid asymmetry in membranes. *Ann. Rev. Biochem.* **48,** 47–71.

Overton, E. (1902) Beiträge zur allgemeinen Muskel und Nervenphysiologie. *Arch. ges. Physiol.* **92,** 115–280.

Owen, J. S., Bruckdorfer, K. R., Day, R. C., and McIntyre, N. (1982) Decreased erythrocyte membrane fluidity and altered lipid composition in human liver disease. *J. Lipid Res.* **23,** 124–132.

Panasenko, O. M., Shalina, R. I., and Azizova, O. A. (1985) Structural changes in erythrocyte membranes in nephropathy. *Biull. Eksp. Biol. Med.* **99,** 434–437.

Parsons, H. C., Hill, R., Pencharz, P., and Kuksis, A. (1986) Modulation of human erythrocyte shape and fatty acids by diet. *Biochim. Biophys. Acta* **860,** 420–427.

Pfeffer, S. R. and Swislocki, N. I. (1976) Age related decline in the activities of erythrocyte membrane adenylate cyclase and protein kinase. *Arch. Biochem. Biophys.* **177,** 117–122.

Pfeffer, S. R. and Swislocki, N. I. (1982) Role of peroxidation in erythrocyte aging. *Membrane Aging Dev.* **18,** 355–367.

Rao, G. A., Siler, K., Larkin, E. C., Howland, J. L., and McAllister, D. J. (1978) Palmitoleic acid in erythrocytes from carriers of Duchenne muscular dystrophy. *Science* **200,** 1416.

Raval, P. J. and Allan, D. (1984) Sickling of sickle erythrocytes do not alter phospholipid asymmetry. *Biochem. J.* **223,** 555–557.

Rimon, G., Meyerstein, N., and Henis, Y. I. (1984) Lateral mobility of phospholipids in the external and internal leaflets of normal and hereditary spherocytic human erythrocytes. *Biochim. Biophys. Acta* **775,** 283–290.

Roelofsen, B., and Schatzmann, H. J. (1977) The lipid requirement of the (Ca^{2+}-Mg^{2+})-ATPase in the human erythrocyte membrane, as studied by various highly purified phospholipases. *Biochim. Biophys. Acta* **464,** 17–36.

Roelofsen, B. and van Deenen, L. L. M. (1973) Lipid requirement of membrane-bound ATPase. Studies on human erythrocyte ghosts. *Eur. J. Biochem.* **40,** 245–257.

Rottenberg, H. (1986) Membrane solubility of ethanol in chronic alcoholism. The effect of ethanol feeding and its withdrawal on the protection by alcohol of rat red blood cells from hypotonic hemolysis. *Biochim. Biophys. Acta* **855,** 211–222.

Ruggiero, F. M., Landriscina, C., Gnoni, G. V., and Quagliariello, E. (1984) Alteration of plasma and erythrocyte membrane lipid components in hyperthyroid rat. *Horm. Metab. Res.* **16,** 37–40.

Schroeder, F. (1984) Role of membrane lipid asymmetry in aging. *Neurobiol. Aging* **5,** 323–333.

Schwartz, R. S., Tanaka, Y., Fidler, I. J., Chiu, D. T., Lubin, B., and Schroit, A. J. (1985) Increased adherence of sickled and phosphatidylserine-enriched human erythrocytes to cultured human peripheral blood monocytes. *J. Clin. Invest.* **75,** 1965–1972.

Sherman, I. W. (1979) Biochemistry of *Plasmodium* (Malarial Parasites). *Microbiol. Rev.* **43,** 453–495.

Shinitzky, M. (1978) An efficient method for modulation of cholesterol level in cell membranes. *FEBS Lett.* **85,** 317–320.

Shinitzky, M. (1984) Membrane fluidity and cellular functions, in *Physiology of Membrane Fluidity*, vol. 1 (Shinitzky, M., ed.) C.R.C. Press, Boca Raton, pp. 1–51.

Shinitzky, M. and Barenholz, Y. (1978) Fluidity parameters of lipid regions determined by fluorescence polarization. *Biochem. Biophys. Acta* **515,** 367–394.

Shinitzky, M. and Henkart, P. (1979) Fluidity of cell membranes: current concepts and trends. *Int. Rev. Cytol.* **60,** 121–147.

Shinitzky, M. and Inbar, M. (1976) Microviscosity parameters and protein mobility in biological membranes. *Biochim. Biophys. Acta* **422,** 133–149.

Singer, S. J. and Nicolson, G. L. (1972) The fluid mosaic model of the structure of cell membranes. *Science* **175,** 720–731.

Sumbilla, C. and Lakowicz, J. R. (1982) Fluorescence studies of red blood cell membranes from individuals with Huntington's Disease. *J. Neurochem.* **38,** 1699–1708.

Taraschi, T. F., Parashar, A., Hooks, M., and Rubin, H. (1986) Perturbation of red cell membrane structure during intracellular maturation of *Plasmodium falciparum*. *Science* **232,** 102–104.

Terranova, R., Alberghina, M., and Carnazzo, G. (1985) Lipid composition and fluidity of erythrocyte membrane in the elderly. *Ric. Clin. Lab.* **15 (Suppl. 1),** 327–334.

Wagner, G. M., Schwartz, R. S., Chiu, D. T., and Lubin, B. A. (1985) Membrane phospholipid organization and vesiculation of erythrocytes in sickle cell anemia. *Clin. Haematol.* **14,** 183–200.

White, D. A. (1973) The phospholipid composition of mammalian tissues, in *Form and Functions of Phospholipids* (Ansell, G. B., Hawthorne, J. N., and Dawson, R. M. C., eds.), Elsevier Scientific Publishing Co., Amsterdam, pp. 441–482.

Wilkerson, L. S., Perkins, Jr., R. C., Roelofs, R., Swift, L., Dalton, L. R., and Park, J. H. (1978) Erythrocyte membrane abnormalities in Duchenne muscular dystrophy monitored by saturation transfer electron paramagnetic resonance spectroscopy. *Proc. Natl. Acad. Sci. USA* **75,** 838–841.

Wing, D. R., Harvey, D. J., Belcher, S. J., and Paton, W. D. (1984) Change in membrane lipid content after chronic ethanol administration with respect to fatty acyl composition and phospholipid type. *Biochem. Pharmacol.* **33,** 1625–1632.

Yawata, Y., Miyashima, K., Sugihara, T., Murayama, N., Hosoda, S., Nakashima, S., Iida, H., and Nozawa, Y. (1984) Self-adaptive modification of red cell membrane lipids in lecithin-cholesterol acyltransferase deficiency. Lipid analysis and spin labelling. *Biochim. Biophys. Acta* **769,** 440–448.

Zwaal, R. F. A., Comfurius, P., and van Deenen, L. L. M. (1977) Membrane asymmetry and blood coagulation. *Nature* **268,** 358–360.

Calcium Fluxes in Pathologically Altered Red Cells

Ole Scharff and Birthe Foder

1. Introduction

Calcium flux is defined as the amount of calcium ions crossing unit area of a membrane in unit time. In red blood cells with no intracellular organelles surrounded by membrane structures, calcium flux refers solely to transport through the plasma membrane, and this transport may be directed inwards (influx) or outwards (efflux). The calcium flux may be passive (i.e., electrodiffusion) or active (i.e., mediated by the calcium pump).

Pathologically altered red cells may be subject to structural changes in membranous or cytoplasmic constituents, or to functional changes that cannot be referred to any known physical cause. In the following, we describe such changes of calcium flux parameters in red cells (for previous reviews, *see* Parker and Berkowitz, 1983; Al-Jobore et al., 1984), but before that we shall deal with tools needed for the study.

2. Tools for Study of Calcium Flux

2.1. Active Ca^{2+} Transport

Ca^{2+} flux mediated by the Ca^{2+} pump can be studied directly by observing the efflux from red cells or reconstituted ghosts loaded with calcium (e.g., ^{45}Ca), or the influx into inside-out vesicles (IOVs) prepared from red cells. Alternatively, the Ca^{2+} pump can be studied indirectly, but often more easily and in greater detail, by examination of its enzymatic counterpart, the calmodulin-stimulated (Ca^{2+} + Mg^{2+})-ATPase, either embedded in the red-cell membrane, or purified and reconstituted in lipid vesicles.

Most studies of the (Ca^{2+} + Mg^{2+})-ATPase have been performed using preparations of red-cell membranes. To obtain maximum enzyme activity, the treatment during preparation should be effective enough to make the membranes permeable to Ca^{2+}, ATP, and calmodulin, and at the same time leave the (Ca^{2+} + Mg^{2+})-ATPase and its environment undamaged. Membranes that are only partly permeable exhibit a low (Ca^{2+} + Mg^{2+})-ATPase activity with decreased affinities for Ca^{2+} and calmodulin.

The importance of the preparation method chosen is illustrated by the work of Hunsinger and Cheung (1986), who studied the (Ca^{2+} + Mg^{2+})-ATPase activity in red cells from cystic fibrosis (CF) patients. The membranes washed in glycylglycine-Mg^{2+}-containing buffer exhibited no difference between patients and controls, contrary to membranes washed in EDTA-containing buffer that showed a 40% reduction of the activity in the CF membranes compared to controls. The EDTA-washed membranes were probably calmodulin deficient. Nevertheless, the first membrane preparation showed lower activity and lower Ca^{2+} affinity, which indicates reduced membrane permeability, making this preparation unfit for a comparison of CF and control (Ca^{2+} + Mg^{2+})-ATPase.

2.2. Passive Ca^{2+} Flux

ATP-independent Ca^{2+} flux can be studied in red cells in which the powerful Ca^{2+}-pump has been inhibited, most often by a preceding depletion of the cellular ATP, achieved, for instance, by preincubation in the presence of iodoacetamide.

2.3. Cellular Calcium Homeostasis

The cellular concentration of total calcium ($[Ca]_i$) is an outcome of Ca^{2+} influx and efflux. Therefore, an increased $[Ca]_i$ in pathological conditions may reflect changes of passive as well as active Ca^{2+} transport. Prolonged exposure of the cell interior to $[Ca]_i$ above 0.1 mmol/L may lead to irreversible changes of membrane proteins mediated by two types of reaction, a transglutaminase-mediated and a proteolytic event (*see* Lorand and Michalska, 1985).

Normally, the rate of Ca^{2+} influx is low in human red cells, but it can be raised in vitro by addition of ionophore, e.g., A23187, and the resultant changes of $[Ca]_i$ may give additional information about the cellular Ca^{2+} homeostasis. For instance, the level of A23187-mediated

Ca^{2+} uptake in normal red cell (Fig. 1, expt., curve N) can be calculated from the measured Ca^{2+} flux parameters (Fig. 1, model) as shown previously (Scharff et al., 1983). However, this is not possible in polycythemia vera red cells, even if the deviating parameters are taken into consideration (Fig. 1, curves P1 and P2), indicating unknown changes in the pathological cells.

2.4. Ionized Cellular Calcium

Owing to the high hemoglobin content in red cells, it has not yet been possible to measure the cellular concentration of ionized calcium ($[Ca^{2+}]_i$) directly by utilizing fluorescent Ca^{2+} chelators, such as quin2 or fura2. Instead, an equilibration technique has been employed (Rhoda et al., 1985; Muallem et al., 1985; Waller et al., 1985). A small elevation of $[Ca^{2+}]_i$ elicits an opening of the Ca^{2+}-sensitive potassium channel (the Gardos effect, *see* Chapter 9), which may lead to loss of cellular K^+ (*see* Table 1) and water.

2.5. Some Pitfalls

Finally, it should be emphasized that, when red-cell preparations from patients and normal donors are compared, a different responsiveness to the method of preparation rather than a genuine difference in vivo should always be considered. In all determinations of ATPase or pump activity, the same degree of calmodulin stimulation should be secured in materials from patients and controls. The variation of Ca^{2+}-ATPase activity in control donors found by different authors (*see* Table 1) demonstrates the great variability of membrane preparations.

In some pathologically altered cells, for instance sickled red cells, intracellular organelle-like structures may occur (Lew et al., 1985), complicating the interpretation of calcium flux data. In the following, we review the available reports on changed calcium flux parameters in red blood cells from patients suffering from various diseases.

3. Sickle Cell Anemia

The alteration of a single amino-acid residue in the β-chain of hemoglobin A, from the highly polar glutamate to the nonpolar valine, inherited mostly by individuals of African origin, leads to the formation of hemoglobin S, which polymerizes during deoxygenation, causing sickle shapes of the affected red cells.

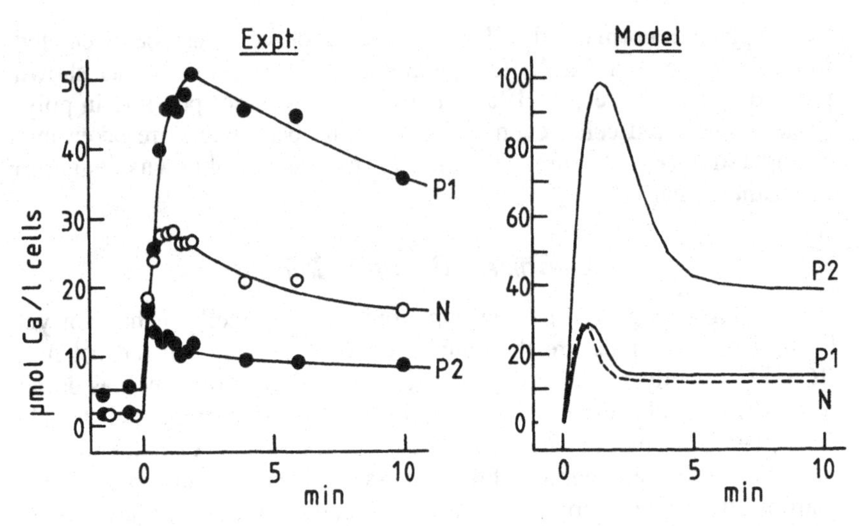

Fig. 1. A23187-induced Ca^{2+} uptake in red cells from a normal donor (N) and two polycythemia vera patients (P1 and P2). Expt. (left): Ca^{2+} uptake was induced at zero time by adding 8 µmol A23187/l cell to a suspension containing 20% red cells in 100 µ*M* $^{45}CaCl_2$, 1 m*M* $MgCl_2$, 75 m*M* KCl, 70 m*M* NaCl, 20 m*M* histidine/imidazole buffer, pH 7.1, 37°C. In parallel experiments (not shown), red cells from the three donors were depleted of ATP by pretreatment with iodoacetamide, and the A23187-mediated Ca^{2+} net influx was determined in the absence of the ATP-dependent Ca^{2+}-pump efflux. Furthermore, (Ca^{2+} + Mg^{2+})-ATPase activity was measured in red-cell membranes prepared as previously (Scharff et al., 1983) from each donor. Model (right): The Ca^{2+} uptake was simulated for each donor by the aid of a mathematical model (Scharff et al., 1983) using the experimental data for A23187-mediated Ca^{2+} net influx (patients: 77% [P1] and 168% [P2] of normal) and pump-mediated Ca^{2+}-efflux based on the ATPase experiments (patients: 51% [P1] and 66% [P2] of normal). It appears that the simulated Ca^{2+} uptake of the normal (dashed curve) fits the experiment reasonably well, whereas the simulated P1 and P2 curves are quite different from the experimental curves. This indicates unknown changes in the polycythemia red cells.

3.1. Ca^{2+} Homeostasis

The main characteristics of the Ca^{2+} homeostasis in sickled cells are: increased $[Ca]_i$ (*see* Table 1), increased passive Ca^{2+} influx during deoxygenation, and impaired pump-mediated Ca^{2+} efflux (Parker and Berkowitz, 1983; Bookchin et al., 1986). The rate of passive Ca^{2+} influx is normal in oxygenated cells, but increases four to five times during deoxygenation (Eaton et al., 1973; Bookchin et al., 1984). At least a part of the increased Ca^{2+} influx seems to occur through specific calcium channels, because the channel blocker nifedipine partially abolishes the calcium uptake (Rhoda et al., 1988). The impaired pump activity has been ascribed to (1) a reduced specific activity of the pump ATPase (Niggli et al., 1982) or (2) a decreased response to calmodulin activation of the (Ca^{2+} + Mg^{2+})-ATPase (Gopinath and Vincenzi, 1979; Dixon and Winslow, 1981). However, the changes of the ATPase activity are small and contradictory (*see* Table 1). Therefore, an impaired Ca^{2+}-pump activity is hardly a main cause of the deviating Ca^{2+} homeostasis in sickled cells.

3.2. Endocytic Vacuoles

More important are the recent findings that the sickled cells contain vacuoles, apparently formed by endocytosis from the plasma membrane as inside-out vesicles, and that these vacuoles store the bulk of the increased cellular calcium (Lew et al., 1985; Rhoda et al., 1985; Rubin et al., 1986). Ca^{2+}-pumps, originating from the plasma membrane, are probably imbedded in the membranes surrounding the vacuoles, which enable the vacuoles to accumulate calcium. In addition, the vesicles that are generated during the endocytosis induced by deoxygenation may envelop extracellular Ca^{2+} (Murphy et al., 1987). However, the endocytosis can only account for about 10% of the total calcium uptake (Rhoda et al., 1988). The total relative volume of the vacuoles varies, with typical values of 0.1–0.4% of the cell volume. The number of vacuoles appears to increase upon deoxygenation and to reduce again by reoxygenation, indicating a reversible mechanism of vesicle formation (Rubin et al., 1986).

The Ca^{2+} pumps in the plasma and vacuolar membranes seem to be able to maintain a low $[Ca^{2+}]_i$, i.e., about 20 n*M* (Rhoda et al., 1985, Murphy et al., 1987) without a permanent activation of the Ca^{2+}-sensitive K^+ channel (Ortiz et al., 1986). However, transient increases in $[Ca^{2+}]_i$ may occur during deoxygenation, and this may activate transiently the K^+ channel, contributing to the loss of K^+ (*see* Table 1) and water of sickle cells.

The compartmentalization of cellular calcium in sickle cells is a very interesting mechanism. The factor(s) releasing the formation of the endocytic vesicles in these cells are, so far, unknown. A prolonged increase of $[Ca]_i$ may be a contributory cause of the vesiculation, which according to Lew et al. (1985) also occur in normal red cells. Vesiculation is stimulated in human red cells when $[Ca]_i$ is increased by addition of ionophore

Table 1
Hemolytic Anemias: Ca^{2+} Parameters in Human Red Blood Cells†

Pathological condition	$[Ca]_i$ µmol/L cells patient	control	$(Ca^{2+} + Mg^{2+})$ − ATPase µmol min^{-1} g^{-1} patient	control	$[K^+]_i$ mmol 1/L cells patient	control	Cell organelles §	Ref.*
Sickle cell anemia (Hb SS)							present	(1)
	189 ± 69	25 ± 8						(2)
Oxygenated	49 ± 20	20 ± 6						(3)
Deoxygenated	102 ± 36	18 ± 6						(3)
Deoxygenated					79 ± 12	105 ± 7		(4)
			26 ± 8	40 ± 9				(5)
			62 ± 12	54 ± 7				(6)
β-Thalassemia							present	(7)
Unsplenectom	26 ± 8	6 ± 6	13 ± 10	11 ± 7				(8)
Splenectomized	85 ± 24	6 ± 3	13 ± 6	11 ± 6				(8)
Fresh blood					76 ± 12	92 ± 11		(9)
24-h incubation					53 ± 23	85 ± 7		(9)

G6PD Deficiency								
No diamide			54 ± 11	56 ± 6				(10)
Diamide			11 ± 4	31 ± 7				(10)
No favism	22 ± 8		29 ± 7		107 ± 8			(11)
Favism	288 ± 158		16 ± 6		77 ± 19			(11)
No favism	17 ± 5	18 ± 6	107 ± 19	107 ± 30				(12)
Favism	143 – 244		41 – 87					(12)
Hereditary spherocytosis							present	(13)
Unsplenectom	40 ± 12	19 ± 5	10 ± 4	10 ± 5				(14)
Splenectomized	31 ± 9	25 ± 6	9.3 ± 4.1	8.7 ± 4.0				(14)
Unsplenectom	15 ± 8	17 ± 8						(15)
Splenectomized	16 ± 4							(15)
					82 ± 8	96 ± 9		(16)

*References: (1) Lew et al., 1985; (2) Eaton et al., 1973; (3) Palek, 1977; (4) Steinberg et al., 1978; (5) Niggli et al., 1982; (6) Luthra end Sears, 1982; (7) Rachmilewitz et al., 1985; (8) Shalev et al., 1984; (9) Vettore et al., 1974; (10) Shalev et al., 1985; (11) De Flora et al., 1985; (12) Turrini et al., 1985; (13) Zipursky et al., 1986; (14) Feig et al., 1974, 1975; (15) Shimoda et al., 1984; (16) Mayman and Zipursky, 1974; † Mean ± S.D. are given. ATPase activities are converted to: μmol/min/g membrane protein, assuming 330 g hemoglobin and 7 g membrane protein/L cells.

†Refers to all values in the table ("Mean ± S.D.").
§The "Cell organelles" are "present" or "not present" (no symbol required).

A23187 or during aging in vitro (Allan et al., 1980). The cells change from discocytes to echinocytes that respond by budding to release small vesicles, but this sort of vesiculation seems to be exocytic rather than endocytic. However, sickle cells may deviate from normal cells in vesiculation. For instance, after treatment with phospholipase A_2, normal cells undergo echinocytic transformation, whereas this phenomenon is suppressed in sickle cells during deoxygenation (Lubin et al., 1981).

4. β-Thalassemia

β-thalassemia is characterized by reduced synthesis of β-chains of globin, leading to precipitating hemoglobin chains and other changes that make the red cell unstable. Anemia is common among these patients (Rachmilewitz et al., 1985).

The affected cells show increased $[Ca]_i$ (*see* Table 1). The $(Ca^{2+} + Mg^{2+})$-ATPase activity, however, is the same in patients and controls, and this raises the question whether the increased $[Ca]_i$ implies a change of $[Ca^{2+}]_i$ or whether calcium is trapped in endocytic vacuoles (Shalev et al., 1984), which are known to occur in thalassemic red cells (Rachmilewitz et al., 1985). Recently, $[Ca^{2+}]_i$ was reported to be within the normal range (about 25 n*M*) in these cells (Rhoda et al., 1987). On the other hand, the increased K^+ permeability and the decreased $[K^+]_i$ found in thalassemic red cells, especially during metabolic stress (Gunn et al., 1972; *see* Table 1), suggest that the potassium channels are activated by, at least, a transiently elevated $[Ca^{2+}]_i$ (cf. sickle cell anemia). Normal $(Ca^{2+} + Mg^{2+})$-ATPase and transiently elevated $[Ca^{2+}]_i$ indicate an increased Ca^{2+} permeability of the plasma membrane, but no change of calcium uptake in thalassemic cells was found (Rhoda et al., 1987).

5. Glucose-6-Phosphate Dehydrogenase Deficiency

Glucose-6-phosphate dehydrogenase (G6PD) is an important enzyme in the pentose phosphate pathway that is the only source of NADPH in human red cells. The deficiency is inherited, and the G6PD activity in the red cells from deficient individuals is below 25% of normal red cells (Vogel and Motulsky, 1986), being lower in the Mediterranean (0–5%) than in the African and Asiatic variants (5–25%).

5.1. Effect of Oxidant Drugs

The affected red cells are characterized by enhanced sensitivity to oxidant agents. Such agents, for instance the antimalarial drug primaquine, may induce severe hemolysis in G6PD-deficient individuals. A reported small inhibitory effect of primaquine on the (Ca^{2+} + Mg^{2+})-ATPase in red cells from deficients (Akoğlu et al., 1984) seems uncertain. In another study, the (Ca^{2+} + Mg^{2+})-ATPase activity was the same in red cells from Mediterranean variants and normal controls (Shalev et al., 1985). After an oxidative stress induced by pretreatment of the red cells with the thiol-oxidizing agent diamide, the (Ca^{2+} + Mg^{2+})-ATPase activity was diminished in both groups, but most in the G6PD-deficient erythrocytes (Table 1).

5.2. Favism

In individuals with very low G6PD activity (the Mediterranean variant), oxidative hemolysis can be provoked by oral intake of fava beans (favism), probably because of two toxic components, divicine and isouramil. During the acute hemolytic crisis, the red cells from such subjects had increased $[Ca]_i$ and, in spite of an elevated reticulocyte count, decreased (Ca^{2+} + Mg^{2+})-ATPase activity when compared to either G6PD-deficient healthy persons or nondeficient controls (*see* Table 1). Furthermore, $[K^+]_i$ was reduced in the red cells from individuals in acute hemolytic crisis (Table 1).

The increase of $[Ca]_i$ could not be explained alone by a reduced calcium-pump activity that corresponded to the found decrease of (Ca^{2+} + Mg^{2+})-ATPase activity (De Flora et al., 1985; Turrini et al., 1985). Other changes probably occurred in the red cells. For instance, a large increase in Ca^{2+} influx has to be assumed in favism (Turrini et al., 1985).

The results above indicate an increased sensitivity of the calcium pump to oxidative stress in G6PD-deficient red cells but, in addition, other membrane components may be influenced. For example, passive Ca^{2+} permeability of the plasma membrane may be increased. Recently, it has been shown that divicine increases the Ca^{2+} uptake of red blood cells or right-side-out vesicles (Turrini et al., 1988). Both normal and G6PD-deficient cells or vesicles were affected, but only the Ca^{2+} permeability of the deficient preparations was increased irreversibly by divicine.

6. Hereditary Spherocytosis

The causes of this inherited disease are unknown, but the genetic defects seem to be intrinsic to the red blood cell. The symptoms are anemia, jaundice, and splenomegaly (Becker and Lux, 1985). $[Ca]_i$ has been reported to be increased (Feig and Bassilian, 1975), and (Ca^{2+} +

Mg^{2+})-ATPase activity to be decreased (Feig and Guidotti, 1974) or increased (Johnsson et al., 1978) in red cells from patients suffering from hereditary spherocytosis (HS). However, the early ATPase results seem rather uncertain, and none of the deviations mentioned were confirmed by others (Kirkpatrick et al., 1975; Zail and van den Hoek, 1976; Shimoda et al., 1984). An increased Ca^{2+} uptake in spherocytic red-cell populations was suggested to be caused by the presence of reticulocytosis in unsplenectomized patients (Shimoda et al., 1984).

Recently, Zipursky et al., (1986) reported a higher percentage of red cells containing vacuoles in splenectomized HS patients (30 ± 18%, mean ± SD) compared to unsplenectomized HS patients (1.5 ± 1.7%) and normal adults (0.4 ± 0.7%). Whether the increased occurrence of vacuoles is related to a changed Ca^{2+} homeostasis in the affected red cells, as in sickled cells (cf above), or whether the phenomenon is the result of other factors remains an open question.

7. Malaria

The widely distributed disease, malaria, caused by Plasmodium parasites occurs in areas housing human populations with a relatively high frequency of the genetical variants: sickle cell disease, glucose-6-phosphate dehydrogenase deficiency, and thalassemia. This coincidence has for a long time been attributed to a positively selective pressure during the human evolution in these areas, probably because of an inhibition of the parasite multiplication in red cells that are either sickled, G6PD deficient, or thalassemic (Vogel and Motulsky, 1986). Since these three types of red cell all exhibit a changed cellular calcium homeostasis (*see* Table 1), it is obvious to focus on the role of cell calcium in the infection of red cells by the malaria parasites.

Plasmodium-infected erythrocytes from various species (human, mice, and rats) are characterized by abnormal transport of Ca^{2+} and K^+, and calcium accumulation (Leida et al., 1981; Bookchin et al., 1981; Tanabe et al., 1982; Krungkrai and Yuthavong, 1983). It has been suggested that the calcium pump in infected rat erythrocytes is damaged by a partial uncoupling of ATP hydrolysis and Ca^{2+} transport, and in these cells more than 90% of the cellular calcium is localized to the parasite compartment (Mikkelsen et al., 1984).

The increased Ca^{2+} uptake in human red cells infected by *P. falciparum* was suggested to contribute to the hemolysis of the cells (Bookchin et al., 1981). Possibly, infected red cells that are, in addition, either sickled, G6PD-deficient, or thalassemic may show increased tendency to hemolysis, which may be detrimental to the malaria parasites at certain stages of proliferation. Contributory factors may be increased viscosity of the host cell cytosol resulting from loss of potassium and water or hemoglobin gelation (Ginsburg et al., 1986).

8. Cystic Fibrosis

Cystic fibrosis (CF) is a recessive disorder with generalized dysfunction of exocrine glands. Clinically the disease is characterized by chronic lung disease, pancreatic insufficiency, and elevated sweat electrolytes. A number of observations indicate that altered calcium homeostasis could be involved in the pathology of CF (Katz et al., 1984).

8.1. (Ca^{2+} + Mg^{2+})-ATPase

About a dozen studies of the (Ca^{2+} + Mg^{2+})-ATPase in red blood cells from CF patients have been published, showing from 0–50% reduction in activity. Calmodulin-stimulated activity was found to be normal (Clough and Hubbard, 1984) or decreased by 15–40% (Foder et al., 1980; Dearborn et al., 1984; Gietzen et al., 1984; Waller et al., 1985; Muallem et al., 1985).

Ca^{2+} sensitivity, calmodulin sensitivity, and trifluoperazine inhibition were unchanged in CF (Katz 1978; Foder et al., 1980; Waller et al., 1985), whereas an increased activation by sodium (Clough and Hubbard, 1984) and altered ATP kinetics (Hunsinger and Cheung, 1986) have been reported. In accordance with these changes the highest reduction in activity (50%) was found in a study using a low ATP concentration in the absence of sodium (Katz, 1978). Phosphoenzyme formation was found to be normal (Waller et al., 1985) or decreased (Muallem et al., 1985; Allen et al., 1986).

(Ca^{2+} + Mg^{2+})-ATPase purified from CF red-cell membranes and assayed in the presence of asolectin had the same activity (and Ca^{2+} sensitivity) as the enzyme purified from control red cells. This suggests that the reduced activity found in some studies may reflect an abnormality of membrane components in association with the (Ca^{2+} + Mg^{2+})-ATPase (Bridges and Katz, 1986). This abnormality may be alterations in the membrane lipids of red cells, especially in CF patients with pancreatic insufficiency, as suggested by Dearborn et al. (1984), who only found decreased (Ca^{2+} + Mg^{2+})-ATPase activity in this type of CF patients.

8.2. Ca^{2+} Fluxes

Reduced Ca^{2+}-pump activity was found in different preparations from CF red cells:

1. In inside-out vesicles (55% reduction; Ansah and Katz, 1980)
2. In cells loaded with calcium by use of an ionophore (29% reduction; Siegal and Settlemire, 1984), and
3. In cells loaded with calcium chelator (75% reduction; Muallem et al., 1985).

The passive Ca^{2+} entry was reported to be normal (Waller et al., 1985) or lower than normal (Muallem et al., 1985), and both CF and control red cells maintained $[Ca^{2+}]_i$ between 20 and 30 n*M* (Waller et al., 1985; Muallem et al., 1985). This indicates that in intact cells the calcium-pump deficit is not large enough to impair the ability of the CF red cell to maintain normal resting levels of $[Ca^{2+}]_i$.

9. Hypertension

Essential (or primary) hypertension (HT) seems to be associated with general alterations of cell membrane function, especially disturbances in cellular Ca^{2+} homeostasis and transport of monovalent cations. Much evidence originates from the study of red blood cells from HT patients or from spontaneously hypertensive rats (SHR) (Postnov and Orlov, 1985).

In red blood cells with a depressed ATP concentration, the accumulation rate of ^{45}Ca in HT patients was twice that in normotensive (NT) controls (Wehling et al., 1983). This might be explained by assuming a residual Ca^{2+}-pump activity, which is lower in the HT cells than in the NT cells (Postnov and Orlov, 1985). A reduced Ca^{2+}-pump activity in HT patients was demonstrated in IOVs prepared from red blood cells (Postnov et al., 1984). Interestingly, differences in pump flux was only found in the presence of calmodulin.

In accordance with these findings, the basal $(Ca^{2+} + Mg^{2+})$-ATPase activity in red-cell membranes from HT patients was normal (Postnov and Orlov, 1985; Lin et al., 1985), but the addition of calmodulin revealed a lower degree of activation in HT membranes (Olorunsogo et al., 1985). These findings, however, have not been confirmed by others (Dagher and Amar, 1986; Vincenzi et al., 1985). According to Postnov et al. (1984), the calmodulin concentration seems not to be changed in HT red cells.

Measurements with a Ca^{2+}-selective electrode in frozen-thawed hemolysates from red cells suggested a higher $[Ca]_i$ in HT than in NT cells (Zidek et al., 1982), and this difference was reduced in patients treated with either the loop diuretic, piretanide, or beta blockers (Zidek et al., 1985). The Ca^{2+} activities measured by these authors (2–10 μmol/L) hardly revealed the $[Ca^{2+}]_i$ values in vivo. However, Okada et al. (1985) recently demonstrated that SHR red-cell populations showed an increased proportion of echinocytic cells compared to those of normotensive rats, suggesting a higher $[Ca^{2+}]_i$ in the hypertensive red cells.

10. Affective Disorders

Alterations in Ca^{2+} homeostasis have been suggested in patients suffering from affective disorders. Low and high serum calcium levels have been reported to be associated with mania and depression, respectively, and antipsychotic drugs, e.g., phenothiazines, are known to interfere with the Ca^{2+}/calmodulin system (for review, *see* Dubovsky and Franks, 1983).

However, the results are ambiguous. Low serum calcium was found in depressed children (Pliszka and Rogeness, 1984), opposite to other findings (*see above*). Ca^{2+} parameters have been determined in red blood cells from patients, and a change of (Ca^{2+} + Mg^{2+})-ATPase activity in manic depressive patients has been reported either to be small (even if the subtle data processing often revealed significant changes) (Choi et al., 1981; Meltzer and Kassir, 1983; Strzyzewski et al., 1984; MacDonald et al., 1984), or to be lacking (Alexander et al., 1986).

11. Other Diseases

11.1. Ca^{2+} Uptake

Changed Ca^{2+} parameters have been found in various diseases because of known or unknown cellular defects. Increased Ca^{2+} uptake in red blood cells has been demonstrated in cases of congenital hemolytic anemia (Wiley and Gill, 1976), iron deficiency, and paroxysmal nocturnal hemoglobinuria (PNH) (Shimoda and Yawata, 1985a,b).

11.2. Active Ca^{2+} Transport

Reduced (Ca^{2+} + Mg^{2+})-ATPase activity was found in red cells in individuals suffering from polycythemia vera (Scharff and Foder 1975, *see also* Fig. 1), familial benign hypercalcemia (Mole and Paterson 1985), hypothyroidism (Dube et al., 1986), protein-energy malnutrition, i.e., kwashiorkor (Ramanadham and Kaplay, 1982), and *Mycoplasma pneumoniae* infection (Tsuchiya and Sugai, 1982). Increased (Ca^{2+} + Mg^{2+})-ATPase or Ca^{2+}-pump activities in red cells have been reported in hyperthyroidism (Dube et al., 1986) and in some studies on Duchenne muscular dystrophy, whereas conflicting results were obtained in studies on other muscular dystrophies (reviewed by Parker and Berkowitz, 1983; Rüdel and Lehmann-Horn, 1985). No changes were found in PNH and autoimmune hemolytic anemia (Asai et al., 1985). These observations are interesting, but represent areas that need more attention before conclusions can be drawn.

12. Summary and Conclusions

In a number of diseases or deficiencies, the Ca^{2+} homeostasis of red cells seems to be disturbed, especially during metabolic stress. Two characteristic features of human red cells are the extremely low Ca^{2+} influx and the potentially high pump-mediated Ca^{2+} efflux. In pathologically altered red cells, increased Ca^{2+} influx and decreased activity of Ca^{2+}-pump or (Ca^{2+} + Mg^{2+})-ATPase are typical findings, resulting in elevated $[Ca]_i$, which may be accompanied by raised $[Ca^{2+}]_i$ and a reduction of $[K^+]_i$, depending on the degree of sequestration of cellular Ca^{2+}. In some diseases, endocytic vacuoles, possibly Ca^{2+}-accumulating, have been found in the pathologically altered red cells.

References

Akoğlu, T., Ozdogu, H., Erdoğan, R., and Özer, F. L. (1984) Erythrocyte membrane ATPase activity of G6PD-deficient individuals and the effect of primaquine metabolite(s) on membrane ATPase enzymes. *J. Tropic. Med. Hyg.* **87,** 219–224.

Alexander, D. R., Deeb, M., Bitar, F., and Antun, F. (1986) Sodium-potassium, magnesium, and calcium ATPase activities in erythrocyte membranes from manic-depressive patients responding to lithium. *Biol. Psychiatry* **21,** 997–1007.

Al-Jobore, A., Minocherhomjee, A. M., Villalobo, A., and Roufogalis, B. D. (1984) Active calcium transport in normal and abnormal human erythrocytes, in *Erythrocyte Membranes 3: Recent Clinical and Experimental Advances* (Kruckeberg, W. C., Eaton, J. W., Aster, J., and Brewer, G. J., eds.), Alan R. Liss, New York, pp. 243–292.

Allan, D., Thomas, P., and Limbrick, A. R. (1980) The isolation and characterization of 60 nm vesicles ('nanovesicles') produced during ionophore A23187-induced budding of human erythrocytes. *Biochem. J.* **188,** 881–887.

Allen, B. G., Bridges, M., Roufogalis, B. D., and Katz, S. (1986) Investigation of (Ca^{2+} + Mg^{2+})-ATPase phospho-protein formation in erythrocyte membranes of patients with cystic fibrosis. *Cell Calcium* **7,** 161–168.

Ansah, T. and Katz, S. (1980) Evidence for a Ca^{2+}-transport deficiency in patients with cystic fibrosis. *Cell Calcium* **1,** 195–203.

Asai, T. Fujioka, S., and Yoshida, S. (1985) Erythrocyte membrane associated enzymes of hemolytic anemias: ATPases, AMP deaminase and GAPD. *Acta Haematol. Jpn.* **48,** 12–21.

Becker, P. S. and Lux, S. E. (1985) Hereditary spherocytosis and related disorders. *Clin. Haematol.* **14,** 15–43.

Bookchin, R. M., Ortiz, O. E., and Lew, V. L. (1984) Silent intracellular calcium in sickle cell anemia red cells, in *The Red Cell: Sixth Ann Arbor Conference* (Brewer, G. J., ed.), Alan R. Liss, New York, pp. 17–28.

Bookchin, R. M., Raventos, C., Nagel, R. L., and Lew, V. L. (1981) Abnormal K and Ca transport and Ca accumulation in red cells infected in vitro with *Plasmodium falciparum. Clin. Res.* **29,** 516A.

Bookchin, R. M., Ortiz, O. E., Hockaday, A., Sepulveda, M. I., and Lew, V. L. (1986) The state of calcium in sickle cell anaemia red cells, in *Intracellular Calcium Regulation* (Bader, H., Gietzen, K., Rosenthal, J., Rüdel, R., and Wolf, H. U., eds.), Manchester University Press, Manchester, pp. 327–334.

Bridges, M. A. and Katz, S. (1986) Isolation and characterization of erythrocyte membrane Ca^{2+}-ATPase in cystic fibrosis. *Pediatr. Res.* **20,** 356–360.

Choi, S. J., Derman, R. M., and Lee, K. S. (1981) Bipolar affective disorder, lithium carbonate and Ca^{++} ATPase. *J. Affect. Disorders* **3,** 77–79.

Clough, D. L. and Hubbard, V. S. (1984) Red cell membrane Ca-ATPase in cystic fibrosis: increased activation by Na. *Clin. Chim. Acta* **138,** 259–265.

Dagher, G. and Amar, M. (1986) Maximal velocity of the Ca pump in RBC of SHR and essential hypertensive patients, in *Proceedings of European Red Cell Club Conference*, Visegrád, Hungary.

Dearborn, D. G., Wityk, R. J., Johnson, L. R., Poncz, L., and Stern, R. C. (1984) Calcium-ATPase activity in cystic fibrosis erythrocyte membranes: decreased activity in patients with pancreatic insufficiency. *Pediatr. Res.* **18,** 890–895.

De Flora, A., Benatti, U., Guida, L., Forteleoni, G., and Meloni, T. (1985) Favism: disordered erythrocyte calcium homeostasis. *Blood* **66,** 294–297.

Dixon, E. and Winslow, R. M. (1981) The interaction between (Ca^{2+} + Mg^{2+})-ATPase and the soluble activator (calmodulin) in erythrocytes containing haemoglobin S. *Brit. J. Haematol.* **47,** 391–397.

Dube, M. P., Davis, F. B., Davis, P. J., Schoenl, M., and Blas, S. D. (1986) Effects of hyperthyroidism and hypothyroidism on human red blood cell Ca^{2+}-ATPase activity. *J. Clin. Endocrinol. Metab.* **62,** 253–257.

Dubovsky, S. L. and Franks, R. D. (1983) Intracellular calcium ions in affective disorders: a review and an hypothesis. *Biol. Psychiatry* **18,** 781–797.

Eaton, J. W., Skelton, T. D., Swofford, H. S., Kolpin, C. E., and Jacob, H. S. (1973) Elevated erythrocyte calcium in sickle cell disease. *Nature* **246,** 105–106.

Feig, S. A. and Bassilian, S. (1975) Increased erythrocyte Ca^{2+} content in hereditary spherocytosis. *Pediat. Res.* **9,** 928–931.

Feig, S. A. and Guidotti, G. (1974) Relative deficiency of Ca^{2+}-dependent adenosine triphosphatase activity of red cell membranes in hereditary spherocytosis. *Biochem. Biophys. Res. Commun.* **58,** 487–494.

Foder, B., Scharff, O., and Tønnesen, P. (1980) Activator-associated Ca^{2+}-ATPase in erythrocyte membranes from cystic fibrosis patients. *Clin. Chim. Acta.* **104,** 187–193.

Gietzen, K., Dick, B., Hauser, U., and Bader, H. (1984) Calmodulin and Ca^{++}-transport ATPase in erythrocytes from cystic fibrosis patients. *Cell Calcium* **5,** 307.

Ginsburg, H., Handeli, S., Gorodetsky, R., Friedman, S., and Krugliak, M. (1986) What is the mechanism of sickle trait cell resistance to malaria?, in *Proceedings of European Red Cell Club Conference, Visegrád, Hungary.*

Gopinath, R. M. and Vincenzi, F. F. (1979) (Ca^{2+} + Mg^{2+})-ATPase activity of sickle cell membranes: Decreased activation by red blood cell cytoplasmic activator. *Amer. J. Hematol.* **7,** 303–312.

Gunn, R. B., Silvers, D. N., and Rosse, W. F. (1972) Potassium permeability in β-thalassemia minor red blood cells. *J. Clin. Invest.* **51,** 1043–1049.

Hunsinger, R. N. and Cheung, H. C. (1986) Probe of the (Ca^{2+} + Mg^{2+})-ATPase in erythrocyte membranes of cystic fibrosis patients. *Clin. Chim. Acta* **156,** 165–178.

Johnsson, R., Santaholma, S., and Saris, N. (1978) Calcium transport and adenosine triphosphatase activities of erythrocyte membranes in congenital spherocytosis. *Scand. J. Clin. Lab. Invest.* **38,** 121–125.

Katz, S. (1978) Calcium and sodium transport processes in patients with cystic fibrosis. I. A specific decrease in Mg^{2+}-dependent, Ca^{2+}-adenosine triphosphatase activity in erythrocyte membranes from cystic fibrosis patients. *Pediatr. Res.* **12,** 1033–1038.

Katz, S., Schoni, M. H., and Bridges, M. A. (1984) The calcium hypothesis of cystic fibrosis. *Cell Calcium* **5,** 421–440.

Kirkpatrick, F. H., Woods, G. M., and LaCelle, P. L. (1975) Absence of one component of spectrin adenosine triphosphatase in hereditary spherocytosis. *Blood* **46,** 945–954.

Krungkrai, J. and Yuthavong, Y. (1983) Enhanced Ca^{2+} uptake by mouse erythrocytes in malarial (*Plasmodium berghei*) infection. *Mol. Biochem. Parasitol.* **7,** 227–235.

Leida, M. N., Mahoney, J. R., and Eaton, J. W. (1981) Intraerythrocytic plasmodial calcium metabolism. *Biochem. Biophys. Res. Commun.* **103,** 402–406.

Lew, V. L., Hockaday, A., Sepulveda, M., Somlyo, A. P., Somlyo, A. V., Ortiz, O. E., and Bookchin, R. M. (1985) Compartmentalization of sickle-cell calcium in endocytic inside-out vesicles. *Nature* **315,** 586–589.

Lin, S., Hong, C., Chiang, B. N., and Wei, Y. (1985) Activities of transport adenosine triphosphatases in erythrocyte membranes of healthy and hypertensive subjects. *Clin. Exp. Hypertens.* **A7,** 1151–1163.

Lorand, L. and Michalska, M. (1985) Altered response of stored red cells to Ca^{2+} stress. *Blood* **65,** 1025–1027.

Lubin, B., Chiu, D., Bastacky, J., Roelofsen, B., and van Deenen, L. L. M. (1981) Abnormalities in membrane phospholipid organization in sickled erythrocytes. *J. Clin. Invest.* **67,** 1643–1649.

Luthra, M. G. and Sears, D. A. (1982) Increased Ca^{2+}, Mg^{2+}, and Na^{+} + K^{+} ATPase activities in erythrocytes of sickle cell anemia. *Blood* **60,** 1332–1336.

MacDonald, E., Rubinow, D., and Linnoila, M. (1984) Sensitivity of RBC membrane Ca^{2+}-adenosine triphosphatase to calmodulin stimulation. *Arch. Gen. Psychiatry* **41,** 487–493.

Mayman, D. and Zipursky, A. (1974) Hereditary spherocytosis: the metabolism of erythrocytes in the peripheral blood and the splenic pulp. *Brit. J. Haematol.* **27,** 201–217.

Meltzer, H. L. and Kassir, S. (1983) Abnormal calmodulin-activated Ca ATPase in manic-depressive subjects. *J. Psychiat. Res.* **17,** 29–35.

Mikkelsen, R. B., Geller, E., Doren, E. V., and Asher, C. R. (1984) Ca^{2+} metabolism of *Plasmodia*-infected erythrocytes. *Prog. Clin. Biol. Res.* **155,** 25–34.

Mole, P. A. and Paterson, C. R. (1985) Calcium-ATPase activity in erythrocyte ghosts from patients with familial benign hypercalcaemia. *Scand. J. Clin. Lab. Invest.* **45,** 349–353.

Muallem, S., Miner, C., and Seymour, C. A. (1985) The nature of the Ca^{2+} pump defect in the red blood cells of patients with cystic fibrosis. *Biochim. Biophys. Acta* **819,** 143–147.

Murphy, E., Berkowitz, L. R., Orringer, E., Levy, L., Gabel, S. A., and London, R. E. (1987) Cytosolic free calcium levels in sickle red blood cells. *Blood* **69,** 1469–1474.

Niggli, V., Adunyah, E. S., Cameron, B. F., Bababunmi, E. A., and Carafoli, E. (1982) The Ca^{2+}-pump of sickle cell plasma membranes. Purification and reconstitution of the ATPase enzyme. *Cell Calcium* **3,** 131–151.

Okada, T., Kuwahara, T., and Nakamura, K. (1985) Ca^{++} and morphological changes in erythrocytes of stroke-prone SHR. *Jpn. Heart J.* **26,** 685.

Olorunsogo, O. O., Okudolo, B. E., Lawa, S. O. A., and Falase, A. O. (1985) Erythrocyte membrane Ca^{2+}-pumping ATPase of hypertensive humans: Reduced stimulation by calmodulin. *Biosci. Rep.* **5,** 525–531.

Ortiz, O. E., Lew, V. L., and Bookchin, R. M. (1986) Calcium accumulated by sickle cell anemia red cells does not affect their potassium ($^{86}Rb^{+}$) flux components. *Blood* **67,** 710–715.

Palek, J. (1977) Red cell calcium content and transmembrane calcium movements in sickle cell anemia. *J. Lab. Clin. Med.* **89,** 1365–1374.

Parker, J. and Berkowitz, L. R. (1983) Physiologically instructive genetic variants involving the human red cell membrane. *Physiol. Rev.* **63,** 261–313.

Pliszka, S. R. and Rogeness, G. A. (1984) Calcium and magnesium in children with schizophrenia and major depression. *Biol. Psychiatry* **19,** 871–876.

Postnov, Y. V. and Orlov, S. N. (1985) Ion transport across plasma membrane in primary hypertension. *Physiol. Rev.* **65,** 904–945.

Postnov, Y. V., Orlov, S. N., Reznikova, M. B., Rjazhsky, G. G., and Pokudin, N. I. (1984) Calmodulin distribution and Ca^{2+} transport in the erythrocytes of patients with essential hypertension. *Clin. Sci.* **66,** 459–463.

Rachmilewitz, E. A., Shinar, E., Shalev. O., Galili, U., and Schrier, S. L. (1985) Erythrocyte membrane alterations in β-thalassaemia. *Clin. Haematol.* **14,** 163–182.

Ramanadham, M. and Kaplay, S. S. (1982) Erythrocyte osmotic fragility in protein-energy malnutrition: cholesterol, phospholipid, and Ca^{2+}, Mg^{2+} adenosine triphosphatase. *Biochem. Med.* **27,** 226–231.

Rhoda, M. D., Beuzard, Y., Apovo, M., Galacteros, F., and Giraud, F. (1988) Ca^{2+} permeability in deoxygenated sickle erythrocytes. *Proceedings of Structure et Fonctions de la Membrane du Globule Rouge* (Cadarache, France).

Rhoda, M. D., Galacteros, F., Beuzard, Y., and Giraud, F. (1987) Ca^{2+} permeability and cytosolic Ca^{2+} concentration are not impaired in β-thalassemic and hemoglobin C erythrocytes. *Blood* **70,** 804–808.

Rhoda, M. D., Giraud, F., Craescu, C. T., and Beuzard, Y., (1985) Compartmentalization of Ca^{2+} in sickle cells. *Cell Calcium* **6,** 397–411.

Rubin, E., Schlegel, R. A., and Williamson, P. (1986) Endocytosis in sickle erythrocytes: a mechanism for elevated intracellular Ca^{2+} levels. *J. Cell Physiol.* **126,** 53–59.

Rüdel, R. and Lehmann-Horn, F. (1985) Membrane changes in cells from myotonia patients. *Physiol. Rev.* **65,** 310–356.

Scharff, O. and Foder, B. (1975) Decreased (Ca^{2+} + Mg^{2+})-stimulated ATPase activity in erythrocyte membranes from polycythemia vera patients. *Scand. J. Clin. Lab. Invest.* **35,** 583–589.

Scharff, O., Foder, B., and Skibsted, U. (1983) Hysteretic activation of the Ca^{2+} pump revealed by calcium transients in human red cells. *Biochim. Biophys. Acta* **730,** 295–305.

Shalev, O., Lavi, V., Hebbel, R. P., and Eaton, J. W. (1985) Erythrocyte (Ca^{2+} + Mg^{2+})-ATPase activity: increased sensitivity to oxidative stress in glucose-6-phosphate dehydrogenase deficiency. *Amer. J. Hematol.* **19,** 131–136.

Shalev, O., Mogilner, S., Shinar, E., Rachmilewitz, E. A., and Schrier, S. L. (1984) Impaired erythrocyte calcium homeostasis in β-thalassemia. *Blood* **64,** 564–566.

Shimoda, M. and Yawata, Y. (1985a) A marked increase of calcium uptake in the ATP-depleted red cells of patients with iron deficiency. *Amer. J. Hematol.* **19,** 55–61.

Shimoda, M. and Yawata, Y. (1985b) An increased calcium accumulation in ATP-depleted red cells of the patients with paroxysmal nocturnal hemoglobinuria. *Amer. J. Hematol.* **20,** 325–335.

Shimoda, M., Miyashima, K., and Yawata, Y. (1984) Increased calcium uptake in the red cells of unsplenectomized patients with hereditary spherocytosis: significant contribution of reticulocytosis. *Clin. Chim. Acta* **142,** 183–192.

Siegal, V. and Settlemire, C. T. (1984) Calcium efflux and cell morphology changes in red blood cells from normal subjects and patients with cystic fibrosis, in *Cystic Fibrosis: Horizons* (Lawson, D., ed.), John Wiley, Chichester, p. 403.

Steinberg, M. H., Eaton, J. W., Berger, E., Coleman, M. B., and Oelshlegel, F. J. (1978) Erythrocyte calcium abnormalities and the clinical severity of sickling disorders. *Brit. J. Haematol.* **40,** 533–539.

Strzyzewski, W., Rybakowski, J., Potok, E., and Zelechowska-Ruda, E. (1984) Erythrocyte cation transport in endogenous depression: clinical and psychophysiological correlates. *Acta Psychiatr. Scand.* **70,** 248–253.

Tanabe, K., Mikkelsen, R. B., and Wallach, D. F. H. (1982) Calcium transport of *Plasmodium chabaudi*-infected erythrocytes. *J. Cell Biol.* **93,** 680–684.

Tsuchiya, Y. and Sugai, H. (1982) The effect of *Mycoplasma pneumoniae* infection on human erythrocytes: changes in osmotic fragility, lipid composition, sialic acid content, Ca^{2+}-ATPase activities, and ATP concentration. *Biochem. Med.* **28,** 256–265.

Turrini, F., Mannuzzu, L., Naitana, A., and Arese, P. (1988) Effect of divicine on passive calcium permeability in G6PD-deficient cells. *Proceedings of Structure et Fonctions de la Membrane du Globule Rouge* (Cadarache, France).

Turrini, F., Naitana, A., Mannuzzu, L., Pescarmona, G., and Arese, P. (1985) Increased red cell calcium, decreased calcium adenosine triphosphatase, and altered membrane proteins during fava bean hemolysis in glucose-6-phosphate dehydrogenase-deficient (Mediterranean variant) individuals. *Blood* **66,** 302–305.

Vettore, L., Falezza, G. C., Cetto, G. L., and DeMatteis, M. C. (1974) Cation content and membrane deformability of heterozygous β-thalassaemic red blood cells. *Brit. J. Haematol.* **27,** 429–437.

Vincenzi, F. F., McCarron, D. A., and Morris, C. D. (1985) Another marker of hypertension in red blood cells? *Fed. Proc.* **44,** 1041.

Vogel, F. and Motulsky, A. G. (1986) *Human Genetics. Problems and Approaches* (Springer Verlag, Berlin).

Waller, R. L., Johnson, L. R., Brattin, W. J., and Dearborn, D. G. (1985) Erythrocyte cytosolic free Ca^{2+} and plasma membrane Ca^{2+}-ATPase activity in cystic fibrosis. *Cell Calcium* **6,** 245–264.

Wehling, M., Vetter, W., Neyses, L., Groth, H., Boerlin, H. J., Locher, R.,

Siegenthaler, W., and Kuhlmann, U. (1983). Altered calcium and sodium metabolism in red blood cells of hypertensive man: assessment by ion-selective electrodes. *J. Hypertens.* **1,** 171–176.

Wiley, J. S. and Gill, F. M. (1976) Red cell calcium leak in congenital hemolytic anemia with extreme microcytosis. *Blood* **47,** 197–210.

Zail, S. S. and van den Hoek, A. K. (1976) Studies on calcium transport and calcium-dependent adenosine triphosphatase activity of erythrocyte membranes in hereditary spherocytosis. *Brit. J. Haematol.* **34,** 605–611.

Zidek, W., Vetter, H., Dorst, K., Zumkley, H., and Losse, H. (1982) Intracellular Na^+ and Ca^{2+} activities in essential hypertension. *Clin. Sci.* **63,** 41s–43s.

Zidek, W., Karoff, C., Baumgart, P., Losse, H., Fehske, K. J., Häcker, W., and Vetter, H. (1985) Intracellular sodium and calcium during antihypertensive treatment. *Klin. Wochenschr.* **63, (Suppl. III),** 147–149.

Zipursky, A., Brown, E. J., and Chachula, D. M. (1986) Erythrocyte vacuoles in hereditary spherocytosis. *Brit. J. Haematol.* **62,** 775–776.

Mechanisms of Red Cell Dehydration in Sickle Cell Anemia

Application of an Integrated Red Cell Model

Robert M. Bookchin, Olga E. Ortiz,
and Virgilio L. Lew

1. Introduction

The erythrocytes from persons with sickle cell anemia (SS cells) exhibit a wide range of volume and density. The lightest (and largest) of these red cells are the young reticulocytes, with densities similar to normal reticulocytes. A large cell fraction (which may include a variable number of denser SS reticulocytes) has a volume and density similar to that of the majority of normal mature red cells. A variable proportion of SS cells, however, has a density range considerably higher than normal; many of these cells, particularly the most dense, retain an elongated, cigar, or spindle shape even when they are fully oxygenated and contain no polymerized hemoglobin S (Bertles and Dobler, 1969), and have been designated "irreversibly sickled cells" (ISCs). The present chapter is concerned with the mechanisms by which SS cells may become dense.

1.1. Cation and Water Content of Dense SS Cells

Several important features of these dense SS cells were first described by Bertles and Milner (1968), who found that the mean cell hemoglobin was normal, whereas the mean cell volume was reduced, so that the increased density, associated with an elevated intracellular hemoglobin concentration, must have resulted from cellular dehydration. Later studies of the composition of dense SS cells showed a marked reduction in K content to less than 30 mmol/L cells, with only a modest increase in Na content, to about 20–30 mmol/L cells (Glader et al., 1978; Clark et al., 1978a), indicating a large net loss of cation.

Calculations from these values and the observed mean cell hemoglobin concentrations indicate that the total monovalent cation concentrations (per vol of cell water) are reduced in these cells, as well as in other dehydrated red cells (homozygous hemoglobin C disease and hereditary xerocytosis) (Brugnara et al., 1984; Glader et al., 1974), all of which have undergone a large loss of K. Recent experiments with K-permeabilized red cells (Freeman et al., 1987) confirmed the predictions made using an integrated red-cell model (Lew and Bookchin, 1986; *also see* chapter 2) that dilution of cell cation under these circumstances can be attributed to the nonideal osmotic properties of hemoglobin. In the dehydrated cells, the increased osmotic coefficient of hemoglobin requires a lower cell content of other osmotic particles to achieve osmotic equilibrium.

1.2. Life Cycle of Dense SS Cells In Vivo

In their early studies, Bertles and Milner (1968) showed by various isotopic labeling techniques that ISCs were generated in the circulation early in the life span of the cells, reaching a maximum in 4–7 d, and then disappearing faster than mature non-ISCs (discocytes). Thus, unlike the normal red cell population, in which the densest cells tend to be the oldest, dense ISCs are relatively young SS cells.

Some interesting observations on the generation of ISCs and other dense SS cells were reported by Embury et al. (1984), who followed the changes in erythropoietin, red cell density, and morphology in sickle cell anemia patients during and after a 5-d nasal administration of oxygen. During oxygen inhalation, which raised the partial pressure of oxygen in the blood, suppressed erythropoiesis was accompanied by a fall in the proportion of ISCs, but there was a sustained elevation in the percentage of non-ISC dense SS cells. When oxygen therapy was discontinued, the number of ISCs rose before an increase in reticulocytes was observed. The number of circulating ISCs and other dense SS cells must reflect the balance between their rates of production and destruction. It is likely that ISCs and other dense SS cells, once formed, have a shortened circulatory survival despite increases in arterial oxygen tension; if so, these data suggest that the extent of deoxygenation of SS cells plays more of a role in formation of ISCs than in the generation of non-ISC dense cells.

1.3. Pathophysiological Significance of Dense SS Cells

Interest in the mechanism of SS cell dehydration has been stimulated by evidence that the dense cells play an important role in the two main clinical consequences of sickle cell anemia—the markedly shortened red-cell survival time and widespread microvascular occlusions. As noted above, ISCs have a particularly short life span in the circulation (Bertles and Milner, 1968), and the severity of anemia in SS disease correlates with the proportion of circulating ISCs (Serjeant et al., 1969).

Fabry et al. (1984) observed that the percentage of circulating dense cells in sickle cell patients was quite stable during the steady state of the disease, but showed a significant decrease during painful crises (presumably because of an abrupt increase in capillary occlusions in several regions of the body over a short period of time). These findings correlate well with those of Kaul et al. (1986), who analyzed the density and morphology of SS cells trapped during perfusion through the isolated rat mesocecum, following microvascular obstruction induced by decreased perfusion pressures. They found that, when the cells were deoxygenated, both dense SS discocytes and ISCs were selectively trapped as compared with SS cells having normal density; when oxygenated cells were perfused, however, only the ISCs were selectively trapped, to the same extent as when they were deoxygenated.

In vitro studies have shown that the high intracellular Hb (S) concentrations of ISCs and other dense SS cells result in polymer formation with minimal deoxygenation and with very short delay times (Coletta et al., 1982; Noguchi et al., 1983), as well as a marked decrease in cell deformability, even when the cells are oxygenated (Clark et al., 1980; Evans et al., 1984; Nash et al., 1984). For these reasons, dehydration of the SS cells appears to contribute to their particularly short circulatory survival times and to capillary occlusion.

2. Mechanisms of Red-Cell Dehydration

A reduction in the water content of red cells implies a loss of osmotic particles from the cells. If a normal hemoglobin content is retained, substantial dehydration must indicate a net loss of cations, with accompanying anions. Such a cation loss could occur by two general mechanisms. One is based on the 3Na-out:2K-in stoichiometry of the sodium pump, as suggested by Glader and Nathan (1978) for red cells that had lost K in excess of Na gain. In fact, this mechanism would apply to any red cells in which the ratio of total passive K efflux to Na influx was more than 2:3. Such a dehydrating effect of the Na-pump was demonstrated in vitro by Clark et al. (1981) on normal human red cells that had been pretreated with nystatin to produce an approximately balanced Na gain and K loss of at least 40 mmol/L cells.

A second general mode of red cell dehydration involves a selective increase in passive K efflux, which could possibly occur by one or several mechanisms. These mechanisms could be activation of the Ca^{2+}-dependent K channel (Gardos, 1958), induction of a selective K leak by oxidant damage to the membrane (Orringer and Parker, 1977; Maridonneau et al., 1983), by lead (Riordan and Passow, 1973), or through the activity of a K:Cl cotransport, as demonstrated in normal human and sheep reticulocytes (Lauf, 1983; Hall and Ellory, 1986; Canessa et al., 1987), as well as in some abnormal red cells under certain conditions (Brugnara et al., 1985; Canessa et al., 1986; Brugnara et al., 1986). The three specific mechanisms for SS cell dehydration studied experimentally to the greatest extent thus far, the Ca^{2+}-dependent K channel, the action of the Na-pump on a balanced increase in Na and K leaks, and the K:Cl cotransport, will be discussed below.

2.1. Activation of the Calcium-Dependent K Channel in SS Cells

Soon after Eaton et al. (1973) and Palek (1973, 1977) found that SS cells had an increased Ca content, especially in the ISCs, and that they accumulated more Ca during deoxygenation-induced sickling, it was suggested that the K channel might be responsible for formation of the dehydrated ISCs (Glader and Nathan, 1978; Eaton et al., 1978). Many studies of Ca and K transport in SS cells over the following years, however, led to an apparent paradox: the sickle cells either did not show the behavior of normal red cells experimentally loaded with μM levels of Ca^{2+}, or they had similar features shown to be unrelated to their Ca content, despite the fact that the excess Ca in these cells was not tightly bound, since it was readily extracted by exposing the cells to the ionophore A23187 and EGTA (Bookchin et al., 1984). Thus, the oxygenated SS cells showed no Ca^{2+}-dependent ^{86}Rb flux, and extraction of their mobilizable Ca did not correct the Na pump inhibition exhibited by the dense SS cells (Ortiz et al., 1986); also, the ISC membranes showed no high mol wt components (Palek and Liu, 1979), as occurs with Ca^{2+}-activated transglutaminase-induced crosslinking of membrane proteins (Smith et al., 1981). Furthermore, despite many observations that SS cells retained significant portions of the Ca taken up during sickling (Palek, 1977; Bookchin and Lew, 1980), no consistent changes were found in the (Ca^{2+} + Mg^{2+}) ATPase activity of SS membranes (Gopinath and Vincenzi, 1979; Litosch and Lee, 1980; Dixon and Winslow, 1981; Niggli et al., 1982; Luthra and Sears, 1982), and ATP concentrations in SS cells (including ISCs) were not reduced (Clark et al., 1978a).

An explanation for these apparent discrepancies came with the demonstration that virtually all of the excess Ca in SS cells is compartmentalized in intracellular vesicles that behave as "inside-out vesicles" (IOVs) in that they can accumulate Ca against an outward gradient in an ATP-dependent manner (Lew et al., 1985). Consistently, the levels of cytoplasmic Ca^{2+} in oxygenated SS cells, about 40 nmol/L cells (Bookchin et al., 1985; Rhoda et al., 1985), were not significantly different from those found in normal red cells (20–30 nmol/L cells) (Lew et al., 1982).

The high Ca content of the intracellular IOVs not only explained many of the paradoxes, but also suggested that the cells were exposed to periods of increased Ca influx in the circulation. There was some reason to believe that transient increases in cytoplasmic Ca^{2+} occurred in deoxygenated SS cells: when these cells were loaded with the Ca chelator Benz2 to suppress the Ca pump by keeping cell Ca^{2+} very low, their mean Ca permeability, which appeared normal in the oxygenated cells, rose about five times on deoxygenation (Bookchin et al., 1984). It must be during such periods of increased Ca permeability, when cell Ca^{2+} levels are somewhat elevated, that the intracellular IOVs accumulate Ca by pumping it inward. It was therefore possible that such transient increases in SS cell cytoplasmic Ca^{2+} in the nanomolar range, during deoxygenation and sickling, might briefly activate the K channels, causing selection K loss and dehydration (Lew et al., 1985).

Ohnishi (1983) had reported that repeated deoxygenation and reoxygenation of the lighter fraction of SS cells, suspended in a plasma-like medium with substrates to maintain ATP levels, resulted in formation of a fraction of cells with increased density only if Ca^{2+} (about 2 m*M*) was present in the medium. To explore further the possibility of Ca^{2+}-induced K channel activation during sickling, we first removed the dense ($\delta >$ 1.118) fraction from SS red cells, resuspended them in their own plasma with added substrates and at 37°C, with or without EGTA, and equilibrated them with alternating nitrogen or oxygen containing 5.6% CO_2. Only with Ca^{2+} (at plasma levels) was present, samples spun on Stractan ($\delta = 1.118$) (Corash et al., 1974) showed the appearance of a small fraction of cells with the density of ISCs, which reached a maximum of about 10% of the cell population after 6 h. Although samples of the total cells showed a slow, progressive Ca^{2+}-independent K loss, which slightly exceeded their Na gain, only the dense cells formed in the presence of Ca had the very low K with moderately increased Na content found in native ISCs. When SS cells from a similar 3-h incubation were placed on a discontinuous Stractan gradient, there was a modest increase in density at each layer, with the largest shift (about 13% of the cells) moving from δ = 1.096 to 1.106, again only in the presence of plasma Ca^{2+}.

A serious problem with such experiments in which SS cells are stirred for hours in the deoxygenated state (or cycled, alternating with oxygenation) is that substantial hemolysis is unavoidable; this raises doubts about whether the results are representative of circulatory conditions in vivo, where intravascular hemolysis is much less.

The long incubation times required to demonstrate cell density shifts in the above experiments suggested that the duration of K channel activation during sickling might be short, and that KCl and water loss might be limited by the diffusional anion permeability. This possibility was explored using a method based on the findings of Lew and Garcia-Sancho (1985) that the diffusional permeability of SCN^- is at least 20 times that of Cl^-, and that at temperatures of 0–5°C, at which the Ca pump is markedly inhibited, there is only moderate K channel inhibition. In a typical experiment, SS discocytes were tonometered in plasma at 37°C and sampled periodically into cold isotonic media containing SCN^- instead of Cl^- anions, and EGTA; the SCN^- rapidly exchanges with cell Cl^-, and the Ca pump is inhibited by low temperature so that the level of cell Ca^{2+} at the time of sampling is maintained; after 30-min incubation, centrifugation of the cells on Stractan (δ = 1.118) demonstrated the formation of any dense cells resulting from K channel activation (with KSCN and water loss). It should be emphasized that the cold SCN^- media is Ca^{2+}-free, so that any Ca^{2+}-dependent K channel activation observed must be the result of an increased level of cell Ca^{2+} in the cells during tonometry in plasma. Under these conditions, sampling into cold SCN^- showed about 8% dense cells formed after a 20-min period of deoxygenation, and about 20% dense cells following a second period of deoxygenation. If the cells were sampled instead into cold Cl^- media, only 2–4% dense cells were seen after five deoxygenation cycles over 150 min. If EGTA (in excess of plasma Ca) was added to the incubation medium, or if the suspension was reoxygenated before sampling, subsequent incubation in cold SCN^- media revealed little or no dense cell formation. Addition to the incubation medium of quinine, an inhibitor of the K channel, reduced subsequent dense cell formation by 60–80%.

These and other experiments indicated that:

1. a single sickling event results in an increase in cell Ca^{2+} ($[Ca^{2+}]_i$) in many SS cells
2. the increase in Ca permeability and $[Ca^{2+}]_i$ is sustained during the period of deoxygenation, but is reversed on reoxygenation and
3. although only a portion of SS cells show these changes during any period of sickling, another group of cells from the discocyte population undergoes comparable changes during a subsequent period of deoxygenation.

The most compelling and direct evidence of K channel activation was obtained when ^{86}Rb uptake was measured in SS discocytes that were deoxygenated in buffered saline with 3 m*M* K, with or without Ca^{2+} (1.6 m*M*) and sampled into cold SCN^- media (free of Ca^{2+} and ^{86}Rb). Only the cells that subsequently became dense in SCN^- showed a large increase in ^{86}Rb uptake against the large outwards gradient, which must have occurred during the tonometry and must reflect activated K channels before sampling into the SCN^- media.

Thus, it appears that during deoxygenation-induced sickling, some SS cells have an increase in cytoplasmic Ca^{2+} sufficient to activate K channels and promote cell dehydration. With reoxygenation (or removal of extracellular Ca^{2+} in vitro), the Ca pump must promptly extrude cell Ca^{2+} to levels below the threshold(s) for K channel activation.

2.2. Role of the Na-Pump in SS Cell Dehydration

Tosteson (Tosteson et al., 1952, 1955; Tosteson, 1955) first showed that, on deoxygenation, SS cells become leaky to Na and K to approximately the same extent, and that the leak is reversed on reoxygenation and unsickling. The modest extent of the increased passive fluxes of Na and K induced by deoxygenation of SS cells limits their detection in some conditions when ^{86}Rb influx against a large outward K gradient is measured in the whole cell population (Clark et al., 1978b; Izumo et al., 1987; Bookchin et al., 1987). In other conditions (Berkowitz and Orringer, 1985), or when net K efflux and/or Na influx across their normal cell-medium gradients were measured (Clark et al., 1978b; Mohandas et al., 1986; Bookchin and Lew, 1978), sickling-related increases in Na and K leaks have been consistently observed. These changes were recently studied in detail by Joiner and coworkers (1986), who reported that, on deoxygenation in a Ca^{2+}-free, high Na, low K medium with ouabain present, SS cells showed a balanced Na gain and K loss ranging from about 2.5–8.5 mmol/L cells/h, which was linear over a 4-h period.

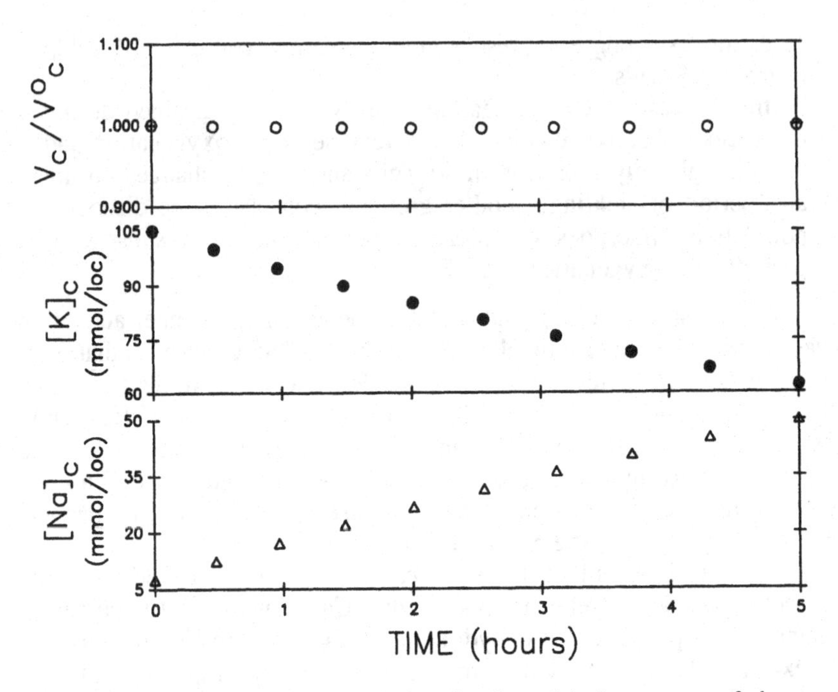

Fig. 1. Computer simulation with red-cell model: time course of changes in cell Na, K, and volume following a balanced increase in passive net Na and K fluxes with Na-pump inhibition, with the red cells "suspended" in a Ca^{2+}-free, plasma-like medium at pH 7.40 (*see* Lew and Bookchin, 1986 and Freeman et al., 1987). A 30-fold increase in Na permeability and a 55-fold increase in K permeability ($P^G_{Na} \times 30$, $P^G_K \times 55$, Na-pump off) produces a balanced cell Na gain and K loss of 10–11 mmol/L cells/h, which remains fairly linear over 4 h. Na and K content are expressed as mmoles/L original cells (mmol/Loc), and relative cell volume is indicated by the ratio of volume at the indicated time to original cell volume. [Note that, in the simulations of Figs. 2–7 which follow, the ordinate scales (as well as the abscissa time scales) are adjusted according to the extent of the changes, so that the spatial shifts in the figures often cannot be compared directly.]

As predicted by Glader and Nathan (1978), and demonstrated in vitro by Clark and coworkers (1981) with nystatin-treated normal red cells, the Na pump, with its 3Na-out/2K-in stoichiometry, acting on a balanced cell Na gain and K loss, should produce a net cell cation loss with dehydration. Using a new integrated model of the mature human red cell (*see* chapter 2 by Lew et al. in this volume), we simulated the experimental conditions used by Clark et al. by first allowing Na and K leaks to produce a balanced Na gain and K loss of 60 mmol/L cell water, with no change in cell volume; all transport functions of the cell were then returned to normal, and the time course of the restorations of the original cell parameters, as predicted by steady-state analyses, was observed. The

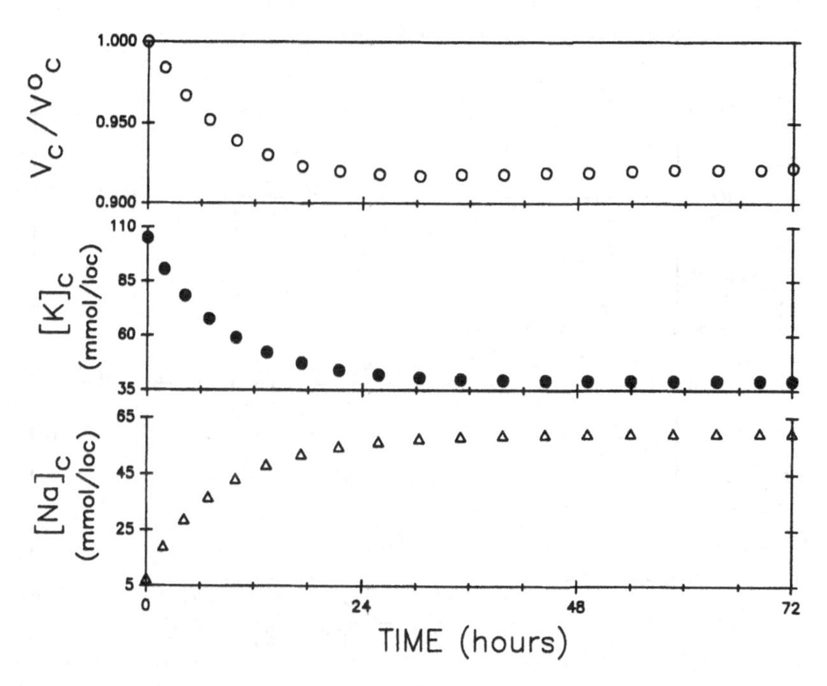

Fig. 2. Computer simulation with red-cell model (*see* Fig. 1 legend). Same leaks as in Fig. 1 ($P^G_{Na} \times 30$, $P^G_K \times 55$); active Na-pump, with normal parameters for mature red cells.

model predicted that cell dehydration would proceed to a maximum (about 10% reduction) in about 15 h, at which time Na and K concentrations in cell water have returned to nearly normal. Cell volume recovery proceeds very slowly over several weeks, however, and is only about 35% complete after one week. These predictions were consistent with the observations of Clark et al. (1981), but deal with experimental conditions different from those applied to SS cells.

Following the report by Joiner et al. (1986) of the extent of Na and K leaks in ouabain-treated doxygenated SS cells (in the absence of external Ca^{2+}), model simulations were performed using these leaks as a guide. Figure 1 shows that increases in the electrodiffusional permeabilities of Na and K by 30- and 55-fold, respectively, with the Na pump suppressed, result in balanced net fluxes of Na inward and K outward of 10–11 mmol/L cells/h initially, which remain fairly linear over a 4-h period, keeping the cells at constant volume. If the same leaks are simulated with the Na-pump functioning (Fig. 2), the initial net K efflux exceeds net Na gain, resulting in a net cation and water loss that is maximal in the first 18–24 h, producing about 8% cell dehydration. When the Na-pump acts on smaller balanced initial Na and K leaks of 6 mmol/L cells/h (Fig. 3), both the extent of the cells' dehydration and the time it takes to reach that

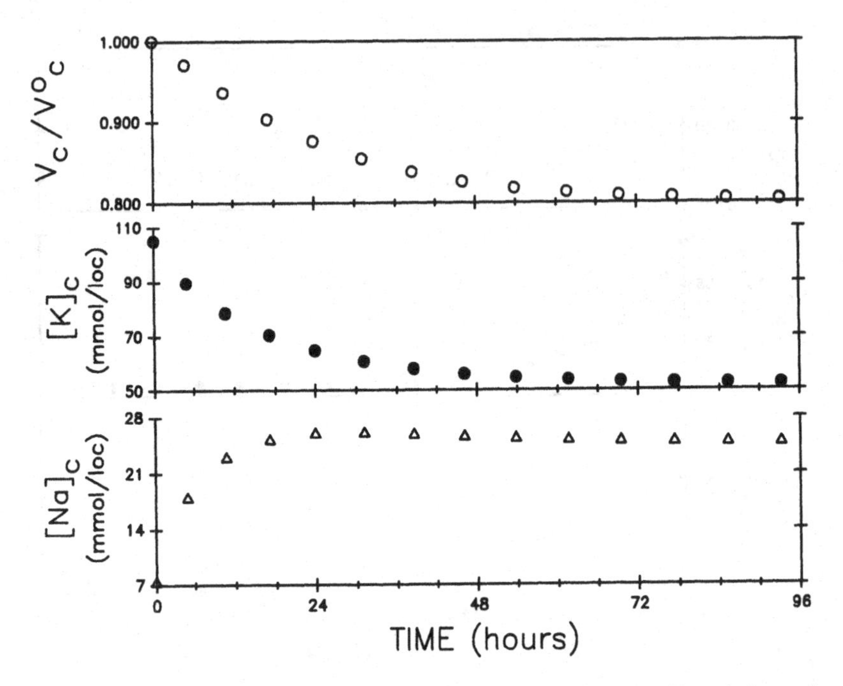

Fig. 3. Computer simulation with red-cell model (*see* Fig. 1 legend); active Na-pump; balanced leaks of about 6 mmol/L cells/h ($P^G_{Na} \times 15$, $P^G_K \times 27.5$).

maximal extent (before leveling off) are increased, with about 10% dehydration in 18 h, leveling off at about 20% dehydration after 2–3 d.

The above leaks and pump-leak effects mimic rather well the experimental observations of Joiner et al. (1986) on SS cells, both in the short incubations with ouabain and in the 28-h incubations under different conditions. It is clear from the present simulations, however, that the dehydration process is not linear, as inferred by Joiner et al. (1986), but levels off to a quasi-steady state after various times; this occurs when the changes in cell contents of Na and K, and the corresponding gradients, result in net passive fluxes that are balanced by the Na-pump.

During circulation in vivo, the majority of SS cells must be repeatedly deoxygenated (partly) and reoxygenated. Figures 4 and 5 illustrate model simulations of possible circulatory cycles in which the two balanced leaks described above are turned on for 60 s (as if deoxygenated) and off for 90 s, repeatedly, while the Na-pump operates continuously. Under both conditions, the model predicts progressive dehydration that is fairly linear over the first 24 h, but gradually tapers off over the next 2–4

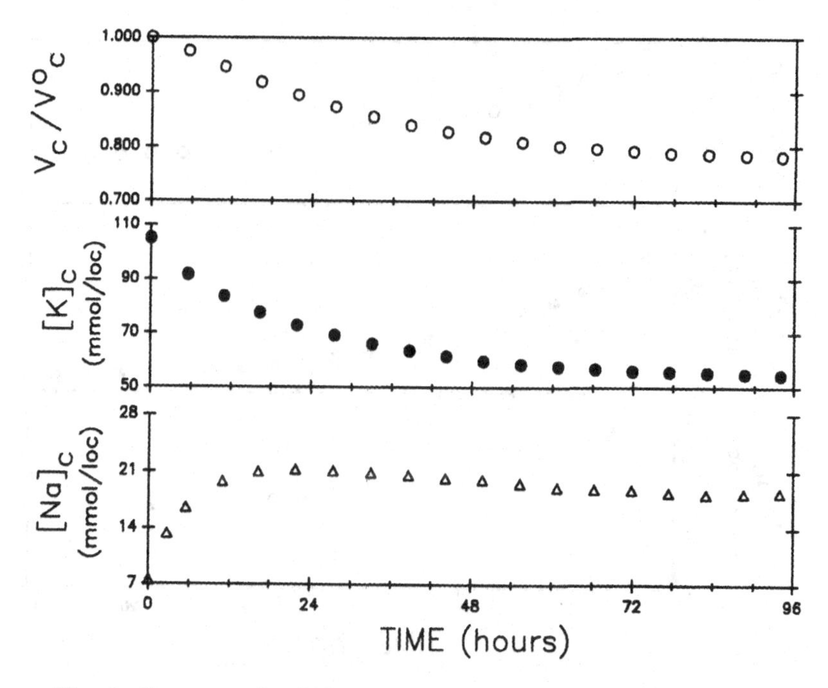

Fig. 4. Computer simulation with red-cell model (*see* Fig. 1 legend), in which the balanced Na and K leaks of 10–11 mmol/L cells/h ($P^G_{Na} \times 30$, $P^G_K \times 55$) are repeatedly cycled on for 60 s and off for 90 s, to simulate the possible effects of repeated deoxygenation and reoxygenation in the circulation, while the Na pump operates continuously.

d, reaching similar volumes (about 22% dehydration) with different relative amounts of cell Na and K. Although the extent of dehydration approaches that seen with ISCs, the cell K contents remain considerably higher than in dense SS cells. In addition, the requirement of several days of continuous deoxy–oxy cycling seems unlikely to be met within the short time during which ISCs are generated and removed from the circulation (Bertles and Milner, 1968).

Since the maximum leaks described by Joiner et al. (1986) for deoxygenated SS cells with ouabain represent mean values for the whole heterogeneous SS cell population, it seems likely that many cells have leaks larger than the mean. Simulation of larger balanced Na and K leaks of 14–15 mmol/L cells/h, however, if run continuously with the Na pump active (Fig. 6), shows only 3.5% maximal cell dehydration after 8 h, followed later by cell swelling—not the pattern observed experimentally (Joiner et al., 1986). If these larger leaks are cycled on and off (Fig. 7),

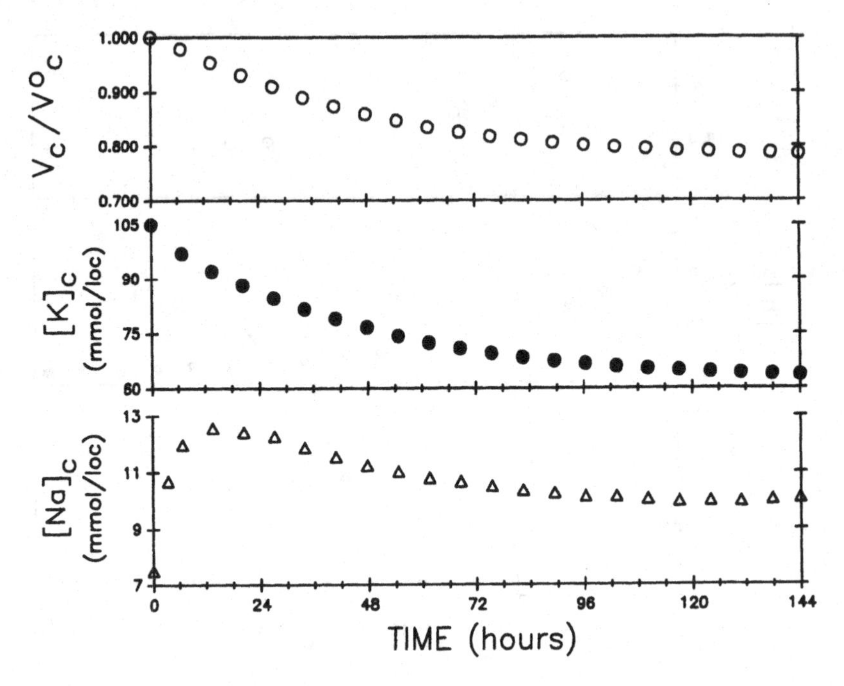

Fig. 5. Computer simulation with red cell model (*see* Fig. 1 legend), in which the smaller balanced Na and K leaks that were run continuously in the simulation of Fig. 3 ($P^G_{Na} \times 15$, $P^G_K \times 27.5$, Na-pump on) are cycled 60 s on–90 s off, as in the simulation of Fig. 4.

on the other hand, cell dehydration occurs to a similar extent as with the smaller cation leaks above. As with the smaller leaks, however, at maximal dehydration, the cell K still remains higher than that found in ISCs. The main effect of "cycling" the leaks, particularly the larger ones (compare Figs. 6 and 7), is that cell dehydration is much greater when the Na-pump, with its 3Na-out–2K-in stoichiometry, has more time to act on the "balanced" gradient dissipation. Variations in the cycling times, from 60 s on –90 s off (as shown in the figures) to 30 s on–60 s off, or 45 s on–90 s off, for example, produced only small quantitative differences from the results of the simulations illustrated.

Some points of caution should be raised about the model simulations. One is that SS (and other abnormal red cells) differ from the normal cells in ways not yet taken into account in the present model; such differences include the lower p*I* of hemoglobin S, a lower cell pH, higher contents of the impermeant anion 2,3-diphosphoglycerate, and the rela-

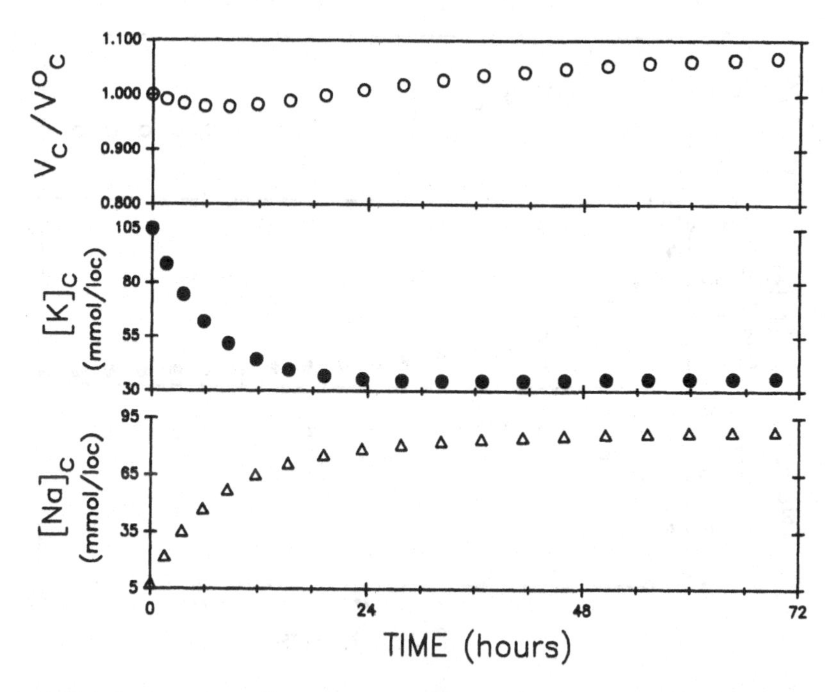

Fig. 6. Computer simulation with red-cell model (*see* Fig. 1 legend), in which larger balanced Na and K leaks of 14–15 mmol/L cells/h ($P^G_{Na} \times 45$, $P^G_K \times 75$, Na-pump on) are run continuously, as with the smaller leaks shown in Figs. 2 and 3.

tive immaturity of the red-cell population. Another caution is that the actual behavior of a variety of passive fluxes, lumped together in the model as facilitated diffusional leaks, may not be represented accurately enough to predict the outcome of small perturbations over extended time periods.

The main conclusion of these simulations, which should not be affected by the caveats above, is that the action of the Na-pump on the observed range of deoxygenation-induced increases in Ca-independent passive Na and K fluxes in SS cells can certainly play a role in the cell dehydration at some stages, but cannot alone account for the observed cation content of dense SS cells. Another consideration in this regard is the unexplained Na-pump inhibition observed in the dense SS cells (Clark et al., 1978b; Ortiz et al., 1986). The extent and time course of Na-pump contribution to SS cell dehydration will obviously depend in part on the stage of dehydration at which significant Na-pump inhibition occurs.

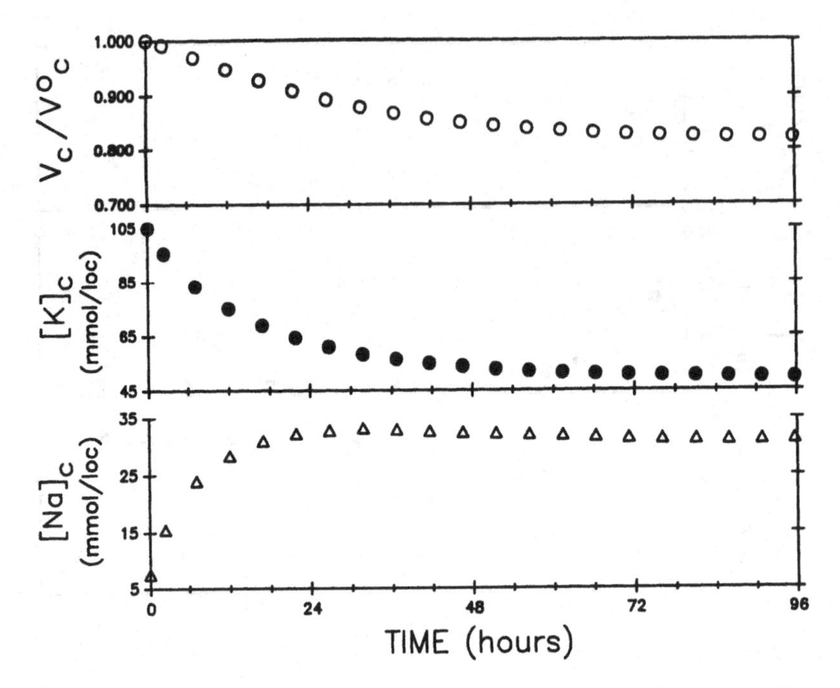

Fig. 7. Computer simulation with red-cell model (*see* Fig. 1 legend), in which the larger balanced Na and K leaks simulated in Fig. 6 ($P^G_{Na} \times 45$, $P^G_K \times 75$) are cycled 60 s on–90 s off, with the Na-pump running, as shown with the smaller leaks in Figs. 4 and 5.

2.3. Possible Role of K:Cl Cotransport in SS Cell Dehydration

Brugnara and coworkers (1985, 1986) and Canessa et al. (1986) have found that substantial fractions of both SS cells and red cells from persons homozygous for hemoglobin C (CC cells) exhibit a large volume and pH-dependent K leak. As studied thus far, this transport system behaves like the K:Cl cotransport pathway described in detail by Lauf and coworkers in low-K sheep red cells and high-K sheep reticulocytes (Lauf, 1983, 1985; Lauf and Theg, 1980), and by Hall and Ellory (1986) and Canessa et al. (1987) in the lightest, reticulocyte-rich fraction of human red cells. By mechanisms not yet clear, this or very similar pathways either persist or are activated in the SS and CC cells. To the extent that this pathway may play a part in SS cell dehydration, it could contribute to the transition of young SS cells (newly released from the bone marrow) to ISCs. The extent of fluxes through this pathway, as well as possible balancing fluxes, in SS cells of different age and volume, and under various conditions that could be encountered in vivo, must be assessed further.

Acknowledgments

This work was supported by grants from the National Institutes of Health (USA) (HL28018 and HL 21016) and from the Wellcome Trust and the Medical Research Council (UK).

References

Berkowitz, L. R. and Orringer, E. P. (1985) Passive sodium and potassium movements in sickle erythrocytes. *Am. J. Physiol.* **249,** 208–214.

Bertles, J. F. and Dobler, J. (1969) Reversible and irreversible sickling: A distinction by electron microscopy. *Blood* **33,** 884–898.

Bertles, J. F. and Milner, P. F. A. (1968) Irreversibly sickled erythrocytes: a consequence of the heterogeneous distribution of hemoglobin types in sickle cell anemia. *J. Clin. Invest.* **47,** 1731–1741.

Bookchin, R. M. and Lew, V. L. (1978) Effects of a "sickling pulse" on the calcium and potassium permeabilities of intact, sickle trait red cells. *J. Physiol. (London)* **284,** 93P.

Bookchin, R. M. and Lew, V. L. (1980) Progressive inhibition of the Ca-pump and Ca:Ca exchange with normal cytoplasmic Ca buffering in sickle red cells. *Nature* **284,** 561–562.

Bookchin, R. M., Ortiz, O. E., and Lew, V. L. (1984) Silent intracellular calcium in sickle cell anemia red cells, in *The Red Cell. Proceedings of the Sixth International Conference on Red Cell Metabolism and Function* (Brewer, G. J., ed.), Alan R. Liss, New York, pp. 17–28.

Bookchin, R. M., Ortiz, O. E., and Lew, V. L. (1987) Activation of calcium-dependent potassium channels in deoxygenated sickled red cells, in *Pathophysiological Aspects of Sickle Cell Vaso-Occlusion* (Nagel, R. L., ed.), Alan R. Liss, New York, pp. 193–200.

Bookchin, R. M., Ortiz, O. E., Somlyo, A. V., Somlyo, A. P., Sepulveda, M. I., Hockaday, A., and Lew, V. L. (1985) Calcium-accumulating inside-out vesicles in sickle cell anemia red cells. *Trans. Assoc. Am. Phys.* **98,** 10–20.

Brugnara, C., Bunn, H. F., and Tosteson, D. C. (1986) Regulation of erythrocyte cation and water content in sickle cell anemia. *Science* **232,** 388–390.

Brugnara, C., Kopin, A. S., Bunn, H. F., and Tosteson, D. C. (1984) Electrolyte composition and equilibrium in hemoglobin CC red blood cells. *Trans. Assoc. Am. Phys.* **97,** 104–112.

Brugnara, C., Kopin, A. S., Bunn, H. F., and Tosteson, D. C. (1985) Regulation of cation content and cell volume in erythrocytes from patients with homozygous hemoglobin C disease. *J. Clin. Invest.* **75,** 1608–1617.

Canessa, M., Spalvins, A., and Nagel, R. L. (1986) Volume-dependent and NEM-stimulated K^+, Cl^- transport is elevated in oxygenated SS, SC and CC human red cells. *FEBS Letters* **200,** 197–202.

Canessa, M., Fabry, M. E., Blumenfeld, N., and Nagel, R. L. (1987) Volume-stimulated, Cl^--dependent K^+ efflux is highly expressed in young human red cells containing normal hemoglobin or HbS. *J. Membr. Biol.* **97,** 97–105.

Clark, M. R., Unger, R. C., and Shohet, S. B. (1978a) Monovalent cation composition and ATP and lipid content of irreversibly sickled cells. *Blood* **51,** 1169–1178.

Clark, M. R., Morrison, C. E., and Shohet, S. B. (1978b) Monovalent cation transport in irreversibly sickled cells. *J. Clin. Invest.* **62,** 329–337.

Clark, M. R., Mohandas, N., and Shohet, S. B. (1980) Deformabililty of oxygenated irreversibly sickled cells. *J. Clin. Invest.* **65,** 189–196.

Clark, M. R., Guatelli, J. C., White, A. T., and Shohet, S. B. (1981) Study on dehydrating effect of the red cell Na^+/K^+-pump in nystatin-treated cells with varying Na^+ and water contents. *Biochim. Biophys. Acta* **646,** 422–432.

Coletta, M., Hofrichter, J., Ferrone, F. A., and Eaton, W. A. (1982) Kinetics of sickle haemoglobin polymerization in single red cells. *Nature* **300,** 194–197.

Corash, L. M., Piomelli, S., Chen, H. C., Seaman, C., and Gross, E (1974) Separation of erythrocytes according to age on a simplified density gradient. *J. Lab. Clin. Med.* **84,** 147–151.

Dixon, E. and Winslow, R. M. (1981) The interaction between (Ca^{2+} + Mg^{2+})-ATPase and the soluble activator (calmodulin) in erythrocytes containing haemoglobin S. *Br. J. Haematol.* **47,** 391–397.

Eaton, J. W., Berger, E., White, J. G., and Jacob, H. S. (1978) Calcium induced damage of haemoglobin SS and normal erythrocytes. *Br. J. Haematol.* **38,** 57–62.

Eaton, J. W., Skelton, T. D., Swofford, H. S., Koplin, C. E., and Jacob, H. S. (1973) Elevated erythrocyte calcium in sickle cell disease. *Nature* **246,** 105–106.

Embury, S. H., Garcia, J. F., Mohandas, N., Pennathur-Das, R., and Clark, M. R. (1984) Effects of oxygen inhalation on endogenous erythropoietin kinetics, erythropoiesis, and properties of blood cells in sickle-cell anemia. *N. Engl. J. Med.* **311,** 291–295.

Evans, E., Mohandas, N., and Leung, A. (1984) Static and dynamic rigidities of normal and sickle erythrocytes. *J. Clin. Invest.* **73,** 477–488.

Fabry, M. E., Benjamin, L., Lawrence, C., and Nagel, R. L. (1984) An objective sign in painful crisis in sickle cell anemia: The concomitant reduction of high density red cells. *Blood* **64,** 559–563.

Freeman, C. J., Bookchin, R. M., Ortiz, O. E., and Lew, V. L. (1987)

K-permeabilized human red cells lose an alkaline, hypertonic fluid containing excess K over diffusible anions. *J. Membr. Biol.* **96,** 235–241.

Gardos, G. (1958) The function of calcium in the potassium permeability of human erythrocytes. *Biochim. Biophys. Acta* **30,** 653–654.

Glader, B. E. and Nathan, D. G. (1978) Cation permeability alterations during sickling: Relation to cation composition and cellular hydration of irreversibly sickled cells. *Blood* **51,** 983–989.

Glader, B. E., Fortier, N., Albala, M. M., and Nathan, D. G. (1974) Congenital hemolytic anemia associated with dehydrated erythrocytes and increased potassium loss. *N. Engl. J. Med.* **291,** 491–496.

Glader, B. E., Lux, S. E., Muller-Soyano, A., Platt, O. S., Propper, R. D., and Nathan, D. G. (1978) Energy reserve and cation composition of irreversibly sickled cells in vivo. *Br. J. Haematol.* **40,** 527–532.

Gopinath, R. M. and Vincenzi, F. F. (1979) (Ca^{2+} + Mg^{2+})-ATPase activity of sickle cell membranes: Decreased activation by red blood cell cytoplasmic activator. *Am. J. Hematol.* **7,** 303–312.

Hall, A. C. and Ellory, J. C. (1986) Evidence for the presence of volume-sensitive KCl transport in 'young' human red cells. *Biochim. Biophys. Acta.* **858,** 317–320.

Izumo, H., Lear, S., Williams, M., Rosa, R., and Epstein, F. H. (1987) Sodium-potassium pump, ion fluxes, and cellular dehydration in sickle cell anemia. *J. Clin. Invest.* **79,** 1621–1628.

Joiner, C. H., Platt, O. S., and Lux, S. E. (1986) Cation depletion by the sodium pump in red cells with pathological cation leaks. Sickle cells and xerocytes. *J. Clin. Invest.* **78,** 1487–1496.

Kaul, D. K., Fabry, M. E., and Nagel, R. L. (1986) Vaso-occlusion by sickle cells: evidence for selective trapping of dense red cells. *Blood* **68,** 1162–1166.

Lauf, P. K. (1983) Thiol-dependent passive K/Cl transport in sheep red cells: II. Loss of Cl^- and N-ethylmaleimide sensitivity in maturing high K^+ cells. *J. Membr. Biol.* **73,** 247–256.

Lauf, P. K. (1985) K+ :Cl- cotransport: Sulfhydryls, divalent cations, and the mechanism of volume activation in a red cell. *J. Membrane Biol.* **88,** 1–13.

Lauf, P. K. and Theg, B. E. (1980) A chloride dependent K+ flux induced by *N*-ethylmaleimide in genetically low K+ sheep and goat erythrocytes. *Biochem. Biophys. Res. Comm.* **92,** 1422–1428.

Lew, V. L. and Bookchin, R. M. (1986) Volume, pH and ion content regulation in human red cells: analysis of transient behavior with an integrated model. *J. Membr. Biol.* **92,** 57–74.

Lew, V. L. and Garcia-Sancho, J. (1985) Use of the ionophore A23187 to measure and control cytoplasmic Ca^{2+} levels in intact red cells. *Cell Calcium* **6,** 15–23.

Lew, V. L., Tsien, R. Y., Miner, C., and Bookchin, R. M. (1982) The physiological (Ca^{2+}) level and pump-leak turnover in intact red cells measured with the use of an incorporated Ca chelator. *Nature* **298,** 478–481.

Lew, V. L., Hockaday, A., Sepulveda, M. I., Somlyo, A. P., Somlyo, A. V.,

Ortiz, O. E., and Bookchin, R. M. (1985) Compartmentalization of sickle cell calcium in endocytic inside-out vesicles. *Nature* **315,** 586–589.

Litosch, I. and Lee, K. S. (1980) Sickle red cell calcium metabolism: Studies on Ca^{2+} + Mg^{2+} ATPase and Ca-binding properties of sickle red cell membranes. *Am. J. Hematol.* **8,** 377–387.

Luthra, M. G. and Sears, D. A. (1982) Increased Ca^{++}, Mg^{++} and Na^{+} + K^{+} ATPase activities in erythrocytes of sickle cell anemia. *Blood* **60,** 1332–1336.

Maridonneau, I., Braquet, P., and Garay, R. P. (1983) Na^{+} and K^{+} transport damage induced by oxygen free radicals in human red cell membranes. *J. Biol. Chem.* **185,** 3107–3113.

Mohandas, N., Rossi, M. E., and Clark, M. R. (1986) Association between morphological distortion of sickle cells and deoxygenation-induced cation permeability increase. *Blood* **68,** 450–454.

Nash, G. B., Johnson, C. S., and Meiselman, H. J. (1984) Mechanical properties of oxygenated red blood cells in sickle cell (HbSS) disease. *Blood* **63,** 73–82.

Niggli, V., Adunyah, E. S., Cameron, B. F., Bababunmi, E. A., and Carafoli, E. (1982) The Ca^{2+}-pump of sickle cell plasma membranes. Purification and reconstruction of the ATPase enzyme. *Cell Calcium* **3,** 131–151.

Noguchi, C. T., Torchia, D. A., and Schechter, A. N. (1983) The intracellular polymerization of sickle hemoglobin: effects of cell heterogeneity. *J. Clin. Invest.* **72,** 846–852.

Ohnishi, S. T. (1983) Inhibition of the in vitro formation of irreversibly sickled cells by cepharanthine. *Br. J. Haematol.* **55,** 665–671.

Orringer, E. P. and Parker, J. C. (1977) Selective increase of potassium permeability in red blood cells exposed to acetylphenylhydrazine. *Blood* **50,** 1013–1021.

Ortiz, O. E., Lew, V. L., and Bookchin, R. M. (1986) Calcium accumulated by sickle cell anemia red cells does not affect their potassium (^{86}Rb) flux components. *Blood* **67,** 710–715.

Palek, J. (1973) Calcium accumulation during sickling of haemoglobin S red cells. *Blood* **42,** 988.

Palek, J. (1977) Red cell calcium content and transmembrane calcium movements in sickle cell anemia. *J. Lab. Clin. Med.* **89,** 1365–1374.

Palek, J. and Liu, S. C. (1979) Membrane protein organization in ATP-depleted and irreversibly sickled red cells. *J. Supramol. Struct.* **10,** 79–96.

Rhoda, M. D., Giraud, F., Craescu, C. T., and Beuzard, Y. (1985) Compartmentalization of Ca^{2+} in sickle cells. *Cell Calcium* **6,** 397–411.

Riordan, J. R. and Passow, H. (1973) The effects of calcium and lead on the potassium permeability of human erythrocytes and erythrocyte ghosts, in *Comparative Physiology* (Bolis, L., Schmidt-Nielsen, K., and Maddrell, S. H. P., eds.), North Holland: Amsterdam, pp. 543–581.

Serjeant, G. R., Serjeant, B. E., and Milner, P. F. (1969) The irreversibly sickled cell: a determinant of haemolysis in sickle cell anaemia. *Br. J. Haematol.* **17,** 527–533.

Smith, B. D., LaCelle, P. L., Siefring, G. E., Jr., Lowe-Krentz, L., and Lorand, L. (1981) Effects of calcium-mediated enzymatic cross-linking of membrane proteins on cellular deformability, *J. Membr. Biol.* **61,** 75–80.

Tosteson, D. C. (1955) The effects of sickling on ion transport II. The effect of sickling on sodium and cesium transport. *J. Gen. Physiol.* **39,** 55–67.

Tosteson, D. C., Carlsen, E., and Dunham, E. T. (1955) The effects of sickling on ion transport I. Effect of sickling on potassium transport. *J. Gen. Physiol.* **39,** 31–53.

Tosteson, D. C., Shea, E., and Darling, R. C. (1952) Potassium and sodium of red blood cells in sickle cell anemia. *J. Clin. Invest.* **31,** 406–411.

Index

Index